T0386962

Sardul S. Guraya

Biology of Spermatogenesis and Spermatozoa in Mammals

With 85 Figures

Springer-Verlag
Berlin Heidelberg New York
London Paris Tokyo

Professor Dr. SARDUL S. GURAYA
ICMR Regional Advanced Research
Centre in Reproductive Biology
Department of Zoology
College of Basic Sciences and Humanities
Punjab Agricultural University
Ludhiana 141004, Punjab
India

Cover illustration: a drawing of a portion of seminiferous epithelium from ram testis

ISBN-13:978-3-642-71640-9 e-ISBN-13:978-3-642-71638-6
DOI: 10.1007/978-3-642-71638-6

Library of Congress Cataloging-in-Publication Data. Guraya, Sardul S., 1930– Biology of spermatogenesis and spermatozoa in mammals. Bibliography: p. 361 Includes index. 1. Mammals-Reproduction. 2. Spermatogenesis in animals. 3. Spermatozoa. I. Title. QL739.2.G88 1987 599'.03'2 86-26310

Softcover reprint of the hardcover 1st edition 1987

Typesetting: K+V Fotosatz GmbH, Beerfelden

2131/3130-543210

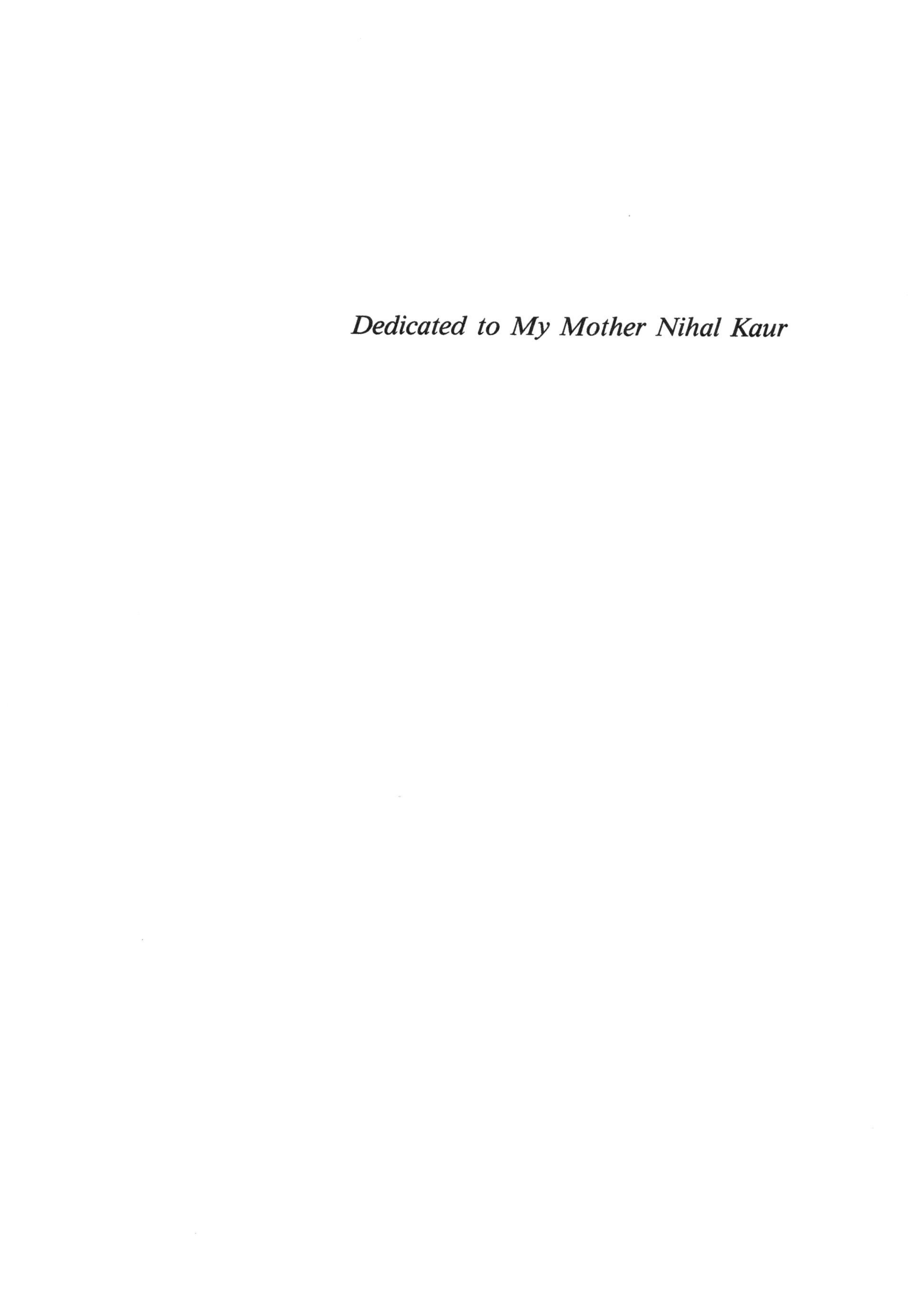

Dedicated to My Mother Nihal Kaur

Preface

The spermatogenesis and spermatozoa of mammals are of fundamental interest to a wide variety of academic and scientific disciplines; zoologists keep up interest in comparative biology of spermatogenesis and spermatozoa in different groups of mammals for determining their phylogenetic interrelationships. Therefore, during the past 15 years, a wealth of reviews and papers have been published on the morphology (including ultrastructure), histochemistry, autoradiography, biochemistry (including immunology), and cell physiology of the seminiferous epithelium including Sertoli cells, spermatogenesis and spermatozoa in mammals. Hormonal regulation of spermatogenesis also forms the subject to numerous studies. Research in this area of reproductive biology continues at a remarkable rate, and new and significant information appears daily in a wide range of journals, published symposia, and specialist reviews. The scattered nature of this information makes it difficult for a scientist, student or andrologist to go through even a small fraction of these publications on the biology of spermatogenesis and spermatozoa and so obtain a general oversight of current activity and new advances. Actually, very little attempt has been made previously to summarize and integrate the vast information which has become available as a result of use of modern techniques of microscopy, surface topography, histochemistry, autoradiography, biochemistry, biophysics, immunology, molecular biology, in vitro systems, etc. Much-needed interdisciplinary approach in biology of seminiferous epithelium, spermatogenesis and spermatozoa is very difficult and thus lacking. Thus this book leads to a better understanding of cellular and molecular aspects of spermatogenesis and spermatozoa. The purpose of this book is, therefore, to present timely thorough reviews on different subcellular and molecular aspects of interrelationships between Sertoli cells and germ cells, spermatogenesis, and spermatozoa in mammals, with special emphasis on modern concepts of structure, chemistry, and function. This book has been designed in such a way as to serve the purpose of the student, the scientist, or the physician, and the veterinarian, and will help them learn the current state of knowledge of mammalian spermatogenesis and its regulation in vivo and in vitro, and spermatozoa on its own as well as other areas of interest. This book is organized into two parts and twelve chapters dealing with cellular, molecular and functional aspects of seminiferous epithelium, spermatogenesis and spermatozoa: Introduction, Seminiferous Epithelium, Spermatogonia, Spermatocytes, Spermatids and Spermiogenesis, Antigens during Spermatogenesis, General Considerations of Spermatozoa, Head, Neck, Cytoplasmic Droplet, Tail, Plasma Membrane and its Surface Components, and Sperm Motility. These various chapters are timely reviews and give extensive bibliographies which will serve as an important source for the investigators on the biology of spermatogenesis and spermatozoa for years to come. An attempt is also made to place basic research in clear perspectives for solving future problems of fertility and sterility in farm animals and humans.

It is hoped that this book will fulfil a long-standing need and serve as an important source of references for research workers engaged in the disciplines of reproductive biology, fertility regulation, andrology including artificial insemination, cell, molecular and developmental biology, biochemistry, immunology, genetics, mammology etc. It has also become clear that great gaps in our knowledge of the biology of spermatogenesis and spermatozoa in mammals still exist, and perhaps this attempt will serve as a stimulus to researchers in various disciplines to fill these gaps.

Completing a book such as this under one cover by a single author is a most difficult job, and I am very grateful to the following internationally known experts in the modern studies of spermatogenesis and spermatozoa in mammals, for critically reviewing/editing the chapters/sections pertaining to their field of specialization: Dr. M. Courot, Dr. M. T. Hochereau-de-Reviers and Dr. J.-L. Courtens for Part One; special thanks are due to Dr. Anthony R. Bellvé, who took very great pains in critically reviewing and editing Part One and Chapter VII (except for Section C) of Part Two twice; Dr. Kenneth L. Polakoski for Section C of Chapter VII; Dr. Leonhard Nelson and Dr. J. A. Volio for Chapters VIII, IX, X (except for Sections B to E) and XII; Dr. C. F. Millette and Dr. R. N. Peterson for Chapter XI; and Dr. I. G. White for Sections B to E of Chapter X. I must thank Dr. S. N. Sewak, Associate Professor of English, for vetting Chapters XI and XII. However, some shortcomings are always expected in this type of adventure involving multidisciplinary approach, for which I assume responsibility. This book is the second volume in the series, the first one entitled "Biology of Ovarian Follicles in Mammals" has also been published by Springer Verlag in 1985. Both books give a complete, up-to-date account of cellular and molecular aspects of development, differentiation and maturation of gametes in mammals which are of great economic importance in human and veterinary medicine, agriculture, wildlife conservation, etc.

Thanks are due to my former students and now colleagues, Dr. K. S. Sidhu, Dr. V. R. Parshad, Dr. G. S. Bilaspuri and Miss Gurinder Randhawa for their help in the preparation of the final version of the manuscript, especially for preparing photomicrographs and checking references to Dr. Charanjit Kaur and Dr. Patwant Kaur for reading the proofs; to Sardar Inderjit Singh for typing the manuscript; and to S. Chain Singh for preparing diagrams. Thanks are also due to the authors and copyright holders for permission to republish some of their illustrations. I must thank Prof. A. F. Holstein for sending a complimentary copy of Atlas of Human Spermatogenesis. The free copies of volumes 383 and 438 of the Annals of the New York Academy of Sciences are acknowledged with thanks. Thanks are particulary due to the Biology Editorial Department of Springer Verlag, Heidelberg, for providing constant encouragement and excellent cooperation during the preparation of this book. Finally, I thank my loving family, Surinder (wife), Gurmeet and Harmeet (sons), and Rupa (daughter), who, through their patience, encouragement and help have made the completion of this book possible.

Ludhiana, January 1987 Sardul S. Guraya

Contents

Part Two Spermatozoa

Introduction

Spermatogenesis is a complex process of differentiation, involving germ-cell proliferation and renewal, meiosis, and spermiogenesis under the influence of Sertoli cells. It is characterized by complex morphological and biochemical transformations that lead to the formation of a highly specialized cell, the haploid spermatozoon. The spermatozoa are genetically diverse owing to the meiotic events of gene recombination and chromosome reassortment. The Sertoli cells forming the critical site for FSH and androgen-mediated biosynthetic events play important roles in the regulation of proliferation and differentiation of germ-cells in the seminiferous epithelium. The considerable progress evident in research on the normal development, differentiation, structure, biochemistry and physiology of spermatozoa in mammals owes much to the perfection of older and the evolvement of new techniques of transmission and scanning electron microscopy, lectin labelling and binding, histochemistry, autoradiography, biochemistry, quantitative research, biophysics, immunology, etc., which have been applied to the study of germ-cells and Sertoli cells both in vivo and in vitro. It is now possible to culture discrete segments of seminiferous epithelium at specific stages of spermatogenesis and separate germ-cells or Sertoli cells, thus allowing the investigator to isolate and monitor specific cell-mediated events. Electron microscopy of thin sections, surface replicas, freeze-cleaving and freeze-etching has made it possible to split the sperm specializations within the plane of the membrane specializations that may be important to the acrosome reaction and to gamete recognition and membrane fusion, which constitute the fundamental events of fertilization (Yanagimachi 1978a, Metz 1978, Shapiro and Eddy 1980, Clegg 1983, Shapiro 1984). The purpose of this book is to summarize the recent information obtained with these modern techniques in order to have a deeper insight into the normal development, differentiation, structure, chemistry and function of various components of mammalian spermatozoa as well as into the morphological, and functional interrelationships between the developing germ-cells and Sertoli cells. In recent years, the ultrastructural and biochemical specializations of Sertoli cells in relation to proliferation and differentiation of germ-cells form the subject of numerous studies, which have suggested that polypeptides of Sertoli cell origin regulate the proliferation and differentiation of germ-cells, and also may act on Leydig cells. Actually, an attempt has been made here to give a comparative account of cellular and molecular aspects of spermatogenesis, Sertoli cells, and spermatozoa in mammals as well as to point out the gaps in our knowledge about these aspects which can be filled in future studies.

Spermatogenesis is sensitive to the effects of various factors, such as heat (Vandemark and Free 1970, Parvinen 1973, R. G. Harrison 1975), light (Hochereau-de-Reviers et al. 1985), ionizing radiation (Ellis 1970, Withers et al. 1974, de Ruiter-Bootsma et al. 1976, Oakberg 1974, 1975, 1978, Huckins 1978b, L. A. Johnson et al. 1984b), nutrition (Leathem 1975, Parshad and Guraya 1984), ageing (Neaves and Johnson 1985), diabetes (Cameron et al. 1985), and a number of chemical agents (Kramer and de Rooij 1970, de Rooij and Kramer 1970, Lobl and Mathews 1978, Scott and Persaud 1978, Tierney et al. 1979, Courtens et al. 1980, Amir 1984, Tanphaichitr et al. 1984, Tanphaichitr and Bellve 1985, Viguier-Martinez et al. 1985), including the drugs used in anticancer chemotherapy (Gomes 1970), the steroid hormones (Barham and Berlin 1974), vitamins (Huang and Hembree 1979) and alcohol (Weignberg et al. 1984). The mechanism(s) involved in the primary action of these drugs have not been resolved, although multiple sites during the development, differentiation and maturation of germ-cells are being subjected to intensive research. It is still to be determined more precisely whether the drugs affect the germ-cells directly, or indirectly by altering the biology of Sertoli cells. Owing to the increasing use of various drugs, prostaglandins, and hormones for the regulation of male fertility (Patanelli 1975, Cunningham et al. 1980, Segal 1985), we need a more thorough knowledge of their action mechanisms at the cellular and molecular levels as well as of their adverse effects on the various components of the testis, and spermatozoa, which affect fertility (Jones 1975). Barham and Berlin (1974) have shown that the morphological (or ultrastructural) effects of injected testosterone in men are most pronounced in the spermatids, whereas the spermatogonia and spermatocytes display progressively fewer morphological defects in response to testosterone propionate. The present detailed, integrated account of the biology of spermatozoa, which has not been available previously, will also be very useful for the better understanding of the effects of long-term storage and ageing of spermatozoa at the subcellular and molecular levels (Mann and Lutwak-Mann 1975, Rowson 1975, Smith et al. 1975). This book will be especially useful for assessing fertility and infertility in mammals including farm animals and humans in relation to spermatid differentiation and sperm structure metabolism and preservation (Jones 1975, Rob and Rezinek 1976, Holstein and Schirren 1979, Amann 1984, Corteel and Paquignon 1984, Garner 1984), as the assessment of semen quality continues to be an important aspect of the clinical practice in male fertility (see Singer et al. 1980a, 1981).

It has also been shown in recent years that gametogenesis in primates and farm animals differs from that in rodents in a number of important respects and the response of germ-cells to such exogenous factors as radiation and drugs is also markedly different (see Baker 1971). Therefore, in this book relatively more emphasis will be laid on the comparative cellular, molecular and biophysical aspects of spermatogenesis, Sertoli cells, and sperm in mammals to reveal species differences. The previous reviews and monographs mostly deal with the morphological differentiation and ultrastructure of germ-cells in various species of mammals, and very little attempt has been yet made to summarize and discuss the results of various new techniques of electron microscopy, histochemistry, biochemistry, immunology, biophysics and physiology (Bishop 1968, Fawcett

1970, 1975a, b, Courot et al. 1970, E. Steinberger 1971, Clermont 1972, Bustos-Obregon et al. 1975, Fritz 1973, Phillips 1975a, Roosen-Runge 1977, Kaczmarski 1978, Gould 1980, Holstein and Roosen-Runge 1981). However, in excellent reviews, Bellvé (1979, 1982) and Bellvé and O'Brien (1983) have discussed the molecular biology of mammalian spermatogenesis and spermatozoa. Such molecular studies on the constituents of germ-cells have become possible as discrete populations of mammalian germ-cells at different stages of their differentiation can be isolated (Bellvé et al. 1977a, b, Meistrich 1972, Romrell et al. 1976).

Besides the potential use of research in reproductive biology, animal production and population control, the comparative studies of subcellular and molecular aspects of spermatogenesis and spermatozoa contribute to our fundamental knowledge of cell, molecular and developmental biology. This is due to the fact that spermatogenesis provides a unique opportunity for the study of both mitotic cell multiplication, for the analysis of cell differentiation, and for the investigation of cell-cell communication. Spermatogenesis also offers an advantage in that all of these aspects can be easily studied in a single developmental lineage. Mammalian spermatogenesis presents little need for indirect means of cell identification and is, therefore, ideally suited for detailed experimental studies undertaken to investigate cell growth, cell differentiation, cell movement, and even cell death. Although the spermatozoa are highly specialized motile cells, they have also many components and characteristics shared by other cells. They form a very useful material for studies of cell motility.

In the previous monographic review, the present author (Guraya 1980) has integrated and discussed the morphological (including ultrastructural), histochemical, biochemical and physiological aspects of development and maturation of mammalian testis. The present book was, therefore, undertaken to deal with similar aspects of spermatogenesis and sperm in the adult. These two timely monographs are intended to provide an insight into the up-to-date data about the cellular and molecular aspects of development and maturation of mammalian testis of subsequent production of spermatozoa and their structure and function. The multidisciplinary approach followed in these two monographs, will certainly permit a much better understanding of the correlation between structure and function in the Leydig cells, spermatogenic cells, Sertoli cells, and spermatozoa.

Part One Spermatogenesis

Chapter I
Seminiferous Epithelium

Mammalian testes form the focal point of the male reproductive system. Their development and differentiation occur during foetal life (Guraya 1980). During the postnatal period, the testes mature to perform two functions, one hormonal (e.g. production of testosterone and other sex steroids, which are also well known to be secreted by the foetal testis) and the other production of spermatozoa (spermatogenesis) in the adult. These processes are localized in two distinct morphological compartments, the vascularized Leydig cells of the interstitium and the avascular seminiferous tubules (Fig. 1). Morphological, histochemical, biochemical, immunological, physiological and endocrinological aspects of development, differentiation and maturation of these two testicular compartments are discussed in a previous monographic review (Guraya 1980). Several other interesting reviews also deal with (1) differentiation and development of the testes (Wartenberg 1981, Ramaswami 1983), (2) endocrinology of the foetal testis (Faiman et al. 1981), and (3) endocrine control of testicular function from birth to puberty (Swerdloff and Heber 1981). The earliest morphological feature by which the foetal testes can be distinguished is the formation of the seminiferous cords. Differentiation of primordial germ-cells in the testes begins at later stages of development, usually after birth. The male-specific antigen, histocompatibility-Y (H-Y) antigen, has been proposed as a testis-determining factor. This hypothesis is based on the finding that the development of testes is related closely with the presence of male-specific antigen regardless of the karyotyped sex. However, the role of these factors during testicular organization is poorly understood. Spermatogenesis proper starts at puberty after a preparatory period of "pre-spermatogenesis" during foetal and postnatal life.

The seminiferous tubules generally constitute 75% of 90% total testicular mass in the adult because of their intense and continuous cell multiplication involved in spermatogenesis (Bellvé and Feig 1984). Their lining is called seminiferous epithelium which is unique among epithelia in the complexity of its organization. It consists of a nonproliferating sessile population of supporting somatic cells, the Sertoli cells, and a proliferating, interdependent population of germ-cells (Figs. 2 and 3) that are displaced centripetally as they progress from being spermatogonia at the base to spermatozoa in the central compartment. Sertoli cells comprise 25% of the normal adult seminiferous epithelium (Dym and Cavicchia 1977). The seminiferous epithelium has a limiting membrane (Figs. 2 and 3), the basement membrane or basal lamina or lamina propria, which consists of flat, myoid cells, fibroblasts and collagen fibres.

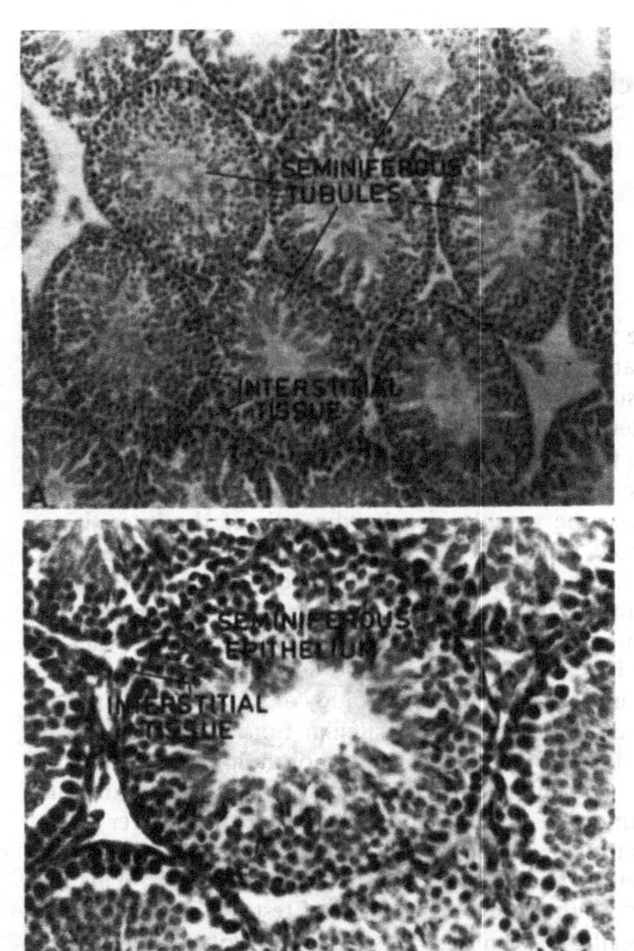

Fig. 1. A Portion of testis of the field rat (*Bandicota bengalensis*) showing two testicular compartments, seminiferous tubules, and interstitial tissue; **B** High-power view of two testicular compartments. Haematoxylin-eosin preparation

Mammalian spermatogenesis is a highly synchronized, regular, long and extremely complex process of cellular differentiation by which a spermatogonial "stem-cell" is gradually transformed into a highly differentiated haploid cell spermatozoon. This differentiation involves three distinct classes of germinal cells – the spermatogonia, the spermatocytes, and the spermatids, which usually are arranged in concentric layers in the seminiferous tubules (Figs. 2 and 3). In the adult mammals, spermatogenesis is a continuous process, which can be divided into three distinct phases (mitosis, meiosis, and spermiogenesis) each characterized by specific morphological and biochemical changes of nuclear and cytoplasmic components (Courot et al. 1970, Clermont 1972, Roosen-Runge 1977, Monesi et al. 1978, Bellvé 1979, 1982, Bellvé and O'Brien 1983). The first phase of spermatogenesis (mitosis), formerly called spermatocytogenesis (spermatogonial stage), now is often referred to as "proliferation and renewal" of sperma-

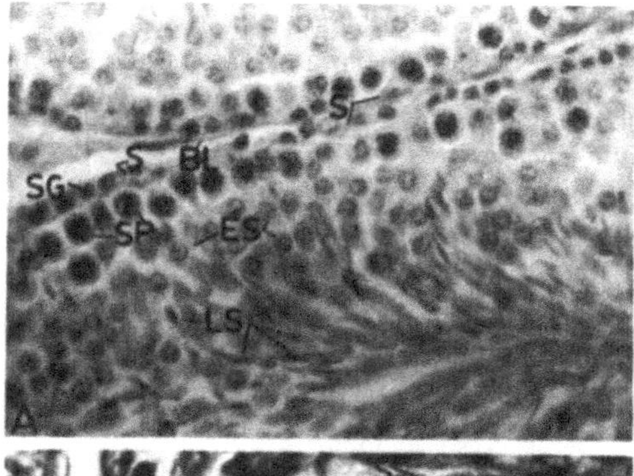

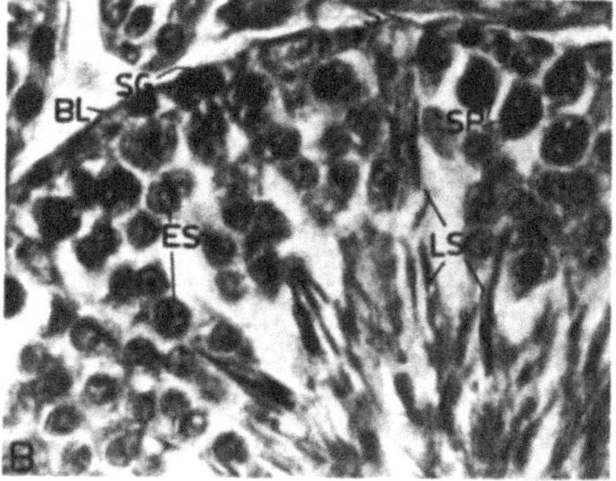

Fig. 2A, B. Portions of the seminiferous tubules in the field rat (*Bandicota bengalensis*) showing the arrangement of spermatogenic cells. The basement lamina (*BL*) is lined by spermatogonia (*SG*) and Sertoli cells (*S*), which towards the tubular lumen are followed by the spermatocytes (*SP*), early spermatids (*ES*) and late spermatids (*LS*) associated with Sertoli cells; maturation of spermatogenic cells is coupled with their shifting towards tubular lumen. Haematoxylin-eosin preparation

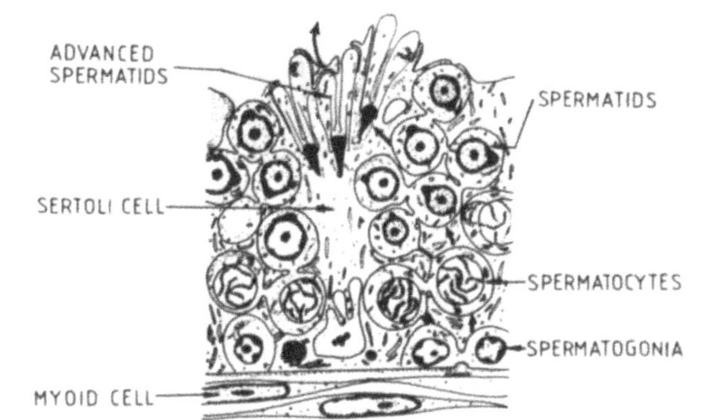

Fig. 3. Arrangement of the cellular components of the seminiferous epithelium (Redrawn from Fawcett 1974)

togonia. During this phase the diploid spermatogonia, which are situated at the periphery of the seminiferous tubule, multiply mitotically to form spermatocytes and also to give rise to new spermatogonial stem cells. The spermatogonial stem cells divide mitotically several times, giving rise to successive generations of type A spermatogonia and subsequently the so-called, intermediate-type spermatogonia which in turn yield type B spermatogonia. In several species there are two generations of type B spermatogonia (two divisions) such as B_1 and B_2. Each one of the last generation of type B spermatogonia divides to form two primary spermatocytes formerly named "resting" but now more appropriately referred to as "preleptotene" spermatocytes, since the nuclei of these cells are by no means resting but actively synthesize DNA in preparation for meiosis (the spermatocytic stage). Meiotic prophase is a very complex process characterized by an ordered series of chromosomal rearrangements which are accompanied by molecular

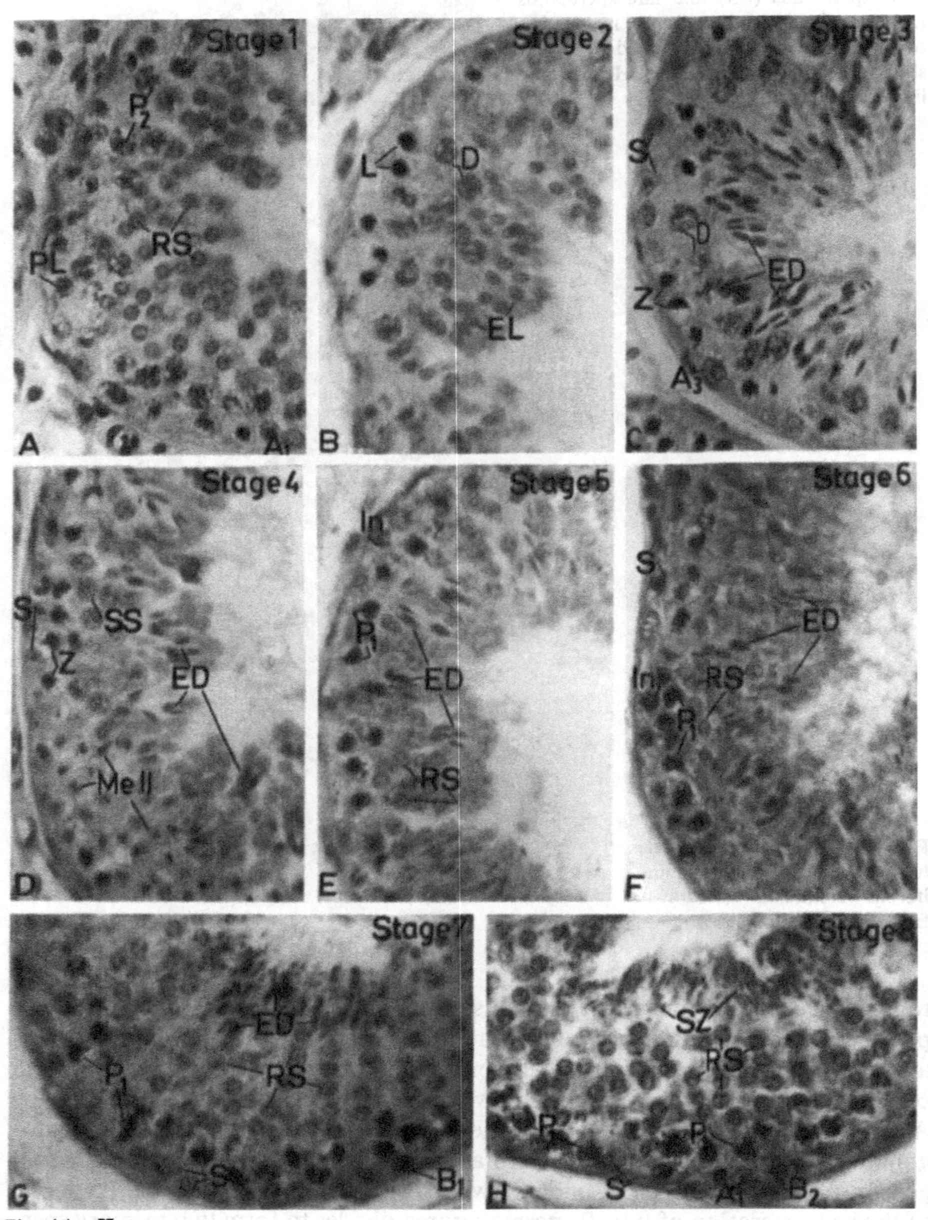

Fig. 4A – H

changes. Meiosis consists of two successive divisions of the primary spermatocyte but is accompanied by only one duplication of chromosomes, thus yielding four spermatids, each one containing only half the somatic number of chromosomes. In the third and final phase of spermatogenesis the spermatids differentiate through complex series of cytological and chemical transformations into spermatozoa; a process usually referred to as spermiogenesis, the stage of spermatids. Four major stages are discernible in spermatid differentiation, designated as the Golgi, cap, acrosomal, and maturation phase, respectively, as will be described in detail in Chap. IV. Each phase can be further subdivided into well-defined steps. All these phases involve very complex morphological, histochemical, biochemical, and biophysical changes in the nucleus and cytoplasm, which are species-specific, as will be discussed in subsequent chapters. The genetic control of spermatogenesis has been demonstrated in mice (Chubb and Nolan 1984, Olds-Clarke 1984).

Various types of evolving germinal cells, i.e. the spermatogonia, spermatocytes, and spermatids at various steps of their respective developments, form well-defined cellular associations or stages (Fig. 4) (Courot et al. 1970, Clermont 1972, E. Steinberger and A. Steinberger 1975, Roosen-Runge 1977, Holstein and Roosen-Runge 1981). It becomes clear on careful examination that these cell as-

Fig. 4A–H. Portions of seminiferous tubules from the testis of goat (*Capra hircus*) after fixation in Zenker-formol and stained with haematoxylin-eosin to show the eight cellular associations or stages of seminiferous epithelial cycle. **A** Stage 1 of cycle which is characterized by the presence of round spermatids towards the tubular lumen. Tubular periphery is lined by A_1 spermatogonia (A_1), Sertoli cells, and preleptotene spermatocytes (*PL*). Towards the tubular lumen, the second generation of preleptotene spermatocytes (*PL*) is followed by the first generation of late pachytene spermatocytes (P_2), which in turn are followed by round spermatids (*RS*). Note the presence of only one generation of spermatids which are round in shape; **B** Stage 2 of cycle in which the preleptotene spermatocytes and late pachytene spermatocytes of previous stage are advanced to leptotene (*L*) and diplotene (*D*) respectively. Towards the tubular lumen are seen elongating spermatids (*EL*) as round spermatids of Stage 1 have started elongating; **C** Stage 3 of cycle in which leptotene spermatocytes of previous stage are advanced to zygotene (*Z*); the elongating spermatids have completed their elongation to form elongated spermatids (*ED*), but spermatocytes of first generation remain at diplotene (*D*). On the tubular periphery are also seen A_3 spermatogonia (A_3) and Sertoli cells (*S*); **D** Stage 4 of cycle which is characterized by the completion of meiotic division of first generation of spermatocytes which were previously at diplotene. Towards the tubular periphery are seen Sertoli cells (*S*) and zygotene spermatocytes (*Z*). Towards the tubular lumen these are followed by the secondary spermatocytes (*SS*), phases of second meiosis (*Me II*), and elongated spermatids (*ED*); **E** Stage 5 of cycle characterized by the presence of second generation of newly formed round spermatids (*RS*) in addition to first generation of elongated spermatids (*ED*). The single generation of spermatocytes which is present is advanced to early pachytene (P_1) in comparison to zygotene phase of Stage 4. In the tubular periphery are also seen intermediate spermatogonia (*In*); **F** Stage 6 of cycle in which the size of the round spermatids (*RS*) has increased. The intermediate spermatogonia (*In*), early pachytene spermatocytes (P_1) and elongated spermatids (*ED*) of Stage 5 remain unchanged. Towards the tubular periphery are also seen Sertoli cells (*S*); **G** Stage 7 of cycle in which the elongated spermatids (*ED*) after their maturation in association with Sertoli cells (*S*) are migrating towards the tubular lumen. Intermediate spermatogonia of Stage 6 are advanced to B_1 spermatogonia (B_1). Spermatocytes remain at early pachytene (P_1) which are followed by round spermatids (*RS*); **H** Stage 8 of cycle in which the testicular spermatozoa (*SZ*) are completely bordering the tubular lumen and these are not intermingled with round spermatids (*RS*). B_1 spermatogonia of Stage 7 have divided to form B_2 spermatogonia (B_2). Also seen in the periphery of the tubule are A_1 spermatogonia (A_1) and Sertoli cell (*S*). The primary spermatocytes remain at early pachytene (P_1)

sociations must succeed one another in time in any given area of the seminiferous tubule. This sequence repeating itself indefinitely has led to the emergence of the notion of a "cycle of the seminiferous epithelium". With cytological methods and [^{3}H] thymidine autoradiography, the details of (1) the seminiferous epithelial cycle and its duration, and (2) the mode of renewal of spermatogonia in various species of mammals are described. All stages of spermatocyte and spermatid differentiation in the mammalian testes are (1) easily defined using routine microscope techniques, (2) readily available in a single anatomical compartment of the body, and (3) usually isolatable in sufficiently high purity to permit both immunological and biochemical analysis (Monesi et al. 1978, Bellvé 1979, Millette 1979a, Meistrich et al. 1981, Wright et al. 1983).

A. Cycle of Seminiferous Epithelium

A "generation" of germ-cells is defined as a group of cells formed at approximately the same time and evolving synchronously through the spermatogenic process (Clermont 1972, E. Steinberger and A. Steinberger 1975, Bilaspuri and Guraya 1980a, 1984b, 1986a). A definite time interval elapses after a syncytium of spermatogonia has entered phase 1 of spermatogenesis and before the next group will start dividing to initiate a second generation. The time interval separating the appearance of germinal cells at the same degree of differentiation between two successive appearances of the same cellular association in a given area of the seminiferous tubule is definite in duration. This interval, called the cycle of the seminiferous epithelium, requires about 16 days in man, 13 in bull and rat, 12 in stallion, and 10 in ram and rabbit (Clermont 1972, Swierstra et al. 1974).

The various generations of spermatogenic cells form cellular associations (stages) of fixed composition (Fig. 4). Due to the precise and regular timing of the steps of spermiogenesis, spermatids at a given step are associated always with spermatocytes and spermatogonia at given steps of their differentiation. Only a limited number of such cell associations can be seen in the various cross-sections of seminiferous tubules (Fig. 1). A close examination of the composition of these various cell associations shows that, in any zone of the seminiferous tubules, they must appear and follow each other in a given and fixed sequence. These cellular groupings, reappearing at regular intervals, represent "stages" of a cycle of the seminiferous epithelium, defined more precisely as "a complete series of the successive cellular associations appearing in any one zone of the seminiferous epithelium" (Courot et al. 1970, Clermont 1972, E. Steinberger and A. Steinberger 1975, Guraya and Bilaspuri 1976a, b, c, Roosen-Runge 1977, Bilaspuri and Guraya 1980a, 1984b, 1986a). These stages represent arbitrary subdivisions through a continuous process and, therefore, the demarcation between the end of a stage and the beginning of the next is often imprecise. For a given mammalian species, the number of cell groupings or stages of the cycle vary depending on the criteria used to characterize or identify them. Such criteria chosen to be helpful in the study of the seminiferous epithelium are arbitrary and vary between investigators.

Recently, two distinct trends have developed in the choice of criteria for identifying the stages of the cycle of the seminiferous epithelium (Roosen-Runge 1962, Clermont 1972, Berndston and Desjardins 1974, E. Steinberger and A. Steinberger 1975). The first method involves the use of the nuclear morphology of spermatids in conjunction with the position of the more mature spermatids within the seminiferous epithelium (Fig. 4). The latter cells, arranged in fascicles, plunge deeply into infoldings within the epithelium, towards the Sertoli cell nuclei. Later the more differentiated cells shift towards the lumen of the tubule for eventual release or spermiation. In this approach various staining methods are used: haematoxylin-eosin being the stain most commonly used (Fig. 4). Such classifications consist of eight stages in rat, ram, bull, rabbit, boar, mink, and monkey (Clermont 1972), dog (Foote et al. 1972), stallion (Swierstra et al. 1974), vole (Grocock and Clarke 1975), and grey squirrel (*Sciurus carolinensis*) (Tait and Johnson 1982). Eight stages of the cycle of seminiferous epithelium have been reported also in the bull (Hochereau-de-Reviers 1970) and buffalo, goat (Fig. 4) and ram (Guraya and Bilaspuri 1976b, Bilaspuri and Guraya 1980a, 1984b, 1986a), but in these studies, stages 1, 2, 3, 4 and 8 have been subdivided. The subdivision of stages of longer duration helps in studying the behaviour of various germinal cells more precisely.

The second method involves the use of the steps in differentiation of spermatids as observed in periodic acid-Schiff (PAS) haematoxylin-stained sections of the seminiferous tubules. This technique, when applied to the sections of testes fixed in Zenker-formol, stains deep purple the developing acrosomic system of developing spermatids and therefore the evolution of this organelle can be used to distinguish the cell groupings to which these spermatids belong. This approach is very useful in providing classifications composed of larger number of stages. With the PAS-haematoxylin technique, 14 stages have been identified in the rat, 13 stages in the hamster and guinea-pig and 12 stages in the mouse and monkey (Clermont 1972). Also 14 stages have been observed in the buffalo (Fig. 5) (Guraya and Bilaspuri 1976b, Bilaspuri and Guraya 1980a) and goat (Bilaspuri and Guraya 1984b). The division of the seminiferous epithelial cycle into 14 stages is useful for studying the behaviour of various germinal cells more accurately.

A close examination of serial sections from biopsies of human testis has revealed six stages in humans (Fig. 6) (Heller and Clermont 1964; Clermont 1972, E. Steinberger and A. Steinberger 1975, Zuckerman et al. 1978). In any one cross-section of seminiferous tubule 3 to 4 stages may be present (Holstein and Roosen-Runge 1981). Moreover, these cell associations are overshadowed often by a mixing of germ-cells at the interface of adjacent cell associations, by the frequent absence of one or more generations of spermatogenic cells, and by the displacement of cells during histological preparations. This form of spermatogenic progression is called irregular and is similar to that reported in the olive baboon (*Papio anubis*) (Chowdhury and Steinberger 1976, Chowdhury and Marshall 1980).

The presence of more than one stage of spermatogenesis in cross-sections of human seminiferous tubules and the inappropriate presence or absence of cell types in typical cell associations ("a typical" or "heterogenous" stages) have made quantitative studies on human spermatogenesis difficult (Chowdhury and

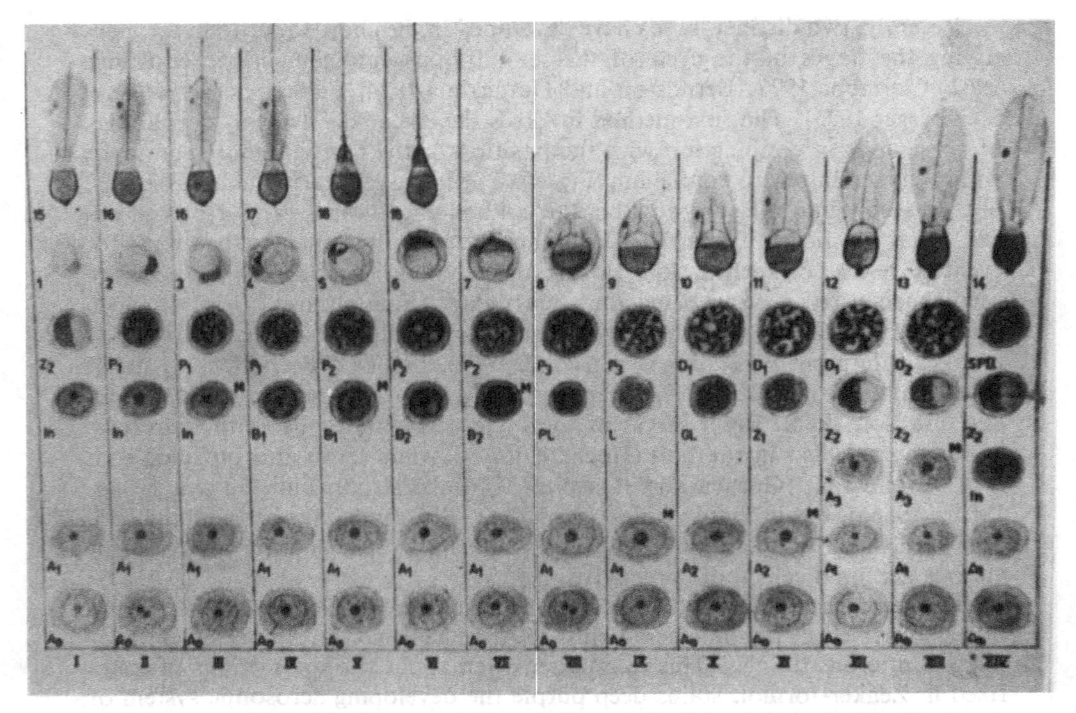

Fig. 5. Cellular composition of the 14 stages of the seminiferous epithelial cycle in the buffalo. Each column (*I* to *XIV*) shows the germinal cells in a given cellular association. The associations or stages are identified by the first 14 out of the 18 steps of spermiogenesis. The latter are defined by the morphological changes in the nucleus (*lightly shaded*) and in the acrosomic structure (*dark grey*) of the spermatids as seen in periodic acid-Schiff/haematoxylin staining. A_0 type A_0 spermatogonia; A_1, A_2 and A_3 represent three types of type A spermatogonia; *In* intermediate spermatogonia; B_1 and B_2 two types of type B spermatogonia; *PL* preleptotene; *L* proper leptotene; *GL* granular leptotene; Z_1 and Z_2 early and late zygotene; P_1, P_2 and P_3 early, mid and late pachytene; D_1 and D_2 early and late diplotene; *SPII* secondary spermatocytes; *M* close to the spermatogonium means mitosis (From Guraya and Bilaspuri 1976b)

Marshall 1980, Schulze and Rehder 1984). In addition, several authors have denied the presence of a wave of spermatogenesis in the human. The factors responsible for the irregular pattern of spermatogenesis in man and baboon also remain to be investigated. However, Chowdhury and Marshall (1980) have studied quantitatively characteristics of the irregular pattern of spermatogenesis in the olive baboon. The frequent occurrence of stages in both cross or oblique sections of seminiferous tubules has been attributed to a patchy distribution of cell associations, rather than to a sectioning artifact. The occurrence of inappropriate cells in various cell associations and spermiation occurring in stages 5 to 7 are believed to be due to asynchronous division of some spermatogonia.

Recently, Schulze and Rehder (1984) have classified various types of human primary spermatocytes by means of morphological and morphometrical studies. Based on this classification, the topographic arrangement of the spermatocyte populations in the longitudinal course of seminiferous tubules is determined. The

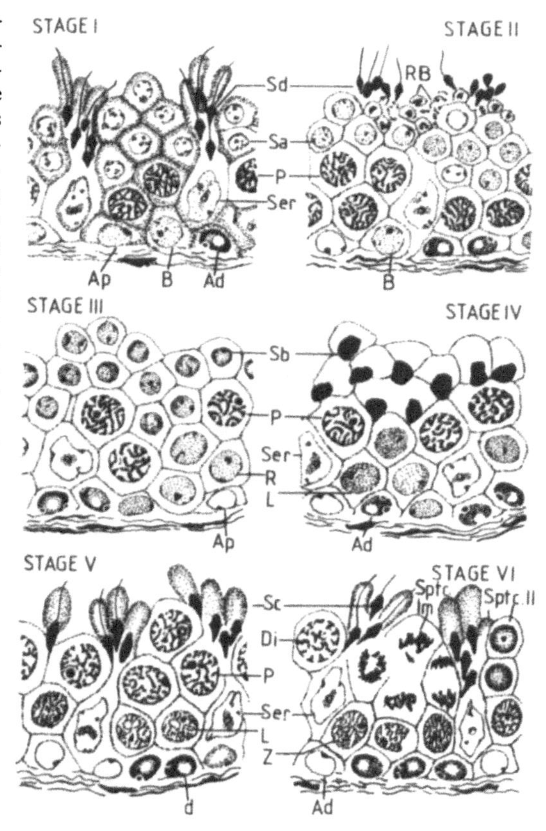

Fig. 6. Representation of cellular composition and topography of the six typical cellular associations found repeatedly in human seminiferous tubules. These cell associations, corresponding to stages of the cycle of the seminiferous epithelium, are numbered with roman numerals, *Stages I–VI. Ser* Sertoli nuclei; *Ap* and *Ad* pale and dark type A spermatogonia; *B* type B spermatogonia; *R* resting primary spermatocytes; *L* leptotene primary spermatocytes; *Z* zygotene primary spermatocytes; *P* pachytene primary spermatocytes; *Di* diplotene primary spermatocytes; *Sptc-Im* primary spermatocytes in division; *Sptc-II* secondary spermatocytes in interphase; *Sa, Sb, Sc, Sd* spermatids at various steps of spermiogenesis; *RB* residual bodies (Redrawn from Clermont 1963)

distribution analysis suggests that spermatocyte populations of successive degrees of development are arranged in helices that are contracted conically to the tubular lumen (Fig. 7). Populations of the same degrees of development are arranged on helices with constant diameter. For each degree of development a "basic helix" proper could be determined. On each of these "basic helices" the centres of gravity of the populations diverge continuously 142.6° ± 14.2°.

Recently, some other techniques have been introduced to study the dynamics of the seminiferous epithelium. Huckins and Oakberg (1978a) have made a morphological and quantitative analysis of spermatogonia in the mouse using whole mounts of seminiferous tubules. Nikkanen et al. (1978) have identified spermatogenic stages in living seminiferous tubules of man. The parts of seminiferous tubules containing late spermatids with condensed nuclei (stage 2) absorb more transmitted light than do parts at other stages. Spermatogenic stages 1, 3, 4 and 6 are identifiable by phase contrast microscopy. Johnson et al. (1979) have made use of scanning electron and light microscopy in the identification of spermatogenic stages in the equine seminiferous tubule. Andersen (1978) has used the fine structure of developing spermatids as a basis for determining the stages of spermateleosis in the blue fox (*Alopes lagopus*). The technique known as transillumi-

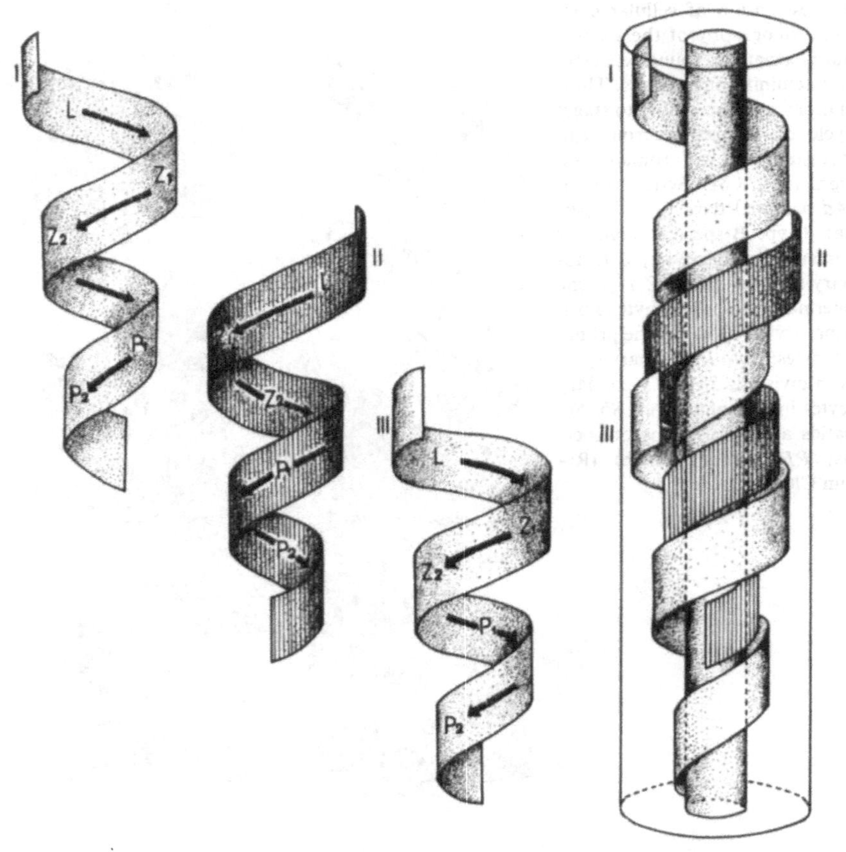

Fig. 7. Construction of a three-dimensional model of the arrangement of primary spermatocytes in the human seminiferous tubule. Populations of subsequent degrees of development occupy helically running strip-shaped areas of the seminiferous epithelium. With advancing degree of development of the spermatocytes the windings of the strip approach the lumen of the tubule. In the longitudinal course of the tubule every 200 – 250 μm a new strip starts, which diverges 142.5° + 1 – 14.2° from the foregoing one. Corresponding points on the subsequent strips can be connected by helices of constant diameters and steep slopes (not shown in the present illustration). *L* leptotene; Z_1 early zygotene; Z_2 late zygotene; P_1 early/mid-pachytene; P_2 late pachytene (Redrawn from Schulze and Rehder 1984)

nation-assisted microdissection has been of great help to isolate tubular segments at defined stages of rat spermatogenic cycle (Parvinen and Ruokonen 1982). The results of all these studies have demonstrated clearly that the cell kinetics of spermatogenesis proceed with a high degree of synchrony. Seasonal changes in spermatogenesis in the blue fox have been quantified by DNA flow cytometry (Smith et al. 1984). Soluble Mn^{2+}-dependent adenylate cyclase activity is also measured.

The classification of cell associations into readily distinguishable stages of a cycle form useful tools in the structural and cytochemical study of germ-cell development and in investigations on the influence of various factors, physical (e.g. X-rays) or chemical (e.g. drugs, hormones), that act on mitosis, meiosis and cell differentiation during spermatogenesis (Bellvé 1979) as well as vasectomy (Hadley and Dym 1983). The classification of cell associations into stages of seminiferous epithelium cycle also has helped us to follow the localization of various enzyme systems in the differentiating germinal cells more precisely (Bilaspuri and Guraya 1982, 1983a, b, e, 1984c).

The definition of the seminiferous epithelial cycle stages, combined with the use of radioactive nucleic acid precursors, has served as a basis of our knowledge of chromosome activity and its variation during spermatogenesis (Monesi et al. 1978, Bellvé 1979). Monesi (1962), using tritiated thymidine, has observed seven distinct periods of DNA synthesis during the seminiferous epithelial cycle in mouse that are localized at stages 8, 10, 12, 2, 3, 5, 7 – 8 and involve different types of spermatogonia (A_1, A_2, A_3, A_4, intermediate and type B) and preleptotene spermatocytes. Corresponding analyses in the rat have revealed the same approximate location of DNA synthesis in relation to the epithelial cycle (Hilscher and Hilscher 1969). Among the various periods of DNA synthesis, L.-M. Parvinen and L. Parvinen (1978) have observed two distinct peaks. The first peak in stages 4 to 6 is due to the intermediate and type B spermatogonia and represents mitotic DNA synthesis. Stages 4 to 6 do not correspond to those defined above, suggesting some differences in distribution beyond stage 5 which may be due to different methods used. The second one in stages 8 to 9 is due principally to replication of DNA in preleptotene spermatocytes; the mitotic cycle DNA synthesis may also be contributing to the second peak. The existence of a distinct mitotic peak in stages 4 to 6, and not in other stages where other spermatogonia are dividing and synthesizing DNA, may be due to diminished degeneration among types A_4 to B spermatogonia in contrast to extensive degeneration of type A spermatogonia (especially $A_1 - A_3$) (Huckins 1978c). However, the bursts of mitotic activity occur at 42-h intervals, the cell-cycle time of each spermatogonial type. Similarly, the preleptotene spermatocytes in rat have not been reported to degenerate, which may explain the second distinct peak in stages 8 to 9. The spermatogonia have a normal diploid amount of DNA. The preleptotene spermatocytes, which are formed by the division of type B spermatogonia, undergo a final phase of DNA replication before progressing further into meiotic prophase. This phase of DNA synthesis takes place in a semi-conservative manner, as reported for somatic cells. However, there exist differences in regard to transcription of myc gene in normal and spermatogenic cells (Stewart et al. 1984). The proliferating spermatogonia express myc at amounts significantly below those for normally dividing somatic cells. The apparent lack of myc RNA in the preleptotene spermatocytes suggests that there is no myc expression in the premeiotic G_1 period. Finally, the spermatocytes have a tetraploid, the secondary spermatocytes a diploid, and the spermatids a haploid amount of DNA (see Courot et al. 1970, Monesi et al. 1978, Bellve 1979, Mann and Lutwak-Mann 1981).

RNA synthesis in relation to the epithelial cycle has been studied by autoradiographic means (Monesi 1964, 1965a, 1971, Utakoji 1966, Loir 1972a, b, Monesi et al. 1978). The main RNA synthesis takes place in spermatogonia, mid-pachytene spermatocytes and, to a small extent, in round spermatids, as will be discussed in detail in Chaps. III and IV.

B. Wave of Seminiferous Epithelium

Each cell association, in addition to being symmetrical radially, shows sequential organization along the length of the seminiferous tubule, there by producing the "wave" of spermatogenesis. The cycle of the seminiferous epithelium therefore differs from the "wave of the seminiferous epithelium" by the fact that it is a dynamic histological phenomenon taking place in time in any one area of the seminiferous epithelium. In contrast to this situation, "wave" does not have any kinetic significance and simply represents, more or less, orderly distribution of the cellular associations along the seminiferous tubules at any one time. In other words, the wave is in space and the cycle is in time. The problems of wave in several species of mammals have been described previously (Clermont 1972, E. Steinberger and A. Steinberger 1975). The presence of the wave of the seminiferous epithelium has been reported to occur in testes of a number of mammals, including the mouse, bull, guinea pig, rabbit, ram, boar, dog, cat, marsupials, etc. Radioautographic studies have provided additional evidence for the concept that the spermatogenic cycle is coordinated in space and time by the mitotic activity of spermatogonia. But the mechanisms involved in synchronization and the origin of the wave need to be determined.

The kinetics of the wave of seminiferous epithelium is difficult to demonstrate in human testes because the associations are highly irregular in their cellular composition and topographic distribution, as already described. Therefore, the clone concept for human spermatogenesis has been suggested, in contrast to the spermatogenic wave so clearly described in other mammalian species. In the baboon, maps of topographic distribution of various stages along the seminiferous tubules have been divided into three categories (Chowdhury and Marshall 1980). The first category includes tubules with a single cell association throughout their entire length. Tubules belonging to the second category show two or more cellular associations arrayed consecutively, but usually with irregular boundaries. The third category includes tubules with relatively small zones occupied by different cell associations. The topographic arrangement of stages in the first and second categories of tubules appears similar to that of rat, whereas the third category of tubules shows a pattern observed only in the human seminiferous tubules. The reasons for these "irregularities", which occur in the seminiferous epithelium of the baboon, the chimpanzee and the human, are poorly understood (E. Steinberger and A. Steinberger 1972). However, Leidl (1968) suggested that the apparent lack of order may reflect a regression of the reproductive capacity of these species. But Schulze and co-workers (Schulze 1982, Schulze and Rehder 1984) have studied the problem of the local organization of human spermatogenesis. These authors have found clear indications that the spermatogenic stages are not distributed at random but are orderly sequences of stages that are fitted helically into the longitudinal course of the seminiferous tubule (Fig. 7).

C. Duration of the Cycle of the Seminiferous Epithelium

The entire process of spermatogenesis or formation of spermatozoa from the least mature differentiating germ-cell (A_1 spermatogonia) needs four cycles of the

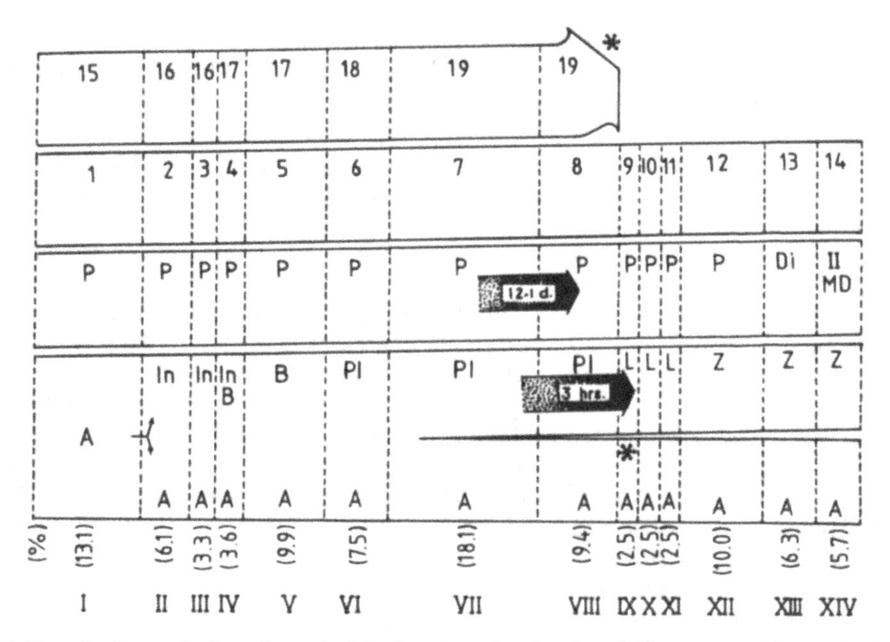

Fig. 8. Quantitative method used to calculate duration of cycle of seminiferous epithelium and of spermatogenesis in Sprague-Dawley rat (details in text). Framework of diagram gives cellular composition of the 14 stages of cycle (14 vertical columns labelled with *Roman numerals*). Width given to various columns is proportional to relative duration of stages expressed in percentages (*numbers in parentheses*). *A, In, B* type A, intermediate-type, and type B spermatogonia; *Pl* preleptotene primary spermatocytes; *L* leptotene spermatocytes; *Z* zygotene spermatocytes; *MD* meiotic divisions; *1 – 19* steps of spermiogenesis. Two *horizontal arrows* show exact location of most advanced labelled spermatocytes in cycle of seminiferous epithelium at 3 h and 12 days + 3 h (12.1 days) after a single injection of [^{3}H]-thymidine. *Asterisks* indicate times of onset (*bottom row*) and of termination (*top row*) of spermatogenesis. With data the duration of both cycle of seminiferous epithelium and of spermatogenesis can be calculated (Redrawn from Clermont 1972 after Clermont and Harvey 1965)

seminiferous epithelium in the rat (Fig. 8). Thus, the duration of the spermatogenic process can be determined by either measuring the time between the division of type A_1 spermatogonia and the release of mature spermatozoa, or by measuring the duration of a single cycle of the seminiferous epithelium and multiplying this value by four. The duration of both spermatogenesis and cycle of the seminiferous epithelium has been studied extensively in different mammalian species (Clermont 1972, E. Steinberger and A. Steinberger 1975, Schuler and Gier 1976, Roosen-Runge 1977, Persona and Bustos-Obregon 1983). Several experimental approaches have been used to evaluate the duration.

With the availability of radioactive tracers such as [^{14}C] thymidine [^{35}S] methionine and [^{32}P] and the development of radioautography, it became possible to determine the duration of the cycle of seminiferous epithelium in intact mammals more precisely. Tritiated thymidine, which is incorporated selectively in the nuclei of spermatogenic cells preparing for mitosis and meiosis, is the most widely accepted label to time the cycle. Quantitative analyses of the tubular cross-sections having labelled cells (especially the most advanced labelled cells originating from the preleptotene spermatocytes, which incorporate [^{3}H] thymidine in their nuclei prior to meiosis) give accurate and reproducible values for the duration of the cycle of different mammalian species studied. Clermont (1972) has illustrated the method employed to calculate the duration of one cycle of the seminiferous epithelium in the Sprague-Dawley rat. The values obtained, which range from 8.6 days (e.g. mouse) to 16 days (man), are constant within a given species, and even within a strain of animal.

With tritiated thymidine the duration of the cycle of seminiferous epithelium can be determined with an appreciable degree of accuracy, but the values obtained for the duration of the whole process of spermatogenesis can only be an approximation for the following reasons. The cycle of seminiferous epithelium shows a sharply demarcated beginning and end point (i.e. the early appearance of step 1 spermatids), but the spermatogenic process has an ill-defined and often debated starting point (i.e. the exact time of the initial spermatogonial stem cell mitosis) and an imprecise end point (i.e. the release of spermatozoa from the seminiferous epithelium). In spite of these problems, the whole process of spermatogenesis in the Sprague-Dawley rats has been calculated to last approximately 51.6 days. In the mouse, the total duration of spermatogenesis, from the stem cell to the mature spermatid, is about 34.5 days (see Monesi et al. 1978). The mitotic phase of spermatogenesis lasts about 8 days, meiosis 13 days, and spermiogenesis about 13.5 days (Monesi 1974). The times of onset and termination of spermatogenesis vary with the species and the criteria employed by the various authors to define them. These may vary also in the mode of stem-cell renewal, as proposed by Huckins (1971a, b) and Oakberg (1971a, b). However, the values obtained by different researchers represent useful estimates of the duration of spermatogenesis. The duration of the cycle of seminiferous epithelium appears to be constant for a given species; 35 days for the hamster, 49 days for the bull, 43 days for the rabbit, and 64 days in man. Apparently, the difference in the duration of spermatogenesis is not only a species-specific characteristic, but also may vary among strains (E. Steinberger and A. Steinberger 1975). No factor is found to influence the rate of development of germ-cells, not even the pituitary gonadotrophins (Clermont and Harvey 1965, Desclin and Ortavant 1963).

D. Coordination of Evolution of Several Superimposed Generations of Germ-Cells

The remarkable constancy in the composition of cell groupings has suggested the presence of some regulatory mechanism that coordinates the rate of development of several generations of spermatogenic cells constituting the seminiferous epithelium. It will be interesting to mention here that although the correlation between the evolutionary steps of the several generations of germ-cells appears to be impressive, it is not absolutely perfect. For example in the rat, the initial spermatogonial mitoses generally appear in association with step 9 spermatids (in stage 9 of the cycle), but these mitoses also may be seen in association with step 8 and step 10 spermatids (Courot et al. 1970, Clermont 1972, E. Steinberger and A. Steinberger 1975). This suggests that there is some fluctuation in the composition of the cell grouping, but such variations appear to be small, and do not disturb the overall composition of the stages of the cycle.

Very divergent views exist about the factors which coordinate the evolution of superimposed generations of germ-cells (Clermont 1972, Mann and Lutwak-Mann 1981). Sertoli cells, the only nongerminal cells in the seminiferous epithelium, are believed to regulate these cyclic changes of germ-cell multiplication by alterations in shape (Leblond and Clermont 1952a), size (Cavicchia and Dym 1977, Ulvik and Dahl 1981), lipid content (Niemi and Kormano 1965, Kerr and de Kretser 1975), and hormone responsiveness (Gordeladze et al. 1982a). Sertoli cells are suggested to play a significant role in this regard as they are in contact with all germinal cells through their long cytoplasmic processes. It has been suggested that these sustentacular cells emit substances (possibly a hormone) to activate (or induce) the simultaneous division of a large group of spermatogonia.

The possible effect of one generation of germ-cells on another has also been ruled out to explain their constant association. This is supported by the fact that the destruction of one or two generations of cells by X-rays or heat does not disturb the specific association of the remaining generations of cells seen within the tubule at variable periods after the treatment (Clermont 1972, E. Steinberger and A. Steinberger 1975).

The residual cytoplasmic bodies also are believed to be involved in regulating the cycle of the seminiferous epithelium (Clermont 1972, Mann and Lutwak-Mann 1981). After being phagocytosed the residual bodies shift to the basal lamina near the spermatogonia at about the time of the first stem-cell mitoses that initiate spermatogenesis. This hypothesis has not been accepted because the spermatogonial stem-cells initiate divisions at fixed intervals even in the total absence of such residual bodies, as occurs in X-irradiated rats. In both situations, the residual bodies are not formed due to the lack of mature spermatids. These observations indicate that the cycle of the seminiferous epithelium is the direct result of the entrance, at fixed intervals, of the spermatogonial stem cells in spermatogenesis and of the fixed duration of the various steps of spermatogenesis. But the mechanisms that control the rate of these phenomena need to be determined. The various proteins which are secreted cyclically by the Sertoli cells (Wright et al. 1983) may be of some significance in this regard.

E. Synchronous Development of Spermatogenic Cells

A characteristic histological feature of the seminiferous epithelium is the synchronous evolvement of groups of spermatogenic cells (Figs. 1 and 2) (Dym and Fawcett 1971, Clermont 1972, E. Steinberger and A. Steinberger 1975, Lok and De Rooij 1983). For example in the rat, groups of germ-cells at approximately the same step of development may occupy large segments of tubules. In other words, one such group may have several thousand germ-cells. But in man, the number of germ-cells observed at the same step of development is relatively small, and such groups of germ-cells occupy restricted areas of the tubular wall.

When a large group of germ-cells in the seminiferous epithelium of the rat is studied very carefully, it can be found that the synchrony may not be perfect as the side-by-side groups of cells may be slightly ahead or behind in their evolution. Lok and De Rooij (1983) have observed that S phase starts and finishes at virtually the same time in adjacent clones of differentiating spermatogonia in the Chinese hamster. But clones of spermatogonia in the S phase are observed intermingled with A_1 cells in other phases of the cell cycles. Adjacent clones of A_2 spermatogonia do not always arise at the same moment. The results of various studies suggest that even if the development of spermatogenic cells is not perfectly synchronized, the parasynchronous evolution of large groups is still remarkable.

Several workers using electron microscope have observed that groups of developing male germ-cells, such as the spermatogonia, spermatocytes, and spermatids, are connected by intercellular bridges (Fig. 9) (Guraya 1980, Holstein and Roosen-Runge 1981, Kleiss and Liebich 1983). These bridges develop as a result of an unusual process of incomplete cytokinesis that leaves patent cytoplasmic channels between daughter cells after the dispersion of spindle microtubules. Fawcett et al. (1959) suggested that the syncytial nature of the germ-cells forms the basis for the synchrony of their differentiation. Although intercellular bridges have been demonstrated in a variety of mammalian species (Kleiss and Liebich 1983), the exact number of cells connected through these bridges was controversial. A careful examination of intercellular bridges has revealed that a group of conjoined spermatids may number in the hundreds (512) (Dym and Fawcett 1971). However, the theoretical numbers of spermatids may not be achieved due to the normal degeneration of spermatogonia and spermatocytes, as will be discussed in Chap. II, thus creating gaps in the long clusters of cells. This will certainly reduce the actual upper size limit of the syncytia, but regardlessly it remains very large. Moens and Hugenholtz (1975), after making a careful study of serial sections oriented parallel to the base of the epithelium, have observed 80 conjoined spermatocytes where 128 may be expected. The presence of such large numbers of conjoined spermatogenic cells further support the original suggestion of Fawcett et al. (1959) that syncytia act as a device for synchronization of their differentiation. Moens and Go (1971) have estimated that some 40 syncytia enter meiosis and spermatid development at the same time in each square millimeter of seminiferous epithelium. Coordination of differentiation throughout these larger areas may be related to the mediation of Sertoli cell communication through their unique junctional complexes (Weber et al. 1983). Clermont (1972) has suggested that in addition to intercellular bridges some other factor may be involved in the synchronous or parasynchronous evolution of germ-cells in the mammalian testes. Lok and De Rooij (1983a) have studied the cell cycle properties and synchronization of differentiating spermatogonia with the "fraction of labelled mitoses technique" and autoradiography of whole-mounted seminiferous tubules at 1 h after injection of [^{3}H] thymidine. Adjacent clones of differentiating spermatogonia start and finish their S phase at virtually the same moment. From the results of this study, a hypothesis is proposed that each generation of differentiating spermatogonia receives a stimulus to divide from outside the spermatogonial compartment. This ensures the synchronous

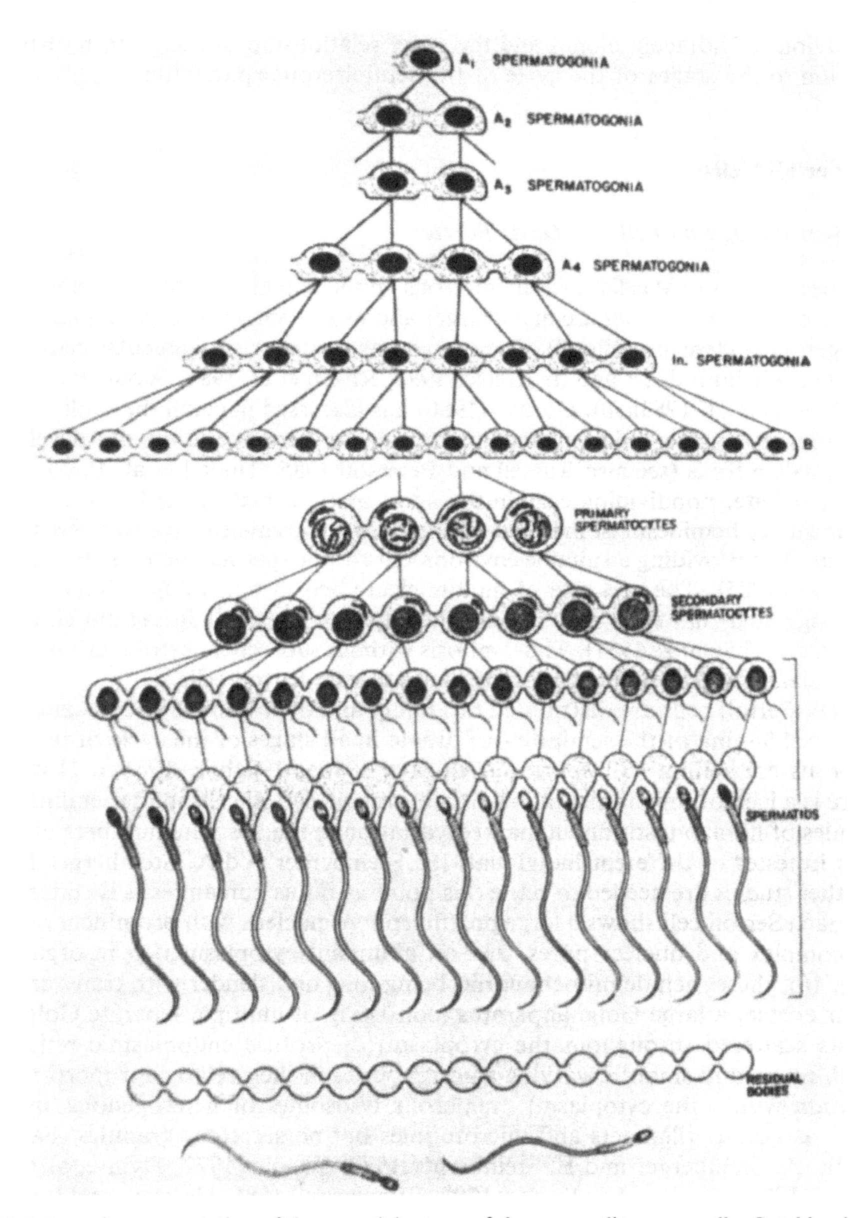

Fig. 9. Schematic representation of the syncytial nature of the mammalian germ-cells. Cytokinesis is incomplete in all but the earliest spermatogonial divisions, resulting in expanding clones of germ-cells that remain joined by intercellular bridges (From Fawcett 1975c, who modified after Dym and Fawcett 1971)

behaviour of adjacent clones and the strict relationship of the pattern of proliferation to the stages of the cycle of the seminiferous epithelium.

F. Sertoli Cells

I. Morphology and Blood-Testis Barrier

Besides the germinal cells, the seminiferous epithelium also contains a stable population of Sertoli cells that are much larger and more complex morphologically than the germinal elements (Fig. 2). They serve as support or sustentacular cells of the whole epithelium (Kerr and de Kretser 1981, Ritzen et al. 1981, Wong and Russell 1983, Vogl et al. 1983a, b). Guraya (1980) has discussed in detail the proliferation, development, differentiation and maturation of Sertoli cells in the developing mammalian testis (see also Russell and Peterson 1985, Tindall et al. 1985). These large, stellate, nondividing cells in the adult are attached to the basal lamina by rudimentary hemidesmosomes and extend between germ-cells to reach the tubular lumen, thus providing a suitable environment for the spermatogenesis (Russell and Peterson 1985). The presence of multinucleate Sertoli cells in aged human testes has suggested that they may resume the capacity to divide in certain situations (Schulze and Schulze 1981). Thus, mitosis without subsequent cytokinesis might be an explanation for the formation of multinucleate Sertoli cells.

The Sertoli cells are distributed randomly and their number is constant along the basal lamina of the seminiferous tubule in all stages of the cycle of the seminiferous epithelium (Bilaspuri and Guraya 1980a, 1984b, c, 1986a). However, there is a lack of uniformity in the distribution of Sertoli cells in the seminiferous tubules of human testis and a marked variation in the absolute numbers of these cells in testes of different individuals (E. Steinberger and A. Steinberger 1975). Further studies are needed to settle this point as far as human testis is concerned.

Each Sertoli cell shows a large polymorphous nucleus with prominent nucleolar complex and nuclear pores, and an abundant cytoplasm rich in organelles (Fig. 10); these include mitochondria, being long and slender with transverse tubular cristae, a large Golgi apparatus (consisting of multiple separate Golgi elements scattered throughout the cytoplasm), a profuse endoplasmic reticulum (both rough and smooth, varying among species in their relative proportion and location within the cytoplasm), numerous lysosomes of heterogenous appearance, as well as filaments and microtubules but no secretory granules (Fawcett 1975a, A. Steinberger and E. Steinberger 1977, Weaker 1977, Dym et al. 1977, Russell 1980a, Kerr and de Kretser 1981, Ritzen et al. 1981, Holstein and Roosen-Runge 1981, Bergh 1983a, Vogl et al. 1983a, b, Russell and Peterson 1985, Tindall et al. 1985). Based on the study of structure of Sertoli cells in the rat, two distinct endosomal compartments have been distinguished (Morales and Clermont 1985). Cruvellier et al. (1985) have observed an increase in volume of mitochondria in Sertoli cells during stages 12–14 of the rat seminiferous epithelium cycle, which coincides with the maximal lipid content of these cells during the cycle, as well as with a marked increase of the endocytotic activity, reflecting the changes of energy requirements of Sertoli cells during the cycle of seminiferous epitheli-

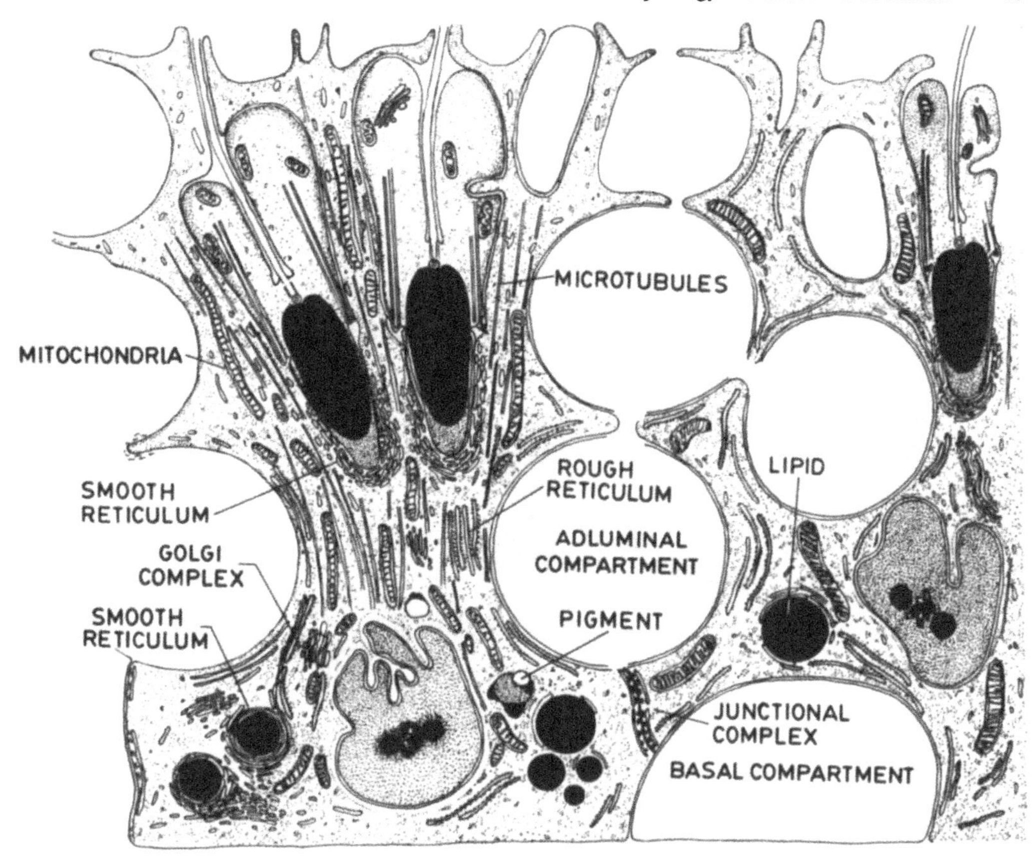

Fig. 10. Diagram of Sertoli cells showing their shape and relationship with germ-cells, as well as form and distribution of their principal organelles and inclusions. The occluding junctions between Sertoli cells divide the seminiferous epithelium into a basal compartment occupied by the spermatogonia and preleptotene spermatocytes and an adluminal compartment containing more advanced stages of the germ-cell population. The occluding Sertoli-Sertoli junctions are the principal component of the blood-testis barrier (Redrawn from Fawcett 1975a)

um. The cyclic nature of endocytotic activity in Sertoli cells has been demonstrated for the rat (Morales et al. 1984). In their amount and distribution, cytoplasmic filaments and microtubules constituting cytoskeleton vary greatly in different regions of the same Sertoli cells (Fig. 11). Filaments of various descriptions have been seen around the nucleus (Dym 1973, 1977, Fawcett 1975a), at the base of the cell (Fawcett 1975a, Russell 1977a), in apical cytoplasm (Fawcett 1975a), and in association with junctions (Flickinger and Fawcett 1967, Dym and Fawcett 1970, Kaya and Harrison 1976, Russell 1977a, b, c). A wreath of filaments close to the Sertoli cell membrane surrounds the human spermatid (Holstein and Roosen-Runge 1981). Sertoli cells of the ground squirrel (*Spermophilus lateralis*) show a remarkably well developed cytoskeleton (Fig. 7) (Vogl et al. 1983a). Microfilaments occur throughout the cell. Intermediate filaments lie around the nucleus,

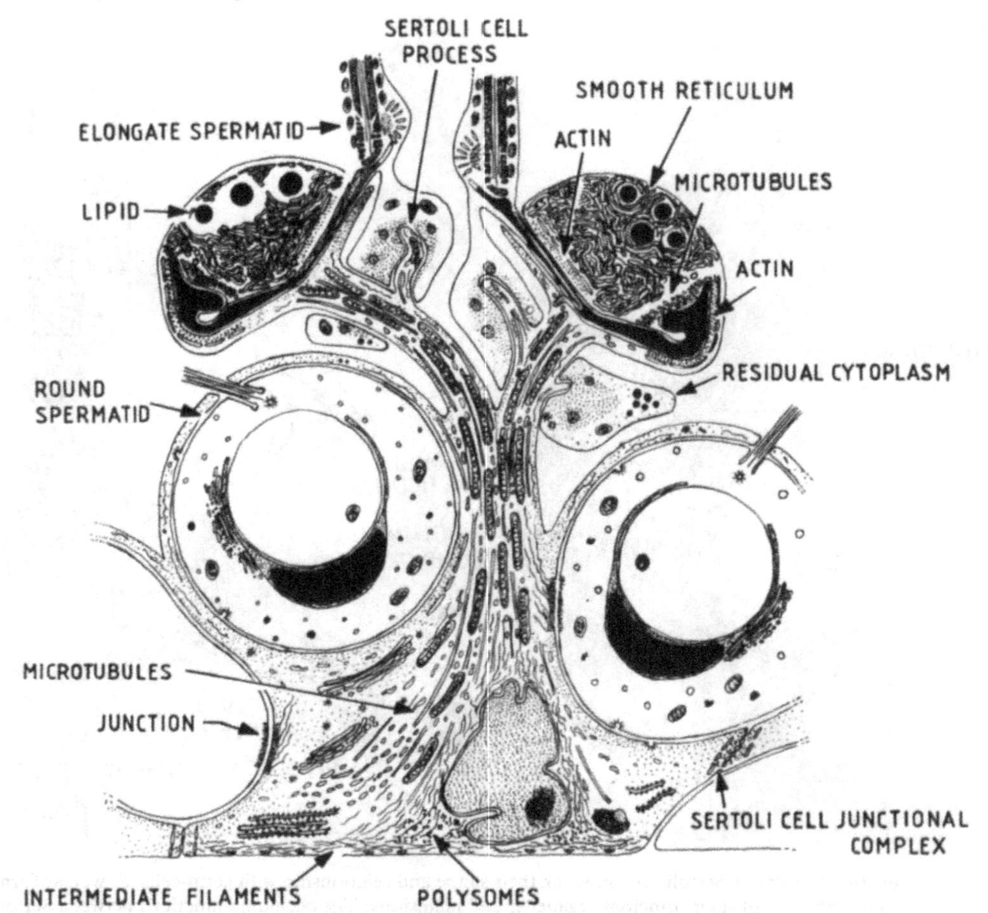

Fig. 11. A ground squirrel Sertoli cell and associated germ-cells at a stage just preceding spermiation. Elongate spermatids and bulbous masses of Sertoli cell cytoplasm, consisting mainly of smooth endoplasmic reticulum, are situated at the ends of stalks that project from the apex of the Sertoli cell. Present also are Sertoli cell processes that extend into the residual cytoplasm of the elongate spermatids. Intermediate filaments occur at the base of the cell and around the nucleus from which similar filaments appear to extend into apical cytoplasm. Intermediate filaments also occur in association with desmosome-like junctions and, like microtubules, are abundant in the Sertoli cell stalks and in their processes. Microtubules occur mainly in cytoplasm apical to the nucleus, where they are oriented parallel to the long axis of the cell. A specialized group of as many as 400 microtubules occurs adjacent to the acrosome of the elongate spermatids. Actin, although present in all areas of the cell, is most concentrated in specialized regions adjacent to junctions (junctional complex) with other Sertoli cells and in areas of adhesion to germ-cells. In both these regions, paracrystalline arrays of actin filaments lie sandwiched between the plasma membrane and a cistern of endoplasmic reticulum (Prepared from Vogl et al. 1983a)

as a layer at the base of the cell, and adjacent to desmosome-like junctions with germ-cells. Intermediate filaments, together with microtubules, are also present in abundance in regions of the cell involved with the transport of smooth endoplasmic reticulum in cytoplasm associated with elongate spermatids, and in processes that extend into the residual cytoplasm of germ-cells. Recently, the filamentous material and the more deeply positioned rough endoplasmic reticulum seen at the cell surface have been termed ectoplasmic (or junctional) specializations which are seen at the level of the blood-testis barrier, as well as in a position facing germ-cells (Russell 1980a, Russell et al. 1980a, 1983c, McGinley et al. 1981, Weber et al. 1983, Vogl et al. 1983a, b). The fate of ectoplasmic specializations adjacent to germ-cells is controversial (Ross 1976, Russell 1977c, Gravis 1979, Russell 1980b, Vogl et al. 1983a, b, Russell and Peterson 1985). The function of ectoplasmic specializations as devices for maintaining a tight association between Sertoli cells and elongated spermatids has been suggested (Russell 1977a, Romrell and Ross 1979, Russell and Peterson 1985). Filaments and subsurface endoplasmic reticulum constitute a cytoskeletal mantle around the spermatid head, thus maintaining the Sertoli cell crypt in which the spermatid is positioned (Vogl and Soucy 1985). Russell (1977c) has suggested that the mantle and the plasma membrane covering the spermatid head may develop a connection to, and be acted upon by other cytoskeletal elements, all of which facilitate the observed movements of the elongate spermatid in preparation for release. The disappearance of ectoplasmic specializations is related to sperm release and is evident in the ground squirrel, where these structures are lost from around spermatid heads in a sequence that has been correlated with the retraction of Sertoli cell cytoplasm from the germ-cells (Vogl et al. 1983a). A graded pattern of loss is also observed in the musk shrew (Cooper and Bedford 1976). The filaments of Sertoli cell appear to possess contractile properties as they consist of actin or actin-like substances. Actin is known to be present in ectoplasmic specializations adjacent to regions of demonstrated adhesion to germ-cells and adjacent to Sertoli cell tight junctions (Toyama 1976, Franke et al. 1978a, Vogl et al. 1985a, Vogl and Soucy 1985), as well as in Sertoli cell region adjacent to the tubulobulbar process of late spermatids. Vogl and Soucy (1985) have studied the arrangement and possible function of actin filament bundles in ectoplasmic specializations of ground squirrel Sertoli cells, which appear to be more skeletal than contractile. Although actin filaments do occur around the nucleus and the base of the Sertoli cell, the most abundant cytoskeletal elements in these locations are intermediate filaments (Franke et al. 1978b, 1979, Vogl et al. 1983a, b). Microtubules together with intermediate filaments, form the most conspicuous cytoskeleton elements in zones of the Sertoli cell associated with the movements of organelles, germ-cells, and residual cytoplasm (Fig. 11) (Vogl et al. 1983a, b). Sertoli cell microtubules appear to be essential for the normal development and translocation of spermatids in the seminiferous epithelium and are involved with positional changes in smooth endoplasmic reticulum of the Sertoli cell (Vogl et al. 1983b). They do not seem necessary for the maintenance of cell junctions. In a variety of physiological and experimental (after hypophysectomy) conditions, electron microscope studies of the Sertoli cells of pig have shown that evolution of the cytoskeleton (microfilaments, intermediate-sized filaments and microtubules) might be regulated directly or indirectly (via testosterone production) by pituitary hormones (see Chevalier and Dufaure 1982). Courot and Courtens (1982) have suggested that the fine structure of ram Sertoli cells is dependent on the hormonal balance. The fine structure of Sertoli cells changes with the maturation of pig testis (Van Vorstenbosch et al. 1984). Seasonal changes in the fine structure of

Sertoli cells are reported for the ground squirrel (Pudney and Fawcett 1985). Means et al. (1982) have discussed the regulation of the cytoskeleton in the Sertoli cells by Ca^2-calmodulin and cAMP. A high ATPase activity near the filaments may be involved in providing an energy source for filament contractility.

Lysosomal structures are scattered throughout the cytoplasm, and are believed to help in the digestion of degenerating germ-cells and residual bodies released by the spermatids during spermiation. They are also involved in the natural autolytic process of degenerating germ-cell cytoplasmic organelles. The arrangement of smooth endoplasmic reticulum around secondary lysosomes and pleomorphic lipofuscin pigment in the human Sertoli cell suggests that they may play a role in the enzymatic digestive process of the Sertoli cell (Tindall et al. 1985). The Sertoli cell cytoplasm constains numerous inclusions (e.g. lipid droplets of different sizes), and in some species (e.g. man) also various crystals and vesicles (Holstein and Roosen-Runge 1981, Tindall et al. 1985). The lipid content of Sertoli cells is variable from species to species both in respect of its total amount and size and of the histochemical nature of the lipid droplets (Bilaspuri and Guraya 1983c, d). It generally is localized near the base of the cells. Human Sertoli cells in culture show aggregations of smooth endoplasmic reticulum often arranged concentrically in multiple layers around lipid droplets (Lipshultz et al. 1982, Tindall et al. 1983, 1985). The significance of species differences in the abundance of lipid remains obscure. Because the lipid content of the cytoplasm varies with stages of spermatogenic cycle (Guraya 1965, Bilaspuri and Guraya 1983c, d), it has been suggested that at least part of this lipid derives from the ingested residual bodies and may be utilized by the Sertoli cells for the synthesis of steroids (Lacy 1962, 1967). But so far, there is no direct evidence that these lipids are used for the synthesis of steroid hormones. However, numerous papers have been published previously on the histochemistry and biochemistry of lipids in the mammalian testis (Guraya 1965, Bilaspuri and Guraya 1980b, 1981a, b, 1983c, d). Conspicuous seasonal variations in the amount and nature of lipids in the testes of different mammalian species have been revealed in these studies, suggesting a role in spermatogenesis. Marzuki and Coniglio (1982) have studied the effect of essential fatty acid deficiency on lipids of rat Sertoli and germinal cells.

The characteristics of the mature Sertoli cell nucleus include an infolded envelope, a relatively homogeneous nucleoplasm, and a single tripartite nucleolus (Fig. 10) whose functional significance is not known; a sphaeridum is present at the side of nucleolus in the human Sertoli cell (Holstein and Roosen-Runge 1981). The nucleolus proper, which incorporates RNA precursors, shows reticular structure accompanied by two electron-dense, smaller structures consisting of heterochromatin; the Sertoli cells synthesize their own ribosomal RNA. These smaller structures, which do not incorporate RNA precursors but stain positively with Feulgen reagent, are usually called heteropycnotic bodies, juxtanuclear bodies, perinuclear spheres, or satellite karyosomes (Fawcett 1975a, A. Steinberger and E. Steinberger 1977, Kerr and de Kretser 1981) but are now known to be centromeric heterochromatin (Jean et aL 1983). The function of these perinucleolar bodies is not known.

The lateral cell membrane of the Sertoli cells forms numerous veil-like processes that extend between spermatocytes and spermatids (Fig. 11) (Fawcett

1975a, Dym et al. 1977, Russell 1980a, Kerr and de Kretser 1981, Vogl et al. 1983a, b, 1985a). Morales and Clermont (1982) studied the evolution of Sertoli cell processes invading the cytoplasm of rat spermatids. According to Wong and Russell (1983), who have made a three-dimensional reconstruction of a rat stage 5 Sertoli cell, lateral processes are of three categories: (1) conical processes extending from the lateral surface near its base; (2) cup-shaped sheet-like processes partially encompassing round germ-cells; and (3) flattened, sheet-like processes extending between round germ-cells. Elongate spermatids occupy deep, irregularly shaped cylindrical recesses directed in the centripetal axis of the Sertoli cell. The tapered luminal extensions of the sheet-like, cylindrical processes are referred to as apical processes. Actually, the interactions between Sertoli cells and differentiating male germ-cells, which certainly occur at the level of their plasma membrane, are dynamic and highly ordered with regard to the timing and spatial arrangements of successive waves of cellular differentiation. Germ-cells multiply at a rapid rate, and grow and shift along the sides and apex of the Sertoli cell. The shape of the latter therefore must alter continually to accommodate cyclic events associated with spermatogenesis. The Sertoli cell, having extensive plasma membrane surface area, shows complex and elaborate surface relationships to the germinal cells, supporting the concept of extensive metabolic exchanges between it and germ-cells, which take place at the cell's surface. Vogl et al. (1985b) have suggested that Sertoli cell processes have an attachment function and that they also may facilitate the movement of residual cytoplasm into the epithelium. These structures also appear to be involved with receptor mediated endocytosis.

Recent electron microscope studies have revealed a great diversity in the morphology and numbers of Sertoli cell junctions in the seminiferous epithelium of mammalian testis, as discussed in an excellent review by Russell and Peterson (1985), who have also described their possible functions. The various types of Sertoli cell-Sertoli cell junctions include (1) tight or occluding junctions, (2) gap junctions, (3) septate-like junctions, (4) desmosomes, (5) close junctions, and (6) Sertoli-Sertoli tubulobulbar complexes to be discussed in Chap. IV. The structure and function of Sertoli-germ-cell junctions, such as desmosomes, gap junctions, tight junctions, ectoplasmic specializations (which are strictly not junctions) and tubulobulbar complexes, are also discussed in detail (Russell and Peterson 1985). Sertoli cells contact and connect with each other by an extensive series of tight or occluding junctions (zonula occludens). These tight junctions, fusing the membranes of adjacent Sertoli cells, form the structural component of the blood-testis barrier, and thereby separate the seminiferous epithelium into two compartments, namely the basal (peripheral) compartment containing spermatogonia and preleptotene spermatocytes, and the adluminal (central) compartment containing meiotic spermatocytes and spermatids (Fig. 11) (Neaves 1977, Bellvé 1979, Guraya 1980, Courot 1980, Connell 1980, Ritzen et al. 1981, Kerr and de Kretser 1981, Holstein and Roosen-Runge 1981, Russell and Peterson 1985). Connell (1980) has observed that meiotic prophase in the dog testis is initiated before the establishment of a complete blood-testis barrier; lumen formation occurs after the initiation of meiosis and the formation of the blood-testis barrier (see also Connell 1984). Sertoli-Sertoli and Sertoli-germ-cell configurational relationships have been demonstrated using morphometric techniques and direct measurements as obtained from micrographs used to reconstruct a model of a rat stage 5 Sertoli cell (Weber et al. 1983). Russell et al. (1983c), using micrographs employed to reconstruct serially a model of a rat stage 5 Sertoli cell, have studied specific Sertoli-Sertoli and Sertoli-germ-cell contacts and/or junctions, their kinds, numbers and position relative to the reconstructed cell and also to adjacent Sertoli cells (see also Russell and Peterson 1985). The Sertoli-Sertoli junctional contact areas exist in a belt-like arrangement near the base of the Sertoli cell. From the results of this study it can be concluded that each Sertoli cell develops contact with the basal lamina, other Sertoli cells, and also germ-cells. All of these junctional contacts act as cohesive elements to hold the epithelium together. The apposing plasma membranes of Sertoli cells studied with transmission electron microscope show up to 40 parallel, linear arrays of particles that course circumferentially near the base of the Sertoli cell (Connell 1980). These particles form an association preferentially with the E-face and show complementary shallow grooves in the P-face. Atypical gap-junction particles also are seen frequently intercalated among the tight junction arrays. Subsurface cisternae and bundles of filaments are present on the inner surfaces of the apposed cell membranes of tight junctions (Kerr and de Kretser 1981, Ritzen et al. 1981, Suarez-Quian and Dym 1984). The junctional complexes also show areas of obliteration of the intercellular spaces. The total complexity of the "Sertoli junction" renders it extremely resistant to hypertonic solutions that dissociate junctions (Russell et al. 1980a, Russel and Peterson 1985). Russell et al. (1983c) have suggested that a belt-like arrangement of the Sertoli-Sertoli junctional contact areas near the base of the Sertoli cell is consistent with their proposed function as sealing element limiting the passage of materials towards the tubular lumen (Russell and Peterson 1985). Pelletier and Friend (1983b), using both freeze-fracture and thin section methodologies, have shown regional differences in the junctional characteristics at the basolateral aspect of the guinea pig Sertoli cell. The most basally localized junctions are uninterrupted parallel strands referred to as "continuous zonules". Further towards the lumen, the junctional strands are meandering and incomplete and are referred to as "discontinuous zonules".

The presence of blood-testis barrier (or Sertoli cell barrier) is further supported by the fact that a number of endogenous or exogenous compounds in serum readily enter the lymphatics and yet are not recovered in rate testis (Fawcett 1975a, A. Steinberger and E. Steinberger 1977, Neaves 1977, Bellvé 1979, Ritzen et al. 1981, Tindall et al. 1985, Russell and Peterson 1985). The impermeable character of the Sertoli junctions has been verified by their ability to exclude electron-dense compounds, such as horseradish peroxidase and lanthanum and sucrose (Dym and Fawcett 1970, 1971, Setchell et al. 1980, Russell and Peterson 1985, Tindall et al. 1985). The blood-testis barrier accounts for the differences in chemical composition of the tubule fluid and the blood plasma or lymph, providing a structural basis for the maintenance of a different milieu in each of the two compartments, which may be crucial during specific stages of germ-cell development. This compartmentation of the epithelium probably has important consequences yet to be determined. However, it is believed that the cells of the basal compartment directly receive hormones supplied by the bloodstream or present in the peritubular fluid (Comhaire and Vermeulen 1978), whereas those of the adluminal compartment receive hormones and their derivatives and nutrients through the Sertoli cells (Ritzen et al. 1981) which establish special junctions with these cells (Russell and Peterson 1985). This suggests that the cells of the basal compartment are affected more rapidly than others by any change in hormonal balance (Courot 1980, Courot and Ortavant 1981). Russell and Peterson (1985) have suggested that transfer of substances across the occluding junctions may be largely dependent on molecular size and physical properties of transported materials. Movement of molecules across the Sertoli cell body is potentially a more selective process. Here, receptors, pumps, exchange mechanisms, enzyme systems, transport and internalization systems may be involved. Sharpe (1984) has made an attempt to study intratesticular factors controlling testicular function. The exceptionally high energy and nutritional demand of a vascular seminiferous epithelium has to be met by local transport in testicular interstitial fluid. The rate of its formation is determined by capillary permeability, and local control of this process is thus a prerequisite for full nutritional and hormonal support for spermatogenesis. Each stage of spermatogenesis appears to have different requirements which only local control mechanisms can ensure.

Gap junctions and septate junctions also interconnect adjacent Sertoli cells and are believed to provide the coordination needed for the translocation of germ-cells from basal to luminal aspect of the seminiferous tubule (Fig. 11) (McGinley et al. 1979, Bellvé 1979, Guraya 1980, Holstein and Roosen-Runge 1981, Russell and Peterson 1985, Kerr and de Kretser 1981, Russell et al. 1983c). The septate desmosome-like complexes are believed to contribute towards inter-Sertoli cell attachment, as do the tight junctions or zonulae occludentes. These transverse septa possibly control the paracellular transepithelial flow of solutes that are required to maintain the atypical ionic composition of tubal fluid. Some forms of cell-to-cell communication (ionic or metabolic coupling) occur through gap junctions. Desmosomes occur between Sertoli cells at the level of occluding junctions (see Russell and Peterson 1985). They function in cell-to-cell adhesion. But their rare occurrence raises some doubt about their precise functional significance.

Besides the Sertoli cell interactions, the supporting cells develop specialized junctional complexes of several types in the plasma membrane, which adhere to developing spermatogenic cells (Fig. 11) (Russell 1980b, Russell et al. 1980a, 1983c, Vogl et al. 1983a, Russell and Peterson 1985). Russell et al. (1983c) have observed that of 37 round germ-cells that are sectioned in their entirety and adjoin the three-dimensionally reconstructed stage 5 Sertoli cell of rat, 23 show desmosome gap junctions with either the reconstructed cell or an adjoining cell. These authors have suggested that since there are multiple junctions connecting some germ-cells to Sertoli cells, the total number of junctions may be much more (35). Desmosome gap junctions of the Sertoli cell are numerous connecting pachytene spermatocytes, less numerous connecting type B spermatogonia, and even less numerous connecting step 5 spermatids; and none is seen joining Sertoli cells with elongate spermatids. Most desmosome gap junctions join germ-cells to the body of the Sertoli cell at its basal aspect. Their numbers and position suggest that they perform an important function in maintaining the integrity of the seminiferous epithelium and may form a route for cell-to-cell communication (Russell and Peterson 1985). These junctional complexes also indicate that various cellular interactions during spermatogenesis may be regulated at the level of the cell surface (Millette 1979a). Electron microscope studies have revealed a greater complexity in the Sertoli junctions. At the focal areas of apparent contact between Sertoli cells and germ-cells, the intercellular space is reduced to 5 – 18 nm. These presumptive junctional specializations in some respects, resemble the typical desmosome (macula adherens) except that the associated fine filaments lie parallel rather than perpendicular to the surface (Kerr and de Kretser 1981, Ritzen et al. 1981, Lingwood and Schachter 1981). Similar rudimentary desmosomes also may bring about a close association between Sertoli cells and spermatids during early stages of differentiation, but in later stages of spermiogenesis, the germ-cell specializations are no longer observed. Russell et al. (1980a, 1983c) have discussed the dynamic aspects of Sertoli ectoplasmic specializations that appear in relation to developing germ-cells (see also Russell and Peterson 1985). Ectoplasmic specializations of the three-dimensionally reconstructed Sertoli cell of rat stage 5 are seen facing only 3 of 37 round cells, and 7 ectoplasmic specializations from adjoining Sertoli cells faced these germ-cells, all of which are step 5 spermatids (Russell et al. 1983c). The absence of ectoplasmic specializations facing pachytene cells suggests that they are needed later in spermatogenesis (Vogl et al. 1983a, b).

The proliferation and renewal of spermatogonia occur in the peripheral (basal) compartment of the seminiferous epithelium, while the spermatogonia and preleptotene spermatocytes are applied closely to the basal lamina, and below the occluding, intercellular Sertoli junctions (Dym et al. 1977). The type B spermatogonia divide to form preleptotene spermatocytes, which undergo a final replication of nuclear DNA before embarking on the prolonged meiotic prophase (see also Chap. III). While in the leptotene and zygotene stages of meiotic prophase, the spermatocytes are translocated across the Sertoli junction into the adluminal (central) compartment of the seminiferous epithelium (see Russell et al. 1983c, Russell and Peterson 1985). Ulvik (1983b) has studied the stage-dependent topographical relationship of spermatogonia and early spermatocytes to Sertoli-Sertoli interspaces in the rat testis. Obviously, the mammalian seminiferous epithelium must have a mechanism that permits the preleptotene spermatocytes to traverse the complex of Sertoli junctions without disturbing the integrity of the permeability barrier. This is believed to be achieved through the formation of a transient intermediate compartment by the Sertoli cells, which envelops early primary spermatocytes during their translocation from the basal to the adluminal (or central) compartment (Russell and Peterson 1985). However, the tracer experiments of Dym and Cavicchia (1977) with lanthanum and peroxidase do not support this concept of a separate intermediate compartment, thus supporting rather the concept of basal and adluminal compartments. The various observations show that the junctional complexes periodically must allow the passage of differentiating germ-cells from the basal to the adluminal compartment by either forming temporary gaps or undergoing a degenerating and regenerating sequence (Fawcett 1974, A. Steinberger and E. Steinberger 1977). Therefore, one can assume that the Sertoli cell junctions are not permanent but open from time to time, possibly in response to the action of some enzyme (Ritzen et al. 1981), to facilitate the passage of the preleptotene spermatocytes (Lacroix et al. 1977, Connell 1980). Abundant actin filaments, present in the Sertoli cell cytoplasm between the subsurface cisternae and the junctional complexes (Fig. 11), are believed to be involved in inducing confirmational alterations in the Sertoli cell and thereby facilitating the intercellular, centripetal movement of spermatocytes. Fawcett (1977) has suggested that germ-cells play an entirely passive role in this process. However, certain indirect and circumstantial observations, as discussed by Bellvé (1979), suggest that this may not be the case. But these observations have suggested that developing germ-cells possess specific surface receptors, possibly glycoproteins, involved in cell-cell adhesion. It will be interesting to mention here that multiple antigenic surface macromolecules occur on spermatogonia, spermatocytes and spermatids, as will be discussed in Chap. V, but whether any of these contribute to intercellular adhesion remains to be determined. Bellvé (1979) has suggested that any array of "adhesive" surface receptors could greatly facilitate the orderly translocation of preleptotene spermatocytes to the adluminal compartment and that this process involves coordinate movement of large syncytial clusters of cells intercalated between numerous Sertoli cells. A depression of spermatogenesis in men of diverse fertility status has been associated with an increase in the incidence of tubules with apparently defective tight junctions (Landon and Pryor 1981). Russell et al. (1983d) have studied the morphological pattern of response after administration of procarbazine in relation to specific cell associations during the cycle of the seminiferous epithelium of the rat. The status of the blood-testis barrier and the integrity of Sertoli occluding junctions after experimental treatments or in animals with genetic disorders form the subject of numerous studies (Russell and Peterson 1985). Actually, the functional integrity of Sertoli junctions in vitro is more difficult to determine than their integrity in the intact animal. However, dissociated Sertoli cells from juvenile animals maintained in culture are observed to develop tight junctions (see Russell and Peterson 1985).

2. *Functional Features*

Sertoli cells are believed to perform several functions in the physiology of the seminiferous epithelium as evidenced by their ultrastructural, histochemical and biochemical features (A. Steinberger and E. Steinberger 1977, 1980, Dym et al. 1977, Guraya 1980, Ritzen et al. 1981, Vogl et al. 1983a, b, Russell and Peterson 1985, Tindall et al. 1985). Fawcett (1975a) has discussed the significance of regional differentiation of the Sertoli cell in relation to its several different functions, which include steroidogenesis, sperm release, maintenance of the blood-testis barrier and exocrine secretion (see also Russell and Peterson 1985, Tindall et al. 1985). These functions are suggested from considerable circumstantial evidence, primarily of a morphological nature (Russell 1980a). Therefore, these functions still continue to be speculative to a great extent. The precise roles of Sertoli cells in spermatogenesis need to be resolved at the molecular level (Tindall et al. 1985). However, the results of recent correlative, biochemical and morphological studies on Sertoli cells alone or on seminiferous tubules cultured in vitro have suggested that the cyclically secreted Sertoli cell proteins may be one means by which these cells influence spermatogenesis. Actually, both hormones and cell-cell interactions are needed for the completion of spermatogenesis. There is now increasing evidence that Sertoli cells may secrete their products into the interstitial space where they enter the lymphatics and the bloodstream, as well as secreting products into the lumen of the seminiferous tubules. This bidirectional secretion has brought new implications to the hypothesized physiological roles of Sertoli cell products in male reproduction.

Sertoli cells are phagocytic, eliminating from the seminiferous epithelium residual bodies and degenerating germ-cells as well as gesting particulate matter (E. Steinberger and A. Steinberger 1975, A. Steinberger and E. Steinberger 1977, Russell 1980a). The residual bodies containing lipid droplets, ribosomes, clusters of small mitochondria and various cytoplasmic membranes are taken up by the Sertoli cells into the cytoplasm and transported from the luminal into a more peripheral position, as will be discussed in Chap. IV. The phagocytic activity is enhanced greatly following epithelial damage and massive degeneration of germ-cells. The Sertoli cells also retain their phagocytic activity in vitro under organ and cell-culture conditions (A. Steinberger and E. Steinberger 1977).

The Sertoli cells play an active role in the endocrine regulation of tubule function, and the release of spermatozoa in the lumen. It has been proposed that the material conveyed to the periphery of the Sertoli cell cytoplasm is utilized subsequently as a substrate for the elaboration of a promoting factor needed for further spermatogenesis (Lacy 1962). That may be so, but whether such a factor is a Sertoli cell hormone is a moot point. In this regard, it has been proposed that Sertoli cells utilize the lipids from ingested residual bodies shed by the spermatids for hormone synthesis (E. Steinberger and A. Steinberger 1975, A. Steinberger and E. Steinberger 1977). The evidence for this is based primarily on changes in cytochemical reaction for lipids, or free or esterified cholesterol in the Sertoli cells (Bilaspuri and Guraya 1983c, d). So far, there is no direct evidence that these lipids are utilized for steroid biosynthesis, but the Sertoli cells have been shown to form testosterone from exogenous progesterone, and to aromatize exogenous testosterone to oestradiol-17β in vitro under the influence of FSH (Dorrington et al. 1976, 1978a, b, Van Damme et al. 1980, A. Steinberger and E. Steinberger 1980, Van der Molen and Rommerts 1981, Ritzen et al. 1982). This occurs only in young age (i.e. before 20 days in the rat) but not in the adult. Actually, very divergent views exist in regard to steroidogenic activity of Sertoli cells (A. Steinberger and E. Steinberger 1977, 1980, Van der Molen and Rommerts 1981, Ritzen et al. 1981). The presence of smooth endoplasmic reticulum (Fawcett 1975a, Tindall et al. 1985) and various steroid dehydrogenases demonstrated histochemically (Bilaspuri and Guraya 1984a) and biochemically (Van der Molen and Rommerts 1981) in Sertoli cells indicates that they are equipped, morphologically, cytochemically, and biochemically, for steroid biosynthesis (Guraya 1980). However, Ritzen et al. (1981) have suggested that the smooth endoplasmic reticulum of Sertoli cells may be involved rather in testosterone metabolism than in synthesis. Van der Molen and Rommerts (1981) have suggested also that no convincing evidence for de novo testosterone synthesis in testicular cell types other than the Leydig cells has been reported as yet. But this does not rule out a function for the seminiferous tubular cells, which contain dehydrogenases and other enzyme systems related directly or indirectly to steroid biosynthesis and metabolism (Bilaspuri and Guraya 1983f, 1984a), in utilizing steroid hormones produced in the Leydig cells for further steroid synthesis. Ludvigson et al. (1982) have demonstrated immunocytochemical localization of aldose reductase in rat Sertoli cells. The physiological significance of possible testosterone synthesis and testosterone metabolism by the Sertoli cells is not clear (A. Steinberger and E. Steinberger 1980, Van der Molen and Rommerts 1981, Ritzen et al. 1981). But it is possible that either testosterone or its metabolites may effect some biological activities within the Sertoli cell, be made available to the developing germ-cells, or exert their effect outside the seminiferous tubules.

Various workers have stressed the usefulness of regarding the Sertoli cells and their associated germ-cells as symbiotic units (Fawcett 1975a, Russell 1980a). This physiological interdependence of the two populations is supported by various recent studies (Wright et al. 1983, Galdieri et al. 1983, Boitani et al. 1983, Welsh et al. 1985, Tindall et al. 1985, Russell and Peterson 1985). Both hormones and cell-cell interactions are of great significance for the completion of spermatogenesis. Ulvik (1983a) has reported the selective uptake by the Sertoli cells of cytoplasm from normal spermatogonia in the rat testis. The functional significance of this uptake is not known. The Sertoli cells act as a supporting matrix for the germ-cells without, however, a permanent attachment to the germ-cells, as that would be incompatible with the translocation of the differentiating germ-cells up the sides of the supporting Sertoli cells towards the tubular lumen (Russell et al. 1983c). Grootegoed et al. (1982a) have studied Con A-induced attachment of spermatogenic cells to Sertoli cells in vitro. This suggests that the influence Sertoli cells may exert upon the germ-cells, by providing them with nutrients (Ritzen et al. 1981) or possibly a substance needed for differentiation, presumably depends upon a two-step process, i.e. formation of a secretory product by the Sertoli cells and uptake of this material by the germ-cells. Wright et al. (1983) have suggested two possible mechanisms by which germ-cells may be coordinating Sertoli cyclic activity. One is through a germ-cell product, but at present there are no recognized secretory substances of germ-cells that are known to act directly on Sertoli cells. The Sertoli cells ingest residual bodies released from the maturing spermatids. Another possible mode of communication between germ and Sertoli cells is by the gap junctions described already. Galdieri and Monaco (1983) have claimed to produce an evidence of protein secretion by cultured pachytene spermatocytes. Its physiological significance needs to be determined. But Galdieri et al. (1983) have observed changes of Sertoli cell glycoproteins induced by removal of the associated germ-cells. These connections, along with the cytoplasmic bridge between spermatocytes and the gap junctions between Sertoli cells, permit metabolic and electrical coupling of all the cells of the seminiferous epithelium (Loewenstein 1981). Ritzen (1983) has suggested the presence of chemical messengers between Sertoli cells and neighbouring cells. There is now increasing evidence for communication between the various cell types of the gonad (Wright et al. 1983). The nature of nutrients provided by the Sertoli cells remains to be defined. It may be glycogen (Fouquet 1968). In response to FSH stimulation, Sertoli cells secrete considerable amounts of lactate (Robinson and Fritz 1981), a metabolic substrate preferred by advanced spermatogenic cells

(Nakamura et al. 1981). Ziparo et al. (1982) have shown for the first time that Sertoli cells in vitro are capable of transferring [^{3}H] choline to germ-cells. The deprivation of pituitary gonadotrophins by hypophysectomy, or blockade of secretion or of actions of hormones and seasonal changes in gonadotrophins (FSH and LH) are known to affect the development, differentiation, morphology and metabolic activities of Sertoli cells (A. Steinberger and E. Steinberger 1977, Courot 1980, Guraya 1980, Courot and Ortavant 1981, Ritzen et al. 1981, Tindall et al. 1985). The vitamin-A depletion and subsequent repletion affect the morphological characteristics of Sertoli cells and germ-cells in the rat testis (Unni et al. 1983). The cryptorchidism also affect the morphology and function(s) of Sertoli cells in rat (Bergh 1983a, b, 1985, Bergh and Damber 1984, Bergh et al. 1984).

The probable importance of Sertoli cells in the process of spermatogenesis has been emphasized by numerous investigators (Fawcett 1975a, Dym et al. 1977, Russell 1980a, Kerr and de Kretser 1981, Ritzen et al. 1981, Tindall et al. 1985), although their precise function and significance in relation to the developing germ-cells still need to be determined at the molecular level (Tindall et al. 1985). Stimulated by FSH and androgens, Sertoli cells play important roles in the maintenance of spermatogenesis (Guraya 1980, Ritzen et al. 1981). These hormones may, in turn, make the Sertoli cells responsive to regulatory signals from differentiating spermatogenic cells. Through the recognition of these hypothetical signals, the Sertoli cells appear to be coordinated over space and time in concert with the spermatogenic cycle. Jutte et al. (1983) have demonstrated Sertoli cell maintenance of germ-cell viability. FSH and testosterone receptors but not LH receptors have been demonstrated in Sertoli cells (Dufau et al. 1973, A. Steinberger and E. Steinberger 1977, Courot et al. 1979, Courot 1980, Guraya 1980, Risbridger et al. 1981, Ritzen et al. 1981, Reichert et al. 1982, Parvinen et al. 1983, Tindall et al. 1985). FSH binds to specific receptors on the Sertoli cell plasma membrane, stimulates adenylate cyclase, activates soluble cAMP-dependent protein kinase, and stimulates RNA and protein synthesis. Tindall et al. (1985) have discussed the biochemical aspects of hormonal regulation of Sertoli cell function(s) by FSH. The biochemical studies have also suggested that androgen receptor in the Sertoli cell plays a pivotal role in the hormonal regulation of spermatogenesis. Both the cytoplasmic and nuclear receptors for androgen have been described. But their physicochemical and biological properties still need to be understood. One characteristic common to all androgen receptors described to date is their charge properties. The most unique property of the androgen receptor in the Sertoli cell is its specificity of binding to the steroid. The suggestion that androgens act through Sertoli cells is supported also by the fact that their nuclear area is depressed after anti-androgens (Aumüller et al. 1975, Viguier-Martinez and Hochereau-de-Reviers 1977). Tindall et al. (1985), after discussing the cellular responses of Sertoli cells to FSH, LH and testosterone, have suggested that their ultrastructure appears to be more highly regulated by LH than FSH. Much of this control is presumably via the stimulatory effect of LH on testosterone production by the testis and the subsequent stimulation of the Sertoli cell by testosterone. But the addition of testosterone to LH-depleted animals does not completely restore the fine structural characteristics, suggesting the possibility of a direct effect of LH on Sertoli cell morphology.

FSH has a conspicuous effect on testicular levels of androgen-binding protein (ABP) which is secreted by Sertoli cells (Means 1977, A. Steinberger and E. Steinberger 1977, 1980, Ritzen et al. 1981, diZerega and Sherins 1981, Tindall et al. 1985). Direct evidence for ABP production by the Sertoli cell was derived from experiments carried out in vitro with cultures of isolated Sertoli cells (A. Steinberger and E. Steinberger 1977, Ritzen et al. 1981, De Philip et al. 1982, Tindall et al. 1985). Cultured Sertoli cells were shown to secrete ABP into the medium and to respond to FSH stimulation with increased ABP production. Other pituitary hormones did not show any effect on the secretion of ABP in culture. The presence of ABP in rodents (rats and mice) testes, where spermatogenesis or germ-cell populations are depleted due to various genetic defects, has provided further evidence for its production by the Sertoli cells (Tindall et al. 1985). The concentration of ABP differs in various species studied (Ritzen and French 1974, Hansson et al. 1975, Jegou 1976, Ritzen et al. 1977, 1981, Danzo and Eller 1978). The existence of ABP in the human testis was not proved conclusively in earlier studies (Troen 1978). But the secretion and characterization of ABP-like proteins by the human Sertoli cells have been investigated in some recent studies (Lipshultz et al. 1982, Tindall et al. 1983, 1985). These data are the first to provide evidence for the synthesis and secretion of ABP by the human Sertoli cells. The physicochemical properties of ABP have been characterized (Ritzen et al. 1973, Weddington et al. 1974, 1975, Musto et al. 1978, 1982, Tindall et al. 1985) but its physiological role is still obscure. Ritzen et al. (1981) have discussed its three possible functions: (1) an intracellular carrier of testosterone or dihydrotestosterone in the Sertoli cell itself; (2) a large store of androgenic hormones in the seminiferous epithelium and epididymis; and (3) a carrier of testosterone from the testis into the epididymis. Since the γ-glutamyl transpeptidase (a membrane-bound enzyme) activity of Sertoli cells is believed to play an important role in protein synthesis, it may be involved in some regulatory function in the synthesis of ABP and the Sertoli Cell Factor (A. Steinberger and E. Steinberger 1977). It catalyzes the transfer of the γ-glutamyl moiety of glutathione and other γ-glutamyl compounds to a large number of amino acid and peptide co-factors. It is believed that this enzyme plays a role in the transport of the amino acid via the γ-glutamyl cycle. Actually, the precise molecular mechanism of ABP secretion is not yet known. Welsh et al. (1980) have suggested the possible involvement of cAMP, calcium, and cytoskeleton. The effect of FSH on ABP production by Sertoli cells on causing local accumulation of androgen may be important in the initiation of spermatogenesis. Similarly, the effect of FSH on the conversion of testosterone to oestrogen in young age but not in the adult also may be significant in the regulation of rat spermatogenesis.

It has been suggested that the Sertoli cells are strategically located to create, by their metabolic and secretory activities, a special environment favouring germ-cell differentiation. The blood-testis barrier may regulate the molecular composition of the intercellular fluid present in the adluminal compartment (Russell and Peterson 1985). But there is still no direct evidence for this suggestion. Actually, relatively very little evidence is available for the secretory processes of the Sertoli cells (Fawcett 1975a, A. Steinberger and E. Steinberger 1977, Russell 1980a, Ritzen et al. 1981, Tindall et al. 1985). The seminiferous tubules of adult animals secrete a fluid differing in composition from that of blood plasma or lymph (Setchell et al. 1969, Ritzen et al. 1981). Since the secretion of tubule fluid occurs in the absence of germinal cells, this function also has been assigned to the Sertoli cells (Ritzen et al. 1981). Waites (1977) reviewed the chemical composition of the seminiferous tubule fluid. Compared to blood plasma, it is very rich in potassium and bicarbonate, but low in sodium and chloride. In this regard, it resembles an intracellular ionic composition. The maintenance of this ionic gradient is a function of the Sertoli cell cytoplasm (Muffly et al. 1985) and the occluding inter-Sertoli cell junctions; however, regulation of the tubular fluid has not been studied in any detail (Ritzen et al. 1981). Relatively more information is available on the characterization of fluid collected from the rete testis rather than directly from the seminiferous tubules (Waites 1977). However, it is estimated that about nine-tenths of the rete fluid actually is secreted directly through the epithelium of the rete testis (Setchell 1974). The latter fluid more closely resembles blood plasma in its ionic composition than the "primary" seminiferous tubule fluid. Thus, it appears that the primary seminiferous tubule fluid must be analyzed if the Sertoli cell contribution is to be understood as recently studied by Muffly et al. (1985). Little is known about the ionic composition of tubular fluid under various pathophysiological situations (Ritzen et al. 1981).

Besides the production of ABP and secretion of fluid, several other polypeptides are secreted by Sertoli cells after hormonal stimulation (Wilson and Griswold 1979, diZerega and Sherins 1981, Ritzen et al. 1981, De Philip et al. 1982, Hedger et al. 1984, Tindall et al. 1985). Recent studies have shown that the synthesis of both cellular and secreted proteins increases during specific stages of the seminiferous epithelium (Wright et al. 1983). All other cyclically secreted proteins such as plasminogen activator, the somatomedin-like molecule, and proteins S_2, S_5, T_3, T_8, T_9 are also recognized as Sertoli cell products (Lacroix et al. 1981, Ritzen et al. 1982, Johnsonbaugh et al. 1982, Fritz et al. 1982). FSH stimulates the production of plasminogen activator which, like ABP, is also secreted into the rete testis fluid. Isolated Sertoli cell membranes contain a membrane-associated form of plasminogen activator (Marzowski et al. 1985). The secretion of ABP and plasminogen activator is markedly increased at Stages 7 and 8 of spermatogenesis (Lacroix et al. 1981, Ritzen et al. 1982). FSH also stimulates the synthesis of γ-glutamyl transpeptidase (an important biological marker for Sertoli cells) and a protein kinase inhibitor, which remain in the Sertoli cell (see diZerega and Sherins 1981, Tindall et al. 1985). Sertoli cells produce and secrete proteins, such as transferrin and ceruloplasmin (Mather et al. 1983a, Skinner and Griswold 1980, 1983, Sylvester and Griswold 1984). The latter authors have studied the localization of transferrin and its receptors in rat testes. A large amount of transferrin is observed in the interstitial tissue between the seminiferous tubules. The receptor for this protein was localized in spermatocytes and early spermatids. The secretion of transferrin and accumulation of transferrin-specific mRNA is regulated by hormones and retinol (Huggenvik et al. 1984). Insulin alone or in combination with various treatments produces the most conspicuous increase in transferrin mRNA accumulation. The transferrin is a single glycopeptide of a reported MW of 70,000–80,000. It is an iron-transporting polypeptide that is needed by somatic cells to traverse the G_2 phase of the cell cycle (Bottenstein et al. 1979, Rudland et al. 1977, Tindall et al. 1985). It is believed that the Sertoli cell transferrin functions as a source of iron for the heme proteins or for non-heme metaloproteins in developing germ-cells. Thus, the Sertoli cell may be functioning as an intermediate in the transport of iron from serum transferrin to the germinal cells in the adluminal compartment (Huggenvik et al. 1984). But this suggestion needs to be confirmed. In addition to these known proteins, cyclic protein-2 is secreted in a stage-specific manner (Mather et al. 1983a, Wright et al. 1983). Cyclic protein-2 shows a pronounced cycle of secretion, its peak at Stage 6 being 30-fold greater than at its nadir at Stages 12–14. It could be recovered only from Sertoli cell-enriched cultures prepared from Stage 6 tubules but not from cultures of dispersed germ-cells prepared from the same tubules (Wright et al. 1983). It will be interesting to mention here that stage-specific peaks in both ABP and cyclic protein-2 levels possibly have a role in the developmental stage observed. Cyclic protein-2 is the product of mature Sertoli cells as it has not been observed in media from immature Sertoli cells (Wright et al. 1981, Kissinger et al. 1982, De Philip and Kierszenbaum 1982). Somatomedin-like compound is secreted by the Sertoli cells in vitro (Johnsonbaugh et al. 1982). The production of inhibin has also been attributed to the Sertoli cells (Setchell et al. 1977, A. Steinberger and E. Steinberger 1977, Ritzen et al. 1981, diZerega and Sherins 1981, Tindall et al. 1985). The precise nature of inhibin and its mode of action both await clarification. Murthy and Moudgal (1983) have reported the biological characteristics of a factor suppressing FSH (inhibin from ovine testis). However, it appears to be a protein. That probably is involved in the regulation of FSH secretion by the pituitary (Tindall et al. 1985). But the question of the relative contributions of sex steroids and inhibin in regulating the secretion of FSH by the pituitary in vivo is not resolved fully (see various papers in Bardin and Sherins 1982, Channing and Segal 1982, McCann and Dhindsa 1983). Sarvamangala et al. (1983) have reported the effect of chronic administration of inhibin and testosterone on spermatogenesis in adult male rats. Meiosis preventing substance and Mullerian inhibiting hormones are believed also to be secreted by the Sertoli cells or their precursors during gonadal differentiation in the embryo (Guraya 1980, Ritzen et al. 1981). Byskov (1983) has described the meiosis regulating substances of the rat testes. Sertoli cells in culture also secrete a few other specific proteins and a large number of more or less unidentified proteins (Rommerts et al. 1978, Ritzen et al. 1981, Wright et al. 1983). Sertoli cell products appear to influence the in vitro survival of isolated spermatocytes and spermatids (Boitani et al. 1983). Transferrin secreted by the Sertoli cells binds specifically to pachytene spermatocytes (Holmes et al. 1983). A mitogenic polypeptide designated as seminiferous growth factor (SGF) has been demonstrated in the Sertoli cells of the prepuberal and adult mouse, and in the seminiferous epithelium of several other mammalian species, including the rat, guinea-pig and calf (Feig et al. 1980, 1983, Bellvé and Feig 1984). The levels of SGF are not appreciably decreased in adult mouse testes following hypophysectomy. SGF purified from either the adult mouse or newborn calf seminiferous epithelium has a molecular weight of 15,700 and pH between 4.8 and 5.8, when exposed to denaturing conditions. SGF from these two species of mammals appears to show few exposed hydrophobic domains and a strong propensity to aggregate into multiple, high molecular weight species. Bellvé and co-workers have suggested that SGF is involved in regulating proliferation of Sertoli cells and spermatogenic cells in both developing and adult testes. But clarification of the precise regulatory role of SGF in the mammalian testis will need purification of the polypeptide to homogeneity.

The synthesis and secretion of several proteins by the Sertoli cells (Wright et al. 1982, 1983, De Philip et al. 1982) are supported by their fine morphology as they contain rough endoplasmic reticulum and the prominent nucleolus which usually are believed to be indicative of protein synthesis (Tindall et al. 1985). Further studies are needed to determine more precisely the nature, amounts and regulation of secretory products of Sertoli cells during different stages of spermatogenesis, as well as their regulatory functions in spermatogenesis. Mather et al. (1983b) have suggested the regulation of Sertoli cell function by peritubular myoid cells and extracellular matrix components. The cooperativity between Sertoli cells and testicular peritubular cells in the production and deposition of extracellular matrix components is demonstrated (Skinner et al. 1985).

A technique known as transillumination-assisted microdissection has been introduced to isolate tubular segments at defined stages of rat spermatogenic cycle (Parvinen and Vanha-Perttula 1972, Parvinen and Ruokonen 1982) and to culture with [^{35}S]-methionine under defined conditions. The isolation of these stage-specific tubular segments has permitted a close biochemical investigation of Sertoli cell proteins and factors and their amounts during different stages of spermatogenesis (see also Parvinen 1983, Wright et al. 1983). The myoid cell-depleted tubules isolated from different stages of the epithelial cycle show, at Stages 6 and 12, two distinct peaks of secretion of total radiolabelled proteins in culture (Wright et al. 1983). Two-dimensional gel electrophoresis has revealed that the patterns of secreted proteins from these two stages are remarkably different, while those from other stages are intermediate between those at the peaks. At least 15 proteins are secreted cyclically, many of them previously unrecognized products of the seminiferous epithelium. In this study by Wright et al. (1983) protein secretion by tubular segments is characterized by immunoprecipitation with two polyspecific antisera directed against Sertoli cell products. Five secretory proteins are observed which show cycles different from one another and from cyclic protein-2. Cyclic protein-2 is of great interest as it shows the most pronounced cycle of secretion and is not synchronous with other cyclic proteins described in various studies (Lacroix et al. 1981, Ritzen et al. 1982, Johnsonbaugh et al. 1982, Wright et al. 1983). In contrast to secreted products, the synthesis of most cellular proteins by tubular segments remains relatively constant throughout the cycle. This study has also indicated that the morphology and the protein synthetic capacity of the seminiferous epithelium are coordinated over space and time. Lalli et al. (1984), using radioautography, have demonstrated glycoprotein synthesis in Sertoli cells during the cycle of the seminiferous epithelium of the adult rat.

Short-term culture of seminiferous tubules is a reliable method for studying the effects of cell-cell interactions on proteins secretion in the seminiferous epithelium, as evidenced by both morphological and biochemical observations. The kinetics of protein synthesis and secretion resemble those for cultured Sertoli cells (Wright et al. 1981). Tubular segments neither release their contents because of the cell death nor synthesize novel proteins in response to being placed in culture (Currie and White 1981). The ability of tunicamycin to inhibit protein accumulation in the culture media further supports the suggestion that these proteins are secretory products of the tubules. A significant proportion of germ-cells within tubular segments also survive for many days in culture and progress through several steps of spermatogenesis (Parvinen et al. 1983). Both the morphological and biochemical data support the suggestion that the interaction of the various cell types present in different stages of the seminiferous epithelium can be studied in short-term culture (D'Agostino et al. 1984). Galdieri et al. (1982, 1983) have observed changes in Sertoli cell glycoproteins induced by the removal of the associated germ-cells.

Sertoli cell proteins secreted cyclically under the influence of testosterone and FSH may be performing very important functions in the seminiferous epithelium, about which our knowledge is still very meagre. Some of these proteins may have an autocrine function. Transferrin, ceruloplasmin and Somatomedin are produced by, and in turn stimulate, Sertoli cells (Mather et al. 1983a). The seminiferous growth factor (SGF), a mitogen produced by the sustentacular Sertoli cells, is also suggested to perform autocrine function as it regulates the proliferation of somatic (Sertoli) cells during testicular development and maturation in the mouse (see review by Bellvé and Feig 1984). This period also corresponds to a large number of myc transcripts (Stewart et al. 1984). Some of the Sertoli cell proteins may influence the division, differentiation or movement of germ-cells within the seminiferous epithelium (Lacroix et al. 1981, Johnsonbaugh et al. 1982, Bellvé and Feig 1984). Transferrin is shown to bind specifically to pachytene spermatocytes (Holmes et al. 1983). Lacroix et al. (1981) have suggested that the 100-fold increase in plasminogen activator secretion by the Sertoli cells at Stages 7 and 12 of spermatogenesis is involved in both spermiation and movement of spermatocytes to the luminal side of the blood-testis barrier. The other changes in the morphological characteristics of the tubules, which are associated with the movement of cells, include orientation of the acrosomes of the step 8 spermatids towards the basal lamina and the release of the bundles of spermatids at Stage 6 (Leblond and Clermont 1952a). The exclusive association of any types of cellular motion with specific Sertoli products needs to be determined in future studies. But it will be interesting to mention here that the activity of protein carboxyl methylase is highest during Stages 2–3 of the epithelial cycle (Cusan et al. 1981) when the differentiating spermatids form compact bundles and move towards the basal lamina of the seminiferous tubules. Further studies are needed to define the roles of Sertoli cell proteins in the spermatids differentiation, cellular motion and spermiation. Luteinizing hormone releasing hormone-like substances secreted by Sertoli cells have been shown to affect Leydig cell function (Sharpe et al. 1981).

Positive correlations between the total number of Sertoli cells per testis, the total length of seminiferous tubules per testis, and the total area of the seminiferous epithelium have been described in adult males of different species (rat, ram, bull) (Hochereau-de-Reviers and Courot 1978). The total number of Sertoli cells per testis and of stem-cell stocks per testis have been determined. After hypophysectomy of adult animals (rat or ram), Sertoli cell stocks per testis do not vary, while spermatogonia stock and the efficiency of the spermatogenetic process are affected greatly (Courot et al. 1979, Courot 1980, Courot and Ortavant 1981). The quantitative relationships between the numbers of Sertoli cells and renewing spermatogonia in the adult testes (bull, ram, rat) have provided evidence for the quantitative regulation of adult spermatogenesis by the population of Sertoli cells that is established during the prepuberal period of testicular development by the proliferative capacity of these cells, a feature that may be subject to genetic control (de Reviers et al. 1980) and season of birth (Hochereau-de-Reviers et al. 1984a, b); the rapid proliferation of Sertoli cells correlates to a large number of myc transcripts (Stewart et al. 1984). Significantly, hemicastration of prepuberal rats (Cunningham et al. 1978), cockerels, lambs, and calves (de Reviers et al. 1980), when performed at the time of Sertoli cell multiplication, results in a compensatory hypertrophy of the remaining testis. This effect is due mainly to the fact that there occurs an enhanced hyperplasia of the Sertoli cells, leading to a subsequent commensurate increase in the number of germ-cells. This effect has been related to the higher circulating levels of FSH that occur in response to hemicastration (Cunningham et al. 1978), a conclusion supported by the observation that FSH enhances the mitotic index of prepuberal rat Sertoli cells in vitro (Griswold et al. 1976, 1977, Guraya 1980). Testosterone is observed to determine the number of Sertoli cells in developing testis of rats and reduces the number of immature Sertoli cells with negative relation to meiotic prophase (Kula 1983). The results of these studies indicate that the sperm production in the adult is programmed (or controlled) during the development of testis (see also Hochereau-de-Reviers et al. 1984a, b). The human Sertoli cell population declines with age and there is a significant relationship between sperm production rates and number of Sertoli cells (L. A. Johnson et al. 1984a, Jane and Johnson 1984).

3. Sertoli Cells and the Hormonal Regulation of Spermatogenesis

It is accepted that the physiological function of the testis as a whole, comprising spermatogenesis and steroidogenesis, is regulated through a synergistic action of FSH and LH. The binding of gonadotrophins to specific sites in testicular cells has been shown, in vivo and in vitro, by methods of autoradiography, immunofluorescence, histochemistry and electron microscopy (de Kretser et al. 1969, 1971, Castro et al. 1970, 1972, Lago et al. 1975, Means 1977, Guraya 1980, diZerega and Sherins 1981, Tindall et al. 1985, Ritzen et al. 1981). Labelled FSH binds to Sertoli cells while labelling appears mainly in the Leydig cells and peritubular structures and only to a small degree in the Sertoli cells. The molecular endocrinology of FSH and LH interaction with testis receptors is becoming increasingly understood (Reichert et al. 1982, Payne et al. 1982, Berman and Sairam 1982, Tindall et al. 1985).

Very divergent views exist concerning the hormonal regulation of spermatogenesis and the mechanisms of hormone action (e.g. gonadotrophins and steroids) in the initiation of the first spermatogenic cycle (E. Steinberger 1971, Bergada and Mancini 1973, Courot et al. 1979, de Kretser et al. 1974, 1980, Rodriguez-Rigau et al. 1980, Davies 1981, Courot and Ortavant 1981, Ritzen et al.

1981, diZerega and Sherins 1981, Haneji and Nishimune 1982). However, a picture is gradually emerging of the mechanism of FSH and LH control of spermatogenesis and the part played by the Sertoli cells, although there is no universal agreement in regard to the quantitative requirement for FSH, LH or testosterone. One certainly experiences some difficulty in shifting and interpreting experimental data relating to the functional links between these three hormones, which have been investigated chiefly in intact immature or hypophysectomized rats. The deprivation of pituitary gonadotrophins by hypophysectomy, or blockade of secretion or their action, is for inducing a severe regression of spermatogenesis which occurs rapidly after the start of the treatment and then develops progressively (Courtens and Courot 1980). Supplementation of hypophysectomized animals with hCG and hMG maintains spermatogenesis, thereby indicating a direct or an indirect role of FSH on spermatogonial multiplication and of LH on other stages of spermatogenic cycle (see Courot 1980, Courot and Ortavant 1981). The various hormones (FSH, LH and testosterone) appear to act simultaneously on the ram testis, apparently on the beginning of the spermatogenetic process as revealed by the higher correlation coefficients, with increasing delay between endocrine and testicular size measurements (Courot and Ortavant 1981). Ishihara et al. (1983) have obtained successful induction of spermatogenesis by a combined use of hCG and hMG in patients with hypogonadotropic eunuchoidism associated with pituitary dwarfism.

FSH and LH in the rat and other mammalian species appear to act synergistically in the initiation of the first spermatogenic cycle (Lostroh 1969, de Kreter et al. 1980). In the prepuberal rat, FSH (but not LH) is required to maintain testicular growth and spermatogenesis (Courot et al. 1971) to increase [^{3}H]-thymidine uptake by spermatogonia and leptotene spermatocytes (Ortavant et al. 1972) and to stimulate mitotic activity and differentiation of spermatogonia (Mills and Means 1972, Orth and Christensen 1978) and the synthetic activity of Sertoli cells (see reviews of Means 1977, Welsh et al. 1980, Guraya 1980, diZerega and Sherins 1981, Ritzen et al. 1981, A. Steinberger et al. 1979). There are many reports of the role of FSH in spermatogenesis (Means 1977) but divergent views exist about its function in the very early stages. In the adult hypophysectomized rat, FSH can be effective in maintaining and restoring spermatogonial multiplications (Vernon et al. 1975, Cunningham and Huckins 1979) and germ-cell development between pachytene spermatocytes and step 7 spermatids (Vernon et al. 1975). Libbus and Schuetz (1979, 1980), who used oestradiol-treated adult and prepubertal testes consisting of spermatogonia and Sertoli cells, have observed that PMSG stimulates DNA synthesis in both the germinal and somatic cells, as well as the production of pachytene spermatocytes. Ortavant et al. (1969) and Chemes et al. (1979) have shown a requirement for FSH in the proliferation and differentiation of type A spermatogonia in immature animals. Haneji and Nishimune (1982) have also demonstrated that FSH activates cell division in type A spermatogonia in mouse cryptorchid testes incubated in vitro, and stimulates them to differentiate, while LH shows neither the promotion of differentiation nor a synergistic effect on FSH-mediated germ-cell differentiation. FSH cannot stimulate the germ-cell differentiation in vitro unless insulin and transferrin are present. Sperm production in gonadotropin-suppressed normal men has been reinitiated by administration of FSH (Matsumoto et al. 1983). The results of all these studies differ from many other studies in adult animals, which have shown that LH and/or androgens are more important than FSH in reinitiating and maintaining spermatogenesis (Woods and Simpson 1961, Clermont and Harvey 1967, Davies et al. 1974). The initiation, restoration and maintenance of spermatogenesis by testosterone have been shown in non-human primates (Marshall and Nieschlag 1984).

The in vitro studies have indicated that no hormones are required for rat spermatogenesis to proceed from the early spermatogonial steps through to those of late pachytene spermatocytes and that FSH and testosterone are needed only for the final stages of the spermatogenic cycle (E. Steinberger and Duckett 1965, A. Steinberger and E. Steinberger 1966, 1967, E. Steinberger 1971). This suggests that the initiation of spermatogenesis in the rat probably is mediated by testosterone (Steinberger 1971) or by other androgens (Chowdhury and Steinberger 1975), and does not require gonadotrophins. The other studies have suggested that androgens, LH and other pituitary hormones [growth hormone (GH) and thyroid-stimulating hormone (TSH)] are required for spermatogenesis (Simpson and Evans 1946, Woods and Simpson 1961, Mancini 1968, Mills and Means 1972). But other views consider the spermatogonial renewal and the early steps of meiosis to be regulated by FSH (Greep et al. 1936, Courot 1970, 1976, Courot et al. 1971, Mills and Means 1972, Ortavant et al. 1972, Huckins et al. 1973, Means 1974, Davies 1981). The last generation of spermatogonia are sensitive to FSH control (Courot et al. 1984). Aside from evidence in one case that FSH may bind directly to these germ-cells (Orth and Christensen 1978) concerted evidence on the hormonal regulation of spermatogonial proliferation is not available. The data obtained in man have suggested that FSH maintains normal spermatogenesis quantitatively following hypophysectomy or following the successful induction of spermatogenesis in hypogonadotrophic hypogonadism (see de Kretser et al. 1980, Bremner et al. 1981, Matsumoto et al. 1983, 1986). In the latter situation, withdrawal of FSH after the induction of spermatogenesis leads to a fast decline in spermatogenic function despite continuation of LH/hCG replacement therapy. The role of FSH in the control of spermatogenesis in the adult rat generally is not accepted (Clermont and Harvey 1967) because a number of studies have demonstrated that testosterone can maintain or restore spermatogenesis in hypophysectomized rats (Boccabella 1963, Clermont and Harvey 1967, Matsuyama et al. 1971, Ahmad et al. 1975, Chemes et al. 1976a, Harris et al. 1977, Rivarola et al. 1977, diZerega and Sherins 1981), although the rate of sperm production is below normal (Chowdhury 1979). In the rat, testosterone has been shown to control the differentiation from A_0 to A_1 spermatogonia (E. Steinberger 1971); and antiandrogens can partly block their proliferation, thereby depressing the yield of spermatogonia (Viguier-Martinez and Hochereau-de-Reviers 1977, Hochereau-de-Reviers 1981). Precise quantitative analysis of spermatogonia in hypophysectomized testosterone-treated adult rats has revealed that undifferentiated type A (A_0-A_2) spermatogonia are sustained by testosterone or pituitary hormones but that testosterone only partially restores A_3 to intermediate spermatogonial multiplication (Chowdhury 1979). However, normal spermatogenesis is maintained despite low levels of intratesticular testosterone (Cunningham and Huckins

1979). It will be interesting to mention here that the germ-cells do not show receptors for androgens including testosterone (Courot 1980, Courot and Ortavant 1981).

The effects of human chorionic gonadotrophin (hCG) and pregnant mare serum gonadotrophin (PMSG) for different periods and the effect of testosterone on the immature rat testis have been studied (Chemes et al. 1976b). Short-term treatment with hCG (1 – 3 days) leads to an early meiotic and postmeiotic stimulatory effect but causes a decrease in spermatogonial numbers. hCG treatment for longer periods (10 days) brings about a reduction in numbers of all cell types. Administration of hCG + PMSG alleviates the inhibitory effects, whereas PMSG alone induces the histological and hormonal signs of stimulation of the interstitial tissue and the meiotic and postmeiotic stages; while the numbers of spermatogonia are not affected. Testosterone promotes the number of meiotic and postmeiotic stages and decreases the number of spermatogonia. These results lead to the conclusion that PMSG directly stimulates spermatogonia and hCG exerts its effect through testosterone at the meiotic and postmeiotic stages. The early inhibitory effects of hCG and testosterone on spermatogonial numbers could be assigned to the inhibition of endogenous FSH by androgens. The observation of precocious spermatogenic maturation in a 6-year-old boy with a Leydig cell tumor has suggested that high levels of local androgens are able to initiate spermatogenesis in some cases (see E. Steinberger et al. 1973, Neaves 1975, diZerega and Sherins 1981). However, various hormones, including testosterone, do not exert a stimulatory effect on the initiation of spermatogenesis in vitro (see E. Steinberger et al. 1970).

The maintenance or restoration of spermatogenesis in the adult rat by testosterone, as discussed above, varies between species. In hypophysectomized rams, the administration of steroids sufficient to induce normal concentrations of testosterone and 5α dihydrotestosterone within the seminiferous tubules cannot maintain the number of leptotene spermatocytes (Monet-Kuntz et al. 1976). This has suggested that testosterone is not efficient in maintaining spermatogonial proliferation in the sheep, as it is in the rat. Courot (1970, 1971), has shown that the initiation of spermatogenesis in the lamb is dependent on the presence of anterior pituitary hormones, as its ablation shortly after birth stops the differentiation of germ-cells. This effect can be reversed after treatment with either FSH or LH alone, or synergistically with both gonadotrophins, but not with testosterone. The latter cannot support the transition from reserve (A_0) to renewing (A_1) stem spermatogonia in the ram (Monet-Kuntz et al. 1977, Courot et al. 1979). Courot (1970, 1971) has suggested that the effect of the gonadotrophins is mediated through the Sertoli cells. Marshall et al. (1983) have observed that testosterone alone can initiate spermatogenesis in a nonhuman primate.

Courot et al. (1979) have made a detailed quantitative histological analysis of spermatogenesis in the adult hypophysectomized ram after treatment with various hormones, such as testosterone, hCG and PMSG. This study has shown that (1) the differentiation from A_0 to A_1 spermatogonia can be maintained by PMSG or hCG but not completely by testosterone, (2) the transition from In spermatogonia to primary spermatocytes, including the B_1 and B_2 spermatogonia, is maintained only by PMSG but not by testosterone or hCG; and (3) meiotic prophase and spermiogenesis are maintained by the three hormones (PMSG, hCG or testosterone) but there are qualitative abnormalities in the spermatids. These results have suggested that in the adult ram, the differentiation of renewing stem spermatogonia is under LH control and that the last stages of spermatogonial multiplication, from intermediate to B spermatogonia and to primary spermatocytes, are under the control of the FSH-like activity of PMSG (Hochereau-de-Reviers 1981, Courot and Ortavant 1981). The reserve stem-cells, A_0 spermatogonia, are not hormone dependent as their number does not vary, whatever the treatment (Hochereau-de-Reviers et al. 1980, Courot and Ortavant 1981). The differentiation from A_1 to In spermatogonia is maintained equally by hCG, PMSG and testosterone, suggesting that the effect of gonadotrophins is secondary to testosterone. Obviously, these steps of spermatogonial differentiation are testosterone dependent. Testosterone has only a small effect at the beginning of the spermatogenic cycle (production of leptotene spermatocytes) and quantitatively maintains meiosis and spermiogenesis, but the differentiation of spermatids occurs on the basis of information stored at the beginning of meiosis and needs the support of both testosterone and other factors (Monet-Kuntz et al. 1977). Courtens and Courot (1980) have observed an initial abnormal development of the acrosomes in hypophysectomized and 15–20-day testosterone-supplemented rams, and the recovery of normal differentiation in the 40-day treated animals. This study has suggested that acrosome development is under the control of endocrine-dependent cellular events occurring before the start of spermiogenesis, possibly via Sertoli cell/germ-cell interactions. The role of the Sertoli cell is consistent with its different ultrastructure following the two treatments.

The quantitative maintenance of meiotic prophase and spermiogenesis in the ram by PMSG, hCG or testosterone suggests that their effect is mediated through testosterone secretion. A similar observation has been made for the hypophysectomized rat in which testosterone maintains meiosis and spermiogenesis (Chowdhury and Teholakian 1979). It will be interesting to mention here that LH antibodies and to a slight extent FSH antibodies can modify spermiogenesis in the rat (Dym and Madhwa Raj 1977, Dym et al. 1977, see also review by Hochereau-de-Reviers 1981). Anti-LH is presumed to act through testosterone depletion.

Courot and Ortavant (1981) have observed recently that spermatogenesis in the ram is sensitive to variations in the levels of circulating hormones, as evidenced by the positive correlation between the number of renewing spermatogonia or the efficiency of spermatogonial multiplication and the mean LH value in the peripheral blood of the adult (Courot 1980, Hochereau-de-Reviers 1981). Some of these relationships operate over long periods and probably involve the Sertoli cells. The level of circulating LH in the nonpubertal lamb is related directly to the number of Sertoli cells, and the latter is related to the number of renewing spermatogonia per testis in the adult ram. The importance of FSH, LH and testosterone in the control of ram spermatogenesis is confirmed (see Matsumoto and Bremner 1985). All these observations suggest that both FSH and LH play a role in the differentiation of spermatogonia. Actually, their synergistic effects are essential for the maintenance of spermatogonia and Sertoli-cell function. It is not known whether FSH and/or LH act directly on ram spermatogonia or only indirectly by causing Sertoli cells to secrete a substance that promotes or enhances spermatogenesis. The latter hypothesis is supported by the fact that total numbers of Sertoli cells per testis are not modified, but stem spermatogonia as well as the daily production of round spermatid are significantly increased after hemicastration (Hochereau-de-Reviers et al. 1976). Gray et al. (1981) have suggested a possible relationship between prolactin and spermatogenesis in humans.

The basal compartment of the seminiferous tubules, containing spermatogonia and the leptotene spermatocytes (Fig. 12), can be reached directly by the hormones supplied by the bloodstream or present in the peritubular fluid. However, this does not rule out a possible control of spermatogonia through Sertoli cells. Indeed, positive correlations are found between the total number of Sertoli cells and stem spermatogonia in the adult rat, ram and bull (Hochereau-de-Reviers and Courot 1978). Courot and Ortavant (1981) have observed that the cells of the basal compartment are affected more rapidly than others by any alteration of the hormonal balance, showing that they may represent a direct target to gonadotrophins and testosterone. The presence of FSH receptors has been demonstrated in Sertoli cells (Means 1977, A. Steinberger and E. Steinberger 1977, 1980, Guraya 1980, Ritzen et al. 1981, diZerega and Sherins 1981) and spermatogonia of the rat (Orth and Christensen 1978, diZerega and Sherins 1981). Sertoli cells do not show LH receptors in spite of the fact that they are altered after treatment with antibody to LH in the rat (Chemes et al. 1979) and show an increased number of pinocytotic vesicles on their membrane after treatment with hCG in the dog (Connell 1977). The presence of LH receptors on spermatogonia and a specific role for LH (not operating through testosterone) on seminiferous tubules have been suggested (Courot et al. 1970, Courot and Ortavant 1981). Testosterone receptors have been reported in the rat Sertoli cells (Grootegoed et al. 1977b, A. Steinberger and E. Steinberger 1977, 1980, de Kretser et al. 1980, Ritzen et al. 1981). Binding of testosterone to specific receptors and its conversion to dihydrotestosterone under the catalytic influence of 5α-reductase is now a prerequisite for androgen-induced responses in all target organs including the seminiferous epithelium (Mann and Lutwak-Mann 1981). Sertoli cells and spermatocytes can reduce testosterone, at least in in vitro conditions, but whereas the major product formed by spermatocytes is dihydrotestosterone, in Sertoli cells further conversion (by 3α-hydroxysteroid dehydrogenase) to 5α-androstane-$3,\alpha 17\beta$-diol takes place (Mulder et al. 1974, Dorrington and Fritz 1975, Hansson et al. 1975, Wilson 1975). In the light of present data, it is still not known whether FSH and/or LH act directly on spermatogonia or only indirectly by causing Sertoli cells to secrete a substance that promotes or enhances spermatogenesis. Accurate data on how the hormones present in Sertoli cells interact with the differentiating germcells are lacking, and the global contribution of Sertoli cell secretion in seminiferous tubular fluid formation still needs to be assessed. But the Sertoli cells form the main target for FSH in the testis and their response has been characterized in some detail, at least in the rat (Dufau et al. 1973, Means 1977, Guraya 1980, A. Steinberger and E. Steinberger 1980, Ritzen et al. 1981, Tindall et al. 1985). ABP is an end-product of FSH-stimulated Sertoli cells. Does ABP represent the message maintaining meiosis and spermiogenesis? This is possible because receptors for androgen are present in spermatids of the rat (Wright and Frankel 1980), but

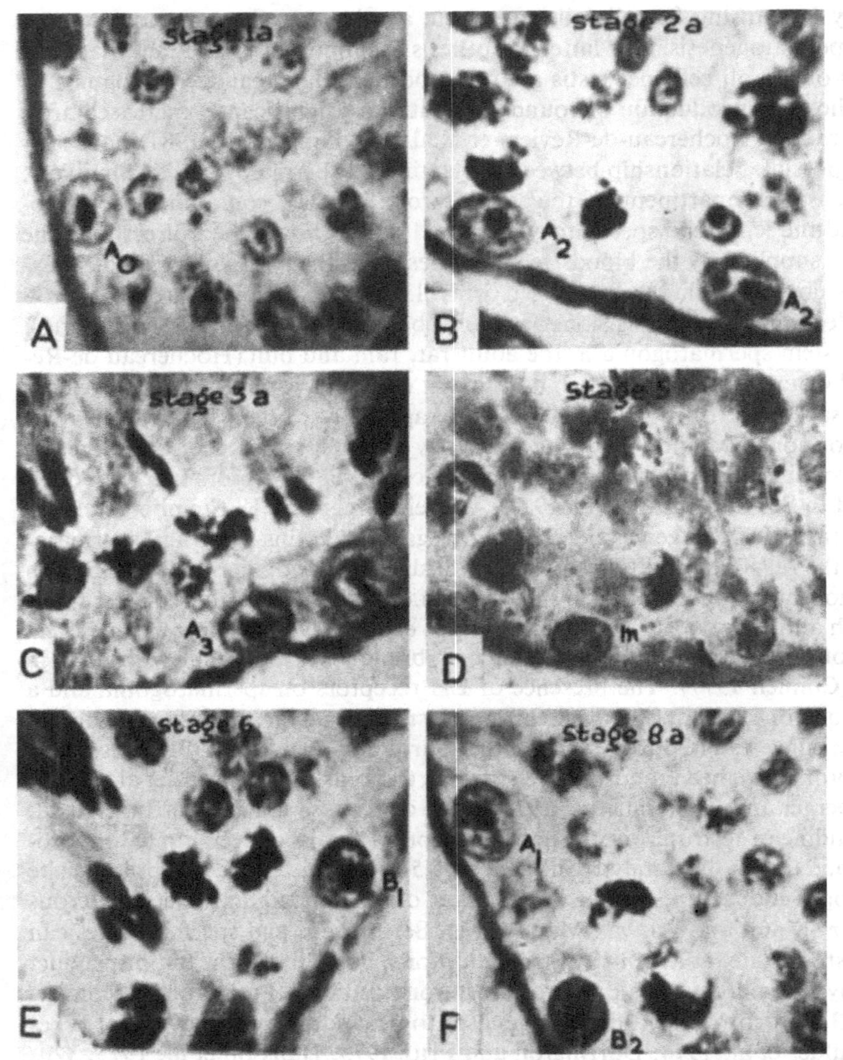

Fig. 12 A – F. Photomicrographs of areas of buffalo seminiferous epithelium showing different types of spermatogonia in different stages. The *number of stage* is indicated *at the top* of each picture on the lumen side of the tubular section. **A** *Stage 1a* showing type A_0 spermatogonia (A_0); **B** *Stage 2a* showing type A_2 spermatogonia (A_2); **C** *Stage 3a* showing type A_3 spermatogonia (A_3); **D** *Stage 5* showing type intermediate spermatogonia (*In*); **E** *Stage 6* showing type B_1 spermatogonia (B_1); **F** *Stage 8a* showing type A_1 spermatogonia (A_1) and type B_2 spermatogonia (B_2) (From Guraya and Bilaspuri 1976c)

several other polypeptides are secreted by the Sertoli cells in response to hormones (Wilson and Griswold 1979, Ritzen et al. 1981, Wright et al. 1983, Tindall et al. 1985) and the possibility cannot be ruled out that one or more of these might control spermatogenesis. The secretion by the Sertoli cells of ABP, which has a high affinity for testosterone and DHT, probably plays a significant role in concentrating androgens within the seminiferous tubule around the Sertoli cells and the germ-cells. The current concepts of hormonal regulation of mammalian spermatogenesis, derived primarily from studies in the rat and lamb, also apply to the human testis (Rodriguez-Rigau et al. 1980, Bremier et al. 1981, Matsumoto et al. 1983, 1984, 1986). Various observations clearly show that both FSH and LH are essential for spermatogenesis (Matsumoto and Bremner 1985), but whereas FSH acts on the Sertoli cells directly and its primary function is to initiate spermatogenesis, the essential quality of LH is ascribable chiefly to its stimulating effect on testosterone biosynthesis in the Leydig cells, thereby providing the germinal epithelium with testosterone needed for spermatogenesis. However, it is debatable whether the Sertoli cells or some other tissue component, such as the outer layer of the boundary tissue, forms the primary target for testosterone (Fritz 1978), but it has been established that before reaching the germ-cells the androgen must be taken up and transported by the Sertoli cells. Uncertainty prevails about the quantitative relationship between the intratesticular concentrations of testosterone and the gonadotrophins, and the rates of sperm production and the secretory function of the Sertoli cell (Kotite et al. 1978, Ritzen et al. 1981). The mechanism enabling the Sertoli cells to secrete ABP into the lumen of the seminiferous tubules, which needs further study, is believed to be controlled by a calcium-binding protein similar to calmodulin (Welsh et al 1980).

The specific effects of FSH, LH and testosterone on mammalian spermatogenesis still need to be elucidated further for the better understanding of hormonal regulation and its qualitative and quantitative aspects. The molecular mechanisms by which these hormones act on the germinal cells also remain to be elucidated (Tindall et al. 1985). A genetic approach is suggested for the study of hormonal requirements of mouse spermatogenesis (Chubb and Lopez 1983). In addition to the secretion of testosterone for peripheral use, the Leydig cells may provide a source of testosterone and other metabolites for use by the seminiferous tubules in the stimulation and maintenance of spermatogenesis. Recent data support the existence of an intercompartment regulatory system (de Kretser et al. 1980, Van der Molen and Rommerts 1981, diZerega and Sherins 1981, Sharpe et al. 1981, Sharpe 1983) which may be of great significance in normal and subnormal testicular function. These data have indicated the existence of a closed-loop feedback system where Leydig cell androgens are aromatized by the Sertoli cell to oestrogens which diffuse back into the interstitial tissue to regulate Leydig cell function. Leydig cells are also known to form the source of peptides and proteins which appear to involve in a paracrine regulation of the Sertoli and germinal cell elements (Melner and Puett 1984a). Some of the peptides produced by the Leydig cells stimulate growth and cAMP production by Sertoli cells (Mather 1980, Mather et al. 1982a, b). There is now increasing evidence that Sertoli cells may secrete their products into the interstitial space where they enter the lymphatics and the bloodstream, as well as secreting products into the lumen of the seminiferous tubules. Melner and Puett (1984a) have suggested that this bidirectional secretion has brought new implications to the hypothesized physiological roles of Sertoli cell products into male reproduction. Bergh and co-workers (Bergh 1985, Bergh and Damber 1983, 1984, Bergh et al. 1984) have suggested the local regulation of Leydig cell function from the seminiferous epithelium in the rat (see Aoki and

Fawcett 1978). The stage-specific inhibition of Leydig cell testosterone secretion by rat seminiferous tubules is observed (Syed et al. 1983). The cyclic regulation of rat Leydig cell testosterone production by seminiferous tubules (Parvinen et al. 1984) and synergistic effects of Sertoli cell and FSH on Leydig cell function (Benahmed et al. 1984) have been demonstrated in in vitro studies. Chen et al. (1983) and Melner and Puett (1984b) have demonstrated that the Leydig cells are the source of synthesis of pro-opiomelanocortin, a precursor polypeptide to several bioactive peptides including β-endorphin (Tsong et al. 1982a, b). Parmentier et al. (1983), using immunohistochemical techniques, have localized renin exclusively in Leydig cells. The application of molecular biology techniques (Bellvé and Moss 1983, Mettler et al. 1983b, Bellvé and Feig 1984) to testicular compartments in the future should be very effective in identifying the regulatory functions of secretions of both Sertoli and Leydig cells at the molecular level.

G. In Vitro Studies

During the past decade several methods have been developed to isolate (1) germ-cells at specific stages of development, (2) Sertoli cells, and (3) segments of seminiferous epithelium at specific stages of spermatogenesis. These protocols have been used for in vitro studies to determine the nature of factors that regulate their survival, surrvival metabolism, and morphological and biochemical characteristics related to germ-cell proliferation and differentiation, as well as the interaction between germ-cells and Sertoli cells and between Sertoli cells and peritubular myoid cells at the molecular level (see Bellvé 1979, Mather et al. 1982b, Wright et al. 1983, Tung et al. 1984, Huggenvik et al. 1984, Suarez-Quian et al. 1984, Sylvester and Griswold 1984, Eikvar et al. 1984, Jutte and Hansson 1984, Heindel et al. 1984, Mather et al. 1984, Welsh et al. 1984, Sanborn et al. 1984, Nakhla et al. 1984, D'Agostino et al. 1984, Tindall et al. 1985, Hadlex et al. 1985, Welsh et al. 1985 for references; see also sect. F of this chapter). The mouse primordial germ-cells have been studied in in vitro culture (DeFelici and McLaren 1983). Burgos et al. (1982) have made in vitro studies on shuttle systems of mouse spermatozoa. Some of these studies have provided newer information that will be discussed here as well as in various chapters of Part Two of this book. Mather et al. (1982b) have studied the problems of culture of testicular cells in hormone supplemented serum-free medium. These studies have shown that the cells can be studied individually and cell-cell interactions can then be investigated by mixing two different cell types in a controlled fashion. Stimulation of rat Sertoli cell adenylate by germ-cells in vitro is studied (Welsh et al. 1985). Grootegoed et al. (1982b) have studied the problems of intercellular adhesion of male germ-cells and Sertoli cells induced by Con A. The formation in vitro of intercellular junctions between isolated germ-cells and Sertoli cells in the rat has been investigated (Ziparo et al. 1982). Vitamin A and FSH function synergistically to induce differentiation of type A spermatogonia in adult mouse cryptorchid testes in vitro (Haneji et al. 1984). Mushkambarov et al. (1982) studied the stability and viability of spermatogenic cells after their isolation and separation. Spermatogenic cells isolated from seminiferous tubules by enzymatic treatment are very sensitive to action of Triton X-100. The cells isolated by mechanical dissociation of spermatogenic epithelium are destroyed during storage. The cell separation by sedimentation velocity technique in humen serum protein gradients leads to viable cells that maintain their specific structural and biochemical properties.

Koulischer et al. (1982) observed with an in vitro system that all meiotic stages are present and normal until the 8th day of culture. During the 9th and 10th day in vitro, the number of the cells in meiosis declines abruptly. By the 14th day, only a few pachytene stages are still observed. Labelling with tritiated thymidine suggests that spermatogonia do not enter meiotic division in vitro. Chromosome preparations reflect cells already in meiosis at the time of initiating the culture. But Parvinen et al. (1983) have claimed to observe the completion of meiosis and early spermiogenesis in vitro. For these studies, segments of rat seminiferous tubules containing primary spermatocytes at late pachynema and diakinesis (Stages 12 and 13 of the cycle) were cultured in a chemically defined medium. After 2 days, most spermatocytes had completed both meiotic divisions and by 6 days the seminiferous epithelium showed the morphologic characteristics of Stage 5, in which the newly formed spermatids show an acrosomic system characteristic of step 5 of spermiogenesis. Rat spermatogenesis in vitro has been traced by live cell squashes and monoclonal antibodies (Toppari et al. 1984). The seminiferous tubules also showed biochemical differentiation as evidenced by increased production of proteins characteristically secreted by Stage 5. These in vitro experiments have also demonstrated that in vitro differentiation of the germinal epithelium occurs in the absence of testosterone and FSH, suggesting that spermatocytes at late pachytene and their associated Sertoli cells have information needed for both meiotic divisions and early spermiogenesis. But this study has demonstrated differentiation only in 10% of the cells in $<10\%$ of the tubules in culture, which is not very high. Tres and Kierszenbaum (1983) have demonstrated that viability of rat spermatogenic cells in vitro is facilitated by their co-culture with Sertoli cells in serum-free hormone-supplemented medium. For this study, spermatogenic cells from 20- to 25-day-old rats were cultured in vitro in the presence of Sertoli cells maintained in serum-free, hormone- and growth factor-supplemented medium, and with either high or low concentrations of FSH in the medium. In the cell reaggregation experiments spermatogenic cells reassociate with Sertoli cells, but not with peritubular cells or cells-free substrate. Autoradiographic experiments using [^{3}H]thymidine as a labelled precursor for DNA synthesis demonstrate that spermatogonia and preleptotene spermatocytes, interconnected by cytoplasmic bridges, have a synchronous S phase . [^{3}H]thymidine-labelled preleptotene spermatocytes progressed until later stages of meiotic prophase. Time-lapse cinematographic studies of Sertoli/spermatogenic cells co-culture demonstrate three major movement patterns. While Sertoli cell cytoplasmic processes between adjacent cells showed tensional forces, spermatogonia were engaged in oscillatory cell movements different from the nuclear rotation observed in meiotic prophase spermatocytes. Results of this study have demonstrated clearly that the multiplicator of the pre-meiotic cells and the differentiation of meiotic prophase cells occurs in vitro in association with Sertoli cells maintained in a medium that allows differentiated cell functions. Gerton and Millette (1984) have studied generation of flagella by cultured mouse spermatids.

Several compounds are known to be secreted by Sertoli cells in vitro, e.g. androgen binding protein (Fritz et al. 1976), transferrin (Skinner and Griswold 1980), plasminogen activator (Lacroix and Fritz 1980, Lacroix et al. 1981, Fritz et al. 1982), glycoproteins (Galdieri et al. 1981, 1982), sulphoproteins (Elkington and Fritz 1980) and myo-inositol (Robinson and Fritz 1979). These studies used either partially pure Sertoli cells or segments of seminiferous epithelium at specific stages of spermatogenesis (Tindall et al. 1985). None of the compounds isolated has been shown to affect the growth and differentiation of germ-cells. Boitani et al. (1983) have demonstrated the influence of Sertoli cell products upon the in vitro survival of isolated spermatocytes and spermatids. Lacroix et al. (1981) studied localization of testicular plasminogen activator in discrete portions (Stages 7 and 8) of the seminiferous tubule in culture, which were isolated by transillumination of Parvinen and Vanha-Perttula (1972). A hypothesis was advanced that plasminogen activator is intimately related to the localized restructuring that takes place at Stages 7 and 8 as the meiotic spermatocytes translocated into the adluminal compartment, and the advanced spermatid into the tubule lumen.

Exogenous lactate has been shown to stimulate RNA and protein synthesis and oxygen consumption of isolated spermatocytes and spermatids (Jutte et al. 1981a, b, 1985). These results are supported by the fact that protein synthesis in round spermatids is also stimulated by lactate (Nakamura and Kato 1981). The isolated germ-cells showed a very low capacity for using glucose as an energy source. But for the maintenance of spermatogenesis in vivo, glucose is necessary (Mancini et al. 1960, Zysk et al. 1975). It is possible that the Sertoli cells in vivo convert glucose to a substrate which is used by germ-cells. Jutte et al. (1982) have investigated whether Sertoli cells in vitro can influence the activity and survival of germ-cells via the secretion of lactate. When fragments of seminiferous tubules are incubated without the addition of glucose, pachytene spermatocytes and round spermatids die within 24 h, while Sertoli cells still remain viable. When the glucose is added in the incubation medium, the germ-cells survive within the seminiferous tubule fragments for at least 72 h. This study has shown also that lactate rather than glucose is necessary for [^{3}H]uridine incorporation and survival of isolated pachytene spermatocytes. However, if the spermatocytes are incubated in the presence of Sertoli cells, glucose maintains the incorporation of [^{3}H]uridine into the germ-cells. Jutte et al. (1985) have recently demonstrated that a stage-specific pattern of protein synthesis can be maintained by pachytene spermatocytes during incubation for a period of 24 h in the absence of Sertoli cells but in the presence of a proper energy source. Sertoli cells secrete lactate in the presence of glucose and the lactate secretion is stimulated two- to fourfold by FSH; FSH can also influence glucose transport by cultured Sertoli cells (Hall and Mita 1984). Mita et al. (1985) have also investigated the influence of insulin and insulin-like growth factor on hexose transport by Sertoli cells. These results have suggested that the activity and survival of pachytene spermatocytes in vitro can be regulated by the supply of lactate from Sertoli cells as also observed by Jutte et al. (1981b). Jutte et al. (1981b) observed that isolated pachytene spermatocytes do not survive in the presence of glucose. But Jutte et al. (1982) have found that RNA synthesis and the integrity of pachytene spermatocytes in cultures of seminiferous tubules and in co-cultures of isolated germ-cells and Sertoli cells are maintained by exogenous glucose. All these observations support the idea that the activity and survival of pachytene spermatocytes are regulated by Sertoli cells via conversion of glucose to lactate. Grootegoed et al. (1984) have observed that spermatogenic cells in the germinal epithelium utilize α-ketoisocaproate and lactate produced by Sertoli cells from leucine and glucose. The lipid oxidation by Sertoli cells is also believed to be to germ-cells (Jutte and Hansson 1984).

Most of the previous studies on metabolic pathways in the testes have been carried out with mixed cell populations and little information is available about metabolic activities of the individual cell types. Sertoli cells in the testis are dependent mainly on lipids for the provision of energy (see Jutte et al. 1982). Jutte et al. (1982) observed that these cells survive incubation for 24 h without glucose, supporting their dependence on intracellular lipids. Study of the metabolism of glucose by cultures of pure Sertoli cells reveals that maximally 2.9% of the glucose utilized is converted to carbon dioxide and 95.8% is converted to anionic compounds, mostly to lactate (Robinson and Fritz 1981). These observations demonstrate the enormous capacity of Sertoli cells to convert glucose into lactate. In contrast to the observations of Jutte et al. (1982), Robinson and Fritz (1981) have not found a stimulatory effect of hormones on lactate production by Sertoli cells. This may be due to differences in the preparations of the Sertoli cells.

Germ-cells may be able to maintain an aerobic metabolism, as the oxygen tension within the seminiferous tubules is as high as the tension in the interstitial tissue (Free et al. 1976). Complete dependence of germ-cells on the activity of the Krebs' cycle has been demonstrated in vivo after feeding the rats with specific inhibitors (Paul et al. 1953, Novi 1968, A. Steinberger and Sud 1970, Sullivan et al. 1979).

Nakamura and Kato (1981) studied the stimulatory effects of glucose and lactate on protein synthesis by spermatids (steps 1 – 8) from rat testis for defining the role of ATP. When the cells are incubated with lactate, the response of protein synthesis in round spermatids is related closely to the intracellular level of ATP. The ATP level in spermatids is increased after incubation of the cells at 34 °C for 60 min in the presence of 20 mM lactate. However, ATP level fell rapidly to undetectable levels during incubation for 30 min at 34 °C without lactate. The ATP level and the rate of protein synthesis in spermatids increased rapidly when lactate (20 mM) was added to the control cells during incubation. The incorporation of $[^{32}P]$ into ATP is increased by treatment with lactate, but glucose (10 mM) has no effect. The rate of utilization of lactate is faster than that of glucose. Mita and Hall (1982) also observed lactate as the preferred substrate. These experiments have shown that ATP is probably a major factor in the stimulation of protein synthesis in spermatids. In a further study, Nakamura et al. (1982) have studied the regulation of glucose metabolism by glycolysis in round spermatids from rat testes. Glyceraldehyde 3-P dehydrogenase, an enzyme believed to play a regulatory role in glycolysis, is inhibited by adenosine, 5′AMP, ADP and 3′ – 5′ cAMP. In each case, the inhibition is competitive with NAD. The results of this study have indicated that the activity of glyceraldehyde 3-P dehydrogenase regulates glycolysis in round spermatids. Hall and Nakamura (1981) observed the influence of temperature on hexose transport by round spermatids of rats.

At successive stages of germ-cell differentiation, there appear to be different requirements for carbohydrate substrates. Round spermatids and pachytene spermatocytes rely on a supply of lactate (Jutte et al. 1981b, Mita and Hall 1982), whereas ejaculated spermatozoa can utilize either glucose or fructose as substrates (Mann and Lutwak-Mann 1981). This suggests that, following a stage of dependence on lactate, germ-cells may become independent of lactate during or after spermatid elongation.

Biosynthesis of lipids by different stages of germ-cell development has been demonstrated in in vitro studies (Grogan and Lam 1982, Grogan and Huth 1983). Grogan and Lam (1982) have observed that all spermatocytes, round spermatids and condensing spermatids from mouse testes possess the various activities necessary for the biosynthesis of fatty acids, but synthesis of fatty acid of >16 carbons declines with progressive stages of differentiation. Complex lipid synthesis is more variable and incorporation into triglycerides generally is much lower in dispersed germinal cells than in whole testis in vitro or in vivo. Isolated germinal cells remain viable throughout the 15 h incubation and synthesize their constituent fatty acids. But the intratubular environment or association with Sertoli cells may be necessary for maintenance of adequate snythesis of complex lipids. In a further study, Grogan and Huth (1983) have demonstrated biosynthesis of long-chain polyenoic acids from arachidonic acid in cultures enriched for mouse spermatocytes and spermatids. The resulting observations suggest that primary spermatocytes, round spermatids and condensing spermatids show some differences in the synthesis of their lipids, which are more pronounced in triglycerides of condensing spermatids.

Gordeladze et al. (1982b) observed maximal Mn^{2+}-dependent adenylate cyclase (AC) activity in Stages 7 – 8 of spermatogenesis (spermiation). FSH-responsive AC activity shows a pattern that coincides with that of the Mn^{2+}-dependent AC. The stage-dependent variation in spermatid AC activity cannot be explained by altered numbers of haploid cells. But the stage-dependent difference in FSH-responsiveness indicates that local influences (from germ-cells) may regulate the response of AC in Sertoli cells to FSH.

Chapter II
Spermatogonia

A. Spermatogonial Types

The earlier germ-cells in the adult testes are the spermatogonia, which are located in the basal compartment of the seminiferous epithelium, close to the basal lamina (Figs. 5 and 12). After a series of mitoses, the spermatogonia give rise to the spermatocytes. The spermatogonial cell divisions are usually incomplete. The daughter cells remain interconnected by cytoplasmic bridges. It generally is accepted that mouse spermatogonia undergo six successive divisions prior to the start of the prolonged meiotic prophase (Clermont and Bustos-Obregon 1968, Oakberg and Huckins 1976). The timing and spatial orientation of each of these mitotic multiplication steps is strictly regulated and can be distinguished from proliferative events in somatic cells by incomplete cytokinesis and the retention of wide intercellular bridges between germ-cells (Fig. 9) (Fawcett et al. 1959, Dym and Fawcett 1971, Schleiermacher and Schmidt 1973, Guraya 1980, Holstein and Roosen-Runge 1981, Kerr and de Kretser 1981). The junctions or desmosome-like structures occur between neighbouring spermatogonia (Holstein and Roosen-Runge 1981).

On the basis of morphology, diameter and volume of the nucleus, number of nucleoli per nucleus, topographic position of spermatogonia with respect to each other and the basal lamina, and the behaviour of their chromosomes during division, three main types of spermatogonia, designated as type A, intermediate (In) and B spermatogonia, have been distinguished in mammals (Figs. 12 and 13) (Nicander and Plöen 1969, Hochereau-de-Reviers 1970, de Rooij 1970, 1973, Rowley et al. 1971, Clermont and Antar 1973, Berndston and Desjardins 1974,

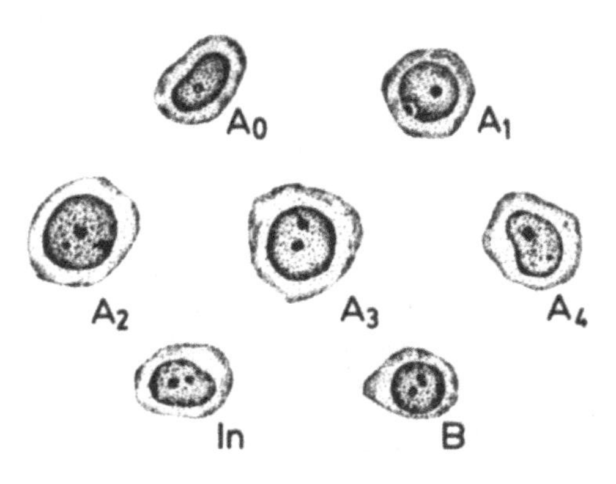

Fig. 13. The various types of spermatogonia as seen in dissected tubules fixed in Carnoy solution, stained with haematoxylin and mounted in toto. These cells are found at the following stages of the cycle of the seminiferous epithelium: A_0 at Stages I–XIV; A_1 at Stages I–IX; A_2 at Stages IX–XII; A_3 at Stages XII–XIV; A_3 at Stages XIV–I; *In* at Stages I–V; *B* at Stages IV–VI (Redrawn from Clermont and Bustos-Obregon 1968)

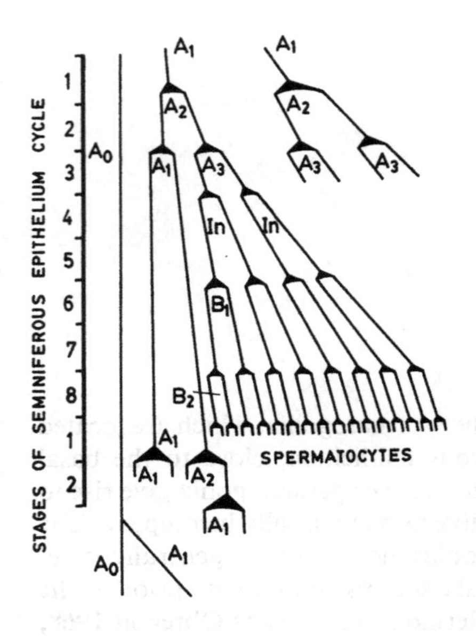

Fig. 14. Spermatogonial divisions in buffalo (From Bilaspuri and Guraya 1980a)

Guraya and Bilaspuri 1976a, b, c, Weaker 1977, Oud and de Rooij 1977, Tait and Johnson 1982, Bilaspuri and Guraya 1980a, 1984b, 1986a). But such a distinction is more difficult in primates including humans (W. Schulze 1978, C. Schulze 1979, 1981, Kerr and de Kretser 1981, Holt and Moore 1984). Holstein et al. (1984) have reported dislocated type A spermatogonia in human seminiferous tubules. The type A spermatogonia divide to form more differentiated spermatogonia (In and B) and for renewing the stem-cell population (Fig. 14) (Hochereau-de-Reviers 1970, Bellvé 1979, Hilscher 1983, Bilaspuri and Guraya 1980a, 1984b, 1986a).

The type A spermatogonia contain slightly elongated or round nucleus with homogeneously distributed fine chromatin granules and one or more nucleoli having well developed nucleonema. These are the largest spermatogonia, and have their longitudinal axis parallel to the tubular wall (Fig. 12); the angle of this axis helps the daughter cells to lie in close association to the basal lamina of the seminiferous tubule. Type A spermatogonia can be further distinguished into three types A_1, A_2, A_3 in the testes of sheep, goat and buffalo (Fig. 12). The A_1 spermatogonia contain a single, large, central nucleolus. The A_2 and A_3 spermatogonia contain one or more nucleoli. The size of the A_3 spermatogonia is smaller than that of the A_2 spermatogonia (Guraya and Bilaspuri 1976a, Bilaspuri and Guraya 1980a, 1984b, 1986a). More than one small nucleolus is present in their darkly stained nucleus which contains coarse chromatin material distributed adjacent to the nuclear envelope. Type B spermatogonia can be distinguished from other spermatogonial types by the presence of chromatin crusts along the nuclear envelope (Figs. 12 and 13), which appears thick under the light microscope. The chromatin crusts are distributed also in the central nucleoplasm. Type B spermatogonia contain two to four small nucleoli and can be further distinguished into B_1 and B_2 spermatogonia in sheep, goat and buffalo (Fig. 12).

Table 1. Nuclear diameter and nuclear volume of various spermatogenic cells in buffalo (From Guraya and Bilaspuri 1976c)

Cell type	Nuclear diameter (expressed in μ)	Nuclear volume (expressed in μ^3)
Spermatogonia		
A_1	7.6	230
A_2	7.48	221
A_3	7.24	191
In	7.01	181
B_1	6.65	160
B_2	6.41	111
Spermatocytes		
Early preleptotene	5.23	75
Late preleptotene	5.7	97
Leptotene	6.18	130
Granular leptotene	6.65	154
Early zygotene	6.89	172
Late zygotene	7.24	200
Early pachytene	7.48	220
Middle pachytene	8.08	276
Late pachytene	8.44	301
Early diplotene	8.55	328
Late diplotene	8.79	357
Secondary spermatocyte	6.65	154
Spermatids 1 (stages 4 and 5)	5.23	75
Late diplotene		
A_1 spermatogonium	1.16	1.55
Late diplotene		
Leptotene	1.42	2.75
Late diplotene		
Spermatid (1)	1.68	4.78

The B_1 and B_2 spermatogonia in sheep, goat and buffalo differ from each other by the fact that nucleus is smaller and more darkly stained in B_2 type spermatogonia.

Generally, the dimensions of spermatogonia decrease from type A through In to B spermatogonia (Table 1), whereas the heterochromatin increases from A_1 to B_2 spermatogonia (Figs. 12 and 13) (Guraya and Bilaspuri 1976a, Bilaspuri and Guraya 1980a, 1984b, 1986a). The nucleus of type A spermatogonia preparing for division is rich in chromatin granules, thus showing their appearance to be similar to that of In spermatogonia, but nuclei of In spermatogonia are smaller than those of A spermatogonia. The chromosomes are elongated during the prophase of type A spermatogonia but become shorter and more contracted during the prophase of type B spermatogonia. The axis of the chromosome spindle in type A spermatogonia is nearly parallel to the basal lamina of the tubule so that the daughter cells remain close to the wall of the seminiferous tubules, whereas in type B spermatogonia it is perpendicular to the tubular wall. A special category of spermatogonia, designated as A_0 or reserve stem-cells, exists in close association with the basal lamina of the seminiferous tubules (Fig. 12). Their nucleus shows a fine granular nucleoplasm, one or more small nucleoli and rare mitoses.

Different types of spermatogonia are seen to divide in specific stages of the seminiferous epithelial cycle (Tait and Johnson 1982; Bilaspuri and Guraya 1980a, 1984b, 1986a). The number of spermatogonial multiplications between A_1 spermatogonia and primary spermatocytes has been determined by different methods (Hochereau-de-Reviers 1981, Hilscher 1983). This includes morphological and morphometrical analyses of spermatogonia; numbering of spermatogonia of each type at each stage of the seminiferous epithelial cylce; analysis of the mitotic index after, or without, blockage by colchicine; analysis of labelling index following the incorporation of [^{3}H]thymidine; and index of labelled mitosis. The number of spermatogonial generations between A spermatogonia and primary spermatocytes varies among different mammalian species (four-six) but is fixed for a given species. It is six in the bull, ram, red deer stag, rat, mouse (Hochereau-de-Reviers 1981) and buffalo, goat and ram (Bilaspuri and Guraya 1980a, 1984b, 1986a). The respective number of type A or B spermatogonial generations also varies among species. In the rat and mouse, there are four generations of type A spermatogonia, one of In, and one of type B spermatogonia (Monesi 1962, Clermont and Bustos-Obregon 1968). The yield of spermatogonial multiplications is calculated from the ratio of the number of primary spermatocytes to that of A_1 spermatogonia per tubular cross-section. The yield is lower than the theoretical one due to the degeneration of cells throughout the multiplication (Huckins 1978c, Bilaspuri and Guraya 1982, 1984b, 1986a).

In the rat, A_1 spermatogonia divide three times to form the successive $A_1-A_2-A_3$ and $-A_4$ spermatogonia (Leblond and Clermont 1952a). The A_4 spermatogonia divide once giving rise to either In spermatogonia or new A_1 spermatogonia (Clermont and Bustos-Obregon 1968, de Rooij and Kramer 1968). For the study of renewal of type A spermatogonia in adult rats, Bartmanska and Clermont (1983) have mapped and scored [^{3}H]thymidine-labelled and unlabelled type A_0, A_1, A_4 and In spermatogonia in tubular segments at Stages I – IV of the cycle of the seminiferous epithelium, before, during, and after the Stage I peak of mitoses. Considering cell numbers and labelling indices, it has been suggested that type A_4 spermatogonia may serve as precursors for both type A_1 and In spermatogonia. After the division of the In spermatogonia the B spermatogonia are formed. These divide once and form the preleptotene primary spermatocytes. In the mouse, six generations of spermatogonia (A_1, A_2, A_3, A_4, In, and B) occur prior to the onset of meiosis (Monesi 1962). Guraya and Bilaspuri (1976a) have made a study of the morphology and dimensions of the spermatogonia and other spermatogenic cells in buffalo testes. Type A, In and B spermatogonia occur in addition to type A_0 spermatogonia (Fig. 12). Types A and B spermatogonia are subdivided into A_1, A_2 and A_3, B_1 and B_2 spermatogonia, respectively (Fig. 12) (Bilaspuri and Guraya 1980a). Similar observations have been made for the goat and ram (Bilaspuri and Guraya 1984b, 1986a). The cyclic release of spermatozoa from a given portion of the mammalian seminiferous

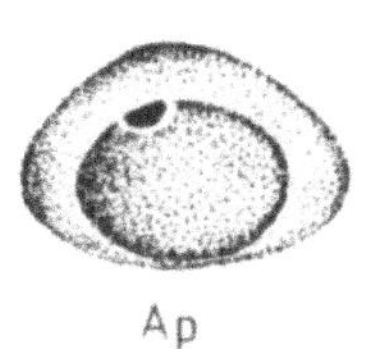

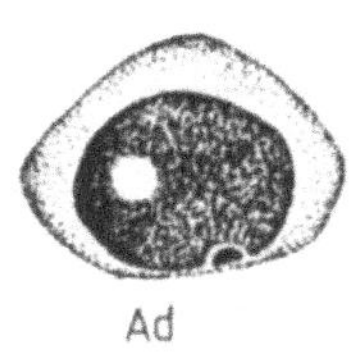

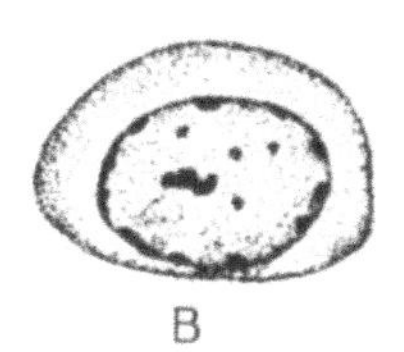

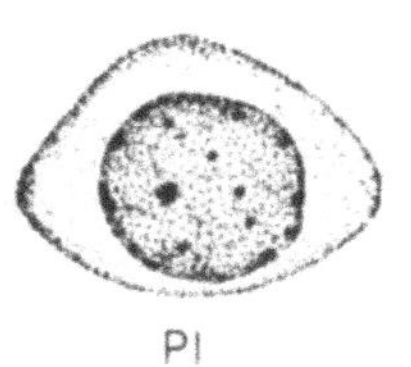

Fig. 15. Main types of spermatogonia and preleptotene spermatocyte in man as seen in Zenker-formol-fixed testes. *Ad* and *Ap* dark and pale type A spermatogonia; *B* type B spermatogonia; *Pl* preleptotene spermatocyte (Redrawn from Clermont 1972)

epithelium is related to the constant delay that occurs between two successive mitoses of A_1 spermatogonia, i.e. the duration of the seminiferous epithelial cycle. These mitoses occur during Stage 1 of the cycle, immediately after the release of the spermatozoa in the tubal lumen.

Types A and B spermatogonia have been identified in the testes of primates including rhesus and man (C. Schulze 1979, 1981, Cavicchia and Dym 1978, Kerr and de Kretser, 1981, Holstein and Roosen-Runge 1981, Hadley and Dym 1983). But the type A spermatogonia in primates and humans have been further distinguished into two main categories, such as the A pale (A_p) and the A dark (Ad) spermatogonia (Fig. 15) (Chowdhury and Steinberger 1976, C. Schulze 1979, 1981, Kerr and de Kretser 1981, Hadley and Dym 1983). Spermatogonium type A cloudy is reported for the human testis (Holstein and Roosen-Runge 1981). It has a nucleus with "cloudy" karyoplasm but no central light area. The human testis also shows a fourth type of spermatogonium designated as spermatogonium type A long which has an elongated ovoid nucleus (Rowley et al. 1971, Holstein and Roosen-Runge 1981). Its karyoplasm is relatively electron-dense without a transparent region. In the *Cercopithaecus* monkey, there are four generations of type B spermatogonia, B_1 to B_4, which apparently originate from type A spermatogonia (Clermont and Antar 1973). Conflicting data have been reported in regard to the morphological criteria of stem cells in the human testis (C. Schulze 1981). But according to Rowley et al. (1971) the long type A spermatogonium constitutes the germinal stem-cell. The cells of type A long become labelled when fresh human seminiferous tubules are incubated in [^{3}H] thymidine (W. Schulze 1978) and thus may represent not a separate cell generation, but a transitional form of A spermatogonia in S phase of DNA synthesis (Holstein and Roosen-Runge 1981). Chowdhury et al. (1975), on the basis of quantitative study of the spermatogonial population in organ culture, suggested that Ap spermatogonia may act as stem cells. This suggestion is supported by the results of C. Schulze (1979, 1981) after chemotherapy and X-irradiation. These spermatogonial types that survive the treatment are considered as stem cells in view of the fact that the stem cells, in contrast to the more differentiated spermatogonia, are radiation-resistant and less sensitive to various noxious agents. The majority of these cells show the characteristic features of Ap spermatogonia, while a few cells may represent variants of this cell type. The Ad spermatogonia are almost completely eliminated from the seminiferous tubules. From these results, C. Schulze (1979, 1981) has proposed a concept that the stem cells of the human testis may be derived from the Ap spermatogonia or variants of this cell type.

Table 2. Cytological criteria of human spermatogonia (According to Rowley et al. 1971)

Classification	AL = long	AD = dark	AP = pale	B
Shape of cell	Very flat	Flat	Rounded	Pear-shaped
Attachment to basal lamina	≈30 μm	Wide	Less wide	<2 μm
Nuclear shape	Oval-irregular	Oval-regular	Spherical	Spherical
Chromatin	Homogeneous	Granular, rarefaction	Homogeneous	Granular
Type of nucleolus	Nucleolonema	Nucleolonema	Nucleolonema pars amorpha	Nucleolonema pars amorpha
Position of nucleolus	Peripheral	Peripheral	Peripheral	Central
Type of mitochondria	Tubular	Tubular	Cristae	Cristae
Arrangements of mitochondria	Groups with ER	Groups with ER and bars	Pairs with bars	Single
Glycogen	Positive	Positive	Negative	Negative

The electron microscopic examination of normal human testes has demonstrated four types of spermatogonia (Rowley et al. 1971, Bustos-Obregon 1975, Bustos-Obregon et al. 1975, Kerr and de Kretser 1981). These are the Ad and Ap, the type B and a new type A (Al), as already stated. Their presence in the human testis has been confirmed by Barham and Berlin (1974) and in monkeys by Cavicchia and Dym (1978). But W. Schulze (1978) and Holstein and Roosen-Runge (1981) have made detailed electron microscope studies of the morphology of various types of spermatogonia (type A dark, type A pale, type A cloudy, and type A long) in men with normal spermatogenesis and in patients treated with anti-androgens. The subcellular criteria used in distinguishing the various types of spermatogonia in humans has included the cellular and nuclear sizes, shape of the cells and their nuclei, the density of their nucleoplasm, the type of nucleolus and its placement within the nucleus, the structure of the mitochondrial cristae, the association of the endoplasmic reticulum and intermitochondrial substance with the mitochondria, the amount of organelles and glycogen present within the cell and the presence of previously undescribed filamentous structures in the cytoplasm of the Al and Ad spermatogonia of the human testes (see also Bustos-Obregon et al. 1975, Kerr and de Kretser 1981, Holt and Moore 1984) (Table 2).

All spermatogonia rest upon the basal lamina of the human seminiferous tubules (Kerr and de Kretser 1981, Holstein and Roosen-Runge 1981). But Holstein et al. (1984) have observed that type A spermatogonia can be found in different locations throughout the germinal epithelium of human seminiferous tubules. Under special conditions, they may be located in the adluminal compartment. They may also occupy an intermediate position between basal and intraluminal locations without showing any signs of cytological damage. Identification of the stem spermatogonia as well as the precise definition of the number of spermatogonial types in the human still remain to be clarified. As demonstrated with the electron microscope, the amount of contact with the tubular/basal lamina progressively decreases from Al, a flat cell lying parallel to the basal lamina, through the Ad and Ap to the B spermatogonia, the latter being a pear-shaped cell with its long axis perpendicular to the basal lamina (Holstein and Roosen-Runge 1981). Vilar and Paulsen (1967) and Vilar et al. (1970) have reported only two types of human spermatogonia (the flat type and round type spermatogonia) lying on the basal lamina and third cell type, a preleptotene spermatocyte, that is surrounded entirely by Sertoli cell cytoplasm.

Cavicchia and Dym (1978) have observed that type Ad spermatogonia in monkeys possess an oblong nucleus containing a deeply stained chromatin in contrast to Ap nuclei which stain very lightly (Hadley and Dym 1983). The staining pattern of some of the type A appears comparable to intermediate spermatogonia. The type B spermatogonia are round cells having increased amount of heterochromatin close to the nuclear envelope.

Preleptotene spermatocytes are smaller both in cellular and in nuclear diameter than the type B spermatogonia, and their nucleoplasm consists of irregular patches of condensed chromatin (Guraya and Bilaspuri 1976a, Bilaspuri and Guraya 1980a, 1984b, 1986a). Bellvé et al. (1977a) have observed that mouse preleptotene cells are smaller than the type B spermatogonia and that the chromatin pattern is more diffuse, although still clumped in a cartwheel fashion.

B. Stem-cell Renewal and Multiplication of Spermatogonia

Spermatozoa are produced by continuous division and renewal of the stem spermatogonia. The quantitative production of spermatocytes and spermatozoa depends on the total number of stem spermatogonia per testis, the scheme of stem-cell renewal, the number of spermatogonial generations between stem cells and primary spermatocytes and the yield of spermatogonial divisions. The process of stem-cell renewal and multiplication of spermatogonia, which occurs throughout the adult life, have been the subject of extensive studies (Monesi 1982, Clermont 1972, Clermont and Antar 1973, Huckins and Oakberg 1978a, b, Guraya and Bilaspuri 1976a, b, c, Hochereau-de-Reviers 1981, Hilscher 1983, Bilaspuri and Guraya 1980a, 1984b, 1986a, Bartmanska and Clermont 1983, Lok and de Rooij 1983). Different nomenclature and schemes of stem spermatogonia renewal have been used during the past 10 years. Actually, very divergent views exist about the stem-cell renewal (Huckins 1971a, b, c, d, Oakberg 1971a, Clermont 1972, De Rooij 1973, Clermont and Hermo 1975, Oakberg and Huckins 1976, Moens and Hugenholtz 1976, Hilscher 1983, Bilaspuri and Guraya 1980a, 1984b, 1986a, Hochereau-de-Reviers 1981, Bellvé and Feig 1984). But the overwhelming body of evidence points towards the principle of a self-renewing stem cell. Clermont (1972) has presented a thorough discussion of the many models, which invariably explain that various types of spermatogonia originate through a sequence of mitotic divisions. But the precise origin of the renewing stem cell is still to be determined.

The early views of bivalent mitosis (differential mitosis), a process whereby each young spermatogonium divides to give rise to one spermatogonium and one spermatocyte (Roosen-Runge 1962), are not supported by the available evidence. Clermont and Leblond (1955) using quantitative techniques developed the concept of stem-cell renewal that involves a new series of divisions being initiated with the onset of each cycle to yield additional stem cells, which lie "dormant" until the subsequent cycle, and to form type A spermatogonia for continuing the spermatogenic process. The dormant spermatogonia multiply to form a new generation of stem cells and a new generation of type A spermatogonia, and the process continues, to repeat in the subsequent cycle. This mode would provide a cyclic replenishment of cells for the spermatogenesis, and meanwhile, would maintain stem-cell line.

In the light of these observations several hypotheses about stem-cell renewal have been presented. Monesi (1962) has assigned the role of a stem cell to the A_1 spermatogonium, whereas Hilscher et al. (1969) have ascribed this function to the intermediate A spermatogonium (see also Hilscher 1983). Some other studies have suggested that A_4 spermatogonium in the rat is the stem cell (Clermont and Bustos-Obregon 1968, de Rooij and Kramer 1968, Clermont and Hermo 1975). The stem-cell renewal hypothesis of Clermont and Bustos-Obregon (1968) is proposed on the basis of presence of five distinct categories of type A spermatogonia such as A_0, A_1, A_2, A_3, and A_4. A series of equivalent divisions originating from type A_1 finally give rise to a population of type A_4 spermatogonia. One of these differentiated type A_4 cells functions as renewing stem cell by producing two dedifferentiated type A_1 as a result of an equivalent mitosis, while the remaining type A_4 spermatogonia divide to form intermediate spermatogonia. By contrast, the type A_0 spermatogonia form reserve stem cells, and remain dormant till there is a need for producing renewing stem cells to repopulate the seminiferous tubules (Dym and Clermont 1970, Courot et al. 1970). This cell type may be analogous to the spermatogonia described by other workers as the primitive type A spermatogonia (E. Steinberger et al. 1964), the immature type A spermatogonia (Sapsford 1962a, b), and the transitional cell (Beaumont and Mandl 1963).

Clermont (1969) has proposed a stem-cell renewal model for the monkey which is similar to that in the rat, except that there are only two classes of type A spermatogonia, such as A_1 and A_2. The A_1 is considered as the reserve stem cell, and A_2 as the renewing stem-cell. Other workers, such as Huckins (1971b) and Oakberg (1971a, b), have criticized the two-stem model put forward by Clermont and his co-workers. Based on detailed analyses of kinetics of spermatogenesis in the rat, Huckins (1971a) has suggested an entirely new model for explaining the renewal of stem cells (Fig. 16), which has also received support from other investigators working with the mouse (Oakberg 1971a, b, de Rooij 1973, van Keulen and de Rooij 1975) and the hamster (Oud and de Rooij 1977). The model of Huckins (1971b) is based on quantitative re-evaluation of whole mounts of seminiferous tubules in the mouse. This hypothesis does not take into account the concept for a reserve stem cell (A_0) and the assumption that type A_1 spermatogonia must be formed in the course of one of the spermatogonial division peaks. But it is suggested that all spermatogonia can be put into one of the following categories: (a) stem-cells (A_s); (b) proliferating cells (A_{pr}, A_{al}); (c) differentiating cells (A_1, A_2, A_3, A_4, In, and B) (Fig. 16) (see Hilscher 1983). It is

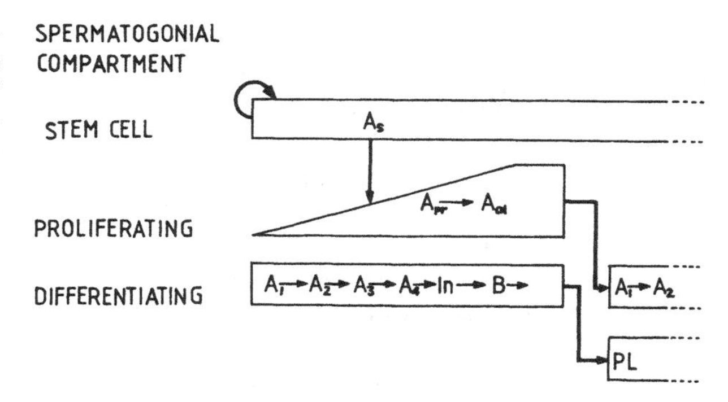

Fig. 16. Model for spermatogonial stem cell renewal in the rat. For the nomenclature of spermatogonia, see the text. *PL* preleptotene spermatocyte (Redrawn from Huckins 1971)

believed that the stem cells (A_s) are isolated rather than forming groups, and are located or distributed randomly throughout the length of the seminiferous tubule rather than lying together in any specific stage of spermatogenesis. These cells are considered as being the "true" stem cells. The evidence suggests the A_s spermatogonia proliferate and either give rise to additional stem cells, which are distributed singly, or to paired daughter cells (A_{pr}). Through a series of synchronous divisions, these cells give rise to chains of cells or aligned spermatogonia (A_{al}), which constitute the proliferative compartment and finally transform into A_1 spermatogonia. The latter enter the differentiating compartment of spermatogonia, as characterized by the series of synchronous divisions, leading to the formation of A_2, A_3 and A_4, In and B spermatogonia. The B spermatogonia divide to form preleptotene spermatocytes. This hypothesis of stem-cell renewal in the mouse has been supported by radioautographic studies on the mouse testis (Oakberg 1971a). These data, obtained by labelling S phase spermatogonia with [^{3}H] thymidine, support the notion of renewing stem-cell compartments. Similar to the rat, the mouse spermatogonia fall into five classes on the basis of their nuclear morphology and appear to undergo mitotic divisions and morphological alterations. A_s cells of Huckins (1971b) can be considered as equivalent to A_0 cells of Clermont and Bustos-Obregon (1968).

De Rooij (1973) has also studied the spermatogonial stem-cell renewal in the mouse. His results have provided evidence for the concept of a population of morphologically undifferentiated A spermatogonia preceding the A_1 spermatogonia in the line of spermatogenesis, as proposed earlier by Huckins (1971b) and Oakberg (1971a, b). Isolated A cells among these A_1 spermatogonia are believed to form the reserve stem cells (A_0) of spermatogenesis. Oud and de Rooij (1977) recently have distinguished four classes of spermatogonia in the Chinese hamster, which include undifferentiated A spermatogonia, A isolated (is), A paired (pr), A aligned (al), and differentiated A spermatogonia (A_1, A_2, A_3), in spermatogonia, and B spermatogonia. $A_{(is)}$ and $A_{(pr)}$ show mitotic activity in several stages of the cycle which serves to generate the $A_{(al)}$ spermatogonia; the latter after one or more divisions transform into A_1 spermatogonia. In more recent studies, the details of cell cycle properties, labelling indices and synchronization have been followed during spermatogonial multiplication in the Chinese hamster (Lok and de Rooij 1983a, b, Lok et al. 1983, 1984). The labelling index (LI) of undifferentiated spermatogonia throughout the whole cycle of the seminiferous epithelium is highest between Stages 11 and 4. The LI of A_{pr} and A_{al} spermatogonia decreases around Stage 3 but remains at a relatively high level for A_s spermatogonia until Stage 7. As calculated from the LI, the proliferative activity of the A_s and A_{pr} spermatogonia is sufficient to yield the necessary number of A_{al}/A_1 spermatogonia for each cycle of the seminiferous epithelium. It is proposed that some of the labelled A_{pr} spermatogonia are false pairs that lose their cytoplasmic bridge following mitosis, thus yielding A_s and/or A_{pr} spermatogonia. Hochereau-de-Reviers (1981) has preferred to use the term A_0 reserve stem spermatogonia (Clermont and Bustos-Obregon 1968) to signify an ensemble of cell types that Huckins (1971b) has subdivided into single, paired, or aligned spermatogonia, and the term A_1 stem spermatogonia for cells that are in the G_1 phase during Stages 6–8 of the seminiferous epithelial cycle. These enter the S phase and participate in mitosis during Stage 1. These cells are at the origin of the cyclic activity of the seminiferous epithelium. According to this hypothesis A_0 spermatogonia are G_0-phase-blocked stem cells, and that other type A spermatogonia are in the G_1, S, and G_2 phases of cyclic activity. For their renewal different types of spermatogonia in buffalo and goat can be classified into three categories: reserve stem cells (A_0), renewing stem cells (A_1, A_2) and differentiating cells (A_3) (Bilaspuri and Guraya 1980a, 1984b).

The two-stem-cell hypothesis is beset with several inherent problems (Oakberg and Huckins 1976), as the origin of A_1 from A_4 spermatogonia has to be explained by taking into consideration various points, such as (1) the method for the selective detachment of a single cell or a small group of cells from the syncytial mass of A_4 spermatogonia; (2) dedifferentiation of the nuclei of type A_4 spermatogonia as evidenced by morphological features; and (3) development of considerably longer cell-cycle kinetics specific to A_1 spermatogonia. Further, A_0 spermatogonium incorporates [^{3}H] thymidine following prolonged exposure, which suggests that it acts as a progenitor stem cell (A_s) instead of having a nonproliferative reserve function (Huckins 1971d, Oakberg 1971b). The progenitor A_s spermatogonia resist the effect of radiation, and the cell-cycle kinetics of this cell type during the recovery phase is normally long (Oakberg 1959, 1964, 1971b, Monesi 1962, Withers et al. 1974).

Based upon three-dimensional configuration made from serial, thin sections of the rat seminiferous epithelium, a variant model has been put forth by Moens and Hugenholtz (1975, 1976) to explain the stem-cell renewal. According to this model, spermatogonia from syncytial clusters, as evidenced by their interconnections through cytoplasmic bridges (Moens and Go 1972, Moens and Hugenholtz 1975). Some of these germ-cell clusters are suggested to be much larger size than those reported in some earlier studies (Dym and Fawcett 1971). Moreover, the occurrence of spermatogonia in the form of single cells and short syncytial chains of one–eight cells is consistent with the model of Huckins (1971b) and Oakberg (1971a). The predominance of odd-numbered chain lengths is believed to be due to either some random cell death or failure of some spermatogonia to divide along with others of the same group (Moens and Hugenholtz 1976). It is significant that the number of spermatogonial syncytia becomes more than that required to replace the advanced spermatocytes, thus indicating the possibility of accumulation of spermatogonial syncytia periodically. At regular intervals, these cells seem to give rise to appropriate amounts of preleptotene spermatocytes. When exhausted, the spermatogonia are filled up again by division from the progenitor stem cells (Moens and Hugenholtz 1975).

C. Cell-cycle Kinetics and Control of Spermatogonial Multiplication

The morphological transformations that occur during the multiplication and differentiation of spermatogonia apparently involve complex changes at the subcellular and molecular levels about which our knowledge is still very meagre. Detailed descriptions of the cell-cycle kinetics of the differentiating spermatogonia, type A_1 through B, are available for the mouse (Monesi 1962), rat (Hilscher and Makoski 1968, Hilscher and Hilscher 1969, Huckins 1971a) and Chinese hamster (Lok and de Rooij 1983a, b). The total cell-cycle time for all types of differentiating spermatogonia is 40 – 42 h in the rat, 27 – 30 h in the mouse and 60 h in the Chinese hamster, representing approximately 14% of the duration of the cycle of seminiferous epithelium. The G_1 phase in the rat and mouse lasts for 11 – 13 h except for the A_1 spermatogonia, in which case this phase is longer and more variable. The duration of S phase, the DNA synthetic phase, in both species becomes considerably longer with the progression of spermatogonial differentiation from A_1 to A_4 (Monesi 1962, Huckins 1971a). The largest increase occurs in the A_2 and A_3 spermatogonia of the rat (Huckins 1971a), and A_4 spermatogonia of the mouse (Monesi 1962, Kofman-Alfaro and Chandley 1970). Unexpectedly, this increase of the S phase takes place at the cost of the G_2 period instead of G_1 which remains rather constant in duration. In the Chinese hamster, with ongoing differentiation, the cell-cycle time of the differentiating spermatogonia increases from 14 to 25 h, while the duration of G_2 phase shortens from 22 to 10 h (Lok and de Rooij 1983a). Lok et al. (1983) have also studied the cell-cycle properties of undifferentiated spermatogonia in the Chinese hamster. The minimum cell-cycle time is approximately 90 h for A_s and 87 h for the A_{pr} and A_{al} spermatogonia, which is appreciably longer than for the differentiating types A_2 – B_2 spermatogonia (60 h; see Lok and de Rooij 1983a). This is mainly accounted for by a longer duration of the G_1 phase. In general, the variability in the duration of the cell-cycle phase is greater than for differentiating spermatogonia. From the shape and position of the second peak of the fraction of labelled mitoses (FLM) curve, Lok et al. (1983) have concluded that the undifferentiated spermatogonia either cycle with a cell-cycle time of approximately 87 – 90 h or become arrested in the G_1 phase. This suggests that the decrease in proliferative activity of the undifferentiated spermatogonia after Stage 4 occurs by the arrest of progressively more cells, i.e. by a gradual decrease of the growth fraction and not by a gradual lengthening of the G_1 phase. The arrested cells either transform into A_1 spermatogonia and then divide at Stage 9, or remain undifferentiated and are stimulated to enter the S phase again during the following seminiferous epithelial cycle. It is suggested also that daughter cells of A_s spermatogonia that become A_{pr} have a greater chance to continue cycling than those that become new A_s cells (see also Lok et al. 1984). Kluin and de Rooij (1981) made a comparison between the morphology and cell kinetics of gonocytes and adult type undifferentiated spermatogonia in the mouse. Hochereau-de-Reviers (1981) has concluded that the type A_0 spermatogonia are G_0-phase-blocked stem-cells and that other type A spermatogonia are in G_1, S and G_2 phases. The passage from the G_1 phase (G_0) to cyclic activity is probably under endocrinological or local control or both, either directly or via Sertoli cell interaction. The molecular mechanisms that promote DNA synthesis in differentiating mammalian spermatogonia are yet to be defined. However, the property of the seminiferous growth factor (SGF) of Sertoli cell origin (Feig et al. 1983) to induce DNA synthesis and cell proliferation among prepuberal Sertoli cells and the fibroblastic BALB/C 3T3 cells suggests a major role for the polypeptide in promoting mammalian spermatogenesis (Bellvé and Feig 1984); significant amounts of SGF also occur in primitive type A spermatogonia and preleptotene spermatocytes. Stewart et al. (1984) have suggested that in somatic cells the level of myc transcription correlates with the rate of division. Such transcription involves the use of two active myc promoters and produces two messenger RNA species that are differentially represented among the transcripts of different tissues. But mitotically and meiotically dividing germ-cells of the mammalian testis show very few myc transcripts and seem to proliferate, at least for a few divisions in the absence of myc transcription. These results have raised important questions in regard to the role of the myc gene products in terminally differentiating cells, especially of the germ line series. With increase in the proportion of heterochromatin that occurs with advancing differentiation of the spermatogonia, the rate of DNA syntheses may decrease (Huckins 1971a). This statement needs to be proven biochemically.

The control of spermatogonial proliferation is poorly understood (Bellvé 1979, Hochereau-de-Reviers 1981). However, the roles of stimulators and endogenous inhibitors in the control of cell proliferation are relatively well recognized in other cell types or tissue (see Nicolson 1976, Edelman 1976, Balázs and Blazsek 1979). Comparable studies need to be made on spermatogenic cells to define control mechanisms involved in their proliferation. The prolonged arrest of A_0 (or A_{is}) and A_1 spermatogonia in the G_1 or G_2 phase indicates the existence of some complex biochemical mechanisms which exert differential control on spermatogonial proliferation. These mechanisms may consist of either negative control in the form of chalones, inhibitors of cell multiplication (Bullough 1975, Irons and Clermont 1979, Balázs and Blazsek 1979) or positive control through growth factors, macromolecules that cause cell multiplication (Gospodarowicz et al. 1978, Balázs and Blazsek 1979, Feig et al. 1983, Bellvé and Feig 1984). The existence of a spermatogonial chalone is supported by the fact that testicular extracts reduce the incorporation of [^{3}H] thymidine into DNA of the repopulating type A spermatogonia, type B spermatogonia, or preleptotene spermatocytes (Clermont and Mauger 1974, 1976, Thumann and Bustos-Obregon 1982, Irons and Clermont 1979; for review see de Rooij 1980, Iversen 1981). As compared with the control rats (saline-injected), the spermatogonia in S phase decrease by about 50% at Stage 1 of the spermatogenic cycle, 74% at Stages 2–3, 84% at Stages 4–5, and 34% at Stages 8–14. But with the injection of a comparable liver extract, there is seen an apparent decrease only in Stages 2–3 (~37%) and 4–5 (~40%). Type A spermatogonia present in these stages show the most pronounced effect of the testicular extracts. These observations are not in accordance with the notion that "chalone" is testis-specific. The apparent difference in the effect of liver and testis extracts may be due to the amount of material injected, since the amount of protein injected in each experiment is not mentioned in either study. Thumann et al. (1981) have described the general properties of the G_1-spermatogonial chalone. It is heat-labile, can be lyopholized and is tissue-specific but not species-specific. As the composition, site of production, and target of the effect of the gonadal chalones have as yet not been satisfactorily determined, the existence of such a substance is still a matter of controversy. The findings of W. Schulze and Rehder (1984) suggest the presence of a local inhibitory effect being involved in the formation of the human seminiferous epithelium. Actually, the isolation and characterization of chalones in somatic cells involve some serious difficulties as various exogenous substances can inhibit cell multiplication in vitro (Balázs and Blazsek 1979). A similar situation can arise in the study of testicular inhibitors of cell multiplication.

Very few attempts have been made to test the presence of testicular growth factors. However, Bellvé and co-workers (Feig et al. 1980, 1983, Bellvé et al. 1983, Bellvé and Feig 1984) have demonstrated the presence of a seminiferous growth factor (SGF) in the mouse and some other mammalian species. SGF is a mitogenic protein resticted primarily to Sertoli cells of the seminiferous epithelium. Thus, the Sertoli cells may direct SGF, as needed, to promote spermatogonial proliferation in the basal compartment or to induce the two meiotic divisions that occur in the avascular, central compartment of the seminiferous epithelium. But in order to resolve the regulatory or physiological functions of the SGF in promoting mammalian spermatogenesis, it will be essential to obtain purification of this polypeptide to homogeneity. In an excellent review, Bellvé and Feig (1984) have discussed the evidence, biological properties, and potential functions of SGF in the mammalian testis. Sertoli cells synthesize transferrin (see Chap. I), an iron-transporting protein that enables cells to traverse the G_2 phase of the cell cycle (Rudland et al. 1977). Therefore, Bellvé and Feig (1984) have suggested that SGF, by inducing cells to proceed through the G_1 phase of the cell cycle, functions coordinately with transferrin to promote spermatogonial proliferation. The partial restoration of spermatogonial divisions by the synergistic effects of transferrin, insulin and FSH (Haneji and Nishimune 1982) is in agreement with this suggestion. Hochereau-de-Reviers (1981) has discussed the role of various factors, such as age, Sertoli cell numbers, genetics, season, temperature, hormonal plasma levels, hypophysectomy and hormonal supplementation, anti hormones, in vitro systems, chalones and inhibin, etc. in the control of spermatogonial multiplication. The total number of type A_1 spermatogonia per testis increases during puberty and in adulthood and is highly correlated with the total number of Sertoli cells per testis, which is established before puberty. However, the endocrinological environment can modify this. There occurs seasonal variations in the yield of spermatogonial divisions as well as in the number of stem cells. Elevation of scrotal temperature or cryptorchidism leads to variable effects on spermatogonia, which are species-specific. Proteinaceous factors (chalones and inhibin) secreted by Sertoli cells may be involved in the control of spermatogonial multiplication (Clermont and Mauger 1974, 1976,

Ewing and Robaire 1978, Lanoiselee-Perrin-Houdon and Hochereau-de-Reviers 1980) but the mechanism of their action, local or central, remains to be determined. Franchimont et al. (1981) have demonstrated the effect of inhibin on rat testicular DNA synthesis in vivo and in vitro. Hormones, by regulating the secretions of various proteins in the proteins in the Sertoli cells, appear to be of great significance in the control of spermatogonial proliferation. Defective spermatogenesis leading to subfertility or complete sterility can be the end result of a multitude of causes, such as malnutrition, radiation, obstruction of the excurrent duct, chromosomal abnormalities, immunological agents, abnormal climatic conditions (low and high temperatures), photoperiod, low atmospheric pressure at high altitude, etc. (see Mann and Lutwak-Mann 1981, Weiner et al. 1984). Blank and Desjardins (1984) have indicated that spermatogenesis is modified by food intake in mice. However, there exist species-specific differences in this regard.

D. Spermatogonial Degenerations and Their Sensitivity to Various Factors

The spermatogenic cells at various stages of their differentiation are known to undergo degeneration in different species of mammals (Roosen-Runge 1955, 1973, 1977, Oakberg 1956, Ortavant 1959, Amann 1962, Clermont 1962, Hochereau-de-Reviers 1970, Gondos and Zemjanis 1970, Barr et al. 1971, Clermont 1972, Russell and Clermont 1977, Johnson et al. 1983c, 1984c). Germ-cell degeneration is observed also in gonads of human foetuses and the new-born (Skrzypczak et al. 1981). Actually, degeneration of a percentage of developing germ-cells is a common feature of developing and mature testes and thus is considered a normal phenomenon (Barr et al. 1971, Roosen-Runge 1973, Russell and Clermont 1977). Germ-cell degeneration has been related generally to spermatogonial mitoses, meiotic divisions and spermiogenesis; germ-cell loss during meiosis has been reported to occur in the human testis (Johnson et al. 1983c). The quantitative aspects of their degeneration, which were poorly understood previously (Clermont 1972, Roosen-Runge 1977), are discussed. Variable percentages of germ-cell degeneration during critical stages of spermatogenesis have been reported for different species of mammals. Degeneration occurs primarily among the types A_2, A_3, and A_4 spermatogonia (Oakberg 1956, Clermont 1962, Clermont and Bustos-Obregon 1968, Huckins 1971b). Approximately 25% of progeny from type A spermatogonia in mice (Oakberg 1956), 10.6% in Sherman rats (Clermont 1962) and 75% in Sprague-Dawley rat (Huckins 1978c) are lost. Ohwada and Tamate (1975) have made a quantitative study of the degenerating spermatogonia at Stage 12 of the cycle of seminiferous epithelium in Holtzman rats. Their frequency is $7.7 \pm 2.8\%$ in average with considerable variation among individuals. These extent of degenerating spermatogonia greatly affect the efficiency of spermatogenesis, as in the rat only half (Clermont and Bustos-Obregon 1968) or less (Huckins 1978c), and in the bull only a third (Hochereau-de-Reviers 1976) of the theoretical numbers of spermatozoa are produced. Huckins (1978c) has studied the morphology and kinetics of spermatogonial degeneration in normal adult rats. Some spermatogonia appear to degenerate selectively during the later spermatogonial divisions so that only 25% of the theoretically possible number of preleptotene spermatocytes finally are produced. During meiotic divisions, a loss of 13% in mice (Oakberg 1956), 22% in Sprague-Dawley rats (Clermont 1962), 25% in rabbits (Swierstra and Foote 1963), and 36% in humans (Barr et al. 1971) has been reported. A_1, A_3, In, B_1 spermatogonia and secondary spermatocytes in the buffalo testis show 20%, 18.75%, 10.2%, 15.5%, and 10% degenerations respectively (Bilaspuri and Guraya 1980a). No degeneration is observed in A_2 and B_2 spermatogonia, primary spermatocytes, and spermatids. In the goat testis types A_3, In, and B_1 spermatogonia show 15.0%, 25.0%, and 25.8% degenerations respectively (Bilaspuri and Guraya 1984b). B_2 spermatogonia and primary spermatocytes do not show degeneration. Secondary spermatocytes in the goat show 10.4% degeneration. Types A_1, A_3, In, and B_1 spermatogonia in the ram testis show 25%, 13.7%, 27.3%, and 21.2% degenerations respectively (Bilaspuri and Guraya 1986a). In the ram, no degeneration is observed during the division of the secondary spermatocytes to form spermatids. But 10% and 23% of the cells degenerate during the divisions of A_1 and In spermatogonia, in addition to 36% and 43% degeneration that subsequently occurs during early pachynema and the reduction divisions, respectively. The extent of degeneration in the ram may be related to normal and long daylight conditions (Ortavant 1959). If the possible degeneration of spermatids during spermiogenesis is not considered, one A_3 spermatogonium in the goat produces 21.1 spermatozoa (Bilaspuri and Guraya 1984b), whereas in the ram it produces 19.4 spermatozoa (Bilaspuri and Guraya 1986a) against the number of 64 reported in other studies (Hochereau-de-Reviers et al. 1980). Significant germ-cell loss occurs during meiosis in human testis (Barr et al. 1971, L.A. Johnson et al. 1983c), possibly due to degeneration of secondary spermatocytes. Whether degeneration of secondary spermatocytes and/or faulty meiotic divisions are responsible, almost half the potential sperm production is lost during meiosis (L.A. Johnson et al. 1983c). The estimated loss of 10% in Sprague-Dawley rats (Roosen-Runge 1955), and an undetermined amount in Sherman rats (Clermont 1962) and mice (Oakberg 1956) occur during spermiogenesis. No significant germ-cell loss appears to occur during spermiogenesis of bulls (Amann 1962) and men (L.A. Johnson et al. 1981). The low numbers of spermatozoa in the grey squirrel testes have been attributed to fewer spermatogonial divisions compared with those of most other mammals, as well as to a low efficiency of spermatogenesis, with only 42% of the germ-cells maturing (Tait and Johnson 1982).

None of the regulatory mechanisms controlling the degenerative processes has been determined (Hochereau-de-Reviers 1981). However, earlier workers (Oakberg 1956, Clermont 1962) suggested that germ-cell loss in the testis is a mechanism to eliminate nuclei having abnormalities of chromosomes. Delayed meiotic development and correlated death of spermatocytes have been observed in male mice with chromosome abnormalities (Speed and deBoer 1983). But this suggestion is not supported by the fact that the relative constancy of the amount of degeneration and the greater susceptibility of certain spermatogonial types to degeneration are not in agreement with simple selection against abnormalities of chromosomes. Huckins (1978a) has proposed that degeneration might be a mechanism to limit germ-cells to a number that can be maintained by the Sertoli cell population. L. A. Johnson et al. (1983c) have suggested that the germ-cell loss during meiosis in the human testis may be related more closely to the functional state of Sertoli cells than to the numbers of these cells in the testis. Ultrastructural and functional data on Sertoli cells from men with varying levels of germ-cell degeneration during meiosis would be helpful to clarify this possibility. A. J. Kaur and Guraya (1982) have observed few differences in the number and thickness of germ-cell layers from Stages 1 – 5 of the seminiferous epithelial cycle during the breeding and nonbreeding seasons of the field rat (*Rattus meltada*). However, quantitative seasonal variations occur in the number of spermatogonia, primary spermatocytes, and round spermatids, indicating the effects of seasons on mitosis and meiosis. Their numbers were greatest during the rainy season, lowest during winter, and again increased during the summer season. There was a significant increase in the percentage degeneration of both preleptotene and secondary spermatocytes during the nonbreeding season. These studies have indicated that spermatogenic cells from B_2 spermatogonia to secondary spermatocytes are more sensitive to seasonal changes than other germ-cells. The number of Sertoli cells, the seasonal alterations in spermatogenesis including germ-cell degeneration in the field rat, may be the result of some metabolic seasonal changes in Sertoli cells which regulate spermatogenesis through the effects of the hormones, FSH and testosterone (Chap. I). The precise nature and hormonal regulation of such Sertoli cell changes must be determined in future studies. L. A. Johnson (1985) has evaluated seasonal variation in the number of spermatogonia and germ-cell degeneration in stallions.

Ohwada and Tamate (1980) have found that in the right and left testes of the same animal, the frequencies of degenerating type A spermatogonia are almost equal. When one testis is removed, there is a negative correlation of the frequency of degenerating spermatogonia in this testis and the other after 25 days. De Reviers and Courot (1976) have made a quantitative study of seminiferous tubules in adult bulls and rams. They have suggested that testicular sperm production is regulated early in the spermatogenetic cycle at the level of the spermatogonia. This is mediated through Sertoli cells which control the seminiferous epithelium development. Sezuki and Withers (1978) have reported exponential decrease during ageing and random lifetime of mouse spermatogonial stem cells. This indicates that with ageing the efficiency of spermatogenesis is also affected. The scrotal insulation affects spermatogenesis in ram (Bayers and Glover 1984).

The effects of various drugs and radiations on the spermatogenic cells of mammalian testes are being studied increasingly to determine their differential sensitivity (L.-M. Parvinen and M. Parvinen 1978, Oakberg et al. 1982, Brezani and Kalina 1982, Oakberg and Crosthwait 1983, Erickson and Hall 1983, Abraham and Franz 1983, C. Schulze 1979, 1981, Hilscher 1983). De Rooij and Kramer (1970) have studied the effects of three alkylating agents on the spermatogonia of rodents. Of all spermatogenic cells, the A_1 spermatogonia in rats and mice are the most sensitive to Myleran. At higher doses this agent also injures later generations of A spermatogonia. In golden hamsters, the A_1 spermatogonia and then subsequent generations show no obvious differences in sensitivity; their B spermatogonia can be killed with higher doses. Two other agents (triethylenemelamine, TEM and isopropylmethane sulfonate, IMS) kill the A_1 and following generations of A spermatogonia in mice and also affect the B spermatogonia if higher doses are given. The A_1 spermatogonia in the G_1 phase are relatively resistant to TEM, making it possible for them to repopulate the seminiferous epithelium within a short time after the administration of this agent. These results indicate that molecular or metabolic differences exist between the spermatogonia of the same species and also between different species. This suggestion is further supported by the observations of Aizawa and Nishimune (1979) who have suggested that serum is necessary for only the early process of spermatogenesis from type A spermatogonia in vitro and not for the later stages. Type A spermatogonia cultured in serum-free medium retain the ability to differentiate for at least 3 days. Tissue culture has provided experimental models in which the environmental conditions can be controlled. Retinoic acid can activate division and induce differentiation of type A spermatogonia of mouse cryptorchid testes in vitro (Haneji et al. 1982, 1983), suggesting that it can play an important role in the control of early stages of spermatogenesis. The metabolic differences between different types of spermatogonia are supported also by the variation in their ultrastructure and possibly in their molecular biology (Monesi et al. 1978, Bellvé 1979, C. Schulze 1979, 1981), as will be discussed below. A combined use of autoradiography and enzyme histochemistry also suggested to reveal the differences between different types of germ cells (B. Hilscher et al. 1985, W. Hilscher et al. 1985).

E. Structure, Cytochemistry and Biochemistry of Spermatogonia

1. Nucleus

The nucleus of A or flat type spermatogonia in marmoset and chimpanzee is elongated with one spherical nucleolus consisting of RNA and protein (Fig. 17). The number of nucleoli varies in different types of spermatogonia of buffalo, goat and ram (Guraya and Bilaspuri 1976a, Bilaspuri and Guraya 1984b, 1986a), as reported for the bull (Hochereau-de-Reviers 1970). Electron microscope studies have shown that the nucleus of A or flat type spermatogonia in monkey, humans and rabbit is slightly elongated with a loose chromatin structure and one to four nucleoli (Fig. 16); the latter may appear homogeneous but generally show a prominent nucleolonema (Tres and Solari 1968, Nicander and Plöen 1969, Vilar et al. 1970, Gondos and Zemjanis 1970, Rowley et al. 1971, C. Schulze 1979, 1981, Cavicchia and Dym 1978, Holstein and Roosen-Runge 1981, Kerr and de Kretser 1981). The spermatogonia of rat contain several nucleoli, which stain for RNA (Daoust and Clermont 1955). The nuclear envelope of A spermatogonia contains pores, and in man two forms of membranous structures known as "stripes" and "blebs" (Rowley et al. 1971, W. Schulze 1978). The intermediary and type B or round type spermatogonia show a decreasing cell size (Table 1), and a more rounded and homogeneous nucleus with several dense granules of chromatin and nucleolar substance (Fig. 13) (Nicander and Plöen 1969, Rowley et al. 1971, Cavicchia and Dym 1978). According to Vilar et al. (1970), the nu-

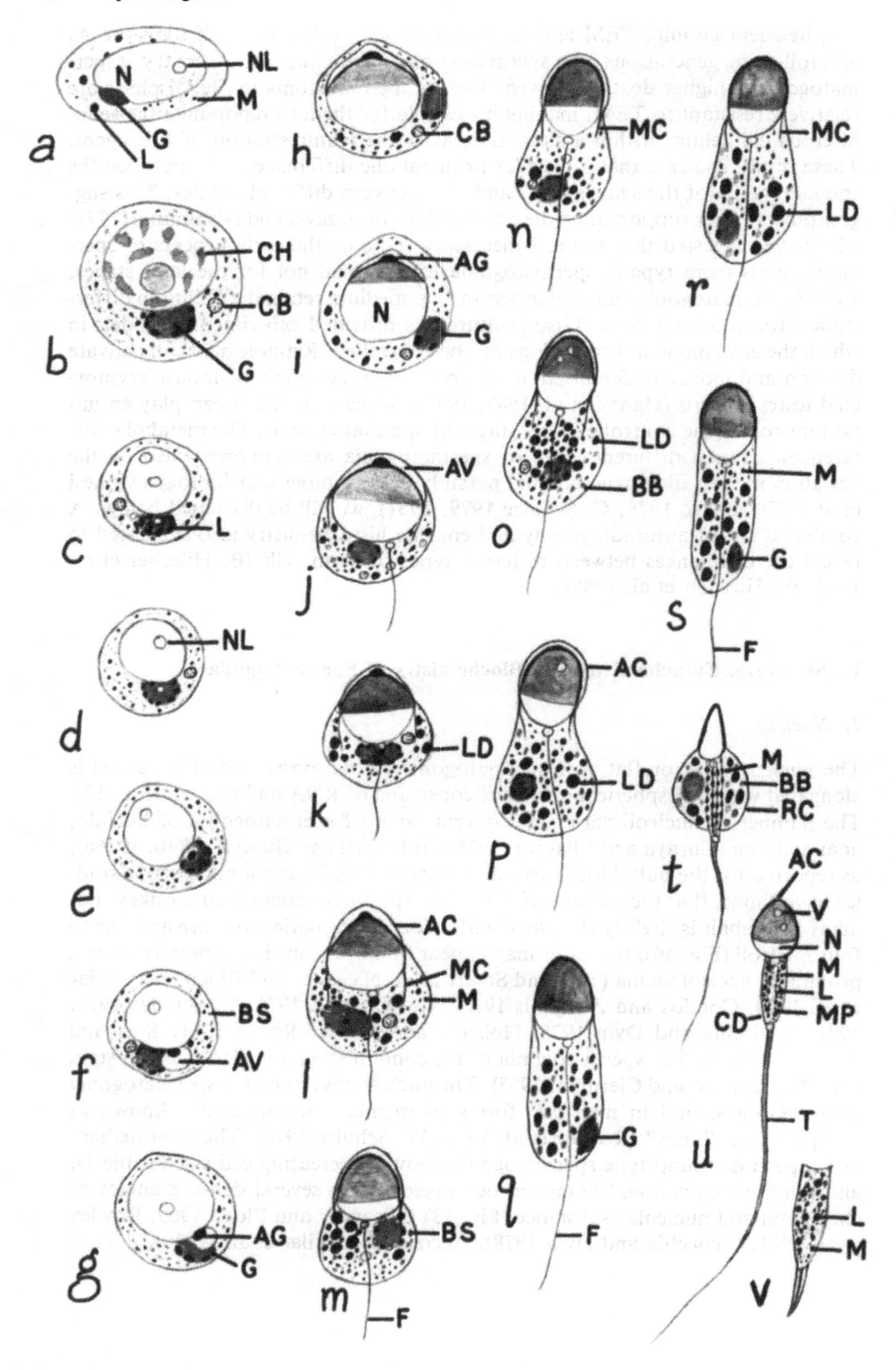
NL
N
M
G
L
a
CH
CB
b
L
c
NL
d
e
BS
AV
f
AG
G
g
CB
h
AG
N
G
i
AV
j
LD
k
AC
MC
M
l
BS
F
m
MC
n
LD
BB
o
AC
LD
p
G
F
q
MC
LD
r
M
G
s
F
M
BB
RC
t
AC
V
N
M
L
CD
MP
T
u
L
M
v

cleolus of round type spermatogonia is of small size. The nucleus of human spermatogonia has been investigated by Tres and Solari (1968) who have distinguished five different types of nuclear bodies in type A (flat) spermatogonia. Type I bodies are typical nucleoli. Types II, III, and IV are considered to be typical nucleoli. Type IV bodies are small chromatin condensations. These different nuclear bodies probably reflect different metabolic activities of spermatogonia, which show varying degrees of nuclear pore clustering (Fawcett and Chemes 1979). Holstein and Roosen-Runge (1981) have illustrated the comparative ultrastructure of karyoplasm in different types of human spermatogonia. The round nucleus of Ap spermatogonia shows a homogeneous granular karyoplasm; a round area of electron-dense, closely packed granules is often seen, it having no connection with the nucleolus. The nucleolus shows a coarse network (nucleolonema), and touches the nuclear membrane. The karyoplasm of Ad spermatogonia is light and floccular with a central, electron-transparent area in which the finely granular material is absent. The spermatogonium type A contains nucleus with "cloudy" karyoplasm resembling the Ad type. But the central light area is absent. The nucleoli are not always applied to the nuclear membrane. The spermatogonium type A long contains an elongated ovoid nucleus. Its karyoplasm is relatively electron-dense without a transparent region. The karyoplasm of type B spermatogonia is floccular. Several noncompact or compound nucleoli are placed near the centre of the nucleus (Rowley et al. 1971), or are associated with the nucleolemina (Holstein and Roosen-Runge 1981). The karyoplasmic densities occur at the inner surface of the nuclear envelope of human spermatogonia A dark (Holstein and Roosen-Runge 1981). A broad zone of granular material may be associated with these densities which, sometimes, contain additional intranuclear cisternae.

Weaker (1977) has reported four generations of spermatogonia, such as stem-cells, type A, intermediate, and type B spermatogonia, in the nine-banded armadillo. The stem-cell shows a highly irregular nucleus and glycogen in the cytoplasm. The type A spermatogonium contains an oblong nucleus with one or two shallow infoldings of the nuclear envelope. The intermediate spermatogonium contains an ovoid nucleus, characterized by one or two nucleoli and heterochromatin scattered in the nucleoplasm. The nucleus of the type B spermatogonium is shaped more spherically with a centrally placed nucleolus and heterochromatin associated with the nuclear envelope.

Fig. 17. Spermatogenic cells, developing spermatids, and spermatozoon from the testicular material of chimpanzee (*Pan troglodytes*), illustrating the morphology and distribution of their various components as revealed by the light microscope and histochemical techniques. *a* A spermatogonium; *b* primary spermatocyte; *c, d* early spermatids; *e – t* successive stages of maturing spermatid illustrating development of acrosomal cap and associated changes in the nucleus and other components; *u* mature sperm; *v* mid-piece of mature sperm; *AC* acrosomal cap; *AG* acrosomal granule; *AV* acrosomal vacuole; *BB* basophilic body; *BS* basophilic substance; *CB* chromatoid body; *CD* cytoplasmic droplet; *F* flagellum; *G* Golgi complex; *L* lipid body; *LD* lipid droplet; *M* mitochondrion; *MC* manchette; *MP* mid-piece; *N* nucleus; *NL* nucleolus; *V* vacuole; *RC* residual cytoplasm; *T* tail

2. *Nucleic Acids and Protein Synthesis*

Nucleic acids, such as DNA and RNA, have been demonstrated cytochemically in the spermatogonia. DNA synthesis in the mammalian testis is an enzymatic process probably dependent on the replicative type of DNA nucleotidyl transferase (DNA polymerase) which has a requirement for all four deoxyribonucleoside triphosphates, i.e. d cytidine, d guanosine, d thymidine, and d adenosine triphosphate, and forms DNA from the four deoxyribonucleoside triphosphate precursors with the concomitant elimination of inorganic phosphate (Calvin et al. 1967). In the seminiferous epithelium, DNA synthesis occurs mainly during the early phase of spermatogenesis and takes place during each cell cycle of spermatogonial multiplication. Much of the fundamental information about DNA synthesis during spermatogenesis has been obtained from autoradiographic and cytochemical analyses in testicular material from the rat (Hilscher 1967), mouse (Monesi 1967), and bull (Hochereau 1967). The final product of spermatogonial division, namely the preleptotene spermatocyte, is the last spermatogenic cell to undergo DNA duplication.

DNA synthesis of the spermatogonia occurs during S phase (Monesi et al. 1978). The molecular mechanisms responsible for increasing the duration of DNA synthesis in differentiating mammalian spermatogonia have not been determined. The length of the S phase increases during development from 19.5 h in A_1 spermatogonia of the rat to 25.5 h in B spermatogonia (Hilscher 1967). A similar trend has been observed in the mouse (Monesi 1962) and the bull (Hochereau 1967). In contrast to the S phase, the G_2 phase decreases from 11.0 h in A_1 spermatogonia to 5.5 h in B spermatogonia in the rat (Hilscher 1967). An increased amount of condensed heterochromatin characterizes the differentiation from A to type B spermatogonia (Courot et al. 1970) which is believed to account for the reciprocal shortening of the G period that occurs during the maturation of the spermatogonia. This is also accompanied by a decrease in the rate of RNA synthesis. Equimolar proportions of somatic, core histones have been reported to occur in chromatin of rat spermatogonia (Chiu and Irvin 1983). Jagiello et al. (1983) have made fiber DNA studies of premeiotic mouse spermatogenesis.

The synthesis of RNA, like that of DNA, occurs during spermatogonial development; the four major ribonucleoside triphosphates needed for the synthesis are guanosine, cytidine, adenosine, and uridine triphosphate, and the enzyme involved is the DNA-directed RNA polymerase. Monesi (1965a) has found a much higher rate of protein and RNA synthesis in type A spermatogonia than in type B spermatogonia of mice. Davis and Firlit (1970) have made similar radioautographic observations of protein synthesis in cells of the seminiferous epithelium in the rat and humans. But Moore (1971, 1972) has demonstrated RNA polymerase activity in the two major types of spermatogonial cells: scant uptake of label occurs in A spermatogonia, while the B spermatogonia are more active, particularly in nucleolar sites. Söderström (1976a), using high resolution autoradiography, has shown that the label remains in the nucleus of the spermatogonia for a long time. After a 2-h pulse the nucleoli of the spermatogonia are labelled heavily (see also Söderström and Parvinen 1976c). This indicates that spermatogonia synthesize RNA actively. Spermatogonia transcribe both nucleolar, that is ribosomal RNA (rRNA), and nonnucleolar RNA species (mainly heterogeneous nuclear RNA). Tijole and Steinberger (1966) have described nucleolar extrusions in the germinal cells of human male gonad. The nuclei of spermatogonia show pyroantimonate deposits which are more dense in areas of condensed chromatin (Gravis 1978). These deposits are indicative of localization of cations which occur in the nuclei of other spermatogenic cells.

3. *Cytoplasmic Organelles and Enzymes*

a) Organelles

Morphological and histochemical techniques have revealed the presence of relatively few cytoplasmic organelles in the spermatogonia of marmoset and chimpanzee (S. S. Guraya, unpublished observations), such as the small Golgi complex, some granular mitochondria and RNA-containing basophilic substance, which are localized mostly in the perinuclear cytoplasm of A spermatogonia (Fig. 17). The Golgi complex is in the form of a small, homogeneous, well-organized mass that is situated in close contact with the nuclear envelope at one pole of the nucleus. Besides these various organelles, some deeply sudanophilic lipid granules consisting of phospholipids also are present (Fig. 17). Gatenby and Beams (1935), using classical techniques of cytology, reported the presence of a crystalloid – the so-called Lubarsch crystal – in the spermatogonia of man. Sohval et al. (1971) and Rowley et al. (1971), using electron microscope, have confirmed its presence in the human A spermatogonia (see also Kerr and de Kretser 1981, Holstein and Roosen-Runge 1981). The significance and chemical composition of the crystalloid are still not known. However, the crystalloids are ultrastructurally composed of closely packed, parallel arrays of dense filaments 8–12 nm wide. Dense granules resembling ribosomes are disposed linearly in the interfilamentous spaces. The complex of filaments and granules is embedded in a finely granular matrix. Such crystalloids also occur in the human Sertoli cells and oocytes (Hadek 1965, Sohval et al. 1971, Guraya 1974, Holstein and Roosen-Runge 1981, Kerr and de Kretser 1981).

The spermatogonial cytoplasm in marmoset, monkey, man, and rabbit is rich in free ribosomes and contains moderate amounts of strands of rough and vesicular smooth-surfaced endoplasmic reticulum, mitochondria lying singly or in groups with a thin layer of electron-dense substance, microtubules, and a simple Golgi apparatus (Fig. 18) (Schmidt 1964, Nicander and Plöen 1969, Gondos and Zemjanis 1970, Vilar et al. 1970, Rowley et al. 1971, Cavicchia and Dym 1978, C. Schulze 1979, 1981, Kerr and de Kretser 1981, Holstein and Roosen-Runge 1981, Holt and Moore 1984). Dym and Pladellorens (1977) have observed that type B spermatogonia in monkey contain a greater amount of smooth endoplasmic reticulum as compared to type A spermatogonia, but their cisternae are scattered randomly and very prominent. According to Vilar et al. (1970), the round or B type spermatogonia in the human testis contain no microtubules, the Golgi apparatus is conspicuous, while the mitochondria, separated and evenly scattered throughout the cytoplasm, have well-developed cristae in a

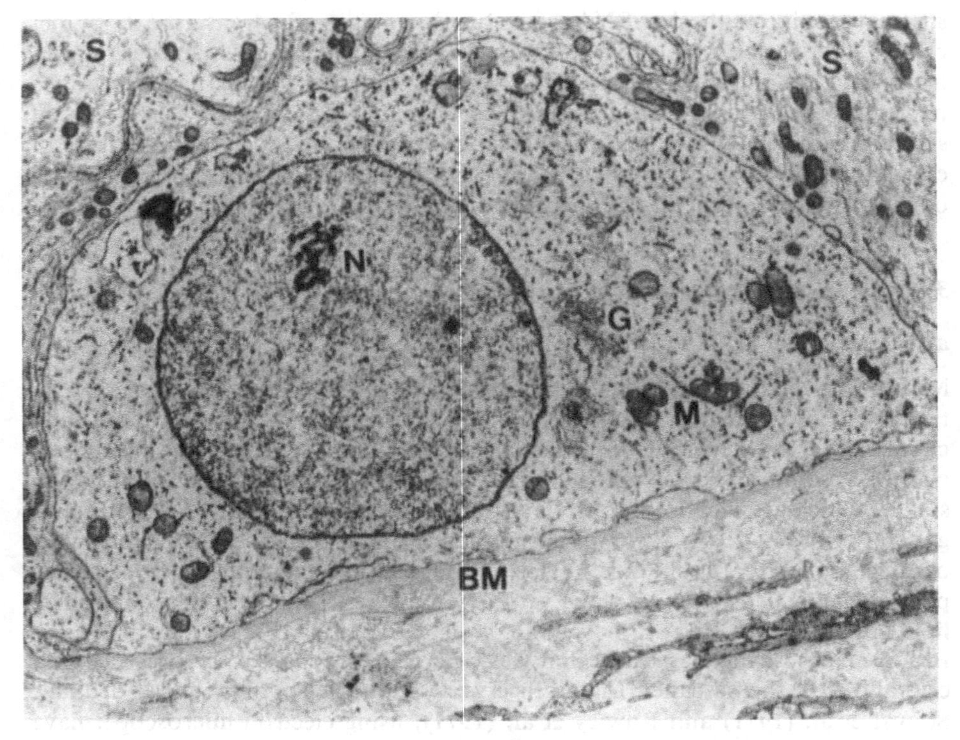

Fig. 18. Electron micrograph of a basally situated pale type A spermatogonium from human testis. The nucleus shows few irregular accumulations of chromatin. The mitochondria (*M*) are interconnected by an electron-dense substance. *N* nucleolus; *G* Golgi complex; *S* Sertoli cells; *BM* basement membrane (From C. Schulze 1981)

dense matrix (see also Holstein and Roosen-Runge 1981). The mitochondria do not congregate, and dense cementing material is absent (Holstein and Roosen-Runge 1981). The spermatogonium type B in the human testis touches the basal lamina either with a broad base or with a more less narrow process (see Holstein and Roosen-Runge 1981). Small crystal-like structures of irregular shape are formed of fibrillar bundless or tubules aligned in the long axis as described by Sohval et al. (1971) and Rowley et al. (1971) in the spermatogonia A of human testis. The fine granular basophilic substance demonstrated with histochemical techniques (Fig. 17) apparently corresponds to the ultrastructural free ribosomes and elements of granular endoplasmic recticulum (Holt and Moore 1984). Kerr and de Kretser (1981) and C. Schulze (1979, 1981) have described the details of ultrastructural differences between different types of human spermatogonia which show considerable amounts of glycogen scattered throughout the cytoplasm and accumulated in membrane-limited areas, as also illustrated by Holstein and Roosen-Runge (1981).

Besides the usual organelles, the spermatogonial cytoplasm may show micropinocytotic vesicles, coated vesicles, occasional multivesicular bodies, vacuoles, lamellar bodies, dense bodies or lysosomes with acid phosphatase activity and glycogen (Nicander and Plöen 1969, Rowley et al. 1971, Barham and Berlin 1974, Barham et al. 1976). Besides these various components, the spermatogonia of the rat (Eddy 1974), rabbit (Nicander and Plöen 1969), and humans (Burgos et al. 1970a, b, Wartenberg et al. 1971, Kerr and de Kretser 1981, Holstein and Roosen-Runge 1981) contain intermitochondrial dense material. Russell and Frank (1978) have demonstrated nuage in the form of spherical particles measuring 70 – 90 nm across in spermatogonia of rat. According to Gondos and Zemjanis (1970), binucleated spermatogonia resemble other spermatogonia in their ultrastructural characteristics, but contain an increased number of lysosome-like structures and degenerating mitochondria. Besides the phospholipid granules, glycogen is another chemical component of which there is an abundance in spermatogonia, but little if any exists in mammalian spermatozoa (Rowley et al. 1971, Mann and Lutwak-Mann 1981).

b) Enzymes

Several enzymes have been reported in the spermatogonia. Bilaspuri and Guraya (1982, 1983b, e, f), using histochemical techniques at the light microscope level, have reported the presence of various enzymes, such as monoamine oxidase, cytochrome oxidase, NADH diaphorase, NADPH diaphorase, glycolytic enzymes, alcohol and secondary alcohol dehydrogenases, and also enzymes of various metabolic pathways in the spermatogonia of buffalo, goat and ram. Nakamura et al. (1983) have observed changes in activities of glucose metabolizing enzymes in germ-cells during spermatogenesis in the rat.

Bilaspuri and Guraya (1983a) have demonstrated histochemically phosphatases (alkaline and acid phosphatases, ATPase, etc.) in the testes of sheep, goat and buffalo. Glucose-6-phosphatase is localized within the rough endoplasmic reticulum and nuclear envelope of spermatogonial cells (Yokoyama and Chang 1977) especially in Al and Ad spermatogonia but not in type Ap or B spermatogonia of human testis (Barham and Berlin 1974, Barham et al. 1976). Glucose-6-phosphatase is an important enzyme in intermediate carbohydrate metabolism and is a good indicator of the general metabolic state of the cell (Wachstein and Meisel 1956). Its presence in the more primitive type of spermatogonia is correlated with the presence of glycogen (Rowley et al. 1971, Barham and Berlin 1974, Barham et al. 1976, Weaker 1977). Inosine diphosphatase activity is present in the endoplasmic reticulum, nuclear envelope, and Golgi complex of all spermatogenic cells except late spermatids (Barham et al. 1976). Adenosine triphosphatase is localized within the plasma membrane of spermatogonial cells of human testis (Barham and Berlin 1974, Barham et al. 1976). But in the rodent testes adenosine triphosphatase occurs in the Golgi cisternae of spermatogenic cells and in the interface between spermatogonia and Sertoli cells, and between late spermatids and Sertoli cells (Tice and Barnett 1963, Chakraborty and Nelson 1974, Chang et al. 1974). Thiamine pyrophosphatase reaction product is present in the Golgi bodies of spermatogenic cells (Barham et al. 1976). Further comparative cytochemical studies at the ultrastructural level should be carried out on the enzymes of different types of spermatogonia for revealing their metabolic differences more precisely, which will provide an insight into the processes of their differentiation at the subcellular and molecular levels.

From the cytological, histochemical and biochemical characteristics of spermatogonia as discussed above, it can be concluded that the spermatogonia are metabolically active and this activity appears to be related to their multiplication and growth processes.

Chapter III
Spermatocytes

The cells in meiosis are called spermatocytes which are formed from the last spermatogonial division and represent the meiotic stage of the male germ-cells. The process of meiosis comprises two divisions, the cells before the first division are called primary spermatocytes (spermatocytes I), after the first division secondary spermatocytes (spermatocytes II). The cells about to enter meiosis usually are designated as resting spermatocytes (Leblond and Clermont 1952b). Since their nuclei synthesize DNA prior to the first meiotic division (Monesi 1962, Hilscher 1964), and thus are active metabolically, they are termed preleptotene spermatocytes. The nuclei of these cells are smaller and darker than those of B_2 spermatogonia (Guraya and Bilaspuri 1976a, c, Bilaspuri and Guraya 1980a, 1984b, 1986a) or B_1 spermatogonia in most species (Clermont 1972, Roosen-Runge 1977). These contain fine chromatin filaments besides the dark chromatin granules lying adjacent to the nuclear envelope and in the central nucleoplasm. The advanced pachytene spermatocytes are the largest spermatogenic cells of the seminiferous epithelium (Table 1). The nucleus and cytoplasm of primary spermatocytes increase markedly in volume during the meiotic prophase. Sertoli cells are just as large in total volume as pachytene spermatocytes (Bellvé et al. 1977a).

A. Meiosis and its Regulation

Meiosis in mammals is a very complex process characterized by an ordered series of chromosomal rearrangements (Figs. 5 and 19), which results in the reduction of the chromosome number to the haploid set by means of two nuclear divisions (1st and 2nd meiotic divisions), involving only a single division of the chromosome. The control mechanisms involved in the initiation and completion of meiosis still need to be resolved. Although intracellular factors may be most important to induce/promote these events of meiosis, recent observations have provided some evidence for the external controls initiating meiosis in mammals (see Guraya 1980). The onset of meiosis at the time of puberty is believed to be controlled by two diameterically opposed testicular factors, a meiosis-inducing and a meiosis-preventing substance (see Grinsted et al. 1979, Parvinen et al. 1982). It will be interesting to mention that the primordial germ-cells present in the mouse foetal ovaries enter meiosis by the 5th day of culture (Taketo and Koide 1981). It is believed that the germ-cells in the developing testis can start meiosis provided they are not surrounded by testis cords (Byskov 1978, Upadhyay and Zamboni 1982). However, Taketo et al. (1984) have observed that

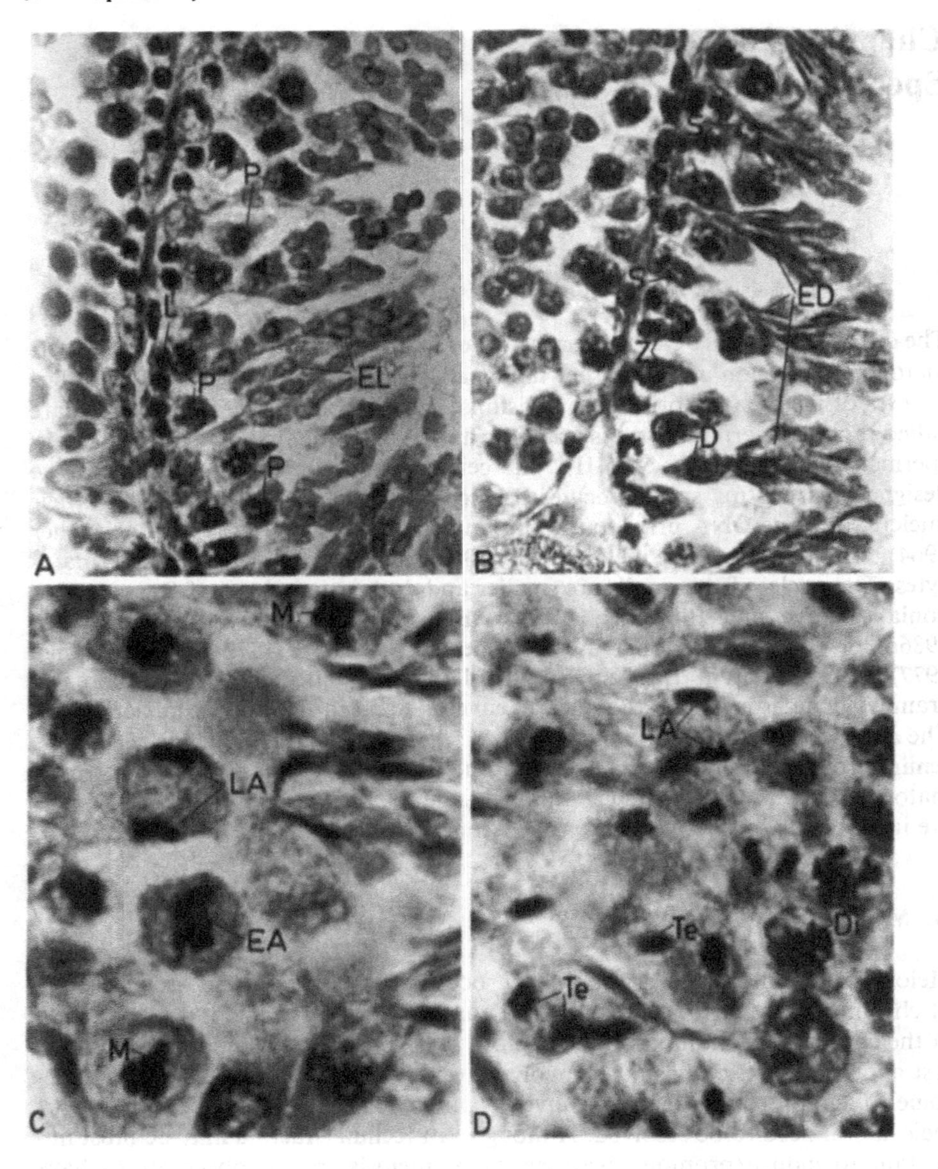

Fig. 19 A – D. Portions of seminiferous epithelium from field rat (*Bandicota bengalensis*) testes after fixation in Bouin's fluid and staining with haematoxylin-eosin to show various phases of meiosis (For diagrammatic representation of these cell types also see Figs. 5 and 27). **A** Cellular associations at Stage 2 of cycle. The tubular periphery is lined by leptotene spermatocytes (*L*) which towards the tubular lumen are followed by pachytene spermatocytes (*P*). Also seen towards the lumen are elongating spermatids (*EL*); **B** Cellular associations at Stage 4 of cycle. The tubular periphery is lined by zygotene spermatocytes (*Z*) which towards the tubular lumen are followed by diplotene spermatocytes (*D*). Also seen are elongated spermatids (*ED*) in association with Sertoli cells (*S*); **C** Figures of meiosis-I at Stage 4 of cycle. The primary spermatocytes in metaphase (*M*), early anaphase (*EA*) and late anaphase (*LA*) are seen; **D** Figures of meiosis-I at Stage 4 of cycle. In addition to primary spermatocytes at diakinesis (*Di*), the spermatocytes at late anaphase (*LA*) and telophase (*Te*) are seen

primordial germ-cells in Bt_2 cAMP-treated explants do not start meiosis even though cords are absent. These germ-cells appear to be morphologically normal and start spermatogenesis by the 14th day after transplantation into adult mice. These results have suggested that primordial germ-cells in foetal testes develop into prespermatogonia independently of testes cord organization, which can be prevented specifically by cAMP analogues, without affecting the differentiation of some cellular components of foetal testes. Millette (1979a) has suggested that, whether mediated by specific receptors for diffusible substances or by direct intercellular communication, the various inductive events may implicate the plasma membrane of seminiferous cells in the regulation of meiosis. But there is no direct evidence to support this contention. Further studies should be carried out on the isolation and characterization of cell surface components which may be responsible for triggering the switch from mitotic cell division to meiotic proliferation in the testis. The meiosis-inducing substance is likely to be a steroid (based on current evidence) and therefore would not require surface receptors – only nuclear receptors. Little is known about the genetic control of meiosis (Guraya 1980). Actually, the various factors regulating meiosis have not been revealed on a biochemical basis, nor have the specific protein changes during meiosis (Bellvé 1979, 1982, Bellvé and O'Brien 1983). Detailed knowledge of these and other molecular events occurring during spermatogenesis will facilitate the development of potential contraceptives for males as well the treatment of infertile cases. In future, this area of research would benefit greatly from the increasing use of molecular biology techniques. The preparation, cloning and sequencing of complementary cDNA's coding for the stage-specific proteins and regulatory factors are going to provide valuable information on the structure of these molecules and their homology with other known proteins (Bellvé and Moss 1983). In addition, information on partial homologies may lead to hypotheses on the potential function of these proteins in regulating germinal cell development including meiosis. There is now increasing evidence to suggest that the development and subsequent day-to-day function of the mammalian testis is controlled to a large extent by local (i.e. intratesticular) factors (Parvinen 1982, Sharpe 1983, 1984, Bellvé and Feig 1984). Most locally acting factors are small peptides, as already discussed in Chap. I.

Whatever may be its regulation, the chromatin material during meiosis passes through various configurations of preleptotene, leptotene, zygotene, pachytene and diplotene during the first meiotic division (Figs. 5 and 19). The two maturation divisions of each spermatocyte result in four haploid cells, the spermatids. The chain of events associated with meiosis is as follows:

1st Meiotic Division: Preleptotene primary spermatocyte → Leptotene → Zygotene → Pachytene → Diplotene → Diakinesis (final stage of prophase) → 1st Metaphase → 1st Anaphase → 1st Telophase → 2 secondary spermatocytes.

2nd Meiotic Division: Secondary spermatocytes → 2nd Metaphase → 2nd Anaphase → 2nd Telophase → 2 Spermatids.

In preparation for the 1st meiotic division, the preleptotene spermatocyte with its nucleus in interphase first changes into the leptotene spermatocyte, which enters zygonema when the pairing of the homologous chromosomes is initiated. As the nucleus progressively increases in size and its chromosomes thicken, the spermatocyte enters pachynema and later the brief diplotene stage, when the chromosomes partially disjunction/separate (Figs. 5 and 19). Holstein and Roosen-Runge (1981) have illustrated the ultrastructural changes in the chromosomes during leptotene, zygotene, and pachytene stages of meiosis in the human testis. The chromatin shows progressive condensation during these stages of meiosis. The reduction of chromatin takes place in the course of two successive divisions: (1) the homologous maternal and paternal chromosomes, already split longitudinally into two chromatids become separated from each other as the nucleus goes through metaphase, anaphase and telophase, and the primary spermatocyte divides into the secondary spermatocytes. Another division follows quickly, leading to the separation of the paired sister chromatids and the formation of two spermatids. In mammals, as a result of the separation of the sex chromosomes, which occurs at the end of the first meiotic division, two categories of haploid spermatids are produced, one having the female X chromosome and the other the male Y chromsome.

On the basis of three-dimensional reconstructions of 29 early pachytene and 28 mid-late pachytene nuclei, Berthelsen et al. (1980) have reported the occurrence and morphology of three ultrastructural markers on meiotic chromosomes of human spermatocytes. A condensed granular structure surrounded by a larger body of uncondensed chromatin is placed near the telomere of the short arm of bivalent 1. A globoid structure with a fine granular substructure is present in the middle of the short arm of bivalent 6. A body of coarse granules is situated near the middle of the long arm of one of the two longest D group bivalents. Guraya and Bilaspuri (1976c), using haematoxylin-eosin and Feulgen reaction techniques, have made a study of the morphology and dimensions of the phases of meiotic division of the spermatocytes in the buffalo testis (Table 1). On the basis of these results, the preleptotene, zygotene and diplotene stages are divided further into early and late stages. Pachynema is divided into early, mid and late pachytene stages. Actually, the stages of meiotic prophase are each based on a progression of distinctive chromosomal events (Figs. 5 and 19), which have been discussed in detail by Bellvé (1979). The chromosomal events that occur during meiotic prophase include a partial compaction of the chromosomes, the precise alignment of the homologues, stabilization of that alignment by the synaptonemal complex, formation of chiasmata, and subsequent disjunction. Each of these processes is closely involved in ensuring successful genetic recombination, as determined by genetic and morphological criteria.

Meiosis in the mammalian testis constitutes a lengthy process as it takes a variable number of days for its completion in different species of mammals (Ortavant 1956, Heller and Clermont 1963, Swierstra and Foote 1965, Utakoji 1966, Kofman-Alfaro and Chandley 1970, Ghosal and Mukherjee 1971). Meiosis occurs more rapidly in males than in females of the same species due to the fact that spermatocytes are not subjected to a·protracted arrest at the diplotene stage (see Bennett 1977).

According to Clausen et al. (1977) microflow fluorometry offers a practical and sensitive technique for monitoring meiosis in the rat. Libbus and Schuetz

(1978) have made an analysis of progression of meiosis in dispersed rat testicular cells by flow cytofluorometry. Very few preleptotene spermatocytes of human testis reach pachytene stage, and essentially none the spermatid stage in cultured material (A. Steinberger et al. 1979). With further improvements in these in vitro methods (see also Chap. I), it may perhaps become feasible one day to reproduce the process of meiosis. This will form an important contribution to the knowledge of the spermatogenic process in the normal testis, and an unique opportunity for studying the effects of antispermatogenic agents to advance knowledge on male contraception. The chromosomes during different stages of meiosis undergo morphological and biochemical changes. Various biochemical changes (or molecular aspects of meiosis) in primary spermatocytes include replication of DNA, formation and function of chromosomal proteins (especially histones, atypical histones and non-histone proteins), transcription (synthesis of various species of RNA), and translation (synthesis of cytoplasmic proteins) which, along with morphological characteristics of different nuclear components, will be discussed here. The peculiar composition of the germ-cell nuclear matrix may be due to the transformation of mitotically dividing spermatogonia into spermatocytes undergoing meiosis.

B. Subcellular and Molecular Aspects of Meiosis

I. Nuclear Components

The nucleus of primary spermatocytes shows various components, such as prominent nucleolus/nucleoli, sex vesicle (or residual XY heterochromatin), some membranous components, etc. besides the chromatin or chromosomes. A large nucleolus consisting of RNA and protein forms the specific feature of the spherical nucleus during the pachytene stage (Fig. 17). It undergoes a cycle during the process of division. It disappears by late prophase and reappears by telophase, in association with certain secondary constrictions (nucleolar organizer) of usually two, defined chromosomes (Roosen-Runge 1977, Bellvé 1979). Daoust and Clermont (1955) studied the behaviour of nucleoli during meiosis in the rat. During leptotene and zygotene stages they increase in number, and during the early pachytene stage in size. At late pachytene stage the nucleoli resume an oval shape and are reduced in number. The nucleolus shows a dense, rounded and granular part and a lighter, finely granular and filamentous, part lying in contact with the nuclear membrane of human spermatocyte 1 (Holstein and Roosen-Runge 1981).

Besides the nucleoli, there is spherical heterochromatic body called "sex vesicle", which is formed by the condensed XY bivalent during the meiotic prophase and is applied to the nuclear membrane (Solari 1970, 1974, Monesi et al. 1978, Holstein and Roosen-Runge 1981). Takanari et al. (1982) have demonstrated the presence of dense bodies in silver-stained spermatocytes of the Chinese hamster, which may be formed also by the condensation of XY bivalents. As the sex vesicle is not a membrane-bound structure, Solari (1970, 1974) has used the term "heterochromatic XY pair" for it. RNA has been observed to be localized exclusively in the granular portion of sex vesicle, whereas the DNA appears only in its chromatin part (Solari 1964, Solari and Tres 1967). Solari (1971), studying the behaviour of chromosomal axes in Searle's X-autosome translocation, has shown that autosomal chromatin becomes heteropycnotic in the proximity of the X-Y chromatin. This effect explains the enlarged volume of the sex vesicle. The X and Y chromosomes in mammals are disproportionate in length and bear extensive regions that are not homologous for gene sequence (Dietrich et al. 1983). Several lines of evidence suggest that the single X chromosome in males is inactivated at meiosis during spermatogenesis (Monesi 1965b, Kofman-Alfaro and Chandley 1970, Lifschytz and Lindskey 1974). Kramer (1981) has suggested that the transcription of the Pgk-1 locus will cease at meiosis and the amount of X-linked form (PGK-I enzyme) will be expected to decline as cells progress through late stages of spermatogenesis. But such a decline in PGK-I could not be established, due to insufficient sensitivity of the immunofluorescence technique used by Kramer (1981).

In mid and late pachytene and diplotene phases of meiosis, rat spermatocytes show two types of unusual membranous structures within the nucleus, which are named the saccule and the bleb (Russell 1977b). The sex vesicle or components of the nucleolus frequently are seen lying adjacent to the saccule which faces the nuclear envelope. A 12 – 16 nm space containing dense material separates the saccule from the nuclear envelope. The bleb, in contrast to the single walled saccule, is bounded by two membranes and causes the nucleus to bulge outwards. Numerous nuclear pores can be seen on its surface which is exposed to the nuclear cap. The contents of small blebs are similar to those of the nucleus, whereas the contents of larger blebs show certain similarities to cytoplasmic constituents. Most large blebs contain chromatoid-like material. Both the bleb and saccule type of membranous structures also occur in humans (Rowley et al. 1971). The functional significance of these nuclear membranous structures of primary spermatocytes must be determined in future studies. The residual XY heterochromatin, the prominent nucleolus and peripheral vesicles present in the pachynema matrix do not exist in spermatid nuclear matrix. But the protein components of these two germ-cell nuclear matrices are remarkably similar (see Ierardi et al. 1983). Only the spermatid matrix shows the major distinct polypeptides (MW ~16,000 and 22,000). The nuclear envelope of pachytene spermatocytes consisting of the bilamellar membranes develops relatively more nuclear pores (Nicander and Plöen 1969). Freeze-fracture preparations of spermatocytes show very striking pore aggregation with close hexagonal packing in pore-rich areas and large pore-free areas (Fawcett and Chemes 1979). The presence of an increasing number of nuclear pores during meiosis is indicative of extensive nucleocytoplasmic exchanges. This suggestion is supported by biochemical studies on the synthesis and transport of various RNA species, as will be discussed later on.

The primary spermatocytes generally are distinguished into electron micrographs by a nucleus containing one or more synaptonemal complexes, homogeneous chromatin, and several large globular and deeply-stained chromatin masses (Fig. 20) (Heller and Clermont 1964, Moses 1968, Barham and Berlin 1974, Dumontier and Sheridan 1977, Holm and Rasmussen 1977a, b, Holstein and Roosen-Runge 1981, Holm et al. 1982, Ierardi et al. 1983). Various electron

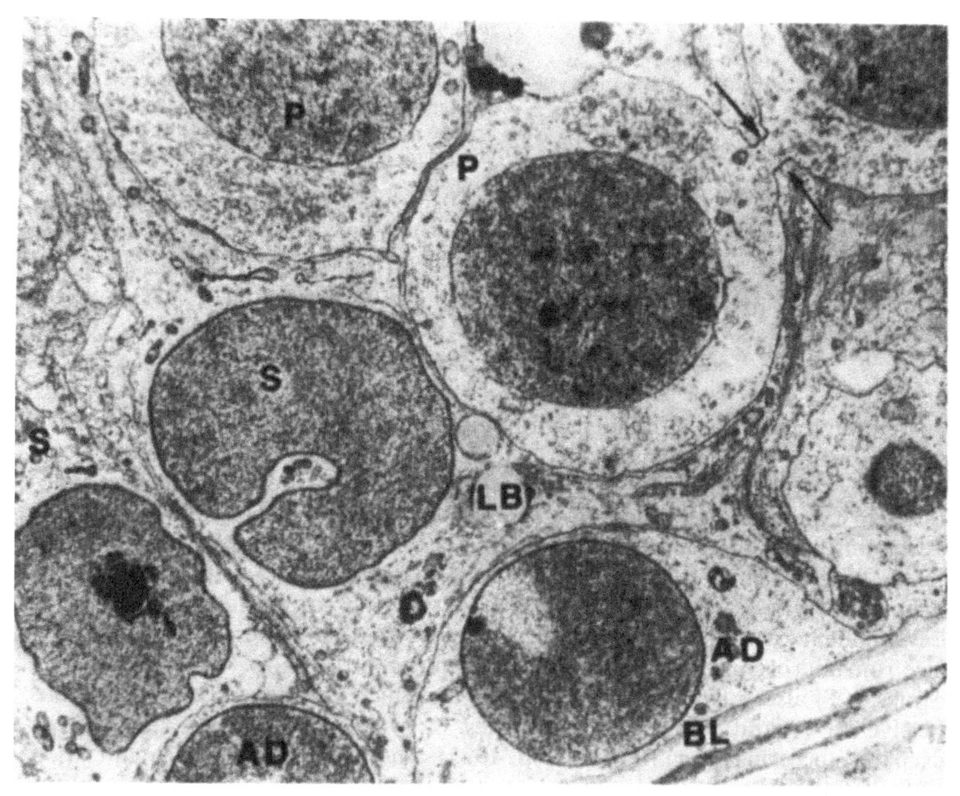

Fig. 20. Electron micrograph of seminiferous tubule from control human testis. Three intratubular cell types are shown. Two type A dark spermatogonia (*AD*) rest upon the basal lamina (*BL*). A nuclear region of rarefaction and extensive contact with the basal lamina characterize these cells. Portions of three primary spermatocytes (*P*), two of which are connected by an intercellular bridge (*arrows*) are evident. Two Sertoli cells (*S*) containing lipofuscin bodies (*LB*) ramify through the germinal cells (From Barham and Berlin 1974)

microscope studies discussed in these papers, reviews and book have revealed that the synaptonemal complexes form axial tripartite structures, as they are composed of two lateral electron-dense, rodlike elements separated by central region containing a medial central element and periodic transverse microfilaments in the primary spermatocytes of different mammals (Fig. 21). Actually, the central element consists of interlacing, microfibrillar elements which arise from the inner aspect of both lateral elements. Lampbrush loops of chromatin radiate from the synaptonemal complexes. Each telomeric end from the synaptonemal complexes is attached firmly to the nuclear envelope by a broad attachment plaque. There is now increasing evidence to show that the synaptonemal complex forms the ultrastructural conterpart of the paired chromosomes or bivalent which develops at the pachytene stage of meiosis (Ierardi et al. 1983). This is further supported by the fact that the number of bivalents for any one species is equal to the number of synaptonemal complexes which may undergo

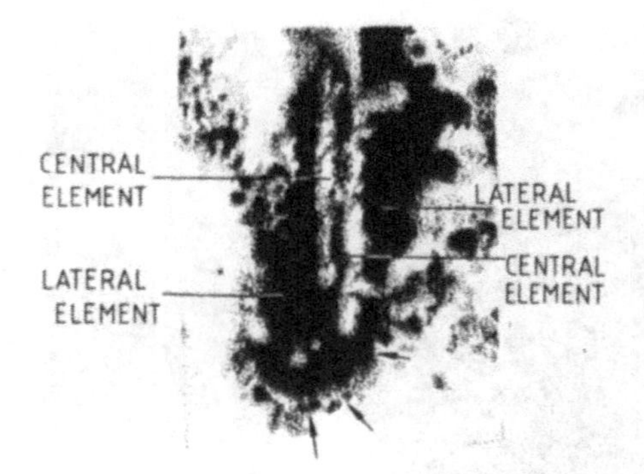

Fig. 21. Ultrastructure of the synaptonemal complex contained within the pachytene spermatocyte matrix. The two lateral elements and one central element are similar in morphology and dimension to those of isolated nuclei. The thickened ends of the lateral elements are associated closely with the attachment plaque (*arrows*). A few transverse microfilaments span the central region (Redrawn from Ierardi et al. 1983)

changes in length due to condensation and contraction of homologous chromosomes during the pachytene stage. The results of studies on whole-mount preparations of chromosomes from spermatocytes have revealed that the synaptonemal complexes course axially through the bivalents (Ierardi et al. 1983). The development of single electron-dense threads, which represent the lateral element of the synaptonemal complex, is helpful in distinguishing the leptotene stage of meiosis. Dietrich and Deboer (1983) have made a sequential analysis of the development of the synaptonemal complex in spermatocytes of the mouse by electron microscopy using hydroxyurea and agar filtration. The sex chromosomes (XY) also form synaptonemal complex which may show species variations (Bellvé 1979) and play significant roles in synapsis, chiasmata and genetic exchange (crossing-over) (Ierardi et al. 1983). The size and complex organization of the synaptonemal complex as reported by Holm and Rasmussen (1977a, b) for the 44 autosomes of human spermatocytes (Fig. 22) suggest that the biochemical processes involved directly or indirectly in crossing-over must be exceedingly complex. The ultrastructural characterization of the meiotic prophase can also be used to assess the radiation damage in man (Holm et al. 1982). Electron microscope examination revealed the coincidental occurrence of synaptonemal complexes and [^{3}H] thymidine label within the nuclei of premeiotic interphase mouse spermatocytes, indicating synapses of homologues had begun during the S (DNA duplication) phase (Grell et al. 1980). Navarro et al. (1981) have introduced a method for the sequential study of synaptonemal complexes by light and electronmicroscopy.

Although the synaptonemal complexes perform a central function in genetic recombination, an understanding of their formation, structure, and function is mostly based on the morphological and histochemical data (Ierardi et al. 1983). The histochemical and enzyme digestion studies have revealed that the synaptonemal complex contains basic protein and trace amounts of DNA and RNA. The proteins form the major component of the lateral and central elements and transverse fibres as evidenced by their digestion with proteolytic enzymes. The protein of the lateral elements may contain one or more species of basic protein specific to meiotic cells. The lateral element also contains a single linear fibril of DNA that passes axially through it. The axial localization of a presumptive DNA fibril in each lateral element appears to be related to the mechanism that brings

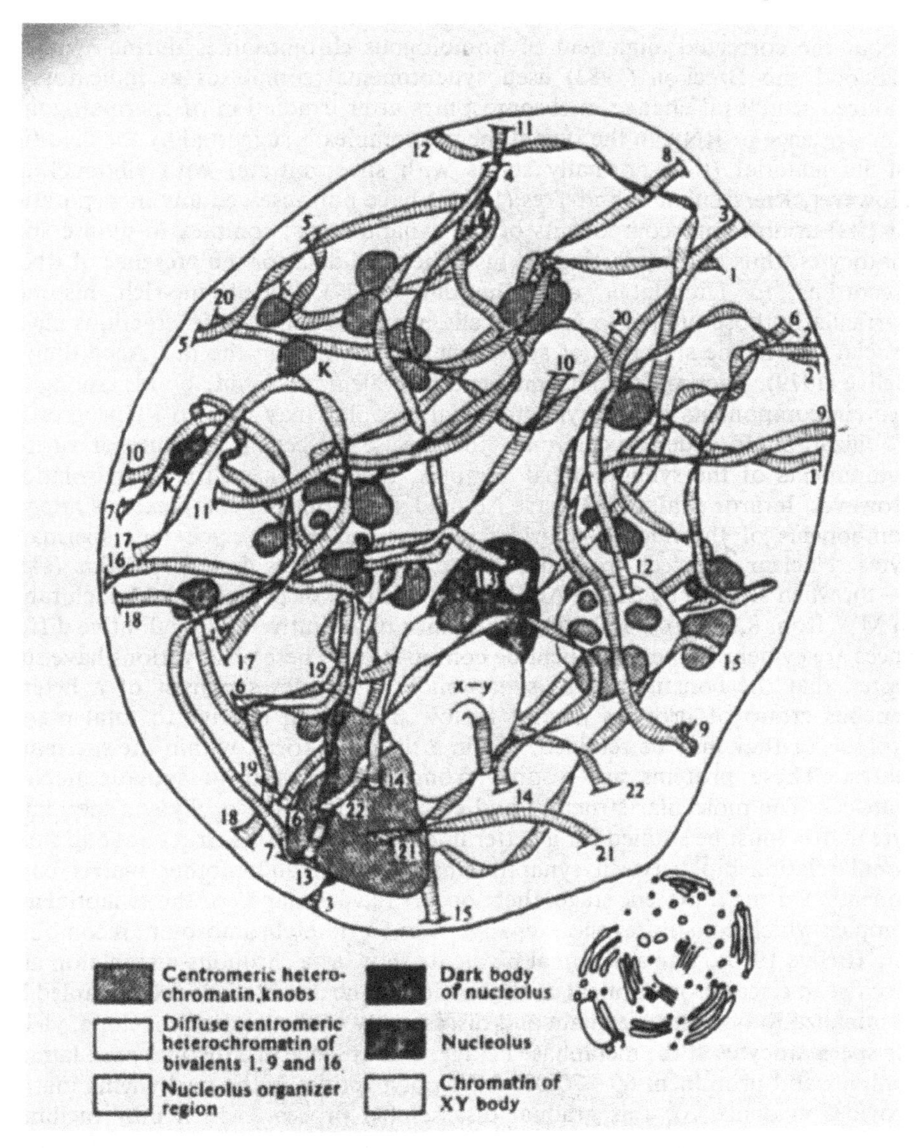

Fig. 22. The spatial organization of the synaptonemal complexes within the nucleus of a human mid-pachytene spermatocyte. The synaptonemal complexes course the entire length of the respective autosomal bivalents (numbered *1* to *22*), and each complex is attached at both ends to the nuclear envelope. The localized block of heterochromatin on each chromosome marks the position of the centromeres. The X and Y chromosomes are paired in the telomeric region only, while the remaining segments are condensed into the heterochromatic XY body. In this case the XY body is associated with the nucleolus (From Holm and Rasmussen 1977b).

about the corrected alignment of homologous chromosomes during synapsis. Cawood and Breckon (1983) used synaptonemal complexes as indicators of induced structural change in chromosomes after irradiation of spermatogonia. The presence of RNA in the synaptonemal complex is suggested by the digestion of its material (that normally stains with silver nitrate) with ribonuclease. However, Kierszenbaum and Tres (1974b) have not observed any incorporation of [^{3}H]-uridine into components of the synaptonemal complex in mouse spermatocytes, thus not supporting the histochemical data for the presence of RNA. According to Dumontier and Sheridan (1977), the lysine-rich histones, particularly those unique to meiotic cells, may through their interactions play a crucial role in the structure of synaptonemal complex in the rat. According to Bellvé (1979), inter and/or intramolecular covalent disulphide bonds among the protein components of the synaptonemal complex may lead to its progressive stabilization. In order to determine the precise molecular structure of various components of the synaptonemal complex, there is a need for their isolation. However, Ierardi et al. (1983) have isolated synaptonemal complexes as integral components of the nuclear matrix from purified mouse pachytene spermatocytes. Nuclear matrices from pachytene spermatocytes and spermatids (steps 1 – 8), when studied by SDS PAGE, show an array of polypeptides which range in MW from 8,000 to ~150,000; only minor quantitative and qualitative differences are evident in their polypeptide composition. These observations have suggested that the constituents of synaptonemal complexes consist of a heterogeneous group of proteins present in low proportion relative to total matrix proteins or they may be retained, but in a different form, within the spermatid matrix. These proteins are distinct from those present in somatic nuclear matrices. The molecular structure and composition of the pachytene spermatocyte matrix must be studied for a better understanding of the structural and functional relationship between synaptonemal complex and mother matrix components. During diplotene stage, there occurs the desynapsis of the synaptinemal complex which constitutes the exchange sites for interchromosomal recombination (Bellvé 1979). The autosomal bivalents show asynchronous disjunction and give rise to circular or elliptical configurations. The diakinesis is accompanied by terminalization of the chiasmata and disassembly of the nuclear envelope, yielding spermatocytes at the metaphase I stage. The absence of a distinct pore-lamina complex and prominent 60 – 70,000 MW polypeptides in the pachynema matrix provides evidence for this gradual disassembly process, which may facilitate movement of the homologous chromosomes during synapsis and disjunction (see Ierardi et al. 1983). The continued absence of a typical lamina structure and its polypeptides in the spermatid matrix is suggestive of the fact that a reassembly of the depolymerized lamina proteins does not take place after meiosis. Further investigations of germ-cell nuclear matrices at other stages of spermatogenesis will be very useful for obtaining a more detailed analysis of matrix transformation during this intricate process of differentiation. Recently Wauben-Penris et al. (1983) have studied the chiasma frequency and nondisjunction in heteromorphic bivalents in different strains of mice.

Recently, lampbrush loops have been demonstrated in the chromosomes of pachytene spermatocytes of the mouse (Kierszenbaum and Tres 1974b). The lack of formation of a partial synaptonemal complex by the axes of the X and Y chromosomes in the marsupial *Monodelphis dimidiata* (Solari and Bianchi 1975) shows an important departure from the observed synaptic pattern known in eutherian mammals (Solari 1974).

2. *Changes in Nucleic Acids and Proteins During Meiosis*

a) DNA

Very divergent views have been expressed about the synthesis of DNA during meiotic prophase. The various studies have shown that DNA synthesis occurs during meiotic prophase (Ortavant 1959, Hotta et al. 1966, Mukherjee and Cohen 1968, Lima-de-Faria et al. 1968, Kofman-Alfaro and Chandley 1971, Meistrich et al. 1975; other references in Bellvé 1979). The preleptotene spermatocytes formed by the division of type B spermatogonia show a final phase of DNA synthesis before entering further stages of meiotic prophase. Because of its semi-conservative nature, this phase of DNA replication resembles that in somatic cells. It will be interesting to mention here that the X and Y chromosomes in the spermatocytes of the hamster and mouse show a differential delay in preleptotene DNA replication (Utakoji 1966, Odartchenko and Pavillard 1970, Kofman-Alfaro and Chandley 1970). This differential delay in DNA replication of sex bivalents, which has been attributed to the greater degree of heterochromatization in their segments, is not specific for pre-meiotic spermatocytes, as temporal difference in their replication pattern has also been reported during DNA formation in spermatogonia and somatic cells. The retarded rate of DNA replication in preleptotene spermatocytes may be indicative of a marked decrease in the number of operative initiation sites for DNA formation as compared to S phase of somatic cells. However, the molecular mechanisms involved in their reduction need to be resolved. A brief G_2 interval is seen after the completion of S phase in the preleptotene spermatocytes. It is immediately followed by a transient condensation of chromatin into a prophase configuration, possibly leading to reorganization of chromosomes for preparation of meiotic prophase.

There exists some difference of opinion in regard to DNA synthesis during different stages of meiotic prophase. Hotta et al. (1966), using biochemical techniques, observed that an exceedingly low amount of DNA is synthesized during zygotene and pachytene stages, constituting 0.3% and 0.1%, respectively, of that synthesized during the S phase of premeiosis. Because of these extremely low levels, it is difficult to localize zygotene DNA with autoradiographic techniques, especially in mammalian species where meiotic synthesis constitutes less than 0.002% of premeiotic DNA synthesis (Meistrich et al. 1975). DNA synthesized during the zygotene stage is believed to be involved in the specific alignment of homologous chromosomes at zygotene and may also play some unspecified role during disjunction (Bellvé 1979). But its precise functions need to be defined. The DNA synthesis during the pachytene stage occurs at levels lower than 0.1% of the premeiotic S phase (Hotta et al. 1966). Söderström (1976a) has demonstrated the presence of trace amounts of DNA synthesis during the meiotic prophase. But the most active DNA synthesis occurs in mid-pachytene spermatocytes in a somewhat later developmental stage than the observed maximum in RNA synthesis of the lampbrush chromosomes. The DNA synthesis during the pachytene stage has been interpreted by Söderström (1976a) as a repair process of damaged DNA, as also discussed by Lee (1983). But the DNA synthesis during leptotene and zygotene stages is believed to be a delayed replication of specific chromosomal regions. Söderström and Parvinen (1976a) have also observed highest synthetic rate of DNA in the mid-pachytene spermatocytes, while moderate or low rate occurs in leptotene, early pachytene, late pachytene and diplotene stages. Chandley (1978) has recently studied the morphological and biochemical aspects of mammalian meiosis. The biochemical events include a small semi-conservative synthesis of DNA during the early prophase and single-strand nicking of DNA coordinated with repair replication of DNA at pachynema. Zygotene DNA synthesis has been assigned a role in synapsis; pachytene DNA synthesis a role in crossing-over. These two forms of DNA synthesis have been shown to occur in mouse spermatocytes (Meistrich et al. 1975, Hotta et al. 1977). The DNA methylation patterns are reported for human pachytene spermatocytes.

The nature and biology of DNA polymerases in the developing and maturing testis have provided further evidence for the synthesis of DNA in primary spermatocytes (Bellvé 1979). The period of major transition in the DNA polymerases is related to the appearance of germ-cells entering meiotic prophase, suggesting that the alteration in specific activity of DNA polymerases α and β may occur in primary spermatocytes. David et al. (1982) have reported developmental changes of DNA ligase during ram spermatogenesis. The uracil-DNA glycosylase has been studied in meiotic and post-meiotic germ-cells of the mouse (Grippo et al. 1982).

Hecht et al. (1979) have studied DNA polymerases in mouse spermatogenic cells separated by sedimentation velocity. DNA polymerase activity continues to be detectable in nonreplicating differentiated pachytene primary spermatocytes. The presence of DNA polymerases has also been shown in the spermatogenic cells of rat. Daentl et al. (1977) have observed that DNA-dependent DNA polymerase activity is maximal in premeiotic and meiotic cells, sequentially declined in progressively more differentiating spermiogenic cells to a minimum value in testicular spermatozoa. Bellvé (1979) has discussed various enzymes needed to effect scission and repair of meiotic DNA, as well as some other proteins performing different functions in relation to DNA. As a result of the last DNA duplication, the primary spermatocyte acquires a tetraploid amount of DNA (Courot et al. 1970). This will be reduced first to the diploid amount when the primary spermatocyte divides into two secondary spermatocytes, and next to the haploid amount when each secondary spermatocyte forms two spermatids.

b) Protein and RNA

The progressive changes in proteins and RNA during meiosis form the subject of numerous studies carried out mostly with autoradiographic techniques. Relatively few attempts have been made to follow their alterations with biochemical techniques which are now being used increasingly with the availability of methods for obtaining homogeneous fractions of spermatogenic cells. Davis and

Firlit (1970) and Loir (1972b) have demonstrated characteristic quantitative radioautographic patterns of protein labelling in the various successive cells of the seminiferous epithelium of rat and humans, and ram respectively, which are consistent with the findings on the nucleic acid labelling in the mouse testis (Monesi 1965a, Monesi et al. 1978). Age differences in proteins at individual stages of differentiation of mouse spermatogenic cells are also revealed (Mushkambarov and Volkova 1983). During late meiotic prophase and/or spermiogenesis additional proteins of the spermatozoon are first detected, including both cytoplasmic (Goldberg 1977) and cell surface constituents or antigens; the latter will be discussed separately in Chap. V. Actually, the relative contributions of meiotic and post-meiotic protein synthesis to the formation of the spermatozoon needs to be resolved (Kramer and Erickson 1982). Unfortunately, autoradiographic techniques do not help us in determining the types of proteins synthesized. Furthermore, estimates of total synthetic activity during spermatogenesis include proteins that are degraded or discarded in the residual bodies in addition to those proteins meant for the mature spermatozoon.

LDH-C and cytochrome ct (E. Goldberg et al. 1977, R.B. Goldberg et al. 1977, Wheat et al. 1977, Meistrich et al. 1977), PGK-1 and PGK-2 isozymes (Kramer 1981, Kramer and Erickson 1982) and the protamines (Loir and Lanneau 1975, 1978b, Bellvé et al. 1975) have been studied for the temporal patterns of their synthesis. Both cytochrome Ct and LDH-X first appear in mid-pachytene spermatocytes, whereas the PGK-2 and protamines exclusively localized to the post-meiotic germ-cells, as their amount increases during the transition between early and late spermatids. Recently, O'Brien and Bellvé (1980b) have studied the temporal synthesis of protein constituents of the mouse spermatozoon during spermiogenesis. The SDS-soluble proteins are synthesized maximally during meiosis and early spermiogenesis. These show both qualitative and quantitative differences at successive stages of spermatogenesis. Their formation has been correlated to the structural and compositional changes of plasma membrane, mitochondria, acrosome, etc. Furthermore, the SDS-soluble fraction, which includes proteins from the plasma membrane, acrosome, axoneme, matrix and cristae of mitochondria from mouse spermatozoa, contains one major 39,000 MW band and numerous minor bands with molecular weight ranging from 30,000 to greater than 100,000 (O'Brien and Bellvé 1980a) Mujica et al. (1978) have also studied the proteins of fractionated human spermatozoa by SDS-polyacrylamide gel electrophoresis. This study is preliminary and thus does not provide convincing evidence for the identification or localization of various protein constituents. Meistrich et al. (1981), by using purified fractions of rat testicular cells, have suggested that testis-specific histones are synthesized during the pachytene stage, as also recently demonstrated for the mouse (Bhatmagar et al. 1985). Similar studies have also been made for the ram (Loir and Lanneau 1975, 1978b). All these studies have demonstrated meiotic histone synthesis. The histones play a fundamental role in the structure and organization of chromatin, as histones 2A, 2B, and 4 constitute the core particle of nucleosome in somatic chromatin (Bellvé 1979). Such nucleosomes are also found in spermatocytes and early spermatids but not during later stages of spermiogenesis. Besides the somatic histones, spermatogenic cells also show several unusual histone species which may regulate the structural function of meiotic chromatin. Some of the atypical histones are testis-specific, as will be discussed in Chap. IV. Actually, the atypical histones show a heterogeneity in their structure. The complex diversity of variants among spermatogenic histones has suggested the possibilty of the presence of a greater number of unique nucleosomes in germ-cells, which may not be present at the same time in any one cell type. Details of stage-specific synthesis of histone species and their functions during meiotic prophase of mammalian spermatogenesis are still to be worked out with biochemical techniques. Bucci et al. (1982) have studied distribution and synthesis of histone subfractions during spermatogenesis in the rat.

Various autoradiographic studies using [^{3}H] uridine have indicated that substantial levels of RNA synthesis occur in spermatogonia, mid-pachytene spermatocytes and, to a small extent, in round spermatids (Henderson 1963, Monesi 1964, 1965a, b, Utakoji 1966, Loir 1972a, b, Geremia et al. 1977b; see reviews by Monesi et al. 1978, Bellvé 1979). Moore (1971, 1972) has observed that RNA synthesis is very low in leptotene spermatocytes but increases during zygotene and early pachytene stages, reaching a peak at the middle of this stage. Further spermatocyte development leads to a decrease in RNA polymerase activity and metaphase I and II chromosomes are completely unlabelled. This indicates that RNA synthesis ceases entirely for the duration of diakenesis and the meiotic reduction divisions.

The results of early studies suggested that ribosomal RNA (rRNA) synthesis did not take place during the meiotic prophase. This suggestion was made from the apparent lack of [^{3}H] uridine incorporation by the prominent nucleolus of mouse pachytene spermatocytes (Monesi 1964, Monesi et al. 1978). But this conclusion was not accepted later on (Stefanini et al. 1974). Recent studies, utilizing high-resolution autoradiography, have produced more convincing evidence for meiotic rRNA synthesis in the mouse (Kierszenbaum and Tres 1974a, b, Tres and Kierszenbaum 1975, Monesi et al. 1978). The synthesis of putative precursor rRNA during leptotene, zygotene, and early pachytene stages was demonstrated for the first time in these studies. Its synthesis occurs at nucleolar organizer regions that are present near the basal knob of certain acrocentric autosomes and, therefore, close to the nuclear envelope.

Besides the main nucleolus other smaller nucleoli associated with the autosomes have been described with electron microscope (Solari and Tres 1967, Solari 1969). The recent ultrastructural autoradiographic studies of the origin and development of nucleoli during meiosis have revealed that the nucleolar organizers in mouse spermatocytes lie close to the paracentromeric heterochromatin of several bivalent autosomes and that, during formation of the nucleolus, nucleolar masses originating from these chromosomal regions become associated secondarily with the sex chromosomes. The more convincing evidence that nucleolus organizer in the mouse spermatocytes does not form association with the sex chromosomes has been provided by DNA-RNA hybridization experiments in cytologicalceprations. The results of this study have revealed that in the mouse the rRNA cistrons are present in the nucleolus organizer regions of three autosomes, 15, 18 and 19 (Henderson et al. 1974). By employing the same method, the ribosomal cistrons have been found in the autosomes 3, 11 and 12 of the rat genome (Kano et al. 1976) and in the acrocentric autosomes of the D and G groups (chromosomes 13, 14, 15, 21 and 22 of the human genome) (Henderson et al. 1972, Evans et al. 1974). It has been clearly shown that the nucleoli incorporate radioactive RNA precursors during meiosis in the mouse (Kierszenbaum and Tres 1974a, b) and in the rat (Stefanini et al. 1974). The pattern of nuclear labelling at different meiotic stages has indicated that during male meiosis there is stepwise increase in the rate of rRNA transcription to a peak in mid-pachytene stage followed by a progressive decrease to a complete arrest in the most advanced stages (Stefanini et al. 1974). This inactivation of rRNA transcription in diplotene spermatocytes has been demonstrated also at the ultrastructural level by the segregation of the fibrillar and the granular components, which is believed to be a morphological expression of nucleolar inactivity (Busch and Smetana 1970). Kierszenbaum and Tres (1974a) have observed that the organization of the nucleolus during meiosis in the mouse starts with the development of the fibrillar component at the zygotene stage followed by the appearance of the granular moiety at the early pachytene stage. The delayed development of the granular component during the early stages of meiosis is believed to reflect a decreased rate of maturation of rRNA precursors.

Recent studies using a quasi-homogeneous population of spermatocytes at the pachytene stage have produced biochemical evidence of synthesis of rRNA during meiosis in the mouse (D'Agostino et al. 1976, Geremia et al. 1978, Meistrich et al. 1981). The formation of rRNA is slow in all stages and it is first seen when a 2-h pulse with [^{3}H] uridine is followed by a 6-h chase (Söderström 1976a, b). The in vivo autoradiographic studies have shown that the nucleolus becomes labelled slowly in pachytene spermatocytes (Kierszenbaum and Tres 1974a, b, Stefanini et al. 1974), which is in agreement with the biochemical findings of the slow rate of rRNA synthesis in the pachytene spermatocytes (Galdieri and Monesi 1974).

The pattern of nonribosomal RNA synthesis during different stages of mouse spermatogenesis was studied formerly with autoradiographic techniques (Monesi 1964, 1965a, 1967, 1971) and more recently with biochemical techniques (Geremia et al. 1976, 1978). The rate of incorporation is much higher during diplotene than in leptotene and zygotene stages. A similar pattern of RNA synthesis has been reported in the hamster (Utakoji 1966) and has been confirmed in the mouse by Moore (1971) and Kierszenbaum and Tres (1974a, b).

The recent detailed studies of Söderström and Parvinen (1976c) and Söderström (1976a, b) have demonstrated the synthesis of some RNA during leptotene, zygotene or early pachytene stages of rat spermatogenesis which, however, could not be demonstrated in the studies of Monesi (1964, 1965a) on the mouse. But a very high rate of RNA synthesis has been found in mid-pachytene spermatocytes of both rat and mouse (see also Geremia et al. 1977b, Monesi et al. 1978). Between the two meiotic divisions, resting secondary spermatocytes show a low level of incorporation. According to Loir (1972a, b) the highest rate of RNA synthesis in the ram takes place in the diplotene primary spermatocytes, suggesting some species differences in the rate of RNA synthesis during the meiotic prophase.

The increase of RNA synthesis in pachytene spermatocytes of rat testes has been related to the development of lampbrush chromosomes (autosomes) (Nebel and Coulson 1962, Monesi 1965a, Kierszenbaum and Tres 1974b, Glätzer 1975, Monesi et al. 1978), which may be special structures that make transcription possible in condensed chromosomes (Söderström 1976a, b, Monesi et al. 1978). The lateral loops projecting from both sides of the chromosomal axis form the sites of RNA transcription in meiosis; this has been shown with light microscopy autoradiography, indicating that the [^{3}H] uridine labelling at the pachytene stage is more concentrated along the margin than over the core of the chromosomes (Monesi 1965a). The low level of RNA synthesis during leptotene, zygotene and early pachytene stages corresponds to the formation of synapsis, the precise point-for-point alignment between homologous chromatid pairs, and the rotation of the chromosomes (Parvinen and Söderström 1976, Söderström 1976a, b). With the formation of synapsis between homologous chromosomes, the lampbrush loops start their development and meanwhile the RNA synthesis increases rapidly. The XY bivalent, which does not form a synapsis along its complete length, remains inactive during pachytene stage (Henderson 1964, Monesi 1965a, b, Monesi et al. 1978).

Actually, two types of RNA synthesis have been reported in the meiotic prophase cells by high-resolution autoradiography: an extranucleolar RNA synthesis of perichromosomal localization and a nucleolar RNA synthetic activity (Kierszenbaum and Tres 1974a, b, Monesi et al. 1978, Bellvé 1979). The nucleolar RNA synthetic activity is related to the formation of rRNA, as already discussed. The perichromosomal parts of bivalent autosomes during late zygotene and pachytene stages synthesize most of RNA, which forms an association with nascent ribonucleoprotein chains, each constituting a fine fibril with granules in the mouse and *Drosophila* (Kierszenbaum and Tres 1974a, b, Meyer and Henning 1974, Henning et al. 1974, Glätzer 1975). These fibrils are labelled with [^{3}H]uridine (Henning 1967, Kierszenbaum and Tres 1975). Kierszenbaum and Tres (1974b), studying the mouse pachytene chromosomes, have suggested that the ribonucleoprotein fibrils represent the primary product of the lampbrush loops of the chromosomes, as already discussed. They contain heterogeneous nuclear RNA (HnRNA) molecules complexed with proteins. The results of these experimental studies have indicated clearly that spermatogonia and spermatocytes are transcriptionally expressing HnRNA and pre-rRNA species (see also Kierszenbaum and Tres 1975).

A remarkable characteristic of the RNA transcribed during meiosis is its stability. Biochemical studies carried out on whole testes of the hamster (Muramatsu et al. 1968) or on isolated seminiferous tubules of the rat (Söderström and Parvinen 1976c) have demonstrated that RNA's synthesized in spermatogenic cells are mostly DNA-like, heterogeneous molecules of high molecular weight with an intranuclear turnover and a remarkable stability in comparison to the high molecular weight nuclear RNA (HnRNA) of somatic cells. Other biochemical studies have shown that RNA synthesized by mid-pachytene spermatocytes is mostly high molecular weight RNA comparable to the HnRNA of somatic cells (D'Agostino et al. 1976, Söderström 1976a, b, Geremia et al. 1976, 1977b); RNA polymerase II is responsible for its synthesis. In contrast to the rapid metabolism of HnRNA in somatic cells the HnRNA in pachytene spermatocytes is stable and remains in the nucleus until diakinesis or prometaphase, when it is released rapidly into the cytoplasm (Monesi et al. 1978). The cytoplasmic RNA present in spermatids is mostly of meiotic origin, as will be discussed in Chap. IV.

Monesi et al. (1978) have discussed in detail the biochemical aspects of RNA synthesis and metabolism in mammalian spermatogenesis, as well as the methodology developed for such studies. The study of the [^{3}H]uridine radioactivity incorporated into acid-insoluble material of the mouse has shown that the middle-late pachytene stage is the meiotic phase of maximum synthetic activity. The discrete peaks of radioactivity have been clearly distinguished in the 4S, 18S and 28S regions of the gradient as well as in the heavier part of the sedimentation gradient. The labelled HnRNA species are probably comparable to nuclear RNA molecules, including rRNA precursors.

Since a poly (A) sequence at the 3′ terminus forms a specific feature of most messenger RNA (mRNA) molecules, the determination of polyadenylated RNA has been considered to provide evidence for the presence of mRNA (Monesi et al. 1978). About 30% of HnRNA contains a poly (A) sequence, although no difference in the rate of metabolism between the poly (A)$^+$ and poly (A)$^-$ RNA has been observed. FSH has been shown to have a marked stimulatory effect on the synthesis of the poly (A)-containing mRNA component of the rat seminiferous tubules (Reddy and Villee 1975). The presence of large amounts of poly (A)$^+$ RNA molecules in the mid-pachytene spermatocytes of the mouse (Monesi et al. 1978) is consistent with the finding that protein synthesis is more active in these cells as compared with earlier and late developmental stages (Davis and Firlit 1965, O‘ Brien and Bellvé 1980b). Geremia et al. (1978) have obtained biochemical evidence of haploid gene activity in spermatogenesis of the mouse. The results have shown that pachytene spermatocytes and round spermatids synthesize both ribosomal RNA (rRNA) and poly (A)$^+$ RNA (presumptive messenger RNA) (mRNA) (see also Loir 1972a, Geremia et al. 1977b, Monesi et al. 1978, Meistrich et al. 1981). D‘Agostino et al. (1978) have observed that mouse pachytene spermatocytes in vitro can synthesize heterodispere poly (A)$^+$ RNA that is engaged in cytoplasmic polysomes. The post-meiotic synthesis of RNA ceases completely in mid-spermiogenesis after nuclear elongation in spermatids has set in, as will be discussed in Chap. IV.

Grootegoed et al. (1977a) using biochemical techniques have characterized the synthesis of RNA in isolated primary spermatocytes (pachytene stage) from rat testes. These cells were used because it was possible to isolate them in a relatively pure form. The rate of incorporation of [^{3}H]uridine into RNA in the isolated cells is constant during the first 8 – 10 h and the labelled RNA appears qualitatively comparable with the RNA synthesized in cells labelled in vivo. The RNA species synthesized from [^{3}H]uridine are mainly ribosomal RNA's and heterogeneous RNA's with a short half-life (see also Meistrich et al. 1981). A predominant synthesis of 32S ribosomal RNA has been observed, which suggests that in spermatocytes cleavage of 32S precursor RNA to 28S RNA is retarded. This 32S ribosomal RNA peak is not in RNA isolated from preparations enriched with Sertoli cells or spermatids.

Rommerts et al. (1978) have studied several biochemical parameters in isolated pachytene spermatocytes which do not show androgen receptors. An incorporation of [^{3}H]uridine into their RNA remains constant for 8 h and decreases thereafter. The relatively high incorporation into 32S rRNA, rather than 28S, is apparent from the electrophoretic profile of labelled RNA and may reflect a delay of the processing of ribosomal RNA in spermatocytes.

The results of various studies, as integrated and discussed above, have clearly indicated that high levels of both transcription and translation occur during meiosis, especially in mid-pachytene spermatocytes. Furthermore, a considerable proportion of the RNA synthesized at the pachytene stage is apparently conserved throughout spermiogenesis (Geremia et al. 1977a, b, Monesi et al. 1978). The diploid genome appears to determine the spermatozoon phenotype by directing meiotic transcription and the subsequent storage of mRNA's and/or the translated protein products, as also supported by the recent observations of O'Brien and Bellvé (1980b).

The various RNA species demonstrated in both in vivo and in vitro biochemical studies may be constituting the different types of nuage described in electron microscope studies of Russell and Frank (1978). Some of the nuage types might be long-lived RNA which is transferred from the nucleus to the cytoplasm. The ultrastructural mechanisms of the nuclear cytoplasmic transfer of substances have been demonstrated in spermatogenesis of both vertebrates and invertebrates (Danilova 1976). The products of gene transcription during meiosis appear to be related to extensive protein synthesis, which is needed during meiotic prophase not only for the biogenesis of new organelles accumulated in primary spermatocytes but also for meeting the demand of their considerable growth, a 10–15-fold increase in volume during their life span. The demand for de novo protein synthesis is supported by the incorporation of radioactive amino acids into cytoplasmic components of primary spermatocytes (Monesi 1965a, 1971). But this autoradiograph study does not provide evidence for the total complexity or the various functions of the different proteins being synthesized. Acutally, the synthetic processes in the primary spermatocyte also contribute a major portion of the protein components that are present in the mature spermatozoon (O'Brien and Bellvé 1980a, b). Until recently most of the previous studies have been concerned with defining the developmental appearance of various enzyme activities, especially enzymes expressing atypical isozymes, that can be distinguished from somatic forms. The first appearance of a wide variety of enzyme activites is closely related to the development of primary spermatocytes. Some of the RNA species may be concerned with protein synthesis at a later time in the spermatogenic process when the RNA synthesis of the spermatids has ceased but protein synthesis is still active, as recently demonstrated by O'Brien and Bellvé (1980b) (see also Loir and Lanneau 1978b). Lactate-dependent protein synthesis has been reported in primary spermatocytes and round spermatids of rat (Jutte et al. 1981a, b, Nakamura and Kato 1981, Nakamura et al. 1981).

3. *Intercellular Bridges and Plasma Membrane*

The spermatocytes are generally connected by intercellular bridges-structures (Fig. 20), as also observed between spermatogonia, and between early spermatids (Holstein and Roosen-Runge 1981). Cytoplasmic bridges result when the cleavage furrow is arrested at the spindle equator (Fawcett et al. 1959, Mahowald 1971, Holstein and Roosen-Runge 1981). Holstein and Roosen-Runge (1981) have illustrated the presence of a bundle of microtubules (diameter approximately 20 nm) within a bridge. Between the thickened plasma membranes the microtubules appear embedded in a finely granular electron-dense substance. These and similar structures represent spindle-fibres which continue to be seen for a time in the intercellular bridges (McIntosh et al. 1979). The cytoplasmic bridges are found connecting groups of germ-cells (Fig. 9). A clone derived from one stem cell forms a syncytium of cells. Syncytial connections are maintained through spermatogonial and spermatocytic stages and are dissolved only in advanced phases of spermatid differentiation. Junctions also develop between a primary spermatocyte and Sertoli cells. Apparently, these are maculae adherentes described as desmosome-like structures or intermediate junctions (see Chap. I).

The cell surface area is increased approximately sixfold concomitantly with extensive cell growth during the first meiotic prophase. Preleptotene spermatocytes are the smallest germ-cells in the mouse, having an average diameter of 7.8 μm after isolation (Bellvé et al. 1977a) with a surface area of 800 μm^2 (Romrell et al. 1976, Millette 1976). The results of these studies suggest that extensive amounts of plasma membrane must be synthesized during the first meiotic prophase. Such assembly would provide a ready mechanism for the insertion of a new surface component not present on earlier spermatogenic cells. Actually, the biogenesis of the spermatocyte plasma membrane has not been investigated with modern techniques. It is believed to undergo a profound transformation during the course of the meiotic prophase. Recently, Mollenhauer et al. (1977) have observed that during spermatogenesis in the rat, the plasma membrane of the germ-cell undergoes an abrupt change which appears as a reduction of electron density and/or a reorganization of the outermost lamella of the unit membrane. This change occurs in the developmental stage between type B spermatogonia and early spermatocytes. The origin of the new type of germ-cell plasma membrane could not be determined. However, a population of thin-membrane cytoplasmic vesicles with asymmetrical stained membranes is present in spermatogonia, before plasma membrane transformation, but is nearly absent in the spermatocytes after the transformation process. The asymmetrical membranes of these cytoplasmic vesicles could be precursor material for the new, or transformed type of germ-cell plasma membrane. The plasma membrane changes described above persist through spermatid development, and are specific to the germinal elements of the seminiferous epithelium. The relationship of these gross morphological changes in the plasma membrane to the insertion of the antigenic determinants to be discussed later in Chap. V remains to be determined more precisely, but the available observations strongly suggest that extensive surface reorganization occurs during late spermatogenesis in mammals. From the observations made for a variety of lectin receptors seen on mouse spermatogenic cells and evidence relating to the selective partitioning of plasma membrane molecules just prior to spermiation, Millette and Bellvé (190) have suggested that molecular rearrangements occur during late spermiogenesis.

McGrady (1979c) has studied the electrophysiological parameters of primary spermatocytes and early and late spermatids isolated from the seminiferous tubules of the mouse. Substantial changes are not observed in membrane potential between developing stages. Membrane potential is dependent on the both K and Na ion concentration gradient, but not on chloride gradients. The ratio of the permeabilities of PN^a/PK varies according to the extracellular concentrations of Na and K Ouabain; a specific inhibitor of $Na^+ - K^+$-activated ATPase produces a maximal reduction in membrane potential of 20%.

By now several techniques have been developed to make isolation of mammalian seminiferous cells possible (Meistrich et al. 1973, Romrell et al. 1976, Bellvé et al. 1977a, b). The detailed biochemical and biophysical analyses of spermatogenic cells in regard to their plasma membranes will greatly increase our knowledge of molecular rearrangements and their functional significance during spermatogenesis. Mouse spermatogenic cells isolated in high-purity by unit-gravity sedimentation are being used extensively to investigate biochemical changes in their various cytoplasmic and nuclear components (Millette 1979a). The correlative biochemical and autoradiographic studies have provided evidence for the localization of adenosine receptor in association with spermatocytes (Murphy et al. 1983). Ziparo et al. (1980) have made a study of surface interaction in vitro between Sertoli cells and germ-cells at different stages of spermatogenesis and discussed the possible regulative role of a somatic-cell – germ-cell interaction.

4. Cytoplasmic Organelles and Enzymes

The amount of various cytoplasmic organelles described for the spermatogonia is increased further during the meiotic prophase (Fig. 17). The Golgi apparatus is enlarged. It is now possible to study the chemistry of the Golgi apparatus of spermatogenic cells as improved methods for its isolation become available (Letts et al. 1978). The mitochondria and RNA-containing basophilic substance are distributed throughout the cytoplasm (Fig. 17). Daoust and Clermont (1955), using histochemical pyronin staining technique, have demonstrated the details of variable distribution of RNA during spermatogenesis of rat, as also recently confirmed by Itagaki and Takahashi (1977) in the mouse. The affinity of the cytoplasm for the stain is decreased gradually during meiotic prophase and is low at the beginning of pachytene. In mid-pachytene the intensity of coloration in the cytoplasm is increased. Thereafter, the cytoplasmic RNA in the cells progressively decreases during the late pachytene and diplotene stages. The increase in the mitochondria of spermatocytes towards the end of meiotic prophase (H. A. Johnson and Hammond 1963, Fawcett and Phillips 1967) corresponds to their considerable oxidative activity (Ambadkar and George 1964). Besides these various cell components, there is present a spherical chromatoid body (Fig. 17) which stains for RNA and protein (Guraya 1965). In the PAS technique, the

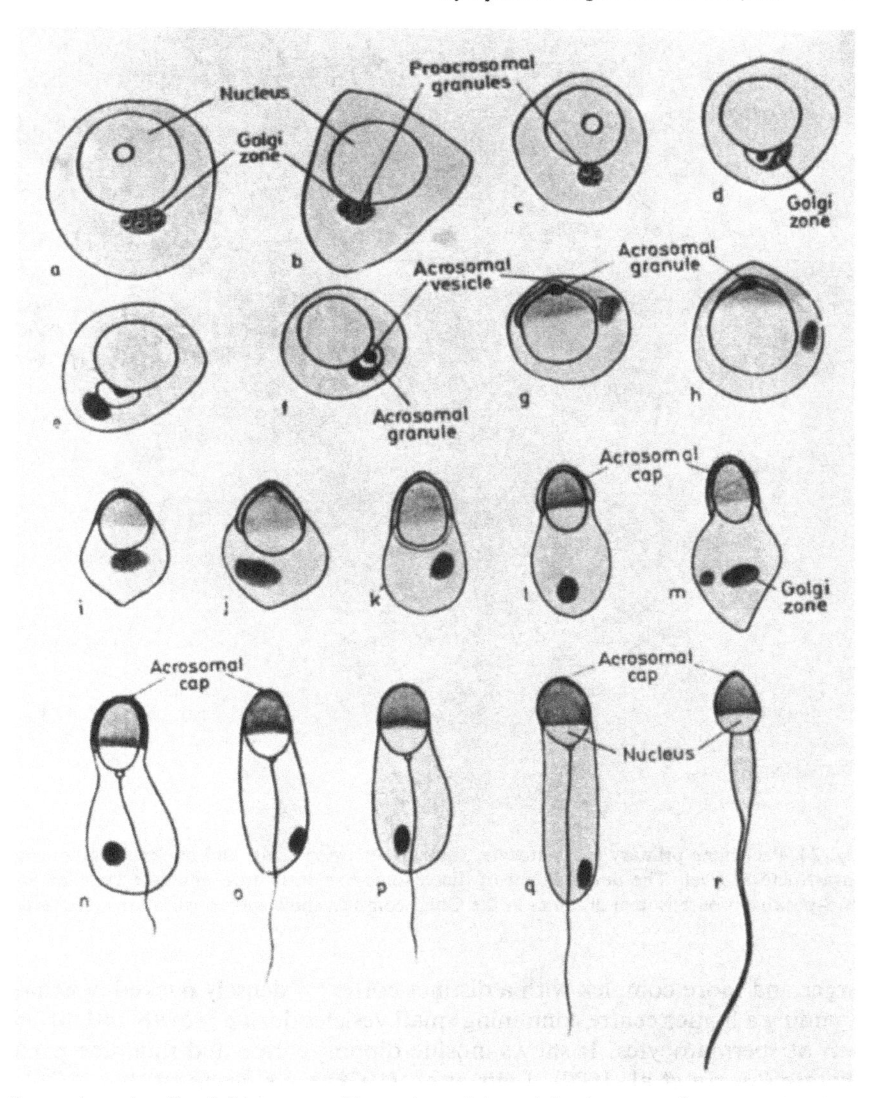

Fig. 23. Spermatogenic cells of chimpanzee, illustrating origin and development of acrosomal vesicle and granule in association with the Golgi complex and their contributions to the acrosomal cap during successive stages of spermiogenesis. Week Bouin's, pyridine extraction, periodic acid-Schiff technique. *a* and *b* primary spermatocytes; *c*–*f* young spermatids; *g*–*q* developing spermatids; *r* mature spermatozoon (From Guraya 1971a)

Golgi complex of spermatocytes shows proacrosomal vesicles/granules which consist of carbohydrate and protein (Fig. 23) (Guraya 1971a). These are associated closely with the PAS-negative proacrosomal vacuoles.

Several electron microscope studies have confirmed the presence of these various organelles (Fig. 24) (Nicander and Plöen 1969, Vilar et al. 1980, Kerr and de Kretser 1981, Holstein and Roosen-Runge 1981). The Golgi region becomes

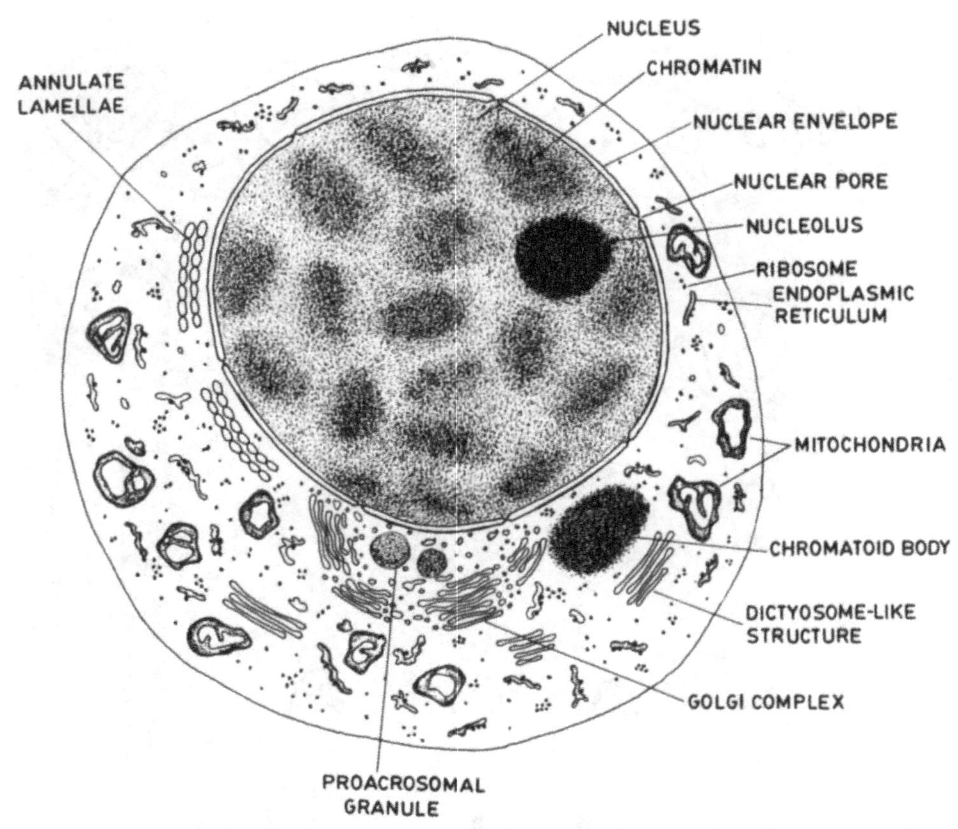

Fig. 24. Pachytene primary spermatocyte, showing its cytoplasmic and nuclear components at the ultrastructural level. The development of dictyosome-like structures, annulate lamellae and some PAS-positive proacrosomal granules in the Golgi complex show species variations (see text)

larger and more complex with a distinct cortex of densely packed lamellae, surrounding a lighter centre containing small vesicles during growth and differentiation of spermatocytes. It shows inosine diphosphatase and thiamine pyrophosphatase (Morre et al. 1980). Letts et al. (1978) have observed that a Golgi-rich fraction from late spermatocytes is enriched 15-fold in N-acetylglucosaminyl transferase, 19-fold in the fucosyltransferase and 3-fold in galactosyltransferase. These observations have suggested that late spermatocytes are highly active in glycoprotein synthesis. This is consistent with the histochemical data that have revealed that the proacrosomal granules consisting of glycoprotein are formed by the activity of the Golgi complex (Figs. 23 and 24) (Guraya 1971a), as also confirmed by electron microscope studies (Holstein and Roosen-Runge 1981). The centrioles move to a position between the Golgi complex and the nucleus. The multivesicular bodies or dense inclusions, which form an anatomical association with the Golgi complex (Mollenhauer and Zebrun 1960, Nicander and Plöen 1969) seem to correspond to the phospholipid-containing bodies of histochemical studies (Guraya 1962, 1965).

The endoplasmic reticulum, which is relatively well developed in spermatogonia, shows great diversity in the spermatocytes of different mammalian species (Fawcett and Ito 1958, Nicander and Plöen 1969, Vilar et al. 1970, Nistal et al. 1980, Kerr and de Kretser 1981, Holstein and Roosen-Runge 1981). The nuclear membrane of spermatocyte I consists of an outer membrane, a perinuclear cisterna and a membrane. In zygotene spermatocytes, flat cisternae of ER studded with ribosomes are applied to the outer membrane (Chemes et al. 1978). In these regions, there are no nuclear pores. Smith and Berlin (1977) have reported the presence of annulate lamellae in the cytoplasm of human spermatocytes (Fig. 24), recently confirmed by Nistal et al. (1980) and Holstein and Roosen-Runge (1981). These are distributed in the perinuclear region, parallel to the nuclear envelope and form single or multiple membranous profiles containing numerous annuli (50 to 60 nm in diameter), which frequently are associated with fibrillar electron-dense material. The numbers of polyribosomes dispersed throughout the ground substance of primary spermatocytes are greater than in any other germ-cell type (Fig. 24). According to Nicander and Plöen (1969), the endoplasmic reticulum in the spermatocytes of rabbit produces numerous small, mainly smooth vesicles, and also might be the source of a new organelle: numerous piles of narrow cisternae with opaque contents. These piles disintegrate late in prophase. Mollenhauer and Morre (1977) have reported the presence of dicytosome-like structures (DLS) (Fig. 24) with cylindrical intersaccular connection (microtubules) in guinea pig spermatocytes which may represent a new organelle unique to spermatocytes. But DLS and Golgi apparatus share cytochemical markers, inosine diphosphatase and thiamine pyrophosphatase (Morre et al. 1980). A distinguishing feature between DLS and Golgi apparatus is in the reaction product distribution in regard to the glutaraldehyde-resistant, NADH-ferricyanide reductase. Golgi apparatus reaction product is concentrated towards the one face of each dictyosome, while DLS reaction product usually is distributed randomly across the stacks. Human primary spermatocytes show a special development of the ER (Nistal et al. 1980) or morula-like formation of ER (Holstein and Roosen-Runge 1981). During the leptotene-zygotene stages a cisterna of ER surrounds a third or half of the nucleus closely associated with the nuclear envelope and connected to it. In the following stages of the 1st meiotic prophase, stacks of these cisternae or morula-like formations appear associated closely to the nucleus and show connections with other cytoplasmic cisternae lying free in the cytoplasm. Ribosomes and elements of endoplasmic reticulum apparently correspond to the RNA-containing basophilic substance of histochemical studies (Guraya 1965).

Mitochondria have highly dilated cristae (Fig. 24) and small particles, interpreted as mitochrondrial ribosomes in the matrix (André 1962, Nicander and Plöen 1969, Vilar et al. 1970, Kerr and de Kretser 1981). The cristae do not show the straight parallel pattern which is seen in the human spermatogonia; instead these follow an irregular course and frequently coalesce with the outer membrane, giving the mitochondria a "vacuolated" appearance. According to De Martino et al. (1979) mitochondria in diplotene spermatocytes, secondary spermatocytes and early spermatids of rat take on a rounded appearance with the inner space containing the matrix flattened against the outer membrane and the

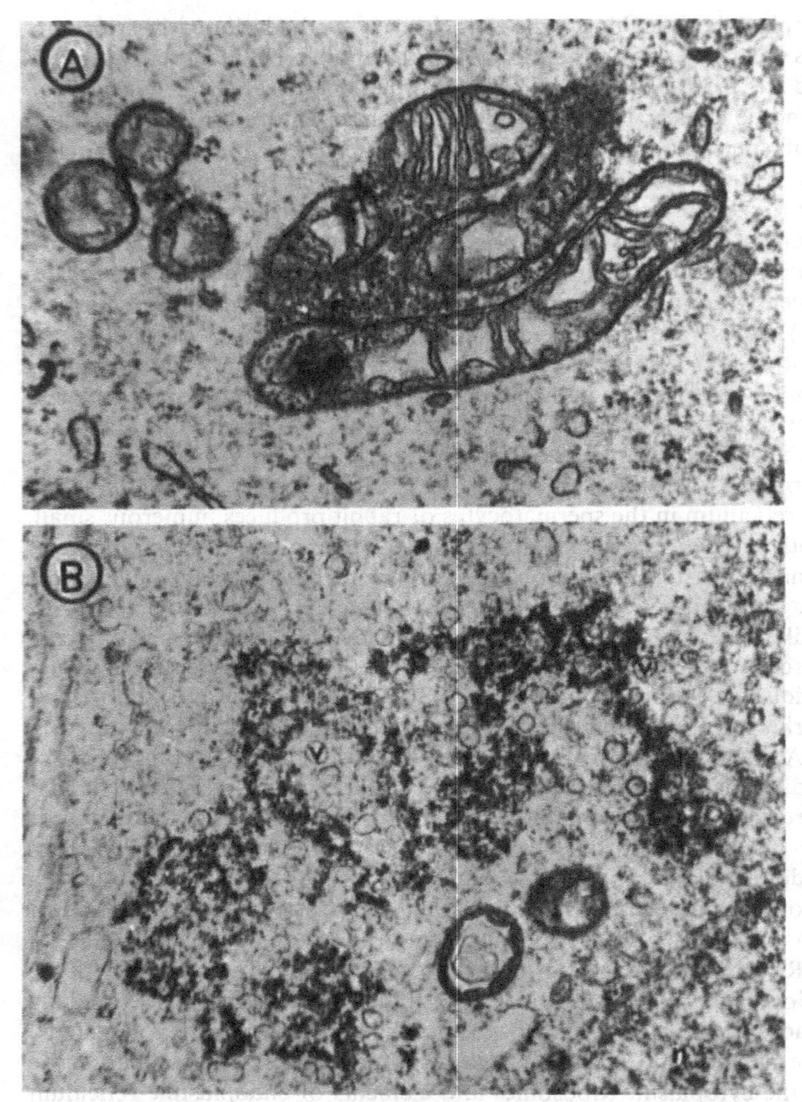

Fig. 25. A Late pachytene spermatocyte of rat, showing two mitochondrial clusters and the intermitochondrial substance (*asterisk*). **B** Late pachytene spermatocyte of rat, showing both nuclear (*n*) and cytoplasmic areas. The chromatoid body appears as irregularly shaped, anastomosing strands of electron-dense material. *Arrowheads* indicate the small fibrils which form these strands. Numerous small vesicles (*v*) are spaced along the strands (From Russell and Frank 1978)

intracistral spaces considerably swollen (condensed mitochondria) (Fig. 24). Biochemical studies have demonstrated a high oxidative capacity coupled with a high level of phosphorylation, suggesting that this condensed state is an expression of functional activity.

Some mitochondria develop an assoication with dense extra-organellar substance (Fig. 25), which has been described in the spermatocytes of different mammalian species (André 1962, Nicander and Plöen 1969, Fawcett et al. 1970, Burgos et al. 1970a, Fawcett 1972, Eddy 1974, Russell and Frank 1978, Kerr and de Kretser 1981). Such mitochondrial clusters associated with the intermitochondrial substance have also been described in the cytoplasm of oocytes (Fawcett 1972, Guraya 1974, 1985).

The presence of some enzymes, such as the sorbitol dehydrogenase, 5′-ribonucleotide phosphohydrolase, uridine diphosphate phosphohydrolase, acid phosphatase isozyme IV, α-glycerophosphate dehydrogenase, t-hexokinase, carnitine acetyl transferase, lactate dehydrogenase (LDH-X), and cytochrome Ct, is believed to be related to the development of primary spermatocytes in the seminiferous epithelium (Vanha-Perttula 1978, Bellvé 1979). Conti et al. (1979) studied adenosine 3′, 5′-cyclic monophosphate-dependent, protein kinase activity in differentiating germ-cells of mouse testes, which include middle-late pachytene spermatocytes, round spermatids (steps 1–8 of spermiogenesis) and elongating spermatids (steps 9–13). It is present primarily in the soluble fraction of the cell homogenate. Various other enzymes have also been studied with regard to their localization in cells constituting testicular tissue and different stages of spermatogenesis (Bilaspuri and Guraya 1982, 1983a, b, e, Redi et al. 1983, Vanha-Pertulla et al. 1985). The precise roles of enzymes in spermatogenesis need to be defined in future studies.

Histochemical studies on alkaline phosphatase in mature mammalian testes have provided inconsistent results, with staining in germinal cells and spermatozoa ranging from absent in hamster, rabbit, and bat to positive in rat, mouse, dog, guinea pig, and humans (see McComb et al. 1979). Kornblatt et al. (1983) have characterized and localized alkaline phosphatase in rat testes. It has the properties of the liver-bone-kidney isozyme. The vast majority of this alkaline phosphatase activity is not present in spermatocytes and early spermatids nor, presumably, in late spermatids. But, it is present in the nongerminal Sertoli cells. The presence of alkaline phosphatase in plasma membranes purified from testes of adult rats is shown. Its form does not change during spermatogenesis and appears to be the same as reported for human testes (C. H. Chang et al. 1980). The decrease in electrophoretic mobility after treatment with neuraminidase has suggested that the testicular alkaline phosphatase is a glycoprotein (Kornblatt et al. 1983), as are most of the previously reported mammalian forms of this enzyme (McComb et al. 1979). The enzyme 5′-nucleotidase is present in both germinal cells and nongerminal cells of the rat testes (Kornblatt et al. 1983). Its specific activity is reported to be higher in nongerminal cells than in germinal cells (Shirley and Schachter 1980, Millette et al. 1980).

Alpha chlorohydrin, a male chemosterilant, affects most of the testicular enzymes severely, as revealed by correlative histochemical and biochemical studies on the normal and treated rat testes (Guraya and Gill 1977, 1978, Gill and Guraya 1980, S. Kaur and Guraya 1981a, b, 1982a, 1983). Enzymes studied include various hydrolytic, glycolytic enzymes, dehydrogenases, oxidases, enzymes of hexose monophosphate shunt pathway, steroid dehydrogenases, NAD and NADP diaphorases, RNAase, proteinase and DNAase etc. Alpha chlorohydrin changes the levels of various dehydrogenases, oxidases and hydrolytic enzymes, but not of glucose-6-phosphate dehydrogenase, 6-phosphogluconate dehydrogenase, $^5\Delta$-3β-HSDH, NADP diaphorase or DNAase.

Some enzymes are associated mainly with the germ-cells and others chiefly with Sertoli cells. γ-Glutamyl transpeptidase, for example, has been suggested as a convenient Sertoli cell marker; this enzyme may be operative in the synthesis of specific proteins secreted by Sertoli cells (Krueger et al. 1974, Lu and Steinberger 1977; Chap. I). Isolated rat spermatocytes and round spermatids utilize lactate, synthesized by Sertoli cells in vivo, as the main source of energy (Nakamura et al. 1981, Robinson and Fritz 1981, Jutte et al. 1981b). Some of the enzymes mentioned are activated by specific hormones, for instance the adenylate cyclase and cyclic AMP-dependent protein kinase by FSH (Kuehl et al. 1970, Means et al. 1974). On the whole, testicular enzymes tend to occur in multimolecular forms, but exceptions are known as in the case of cyclic nucleotide phosphodiesterase. Rat tissues contain seven forms of this enzyme, but the rat testes have two, and only one is testis-specific. In the maturing rat testes the latter appears at 40 days of age, much later than most lysosomal enzymes, coinciding with the elongation of spermatids. Its activity reaches peak level at 50 days, parallel with the formation of spermatozoa and spermiation.

5. Nuage and Chromatoid Body

Fawcett et al. (1970) have drawn attention to the similarity in fine structure of the intermitochondrial material of spermatocytes (Fig. 25) and the substance comprising the chromatoid body of spermatids. From this resemblance, these authors believe that the chromatoid body arises by dissociation of mitochondria from the clumps of dense material and the subsequent aggregation of this material into a single irregular mass forming the chromatoid body. Nicander and Plöen (1969) have observed that the clusters of mitochondria eventually disperse, and their cores of dense intermitochondrial substance (Fig. 25), possibly containing ribonucleoprotein, coalesce into a large chromatoid body, which has been described in spermatocytes of different mammalian species (Nicander and Plöen 1969, Susi and Clermont 1970, Fawcett et al. 1970, Comings and Okada 1972, Eddy 1974, Russell and Frank 1978). Russell (1977a, b) has observed abundant nuage within the cytoplasm of most spermatocytes of adult rats, occurring in at least six different types, as characterized by morphology, distribution, and association with other organelles (Russell and Frank 1978). These include 70 – 90 nm spherical particles, sponge bodies, intermitochondrial substance (Fig. 25), 30 nm particles, chromatoid body (Fig. 25) and definitive chromatoid, etc. The temporal appearance and the dissociation of some nuage types have been followed in conjunction with other events of the spermatogenic cycle. These results have indicated that all types of nuage behave in a dynamic way during the spermatogenic cycle. The morphological heterogeneity of nuage within a single species has been reported previously (Eddy 1974). But it has not been shown previously that six types of nuage may be present in a single cell type, as clearly demonstrated by Russell and Frank (1978). Nuage and chromatoid body are believed to be separate organelles (Söderström 1981).

Actually, very divergent views have been expressed in regard to the origin of the chromatoid body in the spermatogenic cells of mammals and other vertebrates (see Comings and Okada 1972, Yasuzumi 1974). The chromatoid body has been shown to appear first in pachytene spermatocytes at Stage 8 (Leblond and Clermont 1952a) of the rat seminiferous epithelial cycle (Russell and Frank 1978). During the pachytene and diplotene stages of the meiotic prophase it grows, and specific intranuclear membrane modifications may be involved in the transport of material from the nucleus to the chromatoid body (Russell 1977b). From the close relationship of the chromatoid body to the nucleus, its superficial resemblance to a nucleolus and its nucleic acid content (Daoust and Clermont 1955, Sud 1961, Guraya 1965, Yasuzumi 1974), most of the previous workers believe that this structure originates by reaggregation in the cytoplasm of material that passes out of the nucleus through the pores (Maillet and Gouranton 1965, Comings and Okada 1972, Söderström 1979). But Fawcett and co-workers (Fawcett et al. 1970, Eddy 1970, Fawcett 1972) have cautioned against hastily reaching any conclusion about the origin of the chromatoid body from the nucleus. In their view, the finely filamentous material occupying the interstices among mitochondria aggregates to form the chromatoid body (Nicander and Plöen 1969).

Acutally, the problem of the origin of the chromatoid body should be solved with the application of biochemical, cytochemical, autoradiographic and other experimental methods in future studies. However, Russell and Frank (1978) have observed recently that the chromatoid body first appears in mid-pachytene spermatocytes of the adult rat (Fig. 25). By late diplotene and during the first meiotic division it loses its prominence such that its eventual fate could not be determined. Its material has been observed to be dispersed in the cytoplasm as small 30 – nm particles. Large dense bodies (0.5 μm) develop de novo in the cytoplasm of newly formed secondary spermatocytes. These appear scattered throughout the cell and in association with the mitochondria. During early spermiogenesis (Söderström and Parvinen 1978) these bodies appear to coalesce to form the large definitive chromatoid of a type similar to that known to be present during spermiogenesis. According to Söderström (1979) the formation of the chromatoid body can be divided into three stages: mitochondrial, condensing, and maturing. During meiotic prophase pores appear in the nuclear envelope close to the chromatoid body. Material that participates in the formation of the chromatoid body may pass through these pores from the nucleus to the cytoplasm. This substance is initially placed close to mitochondria, where it forms the intermitochondrial chromatoid material of Fawcett and co-workers. During the late pachytene stage the intermitochondrial material begins to accumulate and forms several condensing stage chromatoid bodies that are present in the cells during the meiotic divisions. These appear distributed uniformly within the spermatid forming the mature chromatoid body.

In regard to the chemical nature of the chromatoid body, Sud (1961) reported that it does not stain with the alkaline fast green method for basic proteins, but it does stain with acid dyes and with the Sakaguchi reagent for arginine. He concluded that it contains protamine but no histone. In contrast, according to Vaughn (1966) cytochemical, autoradiographic and microspectrophotometric studies indicate that the chromatoid body contains histone-like, basic proteins rich in lysine and having a lesser amount of arginine. He found no evidence for protamine in this body, and reported that it is rich in RNA but contains no DNA.

Electron microscopic histochemical techniques have failed to confirm the presence of RNA in the chromatoid body (Eddy 1970). Glycoprotein also is believed to be present in the chromatoid body (Susi and Clermont 1970, Yasuzumi et al. 1970a, b).

C. Morphology, Cytochemistry and Biochemistry of Secondary Spermatocytes

Secondary spermatocytes, which are smaller than the primary spermatocytes and larger than young spermatids, show the same components as those described for the primary spermatocytes. Secondary spermatocytes or spermatocytes II exist for only a brief span of time before they undergo the second meiotic division (Ortavant 1958, Heller and Clermont 1964), and thus rarely are seen. The nucleus of secondary spermatocytes in buffalo shows chromatin granules of various sizes which lie closer to the nuclear envelope (Bilaspuri and Guraya 1980a), as also described for the bull (Knudsen 1954, 1958). But in goat and ram, these are distributed throughout the nucleus (Bilaspuri and Guraya 1984b, 1986a), as also reported by Ortavant (1958) for the ram. Fine chromatin granules connected by a network of fine filaments are also seen in the nucleus of secondary spermatocyte. Secondary spermatocyte contains diploid amount of DNA which will be reduced to the haploid amount when each secondary spermatocyte forms two spermatids. According to Vilar et al. (1970) the nucleus of secondary spermatocytes in the human testis is round with a rather homogeneous and finely granular aspect; the nucleolus is very inconspicuous and is absent much of the time. Three or more nucleoli per nucleus may be seen (Holstein and Roosen-Runge 1981).

RNA synthesis has been observed in the secondary spermatocytes of the ram (Loir 1972a, b); some RNA synthesis also occurs in the mouse (Monesi 1964). But no RNA synthesis could be demonstrated in the hamster. These observations suggest some metabolic differences between the secondary spermatocytes of different species of mammals, which must be extended and confirmed in future studies of their enzyme systems. There is no synthesis of DNA in the secondary spermatocytes. In the second meiotic division of buffalo, goat, and ram, the metaphase chromosomes resemble those of the first anaphase (Bilaspuri and Guraya 1980a, 1984b, 1986a), as also reported for the bull (Knudsen 1954, 1958) and ram (Ortavant 1958). However, the size of the metaphase plate of the second meiotic division is smaller than that of the first meiotic division. It is a way for identifying whether a metaphase is from division I or division II. The meiotic prophase-I is relatively slow (or of long duration) division in comparison to other phases of meiosis-II. Intercellular and karyoplasmic bridges can be seen between the secondary spermatocytes (Holstein and Roosen-Runge 1981).

The endoplasmic reticulum of secondary spermatocytes is made of rosaries of small, flat, saccular profiles which are arranged concentrically in relation to the nucleus (Vilar et al. 1970, Holstein and Roosen-Runge 1981). The limiting membranes of these saccules are agranular and tortuous. In comparison to human primary spermatocytes, the number of cisternae of endoplasmic reticulum joined to the nucleus is decreased (Nistal et al. 1980), while the number of free cisternae is increased. Annulate lamellae are conspicuous in secondary spermatocytes.

Holstein and Roosen-Runge (1981) have observed areas of electron-dense granules which are regularly present in secondary spermatocytes of humans between flat cisternae of the ER. The diameter of the granules is 15 to 20 nm. Their nature and function need to be determined. The round mitochondria scattered among the endoplasmic reticulum show the same vesiculization as that described for the primary spermatocytes. Several Golgi complexes are seen around an area containing proacrosomal granules (Holstein and Roosen-Runge 1981).

The diploid secondary spermatocytes, on passing through the second meiotic division (Courot et al. 1970, Holstein and Roosen-Runge 1981) give rise to young spermatids (Figs. 5 and 19). After a short interphase (interkinesis), with no intervening DNA synthesis, the second meiotic division takes place. Its phases are typical of any cell division, which usually are designated as pro-, meta-, ana- and telo-phase II. One chromatid from each pair migrates to each daughter nucleus. The young spermatids resemble the secondary spermatocytes morphologically but have a smaller nuclear diameter.

Chapter IV
Spermatids and Spermiogenesis

The two maturation divisions of each spermatocyte result in four haploid cells, spermatids. These differentiate into spermatozoa, a process called spermiogenesis which ends when the spermatozoa are released from the seminiferous epithelium. After the haploid round spermatids are formed, their nuclear and cytoplasmic components undergo a complex series of morphological, histochemical and biochemical changes ending with the production of highly differentiated and specialized germ-cells, the spermatozoa, which become "free" cells (Fig. 17). The spermatozoon shows a distinct architectural polarity more extensive than any observed in somatic cells. The various alterations of differentiating spermatids occur in the absence of DNA synthesis or cytokinesis and are known to be accompanied by conspicuous changes in plasma membrane components (Chap. V). Actually, mammalian spermatozoa develop many unusual characteristics at the cell surface, which will be discussed in Chap. X.

The recent electron microscope, cytochemical and autoradiographic studies have increased our knowledge about the development and differentiation or behaviour of each component of the spermatid during spermiogenesis when three major developmental processes occur: (1) the condensation of the nucleus, (2) the formation of the acrosome and (3) the development of the flagellum. These processes occur concomitantly and are difficult to show and describe synoptically. Clermont (1963) has classified Sa, Sb_1, Sc, Sd_1, Sd_2 spermatids in relation to developmental stages of the acrosome and the nucleus in man. The evolution of the acrosomal material, as revealed by PAS test, has been one of the criteria used to identify the steps of the spermatid differentiation in light microscopy. Four main phases are distinguishable. These include the Golgi, cap, acrosome and maturation phases, which show the same general pattern but exhibit some variations in different species (Clermont 1967, Guraya and Bilaspuri 1976b, Bilaspuri and Guraya 1984b, 1986a). Based on the development of the acrosome, nucleus and tail structures, Holstein (1976) has distinguished eight stages of spermatid differentiation in man (Fig. 26). The process of spermatid differentiation as well as the structure of mature spermatozoa have been studied extensively with transmission and scanning electron microscopy (see Thibault 1969, Burgos et al. 1970a, Vilar et al. 1970, Fawcett and Phillips 1970, Vilar 1973, Dalcq 1973, Pedersen 1974a, Phillips 1974, Bustos-Obregon et al. 1975, Gould 1980, Menger and Menger 1981a, Holstein and Roosen-Runge 1981, Courtens 1984, Bae 1984). But divergent views have been expressed about the germ-cells' morphological reorganization in the seminiferous epithelium. The normal orientation and transfer of rat spermatids in the seminiferous epithelium during their differentiation have

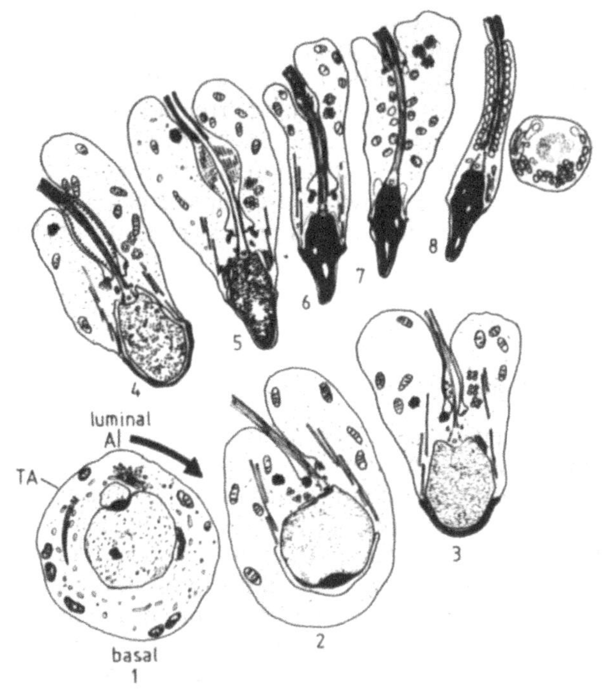

Fig. 26. Semischematic drawing of human spermatid differentiation. According to the development of the acrosome, the condensation of the nucleus, and the development of tail structures eight stages can be distinguished. Drawing based on electron micrographs (Redrawn from Holstein and Schirren 1979)

been attributed to two kinds of motility of the different components in the same cell type, which include primary flagellum and cytoplasmic protrusions at the apical regions of the spermatids (Walt 1981, Walt and Hedinger 1983). The motility due to cytoplasmic protrusions at the apical regions is supported by the presence of myosin in early spermatids of the rat (Walt et al. 1982, Walt and Hedinger 1983). Various electron microscope studies have also revealed that Sertoli cells regularly undergo alterations in shape and internal organization, believed to be partially responsible for many of morphogenetic events of the germ-cells (Fawcett 1975a, 1979, Gravis 1979, Russel 1977c, 1980b). Vogl et al. (1983a) have observed that Sertoli cells of the ground squirrel show very conspicuous changes in structure during spermatogenesis, which are correlated with alterations in the morphology and position of germ-cells within the seminiferous epithelium, and with the presence, in the Sertoli cells, of greatly developed cytoskeleton consisting of microtubules, filaments, smooth endoplasmic reticulum, etc. Newly elongated spermatids are situated in recesses that project deep into the bodies of the Sertoli cells (Fig. 10 and 11). As spermiogenesis continues, the recesses, and hence the attached spermatids, are shifted apically and become positioned at the ends of elongate columns, or stalks of cytoplasm that develop from the apices of the Sertoli cells. These stalks not only suspend the spermatids in the tubule lumens but also seem to be involved in changing spermatid orientation from a vertical to horizontal position relative to the tubule wall. Sertoli cell microtubules are believed to play a role in moving germ-cells through the seminiferous epithelium (Russell 1977c) and determining the shape of spermatid heads (Fawcett 1979). Experiments with microtubule disrupting agents support the suggestion that Sertoli cell microtubules may be essential for the normal development of germ-cell shape and germ-cell's morphological reorganization or orientation (Handel 1979). Vogl et al. (1983b) have also suggested that Sertoli cell microtubules in the ground squirrel are necessary for the normal development and translocation of spermatids in the seminiferous epithelium. This suggestion is supported by the observation that treatment with colchicine appears to prevent spermatids from moving to an apical position. But the mechanism of microtubular involvement in this process needs to be determined.

A. Development of Spermatid Nucleus

1. Chromatin Condensation and Associated Changes in Nucleoproteins

The nucleus of early spermatids, which is smaller than that of secondary spermatocytes, is spherical in shape (Fig. 17). The chromatin bodies of various sizes are scattered on a filamentous network. These stain for DNA and protein (Fig. 27). The early spermatids at steps 1 – 9 of spermiogenesis, in rat (Lalli 1973) and buffalo, goat and ram (Ortavant 1959, Bilaspuri and Guraya 1986b, c) exhibit a fine filamentous network and fine granular material (Fig. 27). Zibrin (1971) using the electron microscope has reported fine filaments in the early spermatids of bull. But he has made no mention of granular material. Although no DNA is synthesized during spermiogenesis, high DNA polymerase activity can be demonstrated in germinal cells (Hecht et al. 1979). There is a possibility of repair DNA synthesis (Bellvé 1979). One or two small spherical nucleoli staining for RNA and protein are present in the spermatids of marmoset and chimpanzee (Fig. 17) (S. S. Guraya, unpublished observations). Krimer and Esponda (1979) have observed that the nucleoli of young spermatids of mice show a very special shape resembling a padlock in which three different areas can be distinguished (Figs. 28 and 29): (a) a compact zone corresponding to the fibrillar component, (b) the granular component, and (c) a fibrillar centre of low density. Fibrillar and granular components usually appear segregated. Both fibrillar and granular components show a positive reaction after silver impregnation (See more details in Czaker 1984, 1985a, b). The fibrillar centres appear to correspond to the nucleolar organizer. Nucleolus-associated structures are reported under various names in male germ-cells (spermatocytes and early spermatids) of several species of mammals (Schultz et al. 1984). Recent studies have shown that nucleolus-associated "round body" in rat spermatocytes and early spermatids consists of non-

Fig. 27. Distribution of DNA in spermatogenic cells during I – XIV stages of seminiferous epithelial cycle from buffalo testes fixed in Zenker-formol and stained with Feulgen reaction/methyl green. Each *column* (*I – XIV*) deals with the distribution of DNA in spermatogenic cells in a given cellular association classified with PAS-haematoxylin technique. The *Arabic letters* show different steps of spermiogenesis. *M* denotes the stage in which spermatogonial mitosis takes place. *Ao* Ao spermatogonia; A_1, A_2, A_3 different types of A spermatogonia; B_1, B_2 B_1 and B_2 spermatogonia; D_1 early diplotene; D_2 late diplotene; *GL* granular leptotene; *In* intermediate spermatogonia; *L* leptotene; *M* stage for mitosis; *PL* preleptotene; P_1 early pachytene; P_2 late pachytene; *1 – 18* different steps of spermiogenesis (Redrawn from Bilaspuri and Guraya 1986c)

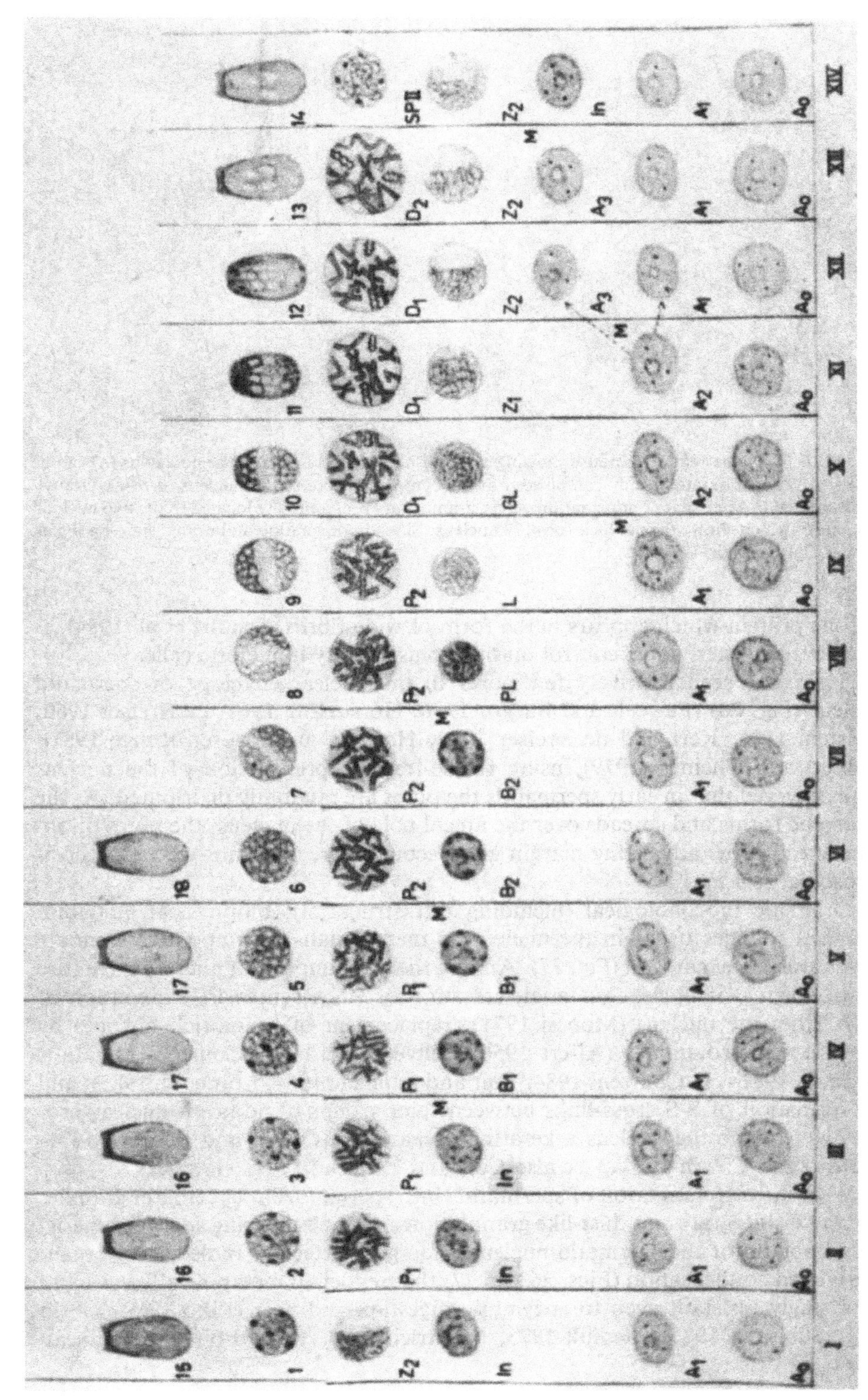

Fig. 27

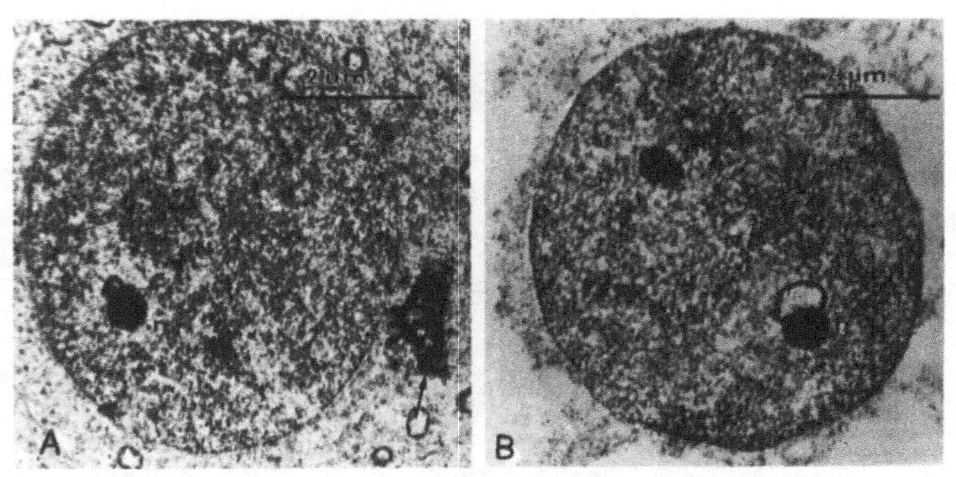

Fig. 28. **A** Early spermatid of mouse showing a round nucleus (*N*) in which the nucleolus (*n*) can be observed. The chromatoid body can be seen in the cytoplasm close to the nucleus (*arrow*). Uranyl-lead staining; **B** After silver impregnation, silver granules are preferentially located in the two nucleoli (*n*), one of which shows the padlock form. *N* nucleus. Silver impregnation and uranyl staining (From Krimer and Esponda 1979)

histone protein which appears in the form of wide fibrils (Schultz et al. 1984). It is believed to exert some control on nucleolus activity in meiotic cells.

There are seen relatively few pores in the nuclear envelope of spermatid nucleus (Fig. 28) (Fawcett and Burgos 1956, Horstmann 1961, de Kretser 1969, Holstein 1976, Kerr and de Kretser 1981, Holstein and Roosen-Runge 1981). Fawcett and Chemes (1979), using freeze-fracture preparation of the nucleus have observed that in early spermatids the pores are randomly distributed. As the acrosome forms and spreads over the apical pole of the nucleus, the pores disappear ahead of its advancing margin and become more concentrated in the post-acrosomal region.

Dramatic morphological (including ultrastructural), biophysical, and biochemical changes occur in the nucleus as mammalian spermatids differentiate into mature spermatozoa (Fig. 27). Among the most important nuclear events are the packaging of nuclear material, cessation of RNA synthesis, elimination of RNA from the nucleus (Monesi 1971), replacement of lysine-rich histones by arginine-rich protamines (Alfert 1956, Bellvé et al. 1975, Courtens and Loir 1975a, b, 1981a, b, Courtens 1984, Loir and Lanneau 1984, Loir et al. 1985), and the formation of S-S cross-links between thiol groups of adjacent nuclear proteins which give the nucleus a keratinoid character (Calvin and Bedford 1971, Bedford and Calvin 1974a, see also Loir and Lanneau 1978a, 1984).

With the differentiation of spermatids into spermatozoa, chromatin granules begin to disintegrate into dust-like granulations. Actually, during spermiogenesis, the chromatin of the spermatid nucleus undergoes extensive molecular reorganization and condensation (Figs. 26 and 27) that renders it metabolically inert and surprisingly resistant even to enzymatic digestion and heat (Pikó 1969, Zibrin 1971, Subirana 1975, Gledhill 1975, Meistrich et al. 1976, Loir and Lanneau

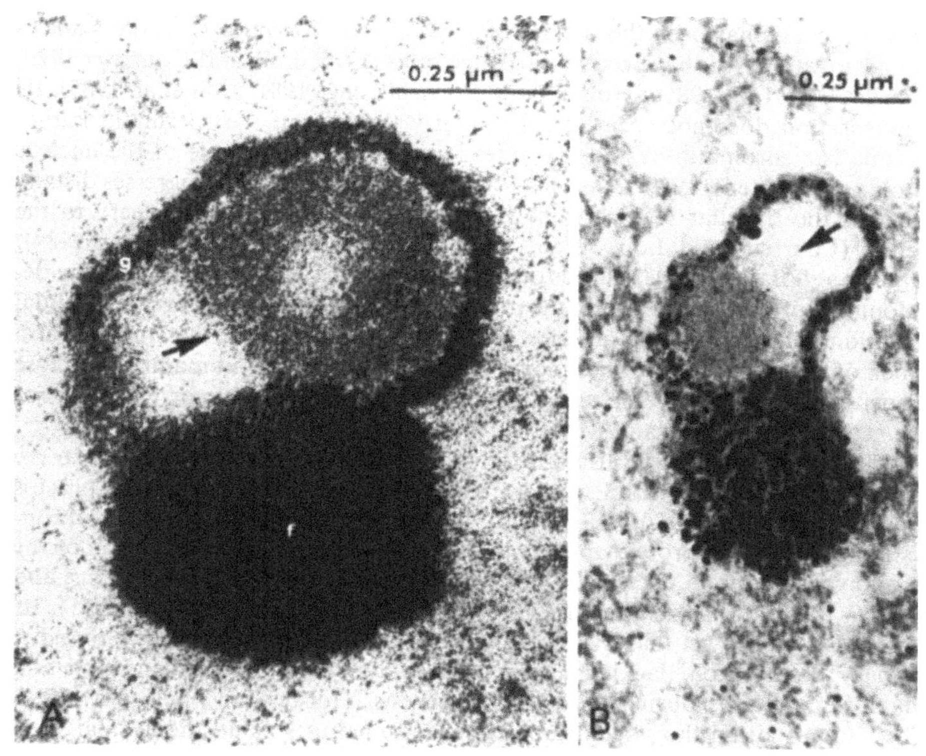

Fig. 29. A A padlock-like nucleolus of mouse spermatid showing the three components; the fibrillar sphere (*f*), the granular area (*g*), and the fibrillar centre (*arrow*). Uranyl-lead staining; **B** Nucleolus of a young spermatid after silver impregnation. The fibrillar centre (*arrow*) is free of granules. Silver impregnation and uranyl staining (From Krimer and Esponda 1979)

1978a, b, 1984, Monakhova and Abbas 1979, Loir and Courtens 1979, Lalli and Clermont 1981, Courtens 1984, Bilaspuri and Guraya 1986c). Meanwhile there also occur changes in basic chromosomal proteins (Kumaroo et al. 1975, Coelingh and Rozijn 1975, Subirana 1975, Gledhill 1975, Loir and Lanneau 1978a, b, Rodman et al. 1979, Mayer et al. 1981, Loir et al. 1985).

The packaging of nuclear material is achieved by aggregation of chromatin granules or threads depending upon the mammalian species (Fig. 26). Lalli (1973), using the electron microscope, has observed the formation of coarsely filamentous material which progressively aggregates to form the electron-dense globular clumps in steps 11–13 of rat spermiogenesis. Following the partial condensation of chromatin from steps 11–14, the nucleus continues to condense during the following steps until the end of spermiogenesis (Lalli and Clermont 1981). But a partial condensation of chromatin in buffalo, goat, and ram occurs during steps 10–12 of spermiogenesis (Fig. 27) (Bilaspuri and Guraya 1986b, c), meanwhile there develops a lightly-stained central zone in the condensing nucleus, which appears to correspond to the equatorial zone of Daoust and Clermont (1955) in the rat. Actually, the course of condensation of chromatin during spermiogenesis has been the subject of great interest for various workers and is found to vary with the species (see Bellvé et al. 1975, Loir and Courtens 1979, Courtens and Loir 1981b, Holstein and Roosen-Runge 1981, Loir et al. 1985). Its condensation does not occur simultaneously throughout the spermatid karyoplasm. The chromatin begins to condense at the acrosomal pole of the nucleus (Loir and Courtens, 1979, Courtens 1984). Then condensation progresses distally towards the flagellar region, while also proceeding from the periphery to the centre (Courtens and Loir 1975a, Loir and Courtens 1979, Courtens 1984). Dooher and Bannett (1973) have found also that the condensation during the development of the sperm head in the mouse proceeds centrifugally and in a caudal direction beginning in the anterior third of the nucleus. The heterochromatin block of one chromosome that moves to the anterior pole and maintains a close association with the developing acrosome is believed to direct morphogenesis of the acrosome (Schmidt and Krone 1976).

The chromosomes establish attachments to the nuclear envelope prior to the onset of chromosomal condensation, which facilitate the condensation process (Schmidt and Krone 1976, Bellvé 1979). Relatively more chromatin material consisting of DNA and protein appears to accumulate in the posterior region of the developing spermatid nucleus of buffalo, goat, and ram (Fig. 27) (Bilaspuri and Guraya 1986b, c), as also reported for the rat (Daoust and Clermont 1955). Its significance is not yet known. Baccetti et al. (1977) have observed that rounded human spermatozoa, lacking an acrosome, show variable condensation and shaping of nucleus during their morphogenesis (Holstein and Schirren 1979). They do not attain the same degree of nuclear maturity as do the normal spermatozoa.

The morphology of the karyoplasmic condensation is species-specific (Horstmann 1970, Courtens 1984, Holstein and Roosen-Runge 1981, Loir and Courtens 1979, Courtens and Loir 1981b) but its relation to the structure of the chromosomes continues to remain unknown. Actually, the density of chromatin in mature mammalian spermatozoa hinders any assessment of chromosome order. Weaker and Cameron (1977) have observed that chromatin fibres of the male gamete nucleus in armadillos (*Dasypus novemcinctus*) contain fibrils. Measurements indicate that chromatin fibre diameter and the number of fibrils within the fibre increase, whereas the diameter of the fibrils decreases during spermiogenesis. The fibrils possibly represent DNA-protein complexes. The increased fibre diameter and the increased number of fibrils possibly correlate with the formation of larger fibres during the condensation of the spermatid nucleus, whereas the decrease in fibril diameter is believed to be correlated with the replacement of histones with new, low-molecular weight, acid-soluble proteins. Loir and Courtens (1979), using electron microscopy, cytochemistry, and treatment of isolated nuclei with chemical and mechanical agents, have investigated nuclear changes as a function of ram spermatid differentiation. In round nuclei, chromatin consists of intertwined knobby fibres ~21 and 12 nm in diameter. In flattening nuclei, these fibres are progressively changed into smooth filaments at least 2.5 – 3.3 nm thick. This change is believed to involve somatic histone removal which occurs in these nuclei. The packaging of chromatin is carried out by aggregation of smooth filaments into large, (30 nm) controlled threads which finally coalesce to form a homogeneous mass, see also Loir et al. 1985. In flattened nuclei a small fraction of chromatin (basal knobs) does not undergo complete packaging and retains both cytochemical properties and resistance to disruption similar to chromatin in flattening nuclei (Courtens 1984). Correlations of changes in nuclear morphology, ultrastructure, and resistance to chemical and physical degradation with changes in nucleoproteins have suggested that the cystine-containing, spermatid-specific proteins may play an important role in chromatin-structure organization, shaping and increasing stabilization of the nucleus (Mayer et al. 1981, Loir et al. 1985). The sperm-specific protein promotes terminal packaging and stabilization of chromatin. From the results of cytochemical studies of nuclear changes in boar, bull, goat, mouse, rat and stallion spermatids, Courtens and Loir (1981b) have suggested that species-specific differences occur in the way the chromatin is condensed, the presence and development of a peripheral layer of dense chromatin, the way lysine-rich nucleoproteins are lost from nuclei, and the staining abilities of nuclei of old spermatids (Courtens 1984).

In the later stages of spermiogenesis, no discrete chromatin bodies are seen as the nucleus becomes more chromophilic and stains intensely and homogeneously for DNA and protein (Fig. 27) (Bilaspuri and Guraya 1986c). Clarke et al. (1980) have made a study of Feulgen-DNA changes in the germ-cells of the male vole (*Microtus agrestis*) during their development histochemically, as also reported by Bilaspuri and Guraya (1986c) for buffalo, goat and ram (Fig. 27). The quantity of Feulgen DNA in round spermatids is one quarter of that in primary spermatocytes. The amount increases slightly in elongated spermatids, and then declines in spermatozoa from the testis, epididymis and vas deferens, below the level of round spermatids. Similar results were also obtained by Esnault et al. (1974) for the ram. This suggests that there is a change in cell DNA content during spermiogenesis rather than in the changing level of Feulgen-stainability of a constant DNA level. Almost complete loss of Feulgen-stainable material occurs in spermiogenic cells and may be due to depurination and elimination of Feulgen-reactant aldehyde groups. Morphologically, nuclear maturation involves condensation of the nucleoplasm, accompanied by moulding to a shape characteristic of the species. However, one or two clear spaces or vacuoles, which give negative reactions with histochemical techniques used for the detection of DNA, RNA, protein, carbohydrate and lipids (S. S. Guraya, unpublished observations), develop in the greatly condensed nucleus of maturing spermatid of the chimpanzee (Fig. 17) and man (Fawcett and Burgos 1956, Horstmann 1961, Holstein 1976). These are not seen in the corresponding differentiating spermatids and spermatozoa of the marmoset (S. S. Guraya, unpublished observations). Czaker (1984, 1985a, b), using electron microscopical silver staining technique, has observed that as chromatin condensation proceeds, the fibrous structures of the nucleolus decrease in size and density and disappear, leaving a space corresponding to the so-called "nuclear vacuole". As the deoxyribonucleoproteins rearrange into a more compact mass, the nucleus gradually elongates, flattens dorsoventrally, and becomes condensed and elliptical in shape (Fig. 17).

The spermatid nucleus gradually takes on the shape characteristic of the species in the course of chromatin condensation (Figs. 17 and 27), suggesting the possibility that the shaping of the nucleus is related to this nucleoplasmic reorganization (Fawcett et al. 1971, Holstein 1976). Fawcett et al. (1971) have suggested that the species-specific shape of the nucleus (or head) must be determined largely from within by a specific, genetically controlled pattern of aggregation of molecular subunits of DNA and proteins during the condensation of chromatin (reviewed by Beatty 1970). This point is also illustrated by the increased incidence of bizarre abnormalities that occur in the nuclear shape during spermiogensis in certain mutant strains of mice (Olds 1971, Hunt and Johnson 1971, Bennett et al. 1971, Dooher and Bennett 1977, Hillman and Nadijcka 1978a, b). But the mechanisms underlying these nuclear abnormalities need to be determined more precisely.

There are other workers who have assigned the shaping role to the caudal sheath or manchette (Illison 1968, Rattner and Brinkley 1971, Courtens 1982b). It is believed that in this condensed form the genome is protected against physical damage or chemical mutagenesis during the sperm transport to the site of fertilization. During this metamorphosis, the nucleic acid molecules in general produce birefringence (Randall and Friedlander 1950) or X-ray diffraction (Wilkins and Randall 1953).

It is accepted generally that the translational events of spermiogenesis are directed primarily by gene products that are transcribed on the diploid genome (Monesi 1971, Monesi et al. 1978). The diploid effect on spermatid development and differentiation is considered to function through the synthesis and accumulation of stable RNA species in the primary spermatocytes, as already discussed in Chap. III. These RNA molecules are translated during spermiogenesis (meiotic control). This suggestion is supported further by the fact that there is an apparent conservation of meiotic RNA in the haploid spermatids (Monesi 1964, 1965a, Loir 1972a, Geremia et al. 1977a, Monesi et al. 1978). The alternative mechanism of diploid regulation of spermiogenesis is that various RNA molecules (or the protein products) formed post-meiotically are exchanged freely between the four daughter spermatids which originate from the two meiotic divisions through the cytoplasmic bridges connecting them (post-meiotic diploid effect). This mechanism, therefore, can prevent the phenotypic expression of the haploid genotype (Monesi et al. 1978). By contrast, see counter arguments by Bellvé (1982). In various genetic, cytological and autoradiographic studies, very conflicting evidence has been produced in regard to the meiotic or the haploid control of spermatid differentiation, as summarized by Monesi et al. (1978) and Bellvé (1982).

Most of the genetic evidence argues against a haploid effect in mammalian spermatogenesis (Beatty 1970, 1972, Brock 1977), although t-allele transmission ratio distortion has been explainable as due to post-meiotic gene action (Gluecksohn-Waelsch 1972, Hammerberg and Klein 1975, Erickson 1978). There is now some well-documented genetic evidence in favour of the expression of the haploid genotype in spermatogenesis, as discussed by Monesi et al. (1978). The occurrence of haploid gene transcription, however, has been revealed by autoradiographic and biochemical evidence (Monesi et al. 1978). Erickson et al. (1980a, b) have provided further evidence for haploid gene expression during spermatogenesis (see also Fujimoto and Erickson 1982). Heterogeneous poly

(A)$^+$-containing RNA is synthesized post-meiotically. Autoradiographic studies have depicted the incorporation of [^{3}H]uridine into RNA of spermatids at steps 1 to 8 of spermiogenesis during the earlier stage of the nuclear shaping and condensation (Monesi 1964, 1965a, 1971, Utakoji 1966, Moore 1971, Loir 1972a, Kierszenbaum and Tres 1975, Söderström and Parvinen 1976c), indicating early cessation of RNA synthesis and its subsequent elimination from the nucleus during early spermiogenesis, as already stated.

Geremia et al. (1977a, 1978), using biochemical techniques, have observed that no haploid gene action is possible after the starting of nuclear elongation in spermatids. These results are in agreement with findings in the mouse (Moore, 1971, Kierszenbaum and Tres 1974b, 1975), rat (Söderström and Parvinen 1976c), and ram (Loir 1970). The RNA synthesis ceases in rat spermatids at step 8 when the chromatin begins to condense (Söderström and Parvinen 1976c, Söderström 1976a). This occurs in a relatively short time because the grains in the [^{3}H] uridine autoradiograms disappear in very closely neighbouring spermatids at step 8. This indicates a precise control of transcription in the spermatids. Transcription in spermatids is predominantly on HnRNA species (Kierszenbaum and Tres 1975, Erickson et al. 1980a, b). According to Moore (1971, 1972), RNA synthesis is reinitiated in round spermatids during spermiogenesis, but is reduced sharply as spermatid nuclei condense, and ceases during an early stage of nuclear elongation. No RNA polymerase activity is demonstrable during the remaining stages of spermatid differentiation, nor is any label incorporation observed in testicular spermatozoa (see Moore 1975).

The nucleoli in the spermatids are unlabelled, indicating a general slowdown of the synthesis and maturation of ribosomal RNA during spermiogenesis (Söderström 1976a). Incorporation of [^{3}H]uridine into the nucleolus-like structure is restricted to its peripheral margins, suggesting that synthesis of rRNA is limited. However, the nucleolus of mouse spermatids gives a positive reaction by the HG-As staining procedure, which is reported to be specific for transcriptionally active nucleolar organizer regions (Schmidt et al. 1977). The autoradiographic studies also have not revealed any RNA synthesis after step 8 of spermiogenesis (Söderström 1976a, Söderström and Parvinen 1976c). Geremia et al. (1978) have observed that round spermatids synthesize both rRNA and mRNA. The postmeiotic transcriptional activity of the rRNA genes has also been reported in other studies (Hofgärtner et al. 1979a, b, Schmidt et al. 1982, 1983). This post-meiotic synthesis of RNA stops completely in mid-spermiogenesis after nuclear elongation in the spermatid has set in (see also Loir 1972a, Geremia et al. 1977b, Monesi et al. 1978, Erickson et al. 1980a, b).

The rate of RNA synthesis per DNA content is the same in round spermatids and pachytene spermatocytes (Loir 1972a, Geremia et al. 1976, 1977a). The round spermatids synthesize both rRNA and polyadenylated RNA (presumably mRNA) (D'Agostino et al. 1976, Geremia et al. 1978). Furthermore, a considerable fraction of rRNA and mRNA produced by pachytene spermatocytes is preserved until late spermiogenesis (see Monesi et al. 1978) when these RNA's are eliminated within the residual bodies (Monesi 1964, 1965a, 1971). As already discussed in Chap. III, the recent biochemical studies in the rat have revealed that the RNA synthesized by pachytene spermatocytes is mostly HnRNA which is more stable than that of somatic cells. This HnRNA may contain precursors of long-lived mRNA molecules needed for directing protein synthesis during late spermiogenesis when RNA transcription stops completely. Monesi et al. (1978), using biochemical techniques, have observed synthesis of 4S, 18S, and 28S RNA in round spermatids as well as pachytene spermatocytes. These cell types synthesize about an equal amount of poly(A)$^+$-containing RNA as well as of nonpolyadenylated RNA per DNA content. The spermatids are considered to synthesize mRNA, favouring the likelihood of continuing mRNA synthesis (Erickson et al. 1980c). cDNA clones encoding cytoplasmic poly (A)$^+$ RNA's first appear at detectable levels in haploid phases of spermatogenesis in the mouse (Kleene et al. 1983, Distel et al. 1984). It appears that much of the protein synthesis in the elongating spermatid occurs under the direction of mRNA's that is stored in round spermatids (Stern et al. 1983b, Gold et al. 1983b). Erickson et al. (1980b) observed that when total cellular RNA prepared from spermatocytes, early spermatids, and middle to late spermatids is translated in vitro in a rabbit reticulocyte system, PGK-2 synthesis is detected only with the RNA prepared from middle to late spermatids, showing that the amount of translatable PGK-2 mRNA is greatly increased in these cells (see also Gold et al. 1983b). However, this does not differentiate conclusively between new mRNA synthesis and "stored" mRNA, as the latter may exist in an untranslatable form prior to the detected increase in PGK-2 synthesis. In contrast to synthesis of PGK-2 during spermiogenesis, the synthesis of LDH-X is highest in pachytene spermatocytes and declines progressively in more mature cells (Meistrich et al. 1977). Thus, it has been proposed that all of the LDH-X mRNA is synthesized in the spermatocytes, and the level declines slowly as the cell matures. The resuls of these studies show that although LDH-X and PGK-2 are both testis-specific isozymes, their patterns of regulation during spermatogenesis are quite different. Alternatively, the appearance and synthesis of protamine-like histone (PLH) do not occur until late-middle stage spermatids (Monesi 1965a, Lam and Bruce 1971, Bellvé et al. 1975) and translatable PLH mRNA can only be detected at this time (Erickson et al. 1980b). This corresponds to the time of increase in PGK-2 synthesis and suggests the synthesis or activation of mRNA for both PGK-2 and PLH may be coordinately regulated. But further studies are needed to determine whether there is coordinate control of multiple genes during spermatogenesis. Recently Distel et al. (1984) have studied the appearance of tubulin messenger RNA (mRNA) transcripts in meiotic and post-meiotic testicular cells of mouse for determining when are specific tubulin genes expressed during the differentiation of the spermatozoon. A complementary DNA (cDNA) clone for an α-tubulin is isolated from a mouse cDNA library. The untranslated 3′ end of this cDNA is homologous to two RNA transcripts seen in post-meiotic cells of the testis, but absent from meiotic cells and from several tissues including the brain. The temporal expression of this α-tubulin cDNA has provided evidence for the haploid expression of a mammalian structural gene.

All these observations, as discussed above, suggest that early spermatids are capable of gene transcription; most of the post-meiotic RNA is nonribosomal, but whether these products are translated to provide specific proteins has not been determined more precisely. However, the appearance of novel α-tubulin transcripts in round spermatids of the mouse coincides with the formation of the manchette and flagellar axoneme (Distel et al. 1984). This suggests that the haploid-specific tubulin mRNA may be involved in these structures, which are specific to spermatogenesis. Protein synthesis in spermatids shows a well-defined pattern. In general, protein synthesis in germ-cells follows a pattern analogous to RNA production. Cytoplasmic protein labelling occurs throughout spermiogenesis, whereas protein labelling of the nucleus is seen only at certain stages (see Monesi et al. 1978). It is found, at a low level, during early spermiogenesis, then stops completely in spermatid stages 8 or 9 to start again from spermatid steps 11 to 14. After this burst in late spermiogenesis, nuclear protein synthesis stops completely in the mature (steps 15 and 16) spermatid (Monesi 1965a). The nuclear labelling in late spermiogenesis of mammals appears to be related to the well-known synthesis of the sperm histone, as will be discussed below.

The early cessation of RNA synthesis and its subsequent elimination from the nucleus during early spermiogenesis, as discussed above, are associated with the turnover of basic chromosomal proteins or extrusion of lysine-containing histone, its replacement by a more basic arginine-rich protein (Grimes et al. 1977, Loir and Lanneau 1978a, b, Bellvé 1979, Mayer et al. 1981, Rodman et al. 1982, Lanneau and Loir 1982, Courtens et al. 1983, Courtens 1984, Bilaspuri and Guraya 1985, 1986d) and coincident incorporation of cysteine (Loir 1970) to such an extent that this amino acid constitutes about 13% of the amino acid residues in the protein of bull sperm nuclear chromatin (Coelingh et al. 1969, Coelingh and Rozijn 1975). Courtens et al. (1983) have studied immunocytochemically the localization of protamine in the spermatids of the ram. In buffalo, goat, and ram, the staining for arginine-rich protein starts in steps 10 to 12 of spermiogenesis, increases in step 13 and becomes maximum in step 14, as revealed by cytochemical techniques (Bilaspuri and Guraya 1985, 1986d). Courtens and Loir (1975a, b, 1981a, b) also have reported a similar sequence of arginine-rich proteins during ram spermiogenesis. This is accompanied closely by condensation of chromatin and a decrease in the intensity of cytochemical reactions for lysine and DNA (Feulgen reaction) (Bilaspuri and Guraya 1986c), as also reported for the bull (Gledhill et al. 1966, Gledhill 1975). The decrease in Feulgen reaction is not due to decrease in DNA content (Esnault and Nicolle 1976).

The sites of synthesis of arginine-rich proteins need to be determined. The dynamics of the displacement of somatic-type histones and the assembly of the sperm-unique nucleoproteins are not understood and there is some question as to whether certain unique basic proteins are associated transiently with the chromatin in the early stages of spermatogenesis (Kistler and Geroch 1975, Kumaroo et al. 1975, Shires et al. 1975, Courtens et al. 1983, Courtens 1984). It is accepted (Bellivé et al. 1975, Loir and Lanneau 1975, Calvin 1976, Balhorn et al. 1977, Bellvé 1979, O'Brien and Bellvé 1980a, b) however that the nucleus of the mature sperm contains none of the complement of the coventional "somatic type" histones detectable by electrophoretic analysis, and that at least 99% of its protein is composed of one or more arginine-rich basic proteins with a high content of cysteine residues (Coelingh et al. 1972, Monfoort et al. 1973, Kistler et al. 1973, Bellvé et al. 1975, Coelingh and Rozijn 1975, Calvin 1976, Tobita et al.

1979, Mayer et al. 1981, Lanneau and Loir 1982), as will also be discussed in detail in Chap. VII.

The nuclear proteins show conspicuous alterations during mammalian spermiogenesis. A cytochemical study of rat spermiogenesis has revealed that histones are displaced from the nucleus and deposited, along with the excess RNA, as the sphere chromatophile bodies that are placed in the caudal lobe of cytoplasm (Vaughn 1966). The first phase of histone displacement corresponds to the onset of nuclear condensation (steps 10 to 12), while the second phase occurs after step 13 when the chromatin condensation is more demarcated. During the first phase, the histones may be modified by acetylation (Grimes et al. 1975c) and phosphorlyation (Marushige and Marushige 1975b) to decrease their net basic charge and thus facilitate their displacement from DNA. Cytochemical studies have suggested the histones are removed from the nucleus in an anterior to posterior direction (Courtens and Loir 1975a, b, 1981a, b, Courtens 1984), implying that the removal of histones and possibly the disassembly of nucleosome substructure proceed in a precise and co-ordinated manner. Courtens and Loir (1975b) have reported the details of ultrastructural changes in cytoplasmic organelles of ram spermatids during somatic type histone transition. The removal of non-histone chromosomal proteins during spermiogenesis is not well characterized (Kolk and Samuel 1975, Balhorn et al. 1977, Mujica et al. 1978, O'Brien and Bellvé 1980a, b). These are synthesized in a defined temporal sequence during spermiogenesis, corresponding to the morphogenesis of the nucleus (O'Brien and Bellvé 1980b). Although their functions need to be determined, they may act as specific regulators of gene transcription by controlling the tertiary structure of chromosomes. The non-histone chromosomal proteins also may play a significant role in establishing or maintaining nuclear shape.

Replacement of the histones involves a series of nuclear protein changes. This is supported by the recent histochemical studies of isolated testicular cells of the rat, which have revealed that at least three histone-like proteins first appear in primary spermatocyte nuclei (Branson et al. 1975, Grimes et al. 1975a, b, Kistler and Geroch 1975, Kumaroo et al. 1975, Shires et al. 1975) but are not retained in late spermatid nuclei (Grimes et al. 1977). These observations are in agreement with the autoradiographic data which have revealed the uptake of [^{3}H]amino acids into the pachytene spermatocyte nuclei of mouse (Monesi 1965a, Kierszenbaum and Tres 1975), rat (Vaughn 1966), and ram (Loir 1972b). Both biochemical and autoradiographic studies have shown that a number of new basic nucleoproteins are synthesized in mammalian spermatids (Monesi 1965a, Lam and Bruce 1971, Loir 1972b, Kistler et al. 1973, 1975a, Kumaroo et al. 1975, Loir and Lanneau 1975, Platz et al. 1975, Shires et al. 1975, Grimes et al. 1975a, b, 1977, Kierzenbaum and Tres 1975, R. B. Goldberg et al. 1977, Bouvier 1977).

Various studies carried out on the spermiogenesis of rat, ram, and mouse have demonstrated that histones of spherically shaped nuclei of young spermatids are replaced by more basic nucleoproteins in elongating nuclei (see references in Courtens and Loir 1981b, Courtens 1982a, 1984, Dupressoir et al. 1985). These basic nucleoproteins then are replaced by even more basic sperm-specific nucleoprotein(s) in mature spermatids. Amino acid incorporation and histochemistry have revealed that the different nucleoproteins have specific amino acid compositions (Bellvé 1979, Courtens and Loir 1981b, Courtens 1982a, 1984, Dupressoir et al. 1985). Loir and Lanneau (1978a) have observed that most striking alterations in protein composition occur during the elongation phase (steps 8 – 12) in the ram. The five histones are displaced from chromatin at the same rate. When they are freed of histones (step 12), the nuclei start to accumulate sperm-specific protein. This stepwise replacement process is accompanied by a reduction in the amount of basic protein bound to DNA. With the disappearance of histones, eight spermatidal protein fractions are present in the nuclei until sperm-specific protein synthesis reaches its maximum rate. Except for one, they all contain cysteine and are partially intermoleculary cross-linked in the chromatin. In in vivo and in vitro labelling experiments, these are synthesized in elongating spermatids (steps 8 – 11). None is a degradation product of histones. An electrophoretic study of mammalian (ram, bull, goat, boars, stallion, rat, cat, hedgehog, European mink and ferret) spermatid-specific nuclear proteins has suggested that differences exist in the total number of these proteins, as well as in the number and amount of cross-linked cystein-containing proteins (Lanneau and Loir 1982). These differences appear to be more family-specific than species-specific. Ram spermatidal nuclear proteins P1, 3 and T have been isolated and characterized (Dupressoir et al. 1985). These are small arginine-and cysteine-rich basic proteins. Proteins PI and T are unusually rich in serine, and histidine content of PI is particularly high. These proteins are present only during the reorganization of the spermatid chromatin.

Bode et al. (1977) have made observations on the competition between protamines and histones for the better understanding of spermiogenesis. Chauviere (1977) has studied the proteolytic activity of chromatin of mammalian testicular cells. The substrate specificity of the protease suggests its probable important functional role during spermiogenesis when histones are replaced by arginine-rich proteins. The stabilization and condensation of spermatid chromatin are promoted through a progressive increase in disulphide bridges.

Marushige and Marushige (1978) have studied the phosphorylation of sperm histone during spermiogenesis in mammals. This phosphorylation is dependent on cyclic -AMP (Geremia et al. 1981) and probably is involved in the proper condensation of spermatid chromatin during spermatogenesis. Grimes and Henderson (1983) studied acetylation of histones during spermatogenesis in the rat. According to Baccetti et al. (1977), three cytochemical findings are consistent with the nuclear immaturity in most of the round-headed human spermatozoa: lower phosphorus concentration, higher zinc content, and a variable amount of lysine. Not all the round-headed spermatozoa show the same degree of nuclear immaturity, some have a lower content of zinc and lysine, and a higher concentration of phosphorus concomitantly with a more compact appearance of the chromatin.

Mayer and Zirkin (1979) recently have made a detailed autoradiographic study of nuclear incorporation and loss of [^{3}H] amino acids in mouse spermatogenesis. Arginine, lysine, valine, and proline are incorporated rapidly into primary spermatocyte nuclei, retained through subsequent spermatocyte divisions and through to step 12 of spermiogenesis, but are lost with spermatid differentiation beyond step 12. Arginine and lysine (not valine or proline) are incorporated rapidly into certain elongated spermatid nuclei but differ strikingly in their distribution and fate. Nuclei of late step 12 through step 15 spermatids initially are labelled with arginine. This label is retained through subsequent spermatid differentiation and sperm maturation in the epididymis. In contrast, lysine is initially incorporated only into late step 12 and step 13 spermatid nuclei, and incorporation of lysine coincides with the initiation of chromatin condensation in late step 13 nuclei and loss of lysine coincides with the completion of condensation in step 14 nuclei (Mayer et al. 1981). The problem of how the successive nucleoproteins enter and leave the nucleus of the mammalian spermatid has been studied by Courtens (1982a), who has suggested that stainable material enters and keeps out of the nucleus via an apparatus made of the nuclear envelope prolongated by the endoplasmic reticulum (see also Courtens 1984). The portamines (or associated stained materials) could enter the nucleus by species-specific ways.

A variety of testis-specific proteins have been demonstrated in several mammalian species and their association with different phases of spermatogenesis has been adequately defined. Some of these nuclear proteins are histones. Many of them show unusual electrophoretic mobility, composition, antigenicity and other properties, and therefore qualify, along with some testicular enzymes, for use as biochemical markers to distinguish specific stages of spermatogenic development (Kistler et al. 1973, 1975c, Kistler and Williams-Ashman 1975, Shires et al. 1975, Chiu and Irvin 1978, Meistrich et al. 1978, Seyedin and Kistler 1979, Wattanaseree and Svasti 1983). The changes in the histone fractions of the rat seminiferous epithelium, which occur during testicular development, are particularly striking in three histone subfractions (designated X_1, X_2 and X_3). Whereas rat spermatogonia contain only histones of the somatic class, after completing the differentiation into primary spermatocytes the germ-cells develop the ability to synthesize the testis-specific histones of the X series. One of them [X_1 (TH_1)] is a variant of the somatic histone H_1 (Russo et al. 1983), another, X_3 (TH_2B), is a direct descendant of the somatic histone H. By the time spermiogenesis has reached the stage of round spermatids, TH_1 and TH_2B will have replaced, largely though by no means completely, their somatic counterparts. No further histone transition occurs in the round spermatid, but during spermatid elongation the somatic- and testis-specific histones and most non-histones are replaced by at least two, low-molecular-weight basic proteins designated TP and TP_2. During the final stages of spermiogenesis these two proteins are replaced again by two new ones, S_1 and TP_3; in the nuclei of mature spermatozoa S_1 predominates (Grimes et al. 1975a, Meistrich et al. 1978). The rat testis actually contains two atypical histone 1 species denoted as TH_1-XA and TH_1-XB (Levinger et al. 1978), which have been isolated and characterized partially by Kumaroo and Irvin (1980). These are lysine-rich histones. Histone TH_1-XB has an unusually high arginine for a H_1 histone. Both TH_1-XA and TH_1-XB are present in nuclei of pachytene spermatocytes and early spermatids. Human testis-specific histone TH_2B has been studied for its fractionation and peptide mapping (Wattanaseree and Svasti 1983). Chiu and Irvin (1978) have presented data to demonstrate the usefulness of two of the unusual basic chromosomal proteins (TH_1-X and TSP) of seminiferous epithelial cells of the testis as "markers" for specific stages of spermatogenesis. Bellvé (1979) has discussed the presence of several atypical histone species that occur in the mammalian testis. The unusual H_2b type histone (designated by Branson et al. 1975 as X_3) and the sperm protamines also could serve as markers for other stages of spermatogenesis.

In mammalian species other than the rat, the variety of specific proteins synthesized in the testicular nuclei appears just as great. In the ram eight spermatidal proteins are described in the nucleus of the advanced spermatid; all but one contain cysteine, and most show overall basic charges similar to those of histones (Loir and Lanneau 1978b, Dupressoir et al. 1985). A highly basic, acid-soluble, low-molecular-weight testis-specific protein has been isolated from human testis; the purified protein is rich in arginine, lysine, and serine but devoid of cysteine, phenylalanine, glutamic acid, isoleucine and tryptophan; it is distinguishable by electrophoretic means from the principal basic chromosomal protein of mature human spermatozoa (Kistler et al. 1975a).

All these marker proteins may be especially helpful to determine the specific stages of spermatogenesis which are blocked by inhibitors, indicating their application in screening studies to develop effective contraceptive drugs. The marker proteins are believed to be especially useful in detecting aspermatogenesis to readily detectable histologic changes.

Very divergent views have been expressed about the functional significance of the sperm-specific basic proteins (Gledhill 1975, Subirana 1975, Coelingh and Rozijn 1975). These are believed to play a role in the rearrangement of condensing male genome (Picheral and Bassez 1971a, b, Walker 1971, MacGregor and Walker 1973, Lung 1972, Bedford et al. 1973, Mayer and Zirkin 1979) and the shaping of the nucleus (Fawcett et al. 1971), inactivation of the genome (MacLaughlin and Turner 1973, Betlach and Erickson 1976), and early embryonic development (Lyon 1971, Lifschytz and Lindskey 1972, Brown and Chandra 1973). Bellvé (1979 has discussed in detail the synthesis, nature and role of spermatogenic histones in chromatin structure (see also Courtens 1984). The diversity of variants existing among them is very complex. The dramatic transcriptional activity changes, which occur during meiotic prophase (Kierszenbaum and Tres 1978), are believed to be indicative of marked alteration among the non-histone chromosomal proteins; the latter appear to function as specific regulators of gene transcription. Bellvé and O'Brien (1979), using new high-resolution two-dimensional gel system, have resolved more than 150 species of non-histone chromosomal proteins of seminiferous epithelial cells. A more definitive approach to this problem also demands using isolated homogeneous populations of germ-cells at successive stages of spermatogenesis. There is now increasing interest to develop various methods for obtaining homogeneous populations of germ-cells in different mammalian species, such as mouse (Romrell et al. 1976, Bellvé et al. 1977a, b, Millette and Moulding 1981a, b, Gerton and Millette 1983), rat (Meistrich et al. 1981), rabbit (O'Rand and Romrell 1977), guinea-pig (Tung et al. 1979), and humans (Shepherd et al. 1981, Narayan et al. 1983).

The various changes in amino acids as described above precede the decrease in Feulgen-staining material (Esnault et al. 1974). Recent studies have demonstrated that the relative amounts of the different histone fractions are changed during the various stages of rat spermatogenesis in an interesting and systematic manner (Kumaroo et al. 1975). The ratio of the trailing (acetylated) to the leading member of the histone 2A doublet was greater in spermatid nuclei than in nuclei of a fraction enriched in primary spermatocytes. Similarly, the ratio X_1/F1 was greatest in the spermatid nuclei. On the other hand, the ratio X_3/F2b was greater in nuclei of pachytene-diplotene primary spermatocytes than in the fraction enriched in nuclei of spermatogonia and preleptotene primary spermatocytes. With the condensation of the chromatin the nucleoli gradually disappear from view and the number of nuclear pores decreases correspondingly (Fawcett and Chemes 1979), suggesting a decrease in the nucleocytoplasmic exchange.

The changes in transcription during spermatogenesis are related closely to physicochemical alterations in the chromatin, as indicated by a progressive decrease in methyl green, Feulgen, and acridine orange staining (Gledhill et al. 1966). Although DNA content of these nuclei remains stable, a decrease occurs in the actinomycin D binding capacity (Brachet and Hulin 1969, Darzynkiewicz et al. 1969, Loir and Hochereau-de-Reviers 1972, Barcellona et al. 1974, Gledhill and Camball 1972, Gledhill 1975), in the number of acidic phosphate groups available for binding cationic dyes, and the sensitivity to thermal denaturation (Ringertz et al. 1970, Gledhill 1975). These alterations are believed not only to reflect changes in the nuclear protein composition, as a result of or as a corollary to cellular differentiation, but also to indicate the secondary structure of nucleic acids. All these changes are doubtless related in some way to chromatin condensation in the later spermatids; it has been suggested that changes are indicative of the progressive increase in the strength of electrostatic binding of nuclear protein to DNA, typical of the final steps of spermiogenesis.

2. *Nuclear Envelope*

The nuclear envelope undergoes conspicuous changes during spermiogenesis (Holstein and Roosen-Runge 1981). Rattner and Brinkley (1971) have followed the fate of the nuclear envelope during spermiogenesis in the cotton top marmoset. Throughout the nuclear shaping and condensation process, excess nuclear membrane is eliminated into the posterior cytoplasm in the form of two continuous sheets at either side of the implantation fossa (Fig. 30). Prior to midpiece formation, these membrane extensions lose their association with the caudal margin of the nucleus. The remaining membrane forms two short extensions at the caudal margin of the nucleus and is morphologically similar to the redundant nuclear membrane described for other mammalian species (Fawcett 1958, 1970). The redundant portion of the excess membrane is formed during the nuclear shaping (Loir and Courtens 1979, Courtens 1984, Holstein and Roosen-Runge 1981). Gordon (1972a) has also reported the presence of extensive out-pockets of nuclear envelope persisting during late spermiogenesis of the guinea pig. These out-pockets may contain portions of DNA that have not been incorporated into the more condensed nuclear mass. This portion of spermatid DNA is believed to remain active to function in the flagellar development when the rest of chromatin has become inert (Fawcett et al. 1970).

Horstmann (1961) observed several phenomena affecting the elimination of excess nuclear envelope in human spermiogenesis. In the early spermatids, stacked lamellae often were observed along the posterior margin of the nuclear envelope. These stacks were believed to be early indications of the elimination of

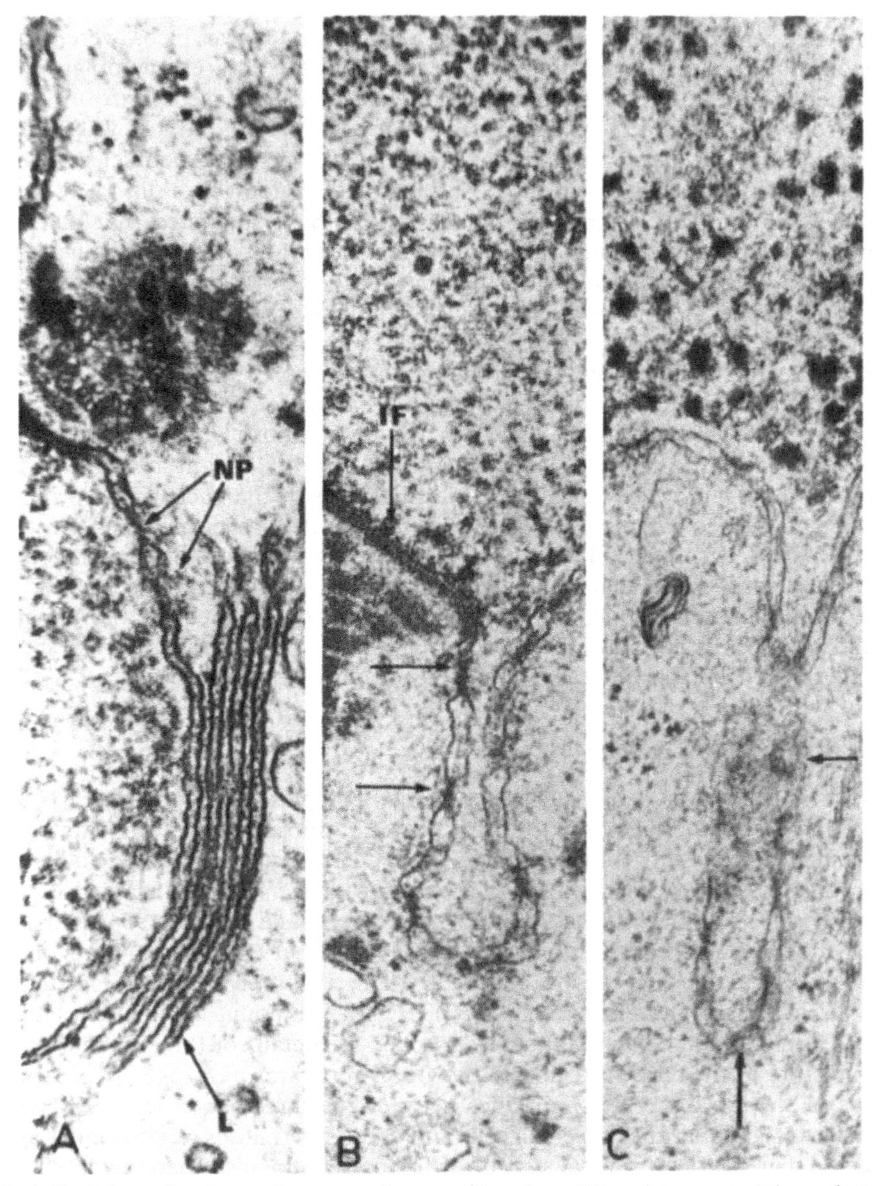

Fig. 30. A Caudal margin of an early marmoset spermatid nucleus; *NP* nuclear pores of the nuclear envelope; *L* lamellar structure associated with the nuclear envelope. **B** Caudal margin of mid-spermatid nucleus from marmoset; *IF* implantation fossa. *Arrows* indicate nuclear pores in membrane sheath extending from the nucleus to the neck region. **C** Caudal margin of marmoset spermatid containing condensing nuclear chromatin. *Arrows* indicate nuclear pores in membrane sheath which extends into the middle-piece region (From Rattner and Brinkley 1971)

the nuclear envelope. In later stages, vesicles or nuclear blebs were observed along the caudal two-thirds of the nucleus and the adjacent cytoplasm. These structures aid both in the elimination of excess nuclear envelopes formed during morphogenesis and in the transmission of nuclear material to the cytoplasm. The transport of nuclear material is supported by the fact that the nuclear pores become more concentrated in the post-acrosomal region (Franklin 1968, Fawcett and Chemes 1979). Loir and Courtens (1979) have suggested that the nuclear posterior space and redundant nuclear envelope are possibly involved in the transfer of proteins leaving the chromatin for the cytoplasm during spermiogenesis. De Kretser (1969) has also observed a budding of vesicles from the redundant nuclear envelope during human spermiogenesis (Kerr and de Kretser 1981), a process which may be involved in the loss of nuclear contents during the gradual reduction in nuclear volume. The suggestions of Horstmann (1961) and deKretser (1969) are in keeping with the observations of Franklin (1968) and Rattner and Brinkley (1971) in marmoset. The redundant nuclear envelope, which forms along the apical and ventral aspects of the nucleus and around the implantation fossa, regresses during steps 17 – 19 of spermiogenesis in the rat (Lalli and Clermont 1981). The nuclear pores do not occur in the acrosomal region, the post-acrosomal lamina and the basal plate of maturing spermatid (Holstein and Roosen-Runge, 1981). Czaker (1985c) has studied the morphogenesis and cytochemistry of the postacromosal dense lamina during mouse spermiogenesis with election microscopical silver-staining technique.

B. Cytoplasmic Components

Various cell components, such as the Golgi complex, mitochondria, RNA-containing basophilic substance (or ultrastructural ribosomes and elements of endoplasmic reticulum), phospholipid granules, multivesicular bodies and chromatoid body, described for the spermatocytes in Chap. III, are also present in the early spermatids (Figs. 17 and 31) (Fawcett and Burgos 1956, Horstmann 1961, deKretser 1969, Rattner and Brinkley 1970, 1971, Bustos-Obregon et al. 1975, Holstein 1976, Hermo and Lalli 1978, Plöen et al. 1979, Holstein and Roosen-Runge 1981, Tang et al. 1982). But their amount is relatively smaller.

Hermo et al. (1979, 1980) have described the details of three-dimensional architecture of the cortical region of the Golgi apparatus of steps 3 – 7 rat spermatids. Although there are similarities between the three-dimensional architecture of the Golgi apparatus in Sertoli cells and young spermatids (e.g. saccular and intersaccular regions), several structural features distinguish the spermatids Golgi apparatus. Clermont et al. (1981) have observed that the Golgi complex of early rat spermatids shows two distinct zones, a cortex made up of flattened saccules and related membranous tubules and a medulla containing various types of vesicular profiles (see also Tang et al. 1982). Mollenhauer and Morre (1977) have distinguished dictyosome-like structures with cylindrical intersaccular connections (microtubules) in late stages of spermatid development in the guinea pig (Mollenhauer and Morre 1983).

Spermatids contain elements of endoplasmic reticulum (ER) (Fig. 31) which show some species variations in development, morphology and distribution dur-

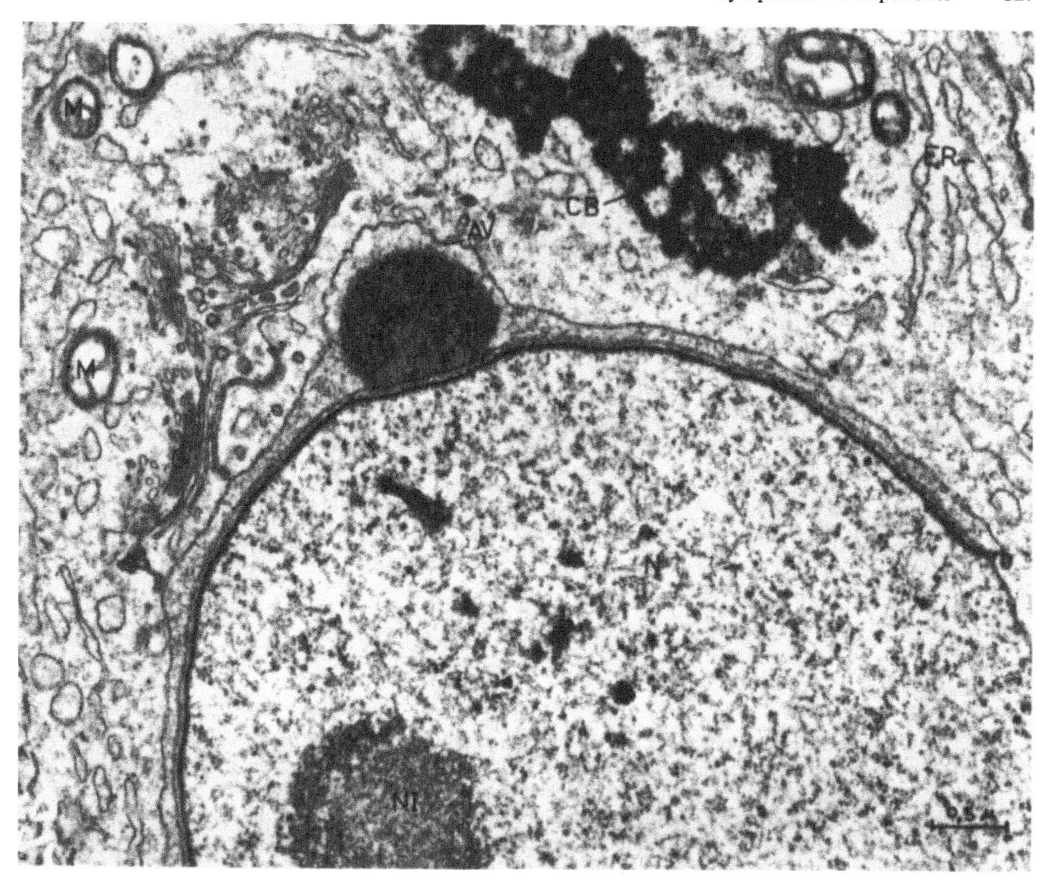

Fig. 31. Electron micrograph of early spermatid of ram, showing nucleus (*N*), nucleous (*NI*), acrosomal granule (*AG*), acrosomal vacuole (*AV*), Golgi complex (*G*), mitochondria (*M*), elements of endoplasmic reticulum (*ER*), chromatoid body (*CB*) and sparsely distributed ribosomes (Courtesy of Dr. M. Courot)

ing spermiogenesis. Hermo and Lalli (1978) have observed that the cisternae of ER in the rat spermatids are associated with the Golgi elements in both the cis and trans faces as well as with the saccules (Hermo et al. 1980). This is in contrast to the previous observations indicating that the numerous cisternae of ER are closely applied to the cis (or cortex) face of the Golgi complex (Fig. 32) (Sandoz 1970, Susi et al 1971, Mollenhauer et al. 1976). Multivesicular bodies usually lie in the Golgi zone (Tang et al. 1982). The detailed analysis of changes in phosphatases of the Golgi apparatus has revealed that during steps 1 – 3 the Golgi apparatus produces multivesicular bodies that become associated with the chromatoid body. Their origin from the Golgi apparatus is further supported by their labelling with [^{3}H] fucose. The function of multivesicular bodies needs to be determined. But these are translocated to the region of the chromatoid body and continue to remain associated with this structure until both components degener-

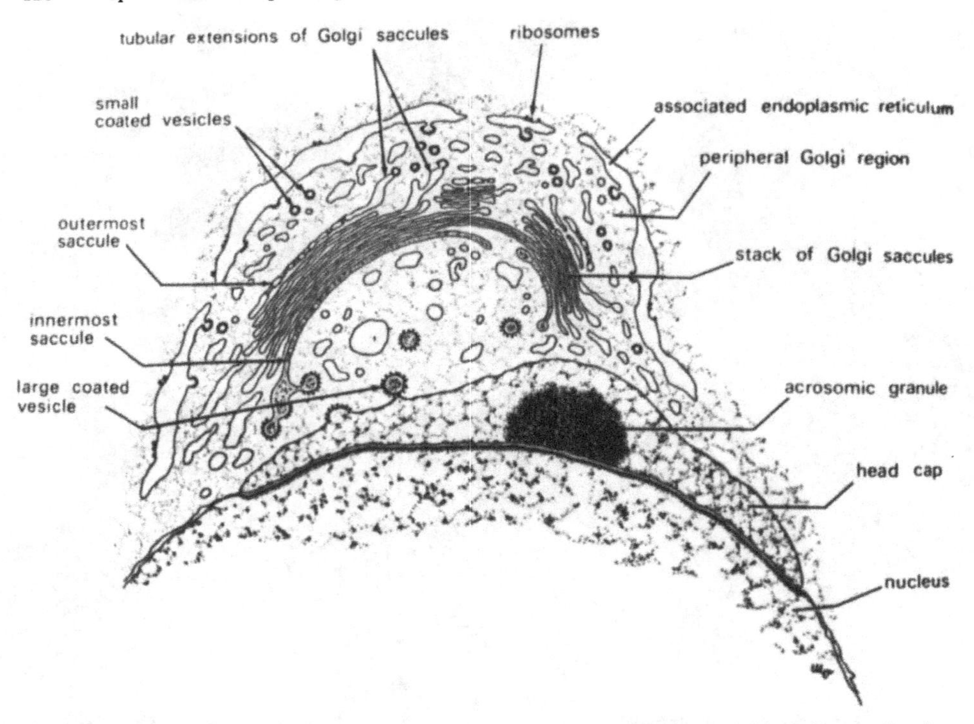

Fig. 32. Characteristic components of the Golgi apparatus in a step 5 spermatid of the rat. *On the right* of the drawing the following major componets of the Golgi region are indicated; the associated layer of endoplasmic recticulum with a few attached ribosomes and small coated buds, the peripheral Golgi region between the stacks of Golgi saccules and layer of endoplasmic reticulum, and finally the acrosomic granule, head cap, and delimiting acrosomic membrane, which together make up the acrosomic system and are applied at the surface of the nucleus. The stacks of Golgi saccules are arranged as a hemisphere around a space referred to as the concavity of the Golgi apparatus (From Susi et al. 1971)

ate (Tang et al. 1982). Clermont and Rambourg (1978) have observed that during the Golgi and cap phases of spermiogenesis the ER is distributed throughout the cytoplasm as a three-dimensional network of spherical and tubular cisternae interconnected by narrow tubules (Fig. 33). In addition, a close network of tubular cisternae is located along the convex surface of the Golgi apparatus and lines the plasma membrane (Fig. 33) where the cell membrane joins that of another spermatid to form an intercellular bridge; this network extends across the bridge. Human spermatids contain numerous layers of annulate lamellae either continuous with the nuclear envelope and caudal to the acrosome or peripherally positioned in the cytoplasm (Smith and Berlin 1977, Nistal et al. 1980, Holstein and Roosen-Runge 1981). Individual lamellae possess therminal dilations and show continuities with ER. The interlamellar space is entirely filled with fine granular electron-dense material. Finally, annulate lamellae and other membranes of the ER are released into the cytoplasmic droplets of the maturing spermatids. The developing spermatids of humans show cytoplasmic granules of 15

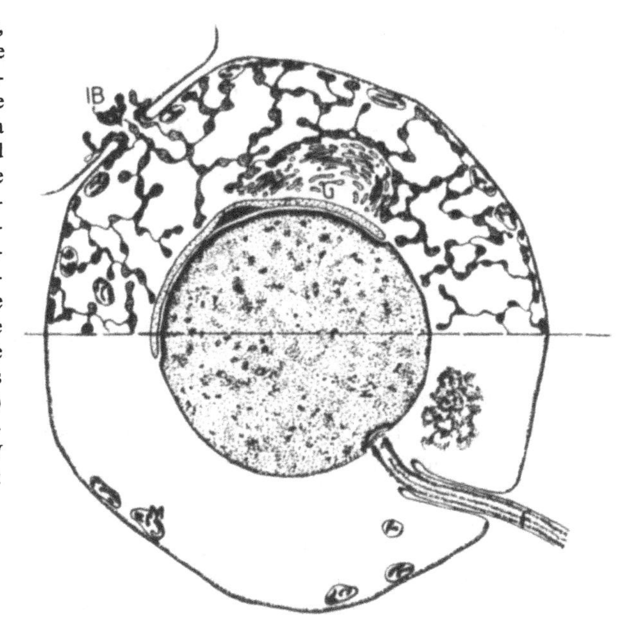

Fig. 33. A step-7 spermatid rat, showing the arrangement of the endoplasmic reticulum throughout the one half of the cell. The endoplasmic reticulum forms a single continuous system and even communicates with the endoplasmic reticulum of adjacent spermatids through intercellular bridges (*IB*). A close network of cisternae is seen subjacent to the plasma membrane and on the convex aspect of the Golgi apparatus (*G*). The dense but unstained chromatoid body is located next to the flagellum (*F*) which is surrounded by the invaginated plasma membrane; *M* mitochondria (From Clermont and Rambourg 1978)

to 20 nm in diameter, which are embedded in a finely granular material in strands and groups (Holstein and Roosen-Runge 1981).

The ribosomes are scattered individually or lie in clusters throughout the cytoplasm of the spermatid (Fig. 31). Nakamura and Hall (1978) have observed that ribosomes of rat spermatids are apparently the most sensitive component to the increased temperature. The inhibition of protein synthesis in spermatids above 34 °C can be partly explained by the breakdown of polysomes in these cells. Nakamura and Hall (1980) in a recent study have demonstrated the mechanism by which the body temperature inhibits protein biosynthesis in spermatids of rat testes.

Besides the various cell components, the developing spermatids show a variety of enzymes such as the hyaluronidase, N-acetyl-β-glucosaminidase III, β-galactosidase II, thiamine pyrophosphatase, adenosine 3′-5′-monophosphate phosphodiesterase F, adenosine triphosphatases, hexokinase, phosphoglycerate kinase B, monoamine oxidase, cytochrome oxidase and diaphorases (Males and Turkington 1970, Monn et al. 1972, Sosa et al. 1972, Turkington and Majumder 1975, Vande Berg et al. 1976, 1981, Rudolph et al. 1982, Bilaspuri and Guraya 1982, 1983a, b). Gordeladze et al. (1982b) have observed stage-dependent variation in Mn^{2+}-sensitive adenylyl cyclase (AC) activity in spermatids. The unilateral cryptorchidism is associated with a rapid decline in soluble Mn^{2+}-dependent AC activity in spermatids (Johnson et al. 1981). But 3 days later there is seen an 82% decrease in germ-cell AC activity. The loss of soluble Mn^{2+}-dependent AC activity is associated with a parallel decrease in Sertoli cell secretion of ABP, showing that Sertoli cell factors may be involved in the maintenance of AC activity of germ-cells. Bilaspuri and Guraya (1983e, f) have made a detailed histochemical localization of glycolytic enzymes, alcohol and secondary-alcohol dehydrogenases, and enzymes of various metabolic pathway in the developing spermatids of the buffalo, goat, and ram. Steroid dehydrogenases are reported during their spermiogenesis (Bilaspuri and Guraya 1984a). But the detailed localization of most of these enzyme systems needs to be made at the ultrastructural level in different components of the developing spermatid. The glucose metabolism in round spermatids of rat testis is regulated by adenine nucleotides (Nakamura et al. 1982).

1. Chromatoid Body

The chromatoid body has been described in the spermatids of different species of mammals (Fig. 17) (Russell and Frank 1978, Stefanov and Penkov 1979, Tang et al. 1982). In its morphology, size and localization, the chromatoid body shows species variations. It is not known whether the chromatoid body of spermatids has some developmental and functional relationship with that of spermatocytes described in Chap. III. Therefore, the origin, morphology, chemistry, behaviour, and function of the chromatoid body during spermiogenesis will be described in detail. The chromatoid body of spermatids consisting of an irregularly shaped, dense mass of fibrillar material is usually placed near the Golgi complex of early spermatid in mammals (Fig. 31) (Brökelmann 1963, Fawcett and Phillips 1967, Eddy 1970, Fawcett et al. 1970, Oura 1971, Susi and Clermont 1970, Yasuzumi et al. 1972, Comings and Okada 1972, Holstein and Roosen-Runge 1981). Associated with it is a spherical body showing a characteristic reticular pattern which has been called the satellite of the chromatoid body (Fawcett et al. 1970, Fawcett 1972). The chromatoid body in the developing spermatids of buffalo stains for RNA and proteins (Guraya 1965).

Some divergent views have been expressed about the origin of the material of the chromatoid body, as also discussed for the spermatocytes in Chap. III. When living young spermatids are studied with phase contrast microscopy, intranuclear dense granules are often seen moving near their chromatoid body (Söderström and Parvinen 1976d). Analysis of living step 1 and step 3 spermatids by time-lapse cinephotomicrography has revealed that the chromatoid body moves in relation to the nuclear envelope in two different ways (M. Parvinen and L.-M. Parvinen 1979), as also reported in other mammals (Stefanov and Penkov 1979). Predominantly in step 1, the chromatoid body moves along the nuclear envelope on a wide area surrounding the Golgi complex and develops frequent transient associations with this organelle. In step 3, the chromatoid body is seen to move perpendicular to the nuclear envelope. It has been observed to be located very transiently at the top of prominent out-pocketings of the nuclear envelope with apparent material continuities through nuclear pore complexes to intranuclear particles. An electron microscope study of the corresponding cells has revealed the accumulation of dense material near the chromatoid body which resembles the intranuclear dense material. In a few cases, a connection between the intranuclear particles and the chromatoid body is seen through a pore in the nuclear envelope (Fawcett et al. 1970, Holstein and Roosen-Runge, 1981).

The vesicle systems of the chromatoid body and of the Golgi complex occasionally are interconnected (Söderström and Parvinen 1976d). The material transported from the nucleus to the chromatoid body may be RNA complexed with proteins. The rapid movements of the chromatoid body have been suggested to play a role in the transport of haploid gene products in the early spermatids, including probably nucleocytoplasmic RNA transport (M. Parvinen and L.-M. Parvinen 1979). In these cells, the chromatoid body apparently is dependent on the function of the haploid genome (Söderström 1977), which continues up to the stage of chromatin condensation (Monesi 1965a, 1971, Utakoji 1966, Söderström and Parvinen 1976c, Geremia et al. 1977a, 1978). This is supported by the fact that the chromatoid body in all the early round spermatids is labelled with [^{3}H]uridine when a 2-h pulse is followed by a 14-day chase in the presence of unlabelled uridine (Söderström and Parvinen 1976b). After step 8, when no RNA synthesis occurs in the spermatids nuclei (Söderström and parvinen 1976c), the chromatoid body remains unlabelled (Söderström and Parvinen 1976b). This also suggests that the label of the chromatoid body originates from the spermatid nucleus. This indicates that the RNA synthesized in the haploid nucleus is transported to the chromatoid body. The mechanisms of this transport have not been worked out in detail, although an increased occurrence of nuclear pore complexes has been reported on an area adjacent to the chromatoid body (Fawcett et al. 1970, Bawa 1975, Reger et al. 1977). Stefanov and Penkov (1979) have suggested the nucleolar origin of the chromatoid body in the spermatids of mammals. The labelling of the chromatoid body with [^{3}H]uridine indicates that it may have a storage function for informational RNA that is utilized during spermiogenesis.

The character of the RNA of the chromatoid body remains to be determined in future studies although it has been shown that α-amanitin, a specific inhibitor of HnRNA synthesis inhibits the labelling of young spermatids (Moore 1972). This indicates that the RNA transported from the spermatid nucleus to the chromatoid body may be messenger RNA which may be playing some important roles during spermiogenesis, especially in the formation of acrosome, tail, etc. The intermitochondrial material in the pachytene spermatocytes (Söderström 1976a) is believed to be the precursor material of the chromatoid body (Fawcett et al. 1970) as already discussed in Chap. III for spermatocytes. Söderström and Parvinen (1976b) have indicated that the chromatoid body of rat spermatids is capable of incorporating [^{3}H] uridine. No mention has been made whether grains occur over the chromatoid body of pachytene spermatocytes; however, occasional labelling was observed over their mitochondrial clusters. Further work should be carried out to determine whether the chromatoid body of spermatocytes and spermatids is actually the same or whether they differ in composition. Indeed, it has yet to be demonstrated whether any of the nuage types (see Russell and Frank 1978) have similar biochemical compositions. However, Söderström (1981) has suggested that the nuage and the chromatoid body of rat spermatogenesis are separate organelles having related functions.

2. The Golgi Complex and Acrosome (or Acrosomal Cap) Formation

a) Golgi complex

Histochemical studies have shown that the spherical or hemispherical Golgi complex of mammalian spermatids contains phospholipids and proteins (Fig. 17)

(Guraya 1962, 1965, Guraya and Bilaspuri 1976b). The Golgi complex in the spermatids of buffalo, goat, and ram also contains phospholipids and lipoproteins (G. S. Bilaspuri and S. S. Guraya, unpublished observations); cerebrosides, protein-bound gangliosides and plasmalogen could not be demonstrated histochemically in their Golgi complex which, however, contains nonacetylated mucosubstances. The various histochemical observations are in agreement with the results of the chemical analysis of Golgi fractions (Favard 1969, Keenan et al. 1972).

Associated with the Golgi complex of spermatids are inclusion bodies that stain for phospholipids (Fig. 17). These bodies may correspond to multivesicular bodies of electron microscope studies (Tang et al. 1982). Corresponding to the formation of acrosomal vesicle and granule (Fig. 17) there is seen some decrease in their number. The Golgi complex is associated with various enzyme activities especially phosphatases (nucleoside diphosphatases, cytidine monophosphatase, thiamine pyrophosphatase and acid phosphatase) (Tice and Barnett 1963, Yasuzumi et al. 1970a, b, Reissenweber 1970, Chang et al. 1974, Dalcq 1965, 1967, Favard 1969, Barham et al. 1976, Clermont et al. 1981, Tang et al. 1982); acid phosphatase is usually found in lysosomes of spermatids. Clermont et al. (1981) and Tang et al. (1982) have studied the detailed localization of phosphatases within the various saccules of the Golgi stacks within the same cells. The nicotinamide adenine dinucleotide phosphatase (NADPase) activity is present within the four or five saccules forming the middle of the stacks (the so-called intermediate saccules). Thiamine pyrophosphatase (TPPase) activity is localized in the last one or two saccules of the stack on its trans (mature) side. Cytidine monophosphatase (CMPase) occurs within one or two thicker saccular structures seen in the trans region of the Golgi apparatus; the latter structures have been named GERL (Novikoff et al. 1977). Some vesicles budding from the GERL element and the acrosomic system give a positive reaction for (CMPase) (Clermont et al. 1981). The multivesicular bodies of the Golgi zone contain CMPase and NADPase (Tang et al. 1982). Neither inosine diphosphatase nor TPPase is observed in the Golgi bodies of spermatids during acrosomal formation (Barham et al. 1976). Their activities decrease in the Golgi complex as spermiogenesis proceeds. Reissenweber (1970) has reported that the loss of TPPase marks the transition from a "young" Golgi apparatus to an "old" one which contains lipofucsin pigments. This may mark the end of synthetic activity in the Golgi apparatus since TPPase is believed to be involved in regulation of co-factor levels and in regulation of Krebs-TCA cycle and lipogenesis (Yasuzumi et al. 1970a). Yokoyama and Chang (1977) have observed that in spermatids of hamster testis, the semicircular profiles of endoplasmic reticulum surrounding the Golgi apparatus show an intense reaction for glucose-6-phosphatase, even after the Golgi has moved away from the acrosome into the redundant cytoplasm. Tang et al. (1982) have demonstrated the presence of a dynamic regional differentiation of enzymatic activities within the Golgi apparatus of the spermatid at 19 steps of spermiogenesis. NADPase is present within the four or five intermediate saccules of Golgi stacks, and TPPase is seen in the last one or two saccules on the trans aspect of the stacks from steps 1 to 17 of spermiogenesis. CMPase lies within the thick saccular GERL elements seen in the trans region of the Golgi apparatus from steps 1 to 7 of spermiogenesis, but CMPase-positive GERL disappear from the Golgi apparatus after its detachment from the acrosomic system at step 8. These data have shown (1) that the Golgi apparatus of spermatids, although it loses its CMPase-positive GERL element in step 8, shows evidence of functional activity until it degenerates in step 17; (2) that in early spermatids the various saccular components of the Golgi region show specializations in regard to enzymatic activities; (3) that each Golgi region may contribute in a coordinated fashion to the formation of the acrosomic system.

b) Development of Acrosome

As studied with the PAS technique, the basic processes involved in the formation and development of acrosome (or acrosomal cap) in various mammals including primates are very similar (Guraya 1965, 1971a, Guraya and Bilaspuri 1976b, Oud and de Rooij 1977, Weaker 1977, Sinowatz and Wrobel 1981, Clermont et al. 1981, Holstein and Roosen-Runge 1981, Bilaspuri and Guraya 1983f). The Golgi complex in early spermatids shows strongly PAS-positive pro-acrosomal granules which are associated closely with the PAS-negative pro-acrosomal vesicles (Fig. 23). It has been shown clearly in electron microscope studies that the pro-acrosomal granules and their associated pro-acrosomal vacuoles are formed by the activity of the Golgi membrane (Fig. 34) (Susi et al. 1971, Bustos-Obregon et al. 1975, Holstein 1976, Plöen et al. 1979, Sinowatz and Wrobel 1981, Holstein and Roosen-Runge 1981, Bilaspuri and Guraya 1983f). Stockert et al. (1975), using light and electron microscopy, have demonstrated the selective staining of the developing acrosome in the spermatids of rodents with phosphotungistic acid, as recently reported by Holt (1979) in golden hamster, guinea pig, bull, boar , and ram. Courtens (1978a) using the periodic acid-thiocarbohydrazide-silver proteinate method and the hydrochloric-phosphotungistic acid stain, have reported the localization of glycoproteins in the developing acrosome of ram spermatids.

Susi et al. (1971) have followed the details of alterations, which occur in the Golgi apparatus during rat spermiogenesis, as also reported by Urena and Malavasi (1978) during meiosis and spermatogenesis in hamster. The various studies have revealed conspicuous morphological and histochemical changes in the Golgi apparatus during spermatogenesis (Tang et al. 1982), which are indicative of physiological alterations. This is further supported by the studies of Mollenhauer et al. (1976) on the membrane transformations in the Golgi apparatus of rat spermatids, which have revealed that dictyosome cisternae proximal to ER resemble ER, and those distal to ER resemble acrosome membrane. The various membrane transformations demonstrated within the Golgi apparatus support the hypothesis that membrane constituents move from the ER to the acrosome via transport vesicles and that the Golgi apparatus is the intermediary component where the transformations occur.

The rapid nonrandom movements of the chromatoid body and associated granules and vesicles in the early rat spermatids have suggested that this organelle may participate from the early stages of spermiogenesis in spermatid maturation (Parvinen and Jokelainen 1974). The morphological association of the chro-

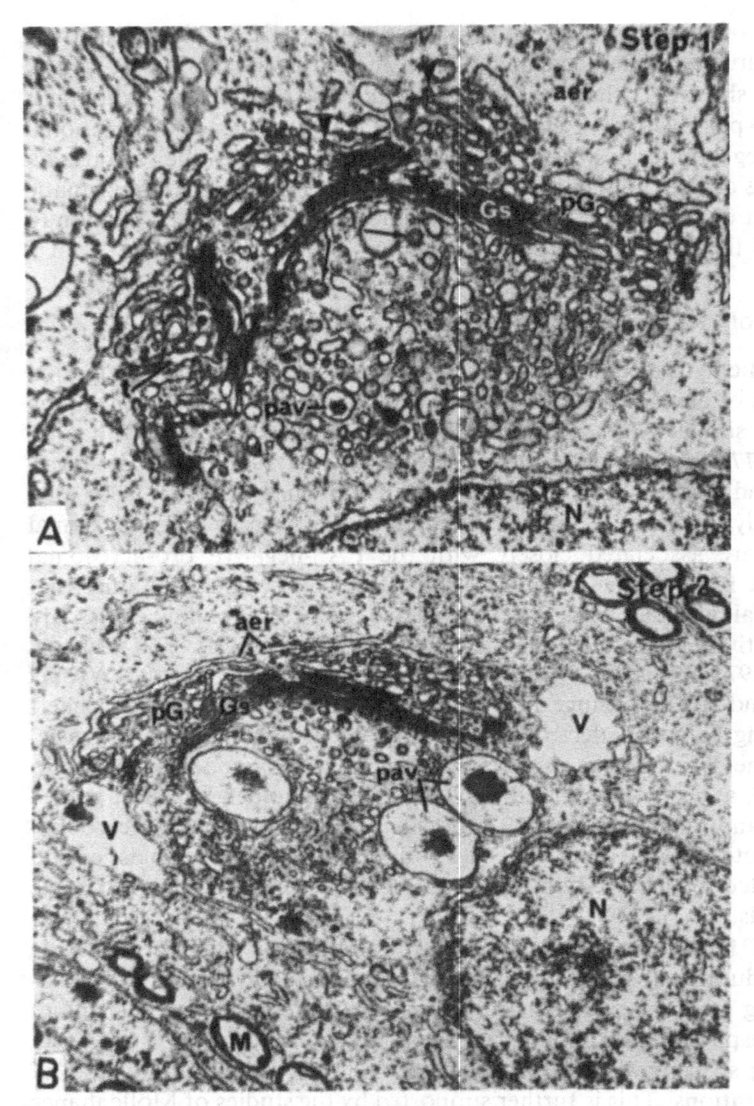

Fig. 34. A Step 1 spermatid of rat. Small coated buds are clearly attached to the extremities of outermost Golgi saccules (*arrowheads*). In addition, tubular extensions (*t*) of the outermost saccule are seen at the *lower left centre* projecting into the peripheral Golgi region between the Golgi stack (*Gs*) and the associated layer of endoplasmic reticulum (*aer*). In the concave zone (*c*) large coated vesicles (*arrows*) and smooth profiles are present, as well as a pro-acrosomic vesicle (*pav*) showing a dense core and a light outer zone. Uranyl acetate and lead citrate. **B** Step 2 spermatid of rat. The enlarged pro-acrosomic vesicles (*pav*) with their dense core and light outer zone lie close to the nucleus (*N*). Two large vacuoles (*V*) are present lateral to the Golgi apparatus. The enveloping layer of endoplasmic reticulum (*aer*), the peripheral region of the Golgi apparatus (*pG*), and the Golgi stacks (*Gs*) are also indicated. Uranyl acetate and lead citrate (From Susi et al. 1971)

matoid body with the Golgi complex during the formation of pro-acrosomal granules or with the developing acrosome itself has provided some evidence for its participation in the formation of the acrosomal system in early spermatids (Bawa 1975, Söderström and Parvinen 1976b, Reger et al. 1977). But the exact nature of its participation or contribution must be determined in future studies.

With the differentiation of spermatids, the pro-acrosomal vesicles and granules fuse with each other to form the large, fluid-filled acrosomal vesicle which surrounds the acrosomal granule (Fig. 23) (Guraya 1971a, Holstein and Roosen-Runge 1981). From the time of formation of pro-acrosomal granules to the step showing a spherical acrosomal granule associated with the nuclear envelope, the spermatids are generally said to be in a Golgi phase (Clermont 1972). The Golgi complex, which contributes glycoprotein material to the acrosomic system throughout its formation and development, eventually separates from it to move back into the caudal region of the cell (Fig. 23). The membranes of the Golgi complex, unlike the mitochondria, have no intrinsic contractility or capacity for independent movement. The caudal migration of the Golgi, therefore, is believed to be a part of general anteroposterior flow of cytoplasm. The accretion of material within the acrosome ends when the Golgi apparatus migrates backwards to the posterior cytoplasm of early elongating spermatid (Susi and Clermont 1970).

With the displacement of cytoplasm including the Golgi complex from the anterior pole of the spermatid, the fluid content of the vesicle is resorbed and its lumen is reduced considerably by collapse of its limiting membrane over the anterior portions of the nucleus (Fig. 23) (Guraya 1965, 1971a, Holstein and Roosen-Runge 1981). Courtens (1978b), using morphometric methods, has produced evidence for a rapid increase in the acrosome volume and area in young spermatids of ram, followed by a decrease in these two parameters in older spermatids. This decrease in volume is in accordance with the observation of a concentration of the acrosome content taking place in the developing spermatids (Courtens 1978a, 1984, Holstein and Roosen-Runge 1981). But the concentrating mechanism is still to be determined, because no material is seen to escape from the acrosome in step 5 and step 6 spermatids. The degree of acrosome spreading appears to be preceded by nuclear and perinuclear substance development (Courtens and Courot 1980, Courtens 1984). The main factors influencing post-acrosomal cytoplasm morphogenesis are believed to be modifications of the spermatid nuclear content as well as changes in structure, composition and position of the envelope (Courtens 1978c, 1984).

During spreading and condensation the substance of the acrosomal vesicle precipitates in the form of a homogeneous, strongly PAS-positive thin layer adjacent to the anterolateral surface of the nucleus (Fig. 23). Among the eutherian mammals there exists species variation in regard to the amounts of material contributed by the acrosomal vesicle and granule to the developing acrosome (Guraya 1971a, Holt 1979, Holstein and Roosen-Runge 1981). The period during which the nucleus and the acrosomic system undergo the most dramatic morphological and physico-chemical transformations has been referred to as the acrosome phase (Leblond and Clermont 1952b). Recent electron microscope studies have revealed that the acrosomic system during rat spermiogenesis splits into two portions early in step 15 to give rise to the main portion with its crest-like acrosome running along the dorsal aspect of the nucleus, and the head cap extending over the lateral surfaces of the nucleus and a smaller head-cap segment, which is seen in steps 15 and 16, along the side of the nucleus at its apical extremity (Lalli and Clermont 1981). This separated head-cap segment reaches the apical-ventral aspect of the head during step 17 and condenses in synchrony with the rest of the acrosomic system in step 19 of spermiogenesis. The large crescentic acrosome in step 15 forms a large fin at the caudal extremity of the acrosomic apparatus, moves anteriorly during steps 16 and 17, while the whole acrosomic system extends farther apically beyond the tip of the nucleus. Factors influencing morphogenesis of the acrosome appear to act in primary spermatocytes and may be controlled by the endocrine balance possibly mediated via the functional relationships between Sertoli cells and germ-cells (Courtens and Courot 1980, Courot and Courtens 1982). In the ground squirrel, the number, time of appearance, and conformation of microtubules adjacent to spermatid heads suggest that, at least in this species, Sertoli cell microtubules facilitate the development of normal acrosome (Vogl et al. 1983a). The results obtained after treatment with colchicine support this suggestion (Vogl et al. 1983b). A wreath of microfilaments close to the Sertoli cell membrane surrounds the head of the developing spermatid in humans (Holstein and Roosen-Runge 1981). These microfilaments may be involved in the morphogenesis of the acrosome. Recently, Welch and O'Rand (1985a) have reported the presence of acrosomally associated actin in rabbit and mouse spermatogenic cells, which may also be involved in the morphogenesis of the acrosome.

Various types of malformations of acrosome have been studied in humans with electron microscope. Baccetti et al. (1977) have studied the development of the acrosome in human testicular biposies from a donor yielding round-headed sperm. An aplasia or hypoplasia of the acrosome has been demonstrated in this study. This aberration has been attributed to the malfunctioning of the Golgi apparatus in producing an acrosomic vesicle, which fails to establish a complete close contact with the spermatid nucleus, and diminishes instead of flattening over the nucleus. Without an acrosome the spermatid nucleus does not attain a conical form at the anterior pole. Vegni-Talluri et al. (1977), using light and electron microscopy, have investigated the acrosome malformations of spermatids and spermatozoa in the testes of two infertile patients. The first visible abnormalities appear at an early spermatid stage. In most cases of malformed spermatids, only the differentiation of the acrosome granule is interfered with as the acrosome vesicles appear either without acrosome granule or with a reduced or malformed granule (Figs. 35 and 36). The lack of the acrosome granule is associated with Golgi vesicles showing no electron-dense material. The fact that almost half of the early spermatids do not develop the acrosomal granule suggests that the original cause is genetic and that the genes are expressed in the haploid phase. Holstein and Schirren (1979) have provided an excellent account of various types of malformations of acrosome in human spermatids, which are the most frequent. They have discussed these malformations under two main categories which include (1) malformations of the acrosome attached to the nucleus of spermatid and (2) malformations of the acrosome, which lacks con-

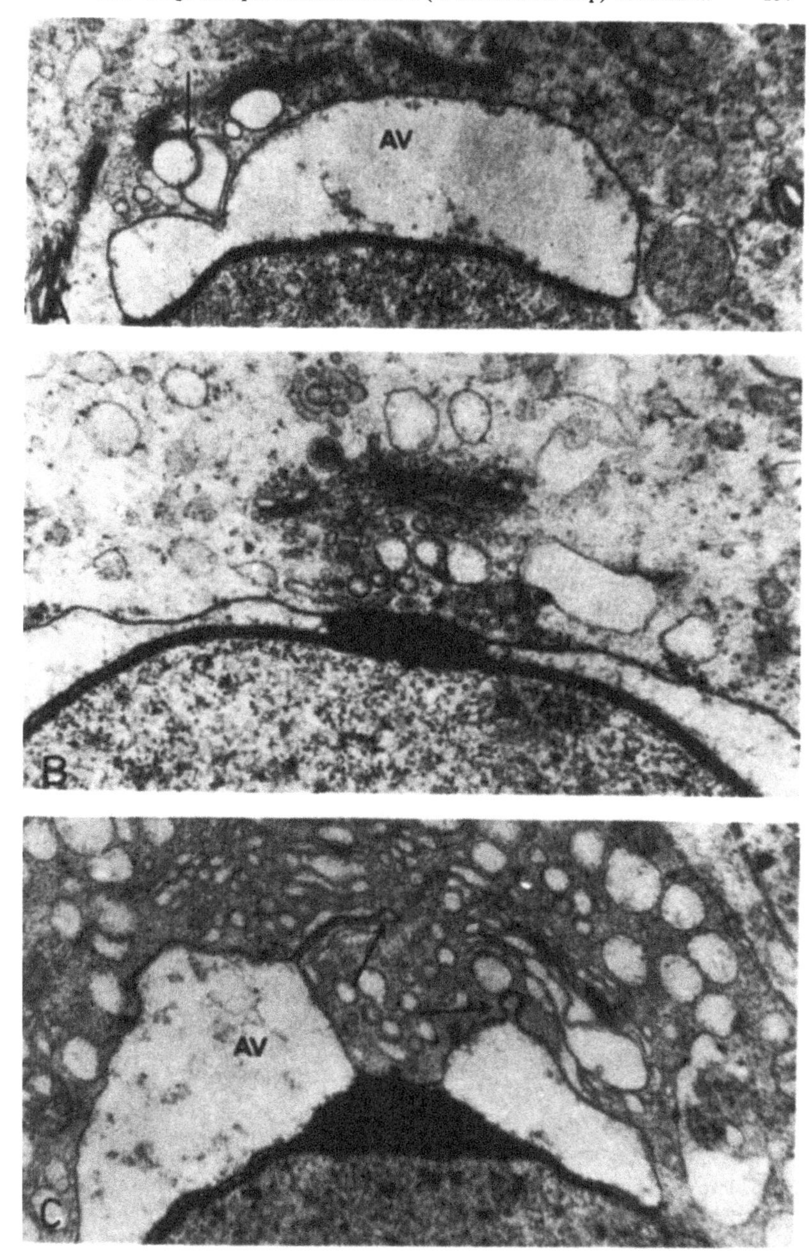

Fig. 35. A Micrograph from serial sections of early human spermatids showing the acrosomic vesicle (*AV*) and Golgi vesicle (*arrow*) devoid of electron-dense material; **B** Early spermatid from serial sections showing moderately electron-dense material in vesicles of Golgi complex. The developing acrosome is slightly more mature than the acrosome of the spermatid; **C** Malformed acrosomic vesicle (*AV*) of early spermatid with membrane protrusion (*arrows*) that surround the Golgi vesicles (*GV*) (From Vegni-Talluri et al. 1977)

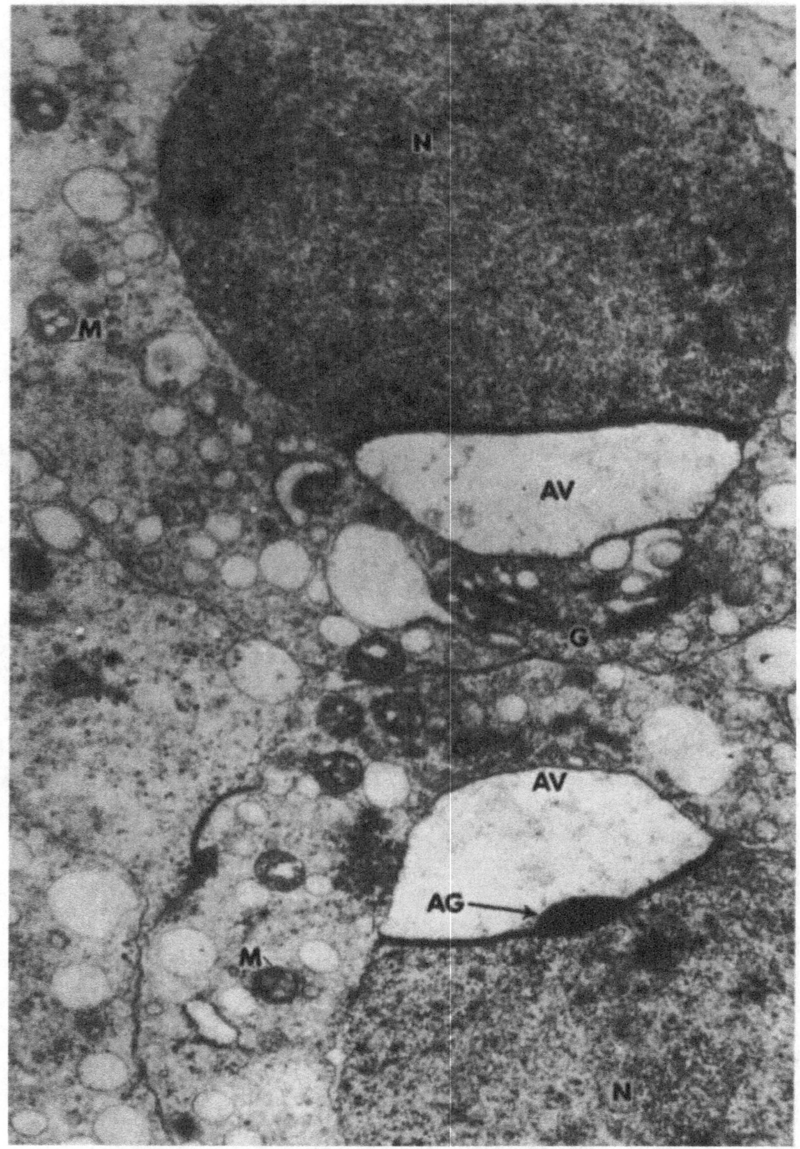

Fig. 36. Two early human spermatids from serial sections showing presence and absence of acrosomic granule (*AG*). Nucleus (*N*), acrosomic vesicle (*AV*), mitochondria (*M*), Golgi complex (*G*) are present (From Vegni-Talluri et al. 1977)

tact with the nucleus of the spermatid. Courtens and Courot (1980) using hypophysectomized and testosterone-supplemented rams have demonstrated that normal acrosome development is under the control of endocrine-dependent cellular events occurring before the onset of spermiogenesis, possibly via Sertoli cell/germ-cell interactions that have been demonstrated in recent in vitro studies

(Ziparo et al. 1980, see Chap. I). The development of the acrosome and the alignment, elongation and entrenchment of spermatids have been studied in rats treated with procarbazine (Russell et al. 1983c). Procarbazine causes various other changes in specific cell association during the cycle of the seminiferous epithelium (Russell et al. 1983d).

The acrosome of testicular spermatozoa in the buffalo, goat, and ram shows PAS-positive mucopolysaccharides (Bilaspuri and Guraya 1983f), as reported in several other species of mammals (Guraya 1965, 1971a, Berndston and Desjardins 1974, Oud and de Rooij 1977, Sinowatz and Wrobel 1981) including Indian mongoose (S. S. Guraya, unpublished observations). Glycogen, acidic mucopolysaccharides and lipids could not be demonstrated. However, Onuma and Nishikawa (1963) have reported the presence of glycolipids in the acrosome. Courtens (1978a), using cytochemical methods at the ultrastructural level, has demonstrated several types of carbohydrate-containing substances in the developing acrosome of the ram. These substances fill the acrosomal granule during the early stages of spermiogenesis and then are evacuated sequentially from it during the cap phase of acrosome development (Holt 1979). These attain their definitive location in the acrosomes of elongating spermatids during the sequential phases, resulting in the subdivision of the organelles into different regions according to their cytochemical affinities. Morphological and cytochemical studies at the ultrastructural level have revealed that the acrosome of the early elongating spermatids of ram emits "post-acrosomic vesicles" containing glycoproteins. Courtens (1979a) has further observed that these post-acrosomal vesicles are carried to the posterior cytoplasm of elongating spermatids following a well-defined route. Finally, they are fused with the plasma membrane which is stained progressively with the cytochemical method used.

Some better techniques should be used to detect the diversity of glycoproteins in the developing acrosome, since numerous acrosomal enzymes are of glycoprotein nature (Dalcq 1967, Borders and Raftery 1968, Brunish and Högberg 1968, Fléchon 1973). Courtens and Loir (1975b) have observed that the protein composition of the acrosomal granule is different from its periphery to its centre. Seiguer and Castro (1972), using electron microscope, have demonstrated the presence of arylsulphatase activity during acrosome formation in the rat. According to Sakai et al. (1979), the specific reactions for acrosomal proteinase occur in the cap but not in the acrosomal collar of developing spermatids and mature spermatozoa of sheep testis. Tang et al. (1982) have observed that the acrosomic system itself is reactive for CMPase and TPPase but is negative for NADPase. Tritiated-fucose is incorporated readily within the Golgi apparatus of steps 1 – 17 spermatids; then in steps 1 – 7 it is incorporated within the acrosomic system.

The behaviour of acrosomal enzymes is of special interest in relation to spermatogenic development. For example, assays of testicular hyaluronidase in young rats have revealed that its activity is barely detectable until about 34 days of age, i.e. throughout the period of spermatócyte development; it then suddenly appears and rapidly increases 400-fold, the rise corresponding chronologically to the formation of acrosomes in the spermatids (Males and Turkington 1970). Hypophysectomy performed at 28 days of age is highly effective in preventing the formation of spermatids and the appearance of hyaluronidase activity, but when the hypophysectomized rats are treated from the age of 30–33 days onwards with LH and FSH or testosterone, the development of spermatids is promoted and hyaluronidase activity increases. These observations have suggested that hyaluronidase activity represents a good biochemical marker for fingerprinting the hormone-dependent cap phase of spermiogenesis. The development of proacrosin and acrosin has been followed during spermiogenesis of different mammals (Florke-Gerloff et al. 1983, Phi-Van et al. 1983, Florke et al. 1983, Mansouri et al. 1983). Arylsulphatase activity also has been used to distinguish between the individual transformations leading from the acrosomal granule and vesicle to the complete acrosome system, as it is evident already in the acrosomal vesicle, and continues to increase during the subsequent stages of the spermiogenesis process (Seiguer and Castro 1972). Arylsulphatase A has been purified to homogeneity from rabbit testes; it is an acid glycoprotein (with 0.8% of N-acetylneuraminic acid) hydrolysing several sulphate esters, including steroid sulphates, cerebroside 3-sulphate and ascorbic acid 2-sulphate (Farooqui and Srivastava 1979a). Another useful lysosomal enzyme is the acrosomal acid phosphatase. Hypophysectomy carried out at 28 days of age produces a decline in activity, but treatment with FSH and LH or with testosterone started on day 30 enables spermatogenesis to progress, and the specific activity of acid phosphatase is restored to control levels (Turkington and Majumder 1975). In the mature spermatozoa, lysosomal acid phosphatase is associated with the subacrosomal space rather than the acrosome proper (Teichman and Bernstein 1971).

3. Mitochondria

The mitochondria of early spermatids in the marmoset, chimpanzee, and buffalo are in the form of granules (Fig. 17) (Guraya 1965). These organelles stain lightly with Sudan black B, indicating low levels of lipids (lipoproteins). The mitochondria in the spermatids of buffalo, goat, and ram also contain lipoproteins (G. S. Bilaspuri and S. S. Guraya, unpublished observations). This is in agreement with the histochemical observations on the mitochondria of other mammalian species (Guraya 1965, 1971b, 1973).

The ultrastructure of mitochondria in the mammalian spermatids is similar (Fawcett 1958, 1979, Fawcett and Burgos 1956, de Kretser 1969, Guraya 1971b, 1973, Bustos-Obregon et al. 1975, Holstein 1976, de Martino et al. 1979, Kerr and de Kretser 1981, Holstein and Roosen-Runge 1981, Courtens 1984). The organelles usually contain a few cristae which generally are folded over and flattened against the inner aspect of the limiting membrane, resulting in the formation of clear vacuoles in the interior of the mitochondria (Fig. 31) (Rattner and Brinkley 1970). Similar vacuoles also have been described in the mitochondria of the developing oocyte of mammals (Wischnitzer 1967, Guraya 1974, 1985). According to André (1962, 1963), the intracristal spaces are filled with clear substance, probably largely aqueous in nature, for which he has proposed the name pseudomatrix. It probably is due to the lesser development of cristae and the greater formation of the pseudomatrix (represented by vacuoles) that the mitochondria show little lipoprotein component demonstrable with histochemical

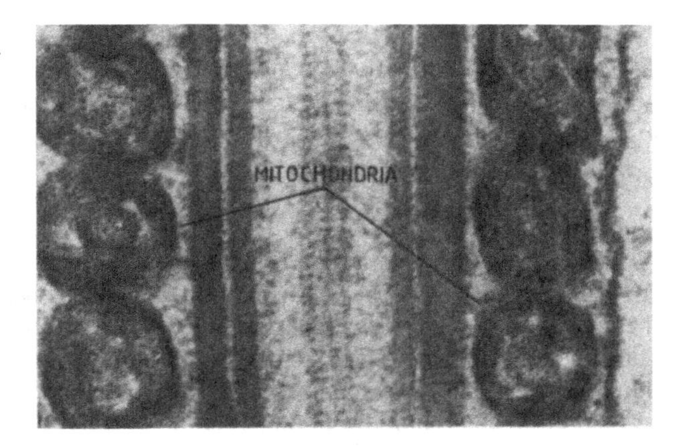

Fig. 37. Ultrastructure of portion of longitudinal section of mid-piece from the testicular spermatozoon of ram, showing mitochondria and various components of axoneme (Courtesy of Dr. M. Courot)

techniques in the early spermatids. Then lipid/protein ratio in general shows a positive correlation with respiratory activity and the development of cristae (Borst 1969). The biochemical studies have suggested that the condensed appearance of spermatid mitochondria is the expression of an active functional state.

The mitochondria during the later stages of spermiogenesis (Fig. 17) stain more deeply with Sudan black B, indicating the development of more lipids, as also described for other mammals including opossum (Guraya 1965, 1971b, 1973). The exact nature of these lipids could not be determined. They may be phospholipids, which usually form more than 90% of mitochondrial lipids (Borst 1969). Guraya (1973) has suggested that the lipids may include choline plasmalogen (phosphatidal choline) in the case of ram, as shown by biochemical techniques (Mann 1964).

The interior of the mitochondria during late spermiogenesis is generally filled with whorls of membranes, and a granular or amorphous substance (Fig. 37) (Hrudka 1968b, Pedersen 1970a, Hughes 1976, Harding et al. 1979); the lipids demonstrated with both histochemical and biochemical techniques may be derived mainly from the mitochondrial membrane (Borst 1969, Guraya 1971b, 1973). As a result of the accumulation of lipid material (or ultrastructural membranes) and granular substance inside the mitochondria, the vacuoles (or pseudomatrix) seen in the mitochondria of early spermatids generally disappear during spermiogenesis. André (1963) has pointed out that the pseudomatrix (corresponding to the vacuoles of Guraya 1973) is expelled from the mitochondria in the middle-aged spermatid. By integrating the results of electron microscope, histochemical and biochemical techniques, Guraya (1971b, 1973) has suggested that the mitochondria during later stages of spermiogenesis in mammals might be functioning in the storage of natural respiratory or oxidative substances (lipids), besides the usual enzyme and co-enzyme systems concerned with respiration, oxidative phosphorylation, and electron transport through the cytochrome system, which form the mitochondrial activity in general (Mohri and Mohri 1965, Farriaux and Frontaine 1967, Borst 1969, Mathur 1971). This is supported further by the biochemical studies which have shown that the available intracellular nutrient reserve of spermatozoa consists, so far as is known, mainly of phospholipids (Mann 1964, Poulos et al. 1973a, Mann and Lutwak-Mann 1981). Besides lipids and enzymes, mitochondria also may accumulate calcium and other ions that are known to play very important roles in mitochondrial physiology (Borst 1969).

4. *Manchette and Shape of the Sperm Head*

As the nucleus and covering acrosomal cap take their new shape, several phenomena can be observed at the caudal end of the nucleus near the point of insertion of the flagellum. A manchette or caudal sheath appears in the cytoplasm of differentiating spermatid and extends from a close association with the nucleus to surround the flagellum for some distance (Fig. 17). It appears prior to the onset of chromatin condensation encircling the caudal pole of the nucleus and extending back into the post-nuclear cytoplasm. Electron microscope studies have shown that the caudal sheath is a transient sleeve-like organelle composed of a layer of related microtubules lying in close parallel array to the long axis of the cell, around the posterior part of the nucleus and the beginning of the flagellum (Figs. 38 and 39). Its microtubules, having a diameter of approximately 20 nm, differentiate from the cytoplasmic, ring-like structure at the base of the acrosome, which constitutes the nuclear ring of manchette, form a cylinder extending from the distal margin of the post-acrosomal dense lamina over the post-acrosomal part of the nucleus, the neck region, to the beginning of the principal piece of the tail (Mackinnon and Abraham 1972, Bustos-Obregon et al. 1975, Holstein 1976, Phillips 1980, Holstein and Roosen-Runge 1981, Courtens and Loir 1981c); proximally, the microtubules arise from a floccular, loosely distributed dense material in the human spermatid (Holstein and Roosen-Runge 1981). Fawcett et al. (1971) have followed the details of origin and morphogenesis of microtubules which polymerize from amorphous precursors. At the same time, the cell membrane near the posterior margin of the acrosomal cap becomes specialized locally by deposition of a layer of dense, fibrillar material on its cytoplasmic surface. The proximal ends of the microtubules, which are of uniform diameter, are embedded in the dense fibrillar material that occupies the concavity of the nuclear ring (Figs. 39 and 40).

Further orientation of the microtubules changes as spermiogenesis proceeds. Fawcett et al. (1971) have described the presence of a dense zone of amorphous material, and vesicular and cisternal elements of the endoplasmic reticulum in the region of microtubules which are connected by slender cross bridges (Figs. 41 and 42), as also reported by Mackinnon and Abraham (1972) and Mackinnon et al. (1973). The authors believe that this dense zone represents precursor protein which polymerizes to form new microtubules. Microtubules from a variety of cell types are now well known to consist of tubulin (Margulis 1973). But it will be interesting to work out the chemistry of isolated microtubules from the manchette of mammalian developing spermatids, as no research in these lines has been carried out previously. Wolosewick and Bryan (1977) have made a study of ultrastructural characterization of the manchette microtubules in the seminiferous epithelium of the mouse. The results of their study have indicated that the manchette microtubules might be intermediate type microtubules as they show the reactions of both cytoplasmic and flagellar-type microtubules.

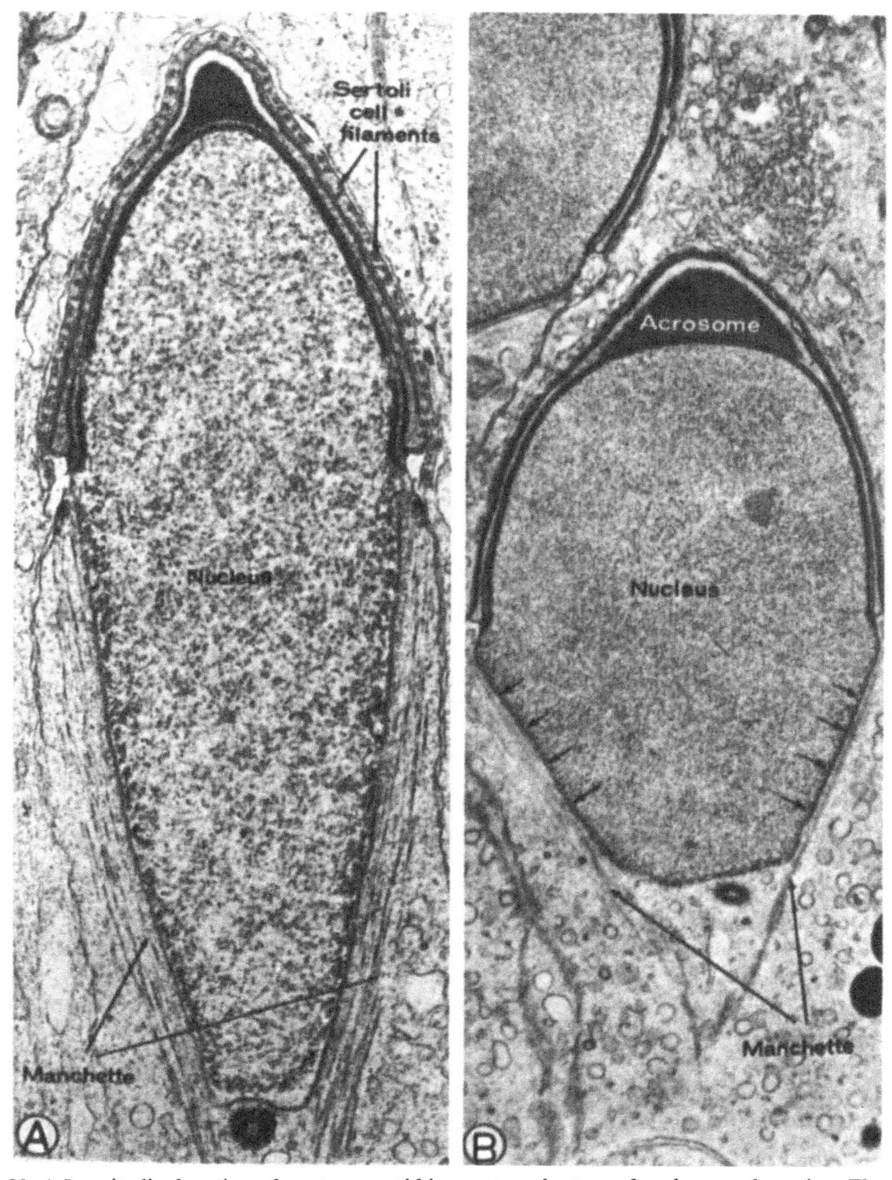

Fig. 38. A Longitudinal section of a rat spermatid in a very early stage of nuclear condensation. The first indications of the nuclear ring are seen as a pair of ill-defined densities subjacent to the cell membrane just behind the acrosomal cap. Microtubules of the developing manchette are tangential to the elongate caudal half of the nucleus and converge towards the base of the flagellum. Although the nucleus appears to be deformed by the microtubules of the manchette, this truncated pyramidal form of the nucleus is very transient and after the manchette assumes parallel orientation extending straight back from a well-developed nuclear ring, the nucleus loses this convergent contour; **B** Longitudinal section of a chinchilla spermatid just before the onset of nuclear condensation. The microtubules of the developing manchette are closely applied to the nuclear envelope in its caudal half (*at arrows*), and their convergent course deforms the posterior half of the nucleus. This transient pyramidal form bears no relation to the final form of the condensed nucleus which diverges slightly towards its caudal surface (From Fawcett et al. 1971)

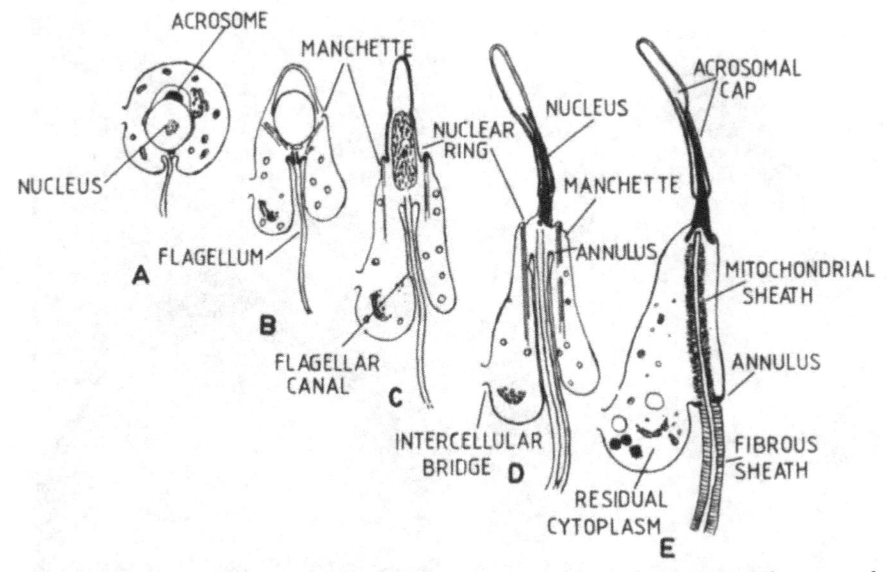

Fig. 39. Successive stages of spermiogenesis in the guinea pig. *A* The nucleus and acrosomal cap of the spermatid are centrally located in the spermatid cytoplasm in the Golgi and cap phases of development; *B* with completion of the acrosomal cap, the Golgi migrates into the caudal cytoplasm, microtubules appear tangential to the caudal half of the nucleus, and spermatid elongation begins; *C* caudal displacement of cytoplasm brings the plasma membrane into close apposition to the acrosomal cap. Nuclear condensation begins: the microtubules become arranged in a cylinder originating in a dense nuclear ring at the caudal margin of the acrosomal cap; *D* nuclear condensation and spermatid elongation continue. Nuclear ring moves back to the level of the base of the flagellum; *E* nuclear ring and manchette disappear; annulus migrates caudally to the anterior margin of the fibrous sheath; mitochondria gather around the base of the flagellum anterior to the annulus and form the mitochondrial helix of the middle piece (Redrawn from Fawcett et al. 1971)

The manchette is a transitory cytoplasmic structure which disappears soon after the formation of the mid-piece and elongation of spermatid (Fig. 39) (Fawcett et al. 1971, Bustos-Obregon et al. 1975, Holstein 1976, Holstein and Roosen-Runge 1981). Its exact function in the formation of spermatozoon is still not known. According to Courot and Fléchon (1966), it could serve as a structural framework within the cell. Others have implicated the microtubules of the manchette in the elongation and shaping of the spermatid nucleus (Dooher and Bennett 1973, 1974). Baccetti et al. (1977) have observed that round-headed spermatozoa in humans derive from spermatids missing the caudal manchette. Only a few sperm show a partially elongated nucleus. These derive from spermatids with differentiated microtubule manchette. These findings are consistent with the view that the caudal manchette plays a role in the nuclear shaping, but conclusive evidence is still required (Courtens 1984).

Some data speak against the view that the manchette is responsible for shaping the nucleus. In marsupials the nucleus is flattened in a plane perpendicular to the long axis of the sperm (Phillips 1970a, Rattner 1972, Harding et al. 1979), whereas the microtubules of the manchette are disposed parallel to the long axis of the cell. Moreover, the microtubules are usually not in actual contact with the

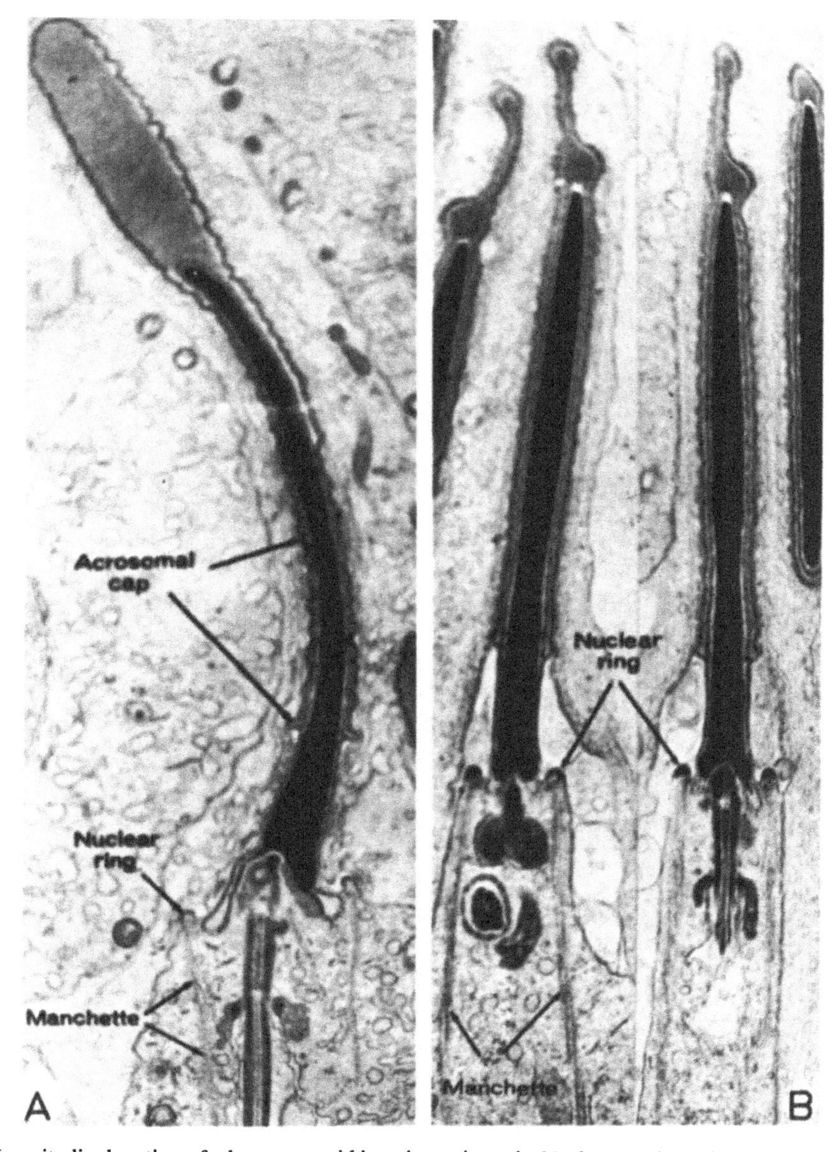

Fig. 40. A Longitudinal section of a late spermatid in guinea-pig testis. Nuclear condensation is essentially complete. Notice that the profile of the caudal third of the nucleus diverges towards the tail. Thus the original convergent deformation of the caudal pole of the nucleus by microtubules is not represented in the final form of the sperm head. In the late stages of nuclear condensation, the nuclear ring moves back to the level of the neck. Therefore, the manchette is not in a position to contribute to the shaping of the nucleus; **B** Longitudinal sections of ram spermatids in a late stage of development. Notice that in this species also the nuclear ring and manchette are caudal to the nucleus. Even in early stages the manchette never extended farther anteriorly than the caudal margin of the acrosomal cap. Being closely related to no more than a fifth of the length of the nucleus, it could not have an important direct influence upon its shape (From Fawcett et al. 1971)

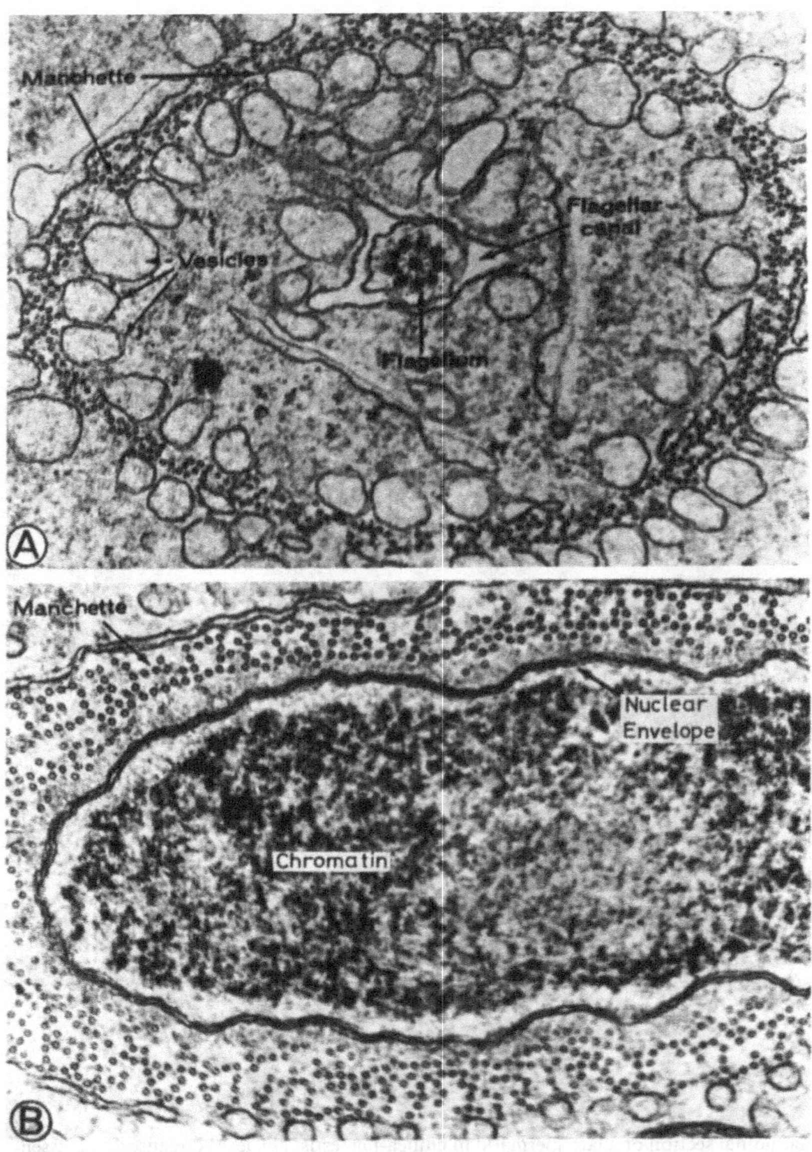

Fig. 41. A Transverse section through the post-nuclear region of a Chinese hamster spermatid, showing the flagellum in its flagellar canal surrounded at a distance by the cylindrical array of microtubules comprising the manchette. There is a striking and intimate association of vesicles with the inside and outside of the manchette. It is suggested that these may be part of an anteroposterior flow of cytoplasm along the manchette during spermatid elongation; **B** Transverse section of the caudal pole of the condensing hamster spermatid nucleus and the associated microtubules of the manchette. It is noteworthy that the microtubules are not in contact with the nuclear envelope, nor is the latter in contact with the condensing chromatin. Were the microtubules compressing the nucleus and underlying chromatin, it is likely that they would be in more intimate contact (From Fawcett et al. 1971)

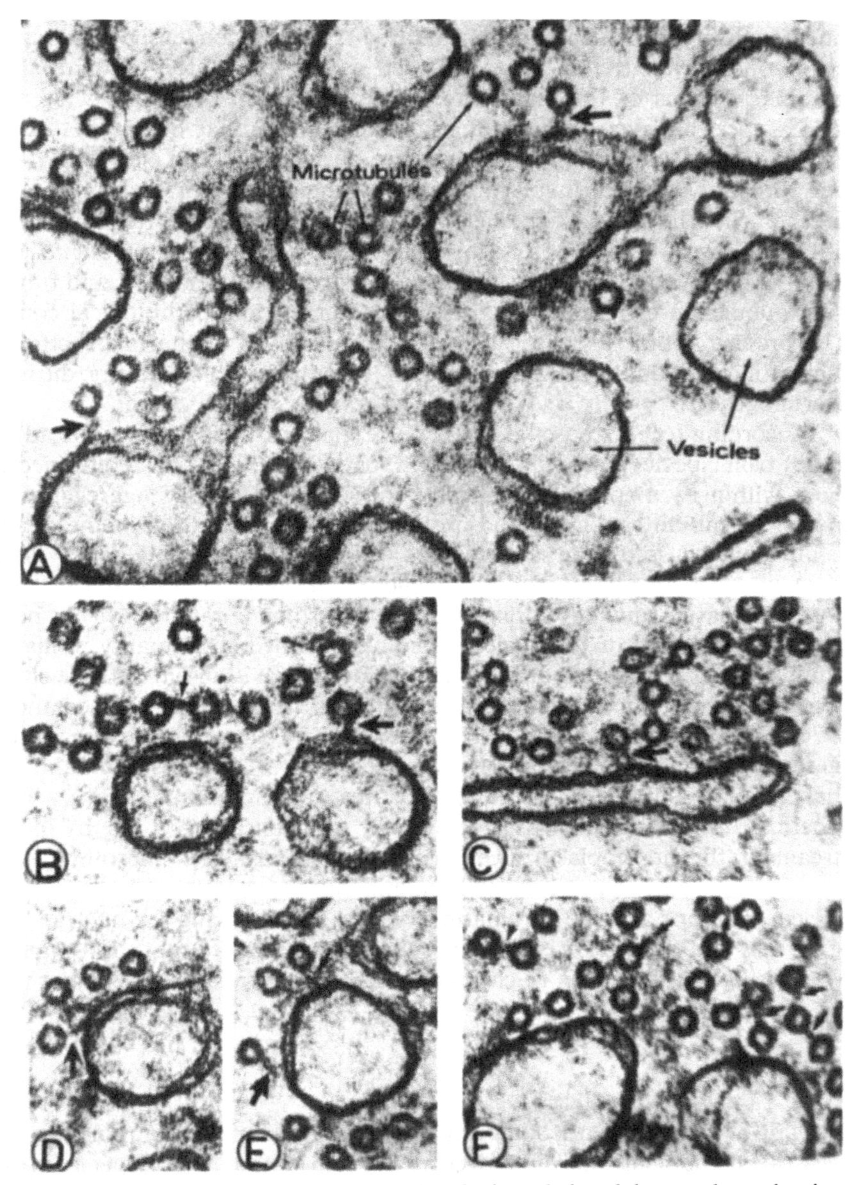

Fig. 42 A – F. High magnification electron micrographs of microtubules of the manchette showing bridgelike linear densities linking adjacent microtubules (*small arrows*). In addition, there are several examples (*at larger arrows*) of similar links between microtubules and the membrane of associated vesicles. Such vesicle-tubule linkages in nerves have been interpreted as the basis of directional movement of vesicle along microtubular tracks in the axoplasm (From Fawcett et al. 1971)

nuclear envelope, which is also separated from the condensed chromatin by a clear zone (Fawcett et al. 1971). Fawcett et al. (1971), after studying the topographic relations of the manchette to the nucleus in the rather wide range of animal species, have concluded that its microtubules may be essential for the redistribution of cytoplasm that takes place during spermatid elongation, but that they probably are not involved directly in the shaping of the nucleus. Several other workers have also suggested that the microtubules are often associated with directed flow of cytoplasm or movement of its particulate components (Fawcett et al. 1971). Dustin (1984) has discussed the structure, chemistry, and function of microtubules in different cell types under normal and pathological conditions. The present author also feels that the microtubules of the manchette may be facilitating the anteroposterior flow of cytoplasm which occurs during early spermiogenesis.

According to Fawcett et al. (1971), the shape of the nucleus (or head), which varies from species to species (Guraya 1965, 1971a), must be determined largely from within by a specific genetically controlled pattern of aggregation of the molecular subunits of DNA and protein during condensation of the chromatin, as already discussed. The same may be valid for the acrosome, which is molded in many cases in absence of a manchette. In addition, Plöen (1973a, b) has observed that normal development of the acrosome in rabbit does not need normal differentiation of the nucleus. The well-defined abnormality in man in the round-headed spermatozoon in which the acrosome is absent and the chromatin is condensed only in a rounded central zone of the nucleus (Pedersen and Rebbe 1974, Holstein and Schirren 1979). This further supports the suggestion that there exists some correlation between chromatin condensation and nuclear shape. Tokuyasu (1974) has suggested that the microtubules have a morphogenetic action on the nucleus, not by their mechanical effect but by inducing a stream of the nucleoplasm away from the nucleus. The microtubules of the manchette in ram, goat, boar, stallion, and bull spermatids are linked to both the nuclear envelope and the chromatin by fibres transpiercing the nuclear envelope (Courtens and Loir 1981c, Courtens 1984). It is suggested that this organization, by allowing redistribution of chromatin prior to spermiation, is similar to half a mitotic apparatus.

5. *Tail*

The tail of the spermatozoon, which consists of various components, namely neck, axial filament, and mid-piece, is formed during spermiogenesis (Figs. 17 and 39). The use of the electron microscope actually has enabled the various workers to learn the details of their development and ultrastructure in various mammalian species (Fawcett and Phillips 1970, Nicander 1970, Bustos-Obregon et al. 1975, Holstein 1976, Holstein and Roosen-Runge 1981, Kerr and de Kretser 1981). But the precise mechanisms in regard to the sites of synthesis of their macromolecules and subsequent morphogenetic changes need to be determined. The connection between the tail and nucleus (or head) and the formation of the mid-piece can be seen at the stage of nuclear condensation and elongation. The neck consists of a proximal centriole and nine cross-striated fibres; four of them fuse 2 by 3 to form the "implantation plate" (Fawcett and Phillips 1970). The axial filament is formed when the acrosomal vesicle appears (Watson 1952).

a) Centrioles and Development of Tail Flagellum (Axoneme)

The tail primordium grows like a cilium at the end of distal centriole in very young spermatids (Fig. 39) (Nicander 1970, de Kretser 1969, Rattner and Brinkley 1970, Fawcett and Phillips 1969a, Fawcett 1972, Gordon 1972a, Bustos-Obregon et al. 1975, Holstein 1976, Holstein and Roosen-Runge 1981, Kerr and de Kretser 1981, Czaker 1985d). The two centrioles gradually move to a position near the nucleus and the proximal part of the cilium then lies within a cytoplasm canal (Fig. 39), which is formed by the fusion of some vesicles (de Kretser 1969, Baccetti et al. 1978). Soon after initiation of axoneme formation at the end of the distal centriole fine filamentous material gathers around the distal end of the proximal centriole to form the centriolar adjunct, which has the structure of a cylinder comprised of nine incomplete triplets (Fig. 43). It is a transient organelle that disappears rapidly from the late spermatid leaving no vestige in the mature spermatozoon (Fawcett and Phillips 1969a). The exact role of this transient organelle is still not known (see also Czaker 1985d). However, Fawcett (1972) believes that it may have a significant role in organizing the development of the neck and flagellum. A study has also been made of in vitro microtubule assembly on centrioles from mammalian spermatids (Esponda and Avila 1983). Next to the distal centriole, a "cup-shaped" lateral junction body is seen in the human spermatids (de Kretser 1969, Holstein and Roosen-Runge 1981).

The mode of formation and differentiation of the basal plate and connecting piece has been considered in various electron microscope studies (Sapsford et al. 1967, 1969, de Kretser 1969, Fawcett and Phillips 1969a, 1970, Gordon 1972a, Yasuzumi et al. 1972, Woolley and Fawcett 1973, Chakraborty 1979, Holstein and Roosen-Runge 1981). These studies have shown that the dense fibrillar material deposited in and around the proximal centriole gives rise to the articular element or basal plate, while similar material emerging from interstices in the wall of the distal centriole polymerizes along its sides to form the primordium of the nine cross-striated columns of the connecting piece (Fig. 44) which surrounds both centrioles. The centrioles do not possess synthetic activity. The precursor materials of the connecting piece are believed to be synthesized elsewhere in the cell and are induced to aggregate or polymerize along the sides of the triplets. Irons (1983) has studied the synthesis and assembly of connecting piece proteins in the rat by electron microscopy and radioautography, following intratesticular injection of radiolabelled amino acids, [^{3}H]proline and [^{3}H]cystine. Early in spermiogenesis (steps 1 – 7), the two centrioles giving rise to the connecting piece essentially remain unmodified. Between steps 8 and 15, the major elements of the connecting piece (striated columns and capitulum) gradually are assembled from an electron-dense material that is deposited around the walls of the centrioles, thus supporting the previous observations. But throughout this period, protein molecules containing proline and cysteine are synthesized by the step 8 – 15 spermatids and incorporated into the developing neck region. These proteins subsequently become permanent structural components of the connecting piece.

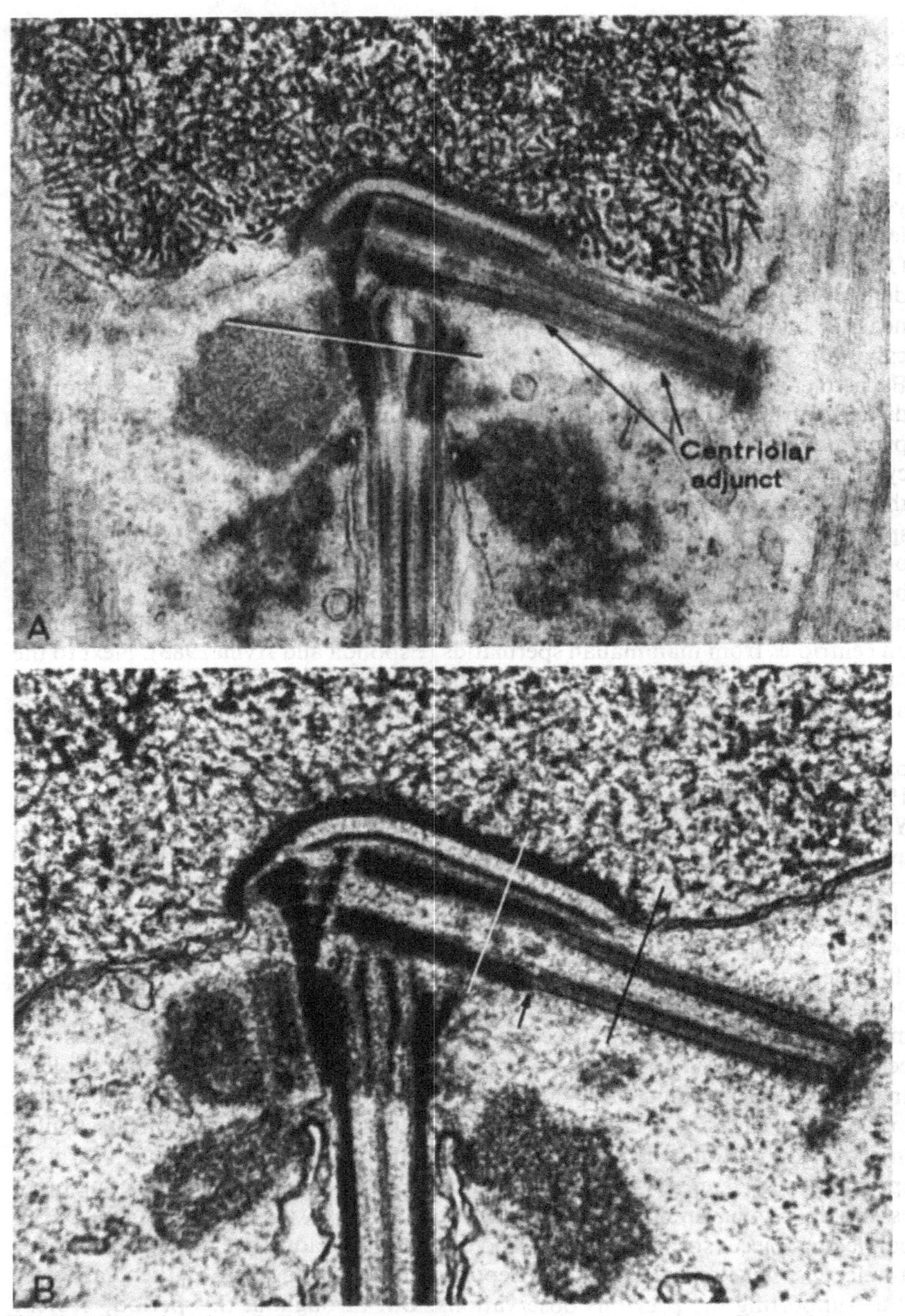

Fig. 43. A Frontal section of chinchilla spermatid at a stage when the condensing chromatin is in the form of coarse dense filaments. The proximal centriole has given rise to a structure called the centriolar adjunct, which at this magnification appears to be simple prolongation of the centriole. The wall of the distal centriole has begun to bow outwards. One of the central pair of microtubules of

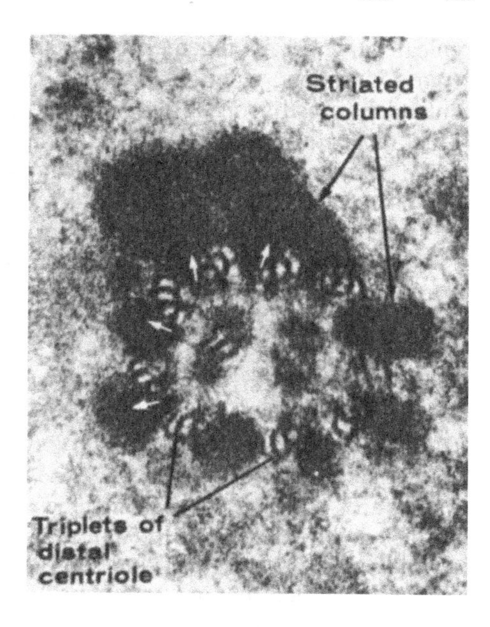

Fig. 44. Transverse section through the distal centriole of Chinese hamster showing the dense matrix of the striated columns arising from between the triplet tubules and extending outwards as indicated by the *arrows*. Material of similar density and texture is associated with the central pair of tubules but evidently in this species does not form discrete rod-like structures in the interior (From Fawcett and Phillips 1969a)

After step 15, few additional proteins are added to the connecting piece during the final steps 16–19 of spermiogenesis. In this process, the regular pinwheel arrangement of the triplets is maintained in the proximal centriole, whereas in the distal centriole the triplets are moved apart due to interstitial accumulation of dense material, resulting in disruption of the integrity of this centriole. Thus according to Fawcett and Phillips (1969a), the proximal centriole persists in the mature mammalian spermatozoon, but the distal centriole no longer exists as a discrete coherent structure. Fawcett (1975b) believes that, concurrent with the development of the connecting piece, the distal centriole disintegrates, while the remnants of the nine triplet microtubules forming its wall may occasionally be found in mature sperm adhering to the inner aspect of some of the nine segmented columns.

Zamboni and Stefanini (1968), studying the neck region of human and macaque spermatozoa, identified persisting triplets or doublets of the distal centriole, which are associated with the inner aspect of the striated columns of the connecting piece. From these observations they concluded that, although the

---

the axonemal complex can be seen extending through the lumen of the centriole to end on the wall of the proximal centriole. Between the central pair of tubules and the wall of the centriole two ill-defined dense rods have appeared. The line transects the distal centriole; **B** Electron micrographs of the neck region of a chinchilla spermatid at an early stage of nuclear condensation. Fine filaments are seen connecting the basal plate to the articular facet of the developing connecting piece. The centriolar adjunct is approaching its maximum length, but accretion of fine filamentous material to its distal end is still in progress. The junction of the proximal centriole with the adjunct is indicated by an *arrow*. The lines are superimposed on the centriole and its adjunct. The substance of the chromatoid body is disposed circumferentially around the site of reflection of the cell membrane into the flagellum, but the annulus is not visible (From Fawcett and Phillips 1969a)

typical pinwheel arrangement of the distal centriole is lost and the triplets are separated widely, it might continue to function as a "transitional connecting centriole" transmitting stimuli from the proximal centriole to initiate propagated waves of bending in the axoneme. They also believed that the persisting triplets of the extensively modified distal centriole are in contact at their anterior end with corresponding elements in the wall of the proximal centriole.

The studies of Fawcett and Phillips (1969a, 1970) have helped to correct the prevalent misconception that the nine striated columns of the neck piece are simply local specializations of the proximal ends of the outer nine dense fibres with which they are continuous in the mature spermatozoon (Nicander and Bane 1962, 1966). Their studies have shown clearly that they are fundamentally different in their fine structure. They have different times and modes of origin, and only secondarily become joined end-to-end with the outer fibres late in spermiogenesis. The proximal centriole is involved actively in the process of formation of the striated neck-piece as the distal centriole.

Towards the end of spermiogenesis the whole tail filament becomes more complex due to the specific differentiation of its different regions (Fig. 39). In the proximal part of the tail filament, the central region now consists of tubular fibrils surrounded by nine "doublets". Each doublet is composed of A-fibre and B-fibre (Fawcett and Phillips 1970, Bustos-Obregon et al. 1975). Nine coarse and thick fibres run close to the fibrillar doublet (Fig. 45A). According to Fawcett and Phillips (1969a, 1970), the outer coarse fibres do not take their origin from the connecting piece and grow distally, but instead, each arises as a lateral outgrowth from the wall of the corresponding doublet. Irons and Clermont (1982a) have followed the details of the sequence of events in the formation of the outer dense fibres as well as the synthesis and incorporation of proteins into them during spermiogenesis in the rat. Their anlagen in the form of nine very fine fibres first develops in association with the most proximal portions of the microtubule doublets in step 8 of spermiogenesis. These gradually increase in length in a proximal-to-distal direction in steps 9 – 14. During steps 15 – 16, the rudimentary fibres suddenly increase in diameter, with the most pronounced growth taking place in step 16, and begin to resemble the mature outer dense fibres. The deposition of electron-dense material along the length of anlagen of the outer dense fibres also corresponds to a period of rapid incorporation of [^{3}H]proline- and [^{3}H]cystine-containing proteins, which are synthesized in the cytoplasm of spermatids during the acrosome and early maturation phases, and form permanent structural components of the outer dense fibres, attaining their definitive form in step 19 of spermiogenesis. The proteins that are incorporated into the forming connecting piece and fibrous sheath are formed in the cytoplasm of step 8 – 17 spermatids as revealed with autoradiography (Irons 1980, 1983).

Once formed, the outer dense fibres lose their attachment to the doublets, except at their caudal extremity (Fawcett and Phillips 1970). In the distal regions of the tail filament, the two circles of fibres and fibrils become less distinct, and the outer thick fibres gradually become thinner and more closely pressed against the corresponding inner circle of fibrils.

Late in the development of the spermatozoon, the central pair of microtubules of the axoneme extends proximally through the interior of the connecting

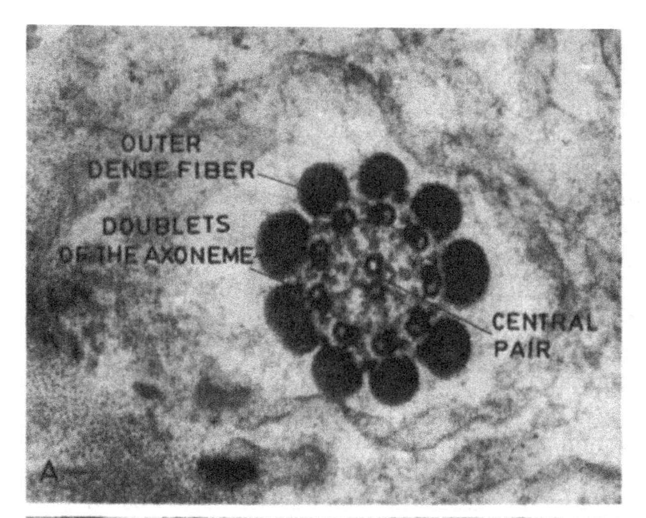

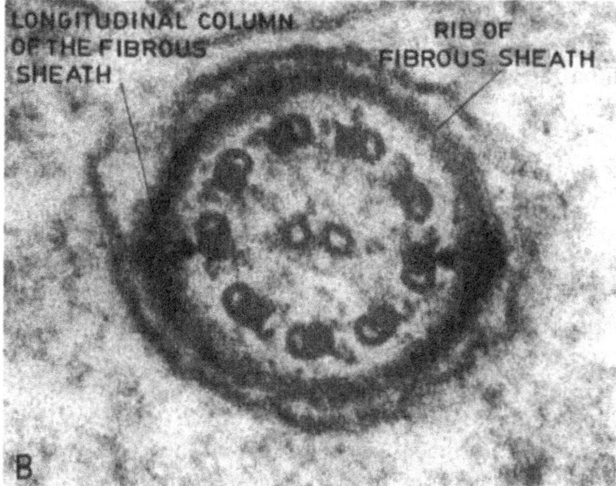

Fig. 45. A Transverse section of flagellum of developing ram spermatid, showing outer dense fibres (9), and 9+2 microtubules of the axoneme; **B** Transverse section of principal piece of testicular spermatozoon of ram showing longitudinal columns of the fibrous sheath and 9+2 microtubules (Courtesy of Dr. M. Courot)

piece as far anteriorly as the proximal centriole (Fawcett 1975b). During the initial development of the flagellum, the microtubules of the axoneme grow by accretion to their distal end. Then in the early mature sperm the central pair elongate at the proximal end, extending through the axis of the connecting piece to contact the juxtanuclear centriole. Finally, the axial filament consists of two central tubules which are surrounded by a central sheath. The spokes are regularly seen between the outer nine double tubules and the central sheath.

b) Development of Fibrous Sheath

According to Nicander (1968), the longitudinal columns of the fibrous sheath, which are present in the principal piece (Figs. 39E and 45B), develop along the filaments 3 and 8. Fawcett and Phillips (1970) have suggested that an attachment of the columns of the fibrous sheath to outer fibres 3 and 8 is a feature common to all mammalian species. The annulus does not play a significant role in initiating fibrous sheath formation. The beginnings of the sheath first are detectable in the free portion of the flagellum that projects beyond the sleeve of spermatid cytoplasm into the lumen of the seminiferous tubule. The deposition of the dorsal and ventral longitudinal columns (Fig. 45 B) between the flagellar membrane and doublets 3 and 8 precedes, by a short time, the development of the circumferential elements of the sheath. The circumferential elements are deposited in the form of very slender rings immediately beneath the plasma membrane of the flagellum. In subsequent development, the slender rings, which are of uniform thickness and very regularly spaced, appear to thicken and become dissociated from the plasma membrane. These also seem to aggregate in groups of two, three and four, fusing with one another to form the much coarser and less regularly spaced ribs of the definitive fibrous sheath, which in the mature sperm is at some distance within the flagellar membrane and is no longer fixed to it. According to Irons and Clermont (1982b), the longitudinal columns of the fibrous sheath develop slowly over a period of 15 days, making their first appearance at the distal end of the principal piece in step 2 of rat spermiogenesis and gradually extending in a proximal direction, ending at the level of the annulus in step 17. All of the ribs develop from anlagen, which are assembled along the length of the principal piece during a much shorter period (4 – 5 days) between steps 11 and 15 of spermiogenesis. New rib anlagen appear to originate from bundles of proteinaceous filamentous material, which is synthesized in the cytoplasm of step 11 – 15 spermatids, and become aligned along the plasma membrane of the principal piece, starting at the distal end. These data have indicated that the two components of the fibrous sheath are formed by means of two independent mechanisms that proceed asynchronously except during an over-lap period of 2 – 5 days between steps 12 and 14 of spermiogenesis.

The radioautographic studies of Irons and Clermont (1982a, b) and Irons (1983) with radiolabelled amino acids [^{3}H]proline and [^{3}H]cystine have demonstrated clearly that the proteins of the outer dense fibres, fibrous sheath and connecting piece are synthesized in the cytoplasm of developing spermatids (steps 8 – 15). These findings have confirmed and extended the results of two recent biochemical studies in which radioactivity has been analyzed in isolated sperm tail proteins (O'Brien and Bellvé 1980b) or intact spermatozoa (Calvin 1981) obtained from the cauda epididymis at various intervals after intratesticular injection of labelled amino acids. It is concluded from both of these studies that sperm tail proteins as a whole are synthesized by the haploid spermatid with a peak of activity during mid-spermiogenesis in the mouse (O'Brien and Bellvé 1980b) and, more specifically, between steps 15 and 17 in the rat (Calvin 1981). But more specific markers for the detection of flagellar proteins are needed for distinguishing them from any other protein within the cytoplasm. The fundamental issue of how proteins incorporated into the outer dense fibres and fibrous sheath are transferred from their site of synthesis to their final destination also needs to be resolved. However, Irons (1983) has suggested that the channel through the centre of the distal and proximal centrioles and the centriolar adjunct forms the possible route for the transport of flagellar proteins from the site of their synthesis.

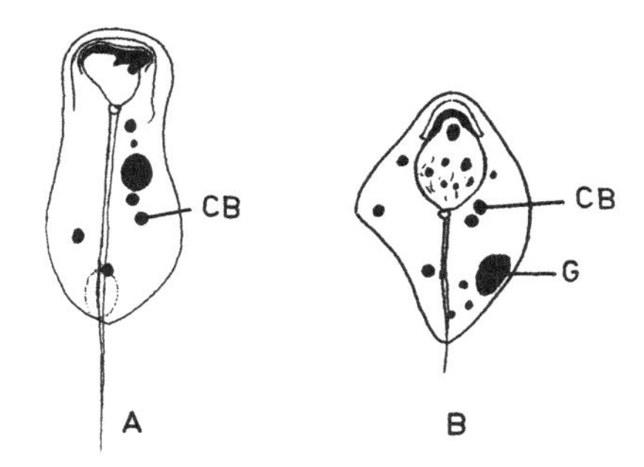

Fig. 46. Developing spermatids from the testis of American opossum, showing several chromatoid bodies (*CB*), and large Golgi complex (*G*)

In some cases of subfertility in man, the absence of the fibrous sheath in connection with an abnormal length of the mid-piece and irregularities in the number of axial fibres have been reported (Ross et al. 1971). In two different cases of oligozoospermic men, the electron micrographs show the contrary: shortening of the mid-piece and an increase in the fibres of the fibrous sheath, which results in an abnormal thickening of the main-piece, or principal piece, of the sperm tail (Ross et al. 1973). Aughey and Orr (1978) have observed a defect affecting 90% of the spermatozoa of an infertile man, which consists of rounded heads and a loss of tails, probably caused by a developmental abnormality at the spermatid stage (Holstein and Schirren 1979). Sperm abnormalities, especially of tails, have been reported in the semen of various mammalian species under normal and experimental conditions (Vale Filho et al. 1976, Uzu et al. 1976, Lobl and Mathews 1978); the abnormalities of other sperm components also occur in bulls (Rob and Rozinek 1976, Uzu et al. 1976, Vale Filho et al. 1976). A thorough understanding of molecular aspects of these abnormalities forms a promising research area for future studies.

c) Mitochondrial Sheath

During the later stages of spermiogenesis, the mitochondria are arranged around the axial filament to constitute the mitochondrial sheath of the mid-piece, which lies between the proximal centriole and the terminal ring or annulus (Figs. 17 and 39E) (de Kretser 1969, Bustos-Obregon et al. 1975, Holstein 1976, Holstein and Roosen-Runge 1981). Most of them arrange themselves helically around the tail (Lung and Bahr 1972). Wartenberg and Holstein (1975) have demonstrated several details of mid-piece formation in man (see also Holstein and Roosen-Runge 1981), which enable a better definition of some steps of spermatid differentiation.

d) Chromatoid Body and Annulus

It is difficult to follow the behaviour of the chromatoid body during the later stages of spermiogenesis with the light microscope (Fig. 17). Several chromatoid bodies of variable size have been observed in the maturing spermatids of opossum (Fig. 46). But electron microscope studies have shown that with the differentiation of spermatid, the chromatoid body draws near to the nucleus and gradu-

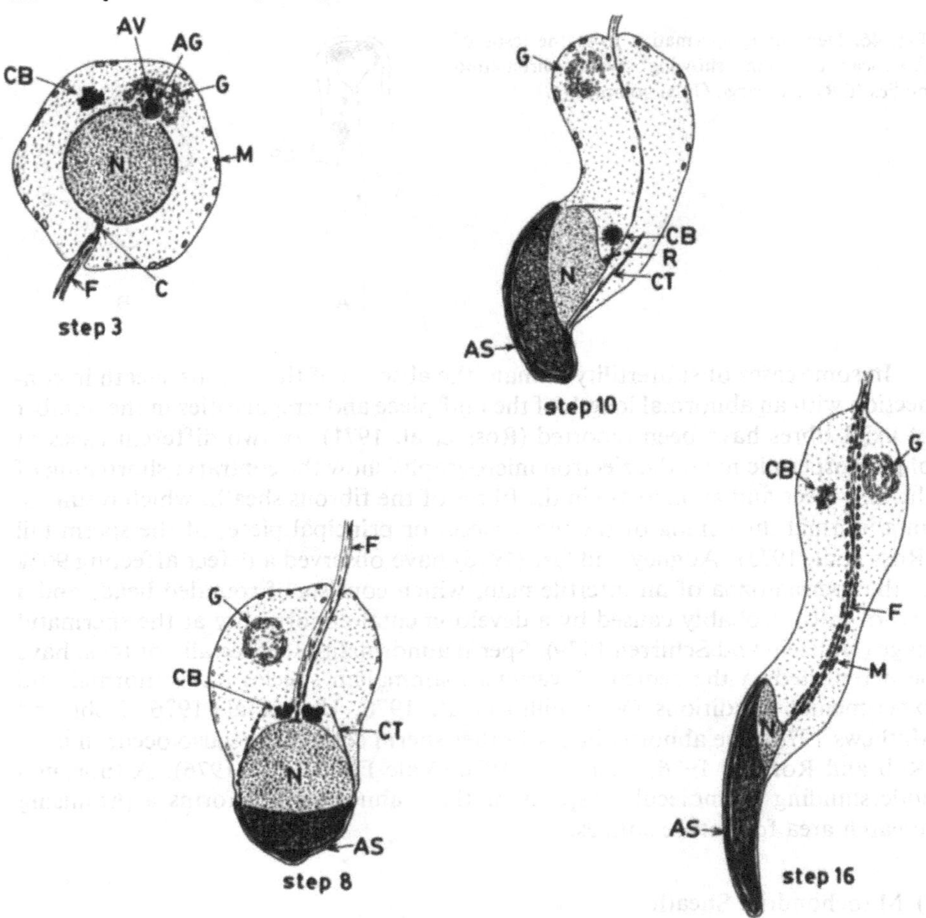

Fig. 47. Series of drawings of rat spermatids at four different steps of spermiogenesis (i.e. steps 3, 8, 10 and 16 of the classification of Leblond and Clermont 1952b). These drawings illustrate the main changes taking place in some of the cytoplasmic components (see description in the text). *AG* acrosomic granule; *AV* acrosomic vesicle; *AS* acrosomic system; *C* centrioles; *CT* caudal tube; *CB* chromatoid body; *F* flagellum; *G* Golgi zone; *M* mitochondria; *N* nucleus; *R* ring. Steps 8, 10, and 16 spermatids are illustrated as seen from the side. The *dense stippling* over part of the nucleus indicates the area covered by the head cap portion of the acrosomic system (Redrawn from Susi and Clermont 1970)

ally moves around it from the anterior to the posterior pole (Figs. 31 and 47) (Fawcett 1972, Yasuzumi 1974). During this caudal migration, streamers of its dense substance extend to pores in the nuclear envelope, which are relatively more numerous in the vicinity of this organelle than elsewhere on the nuclear perimeter. The functional significance of this relationship, which prevails over a considerable span of time during the caudal migration of the chromatoid body (Fig. 47), is still not known. But the pattern of pores evidently changes concurrently with the changing position of the chromatoid body (Fawcett and Chemes 1979). By the time the latter has reached the caudal pole of the nucleus, the

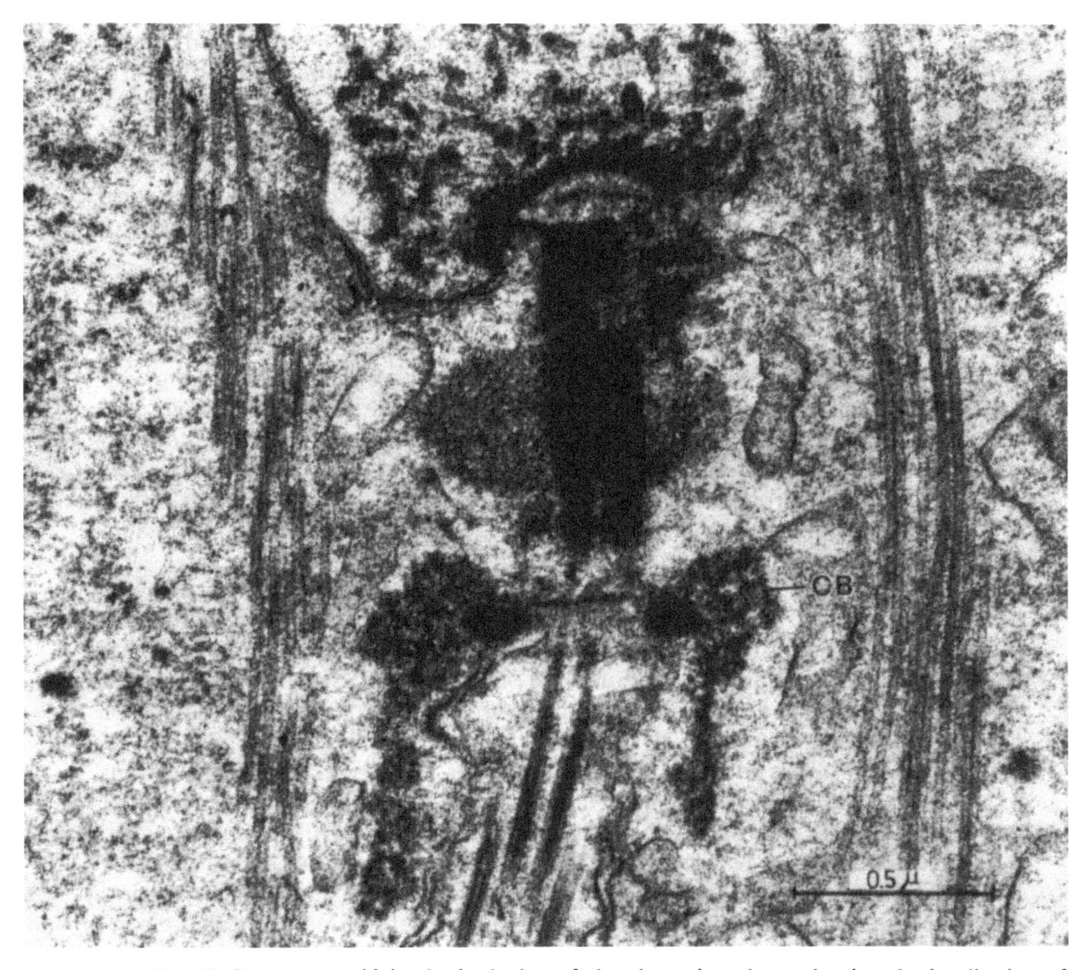

Fig. 48. Ram spermatid in the beginning of the elongating phase, showing the localization of chromatoid body (*CB*) around the axial filament behind the cross-striated columns of the connecting piece. The chromatoid body (*CB*) takes the form of a cylinder, whose posterior end tightens around a homogeneous ring (From Courot and Loir 1968)

centrioles and base of the flagellum have become fixed to the nuclear envelope (Fig. 47) (Fawcett 1972).

According to Courot and Loir (1968), in the beginning of the elongating phase, the chromatoid body is located around the axial filament behind the cross-striated columns (Fig. 48). It takes the shape of a cylinder whose posterior part is wide open, and whose anterior end tightens around a homogeneous ring which is denser to electrons. The cytoplasmic canal membranes attach to the ring. After migration to the posterior part of the mid-piece, the ring called Jensen's ring, remains alone. According to Nicander (1968), the distal ring lying at the bottom of the cytoplasmic canal has two components, i.e. the very dense material and loose material. Fawcett (1972) also has observed that the chromatoid body forms a ring encircling the base of the flagellum contiguous with the annulus which is fixed to the cell membrane along its line of reflection on to the flagellum, as also reported by Holstein and Roosen-Runge (1981). The circumflagellar ring of chromatoid material and the annulus shift caudally to a position that marks the future junction of the mid-piece and principal piece. This migration pattern also permits mitochondria to collect round the flagellum to form the mid-piece (Fawcett et al. 1970, Holstein and Roosen-Runge 1981).

According to Susi (1970), and Susi and Clermont (1970), the chromatoid body in the rat spermiogenesis follows a characteristic path of migration in the spermatid cytoplasm to the region of the developing centriolar apparatus and then into the caudal cytoplasmic lobe where it appears to disintegrate along with the residual cytoplasm (Fig. 47). Fawcett (1972) has observed the gradual dispersal of the chromatoid body in the course of its caudal migration and its final disappearance entirely behind the annulus in the mature spermatozoon. The chromatoid body or bodies, appear to be sloughed off, along with the residual cytoplasm, during the later stages of spermiogenesis in the American opossum (S. S. Guraya, unpublished observations).

From these studies it appears that the chromatoid body may play some role in the development of some components in the neck region of the spermatozoon. But, three alternate suggestions have been made in regard to the functional meaning of the chromatoid body (Sud 1961, Fawcett 1972, Stefanov and Penkov 1979: (1) The chromatoid body is a mass of precursor material required for forming the tail filament; (2) it is involved in the synthesis and introduction of arginine-rich histones into the spermatid nucleus for chromatin condensation; (3) it is an aggregation of long-lived messengers set aside to provide for the protein synthesis essential for differentiation of tail structures that arise after the cessation of transcription (Monesi 1971, Monesi et al. 1978, Bellvé 1979). The specific role of the chromatoid body in the RNA metabolism of spermatogenic cells still is not known. Recent data have further indicated that a considerable proportion of the RNA synthesized in pachytene spermatocytes is preserved through spermatid development until late spermiogenesis (see also Chap. III).

For building up different components of the spermatozoon during spermiogenesis, there is a need for the synthesis of new proteins. Their synthesis suggests the presence of complex control mechanisms involved in the regulation of gene transcription and translation of mRNA's during spermiogenesis. Bellvé (1979) has suggested two possibilities in this regard: (1) The proteins are synthesized during meiotic prophase and are either accumulated and stored or are incorporated directly into cellular components for conservation in the differentiated spermatozoon. (2) There occurs the de novo synthesis of proteins during spermiogenesis corresponding to the formation of various components of the sperm-cell. The autoradiographic studies, do not help us to have a clear cut distinction between these two possibilities. In order to overcome this difficulty, Bellvé and co-workers (Bellvé et al. 1975, O'Brien and Bellvé 1980a) have developed a technique that helps to separate mouse spermatozoon components into three gross fractions. The first fraction that is SDS-soluble contains the plasma membrane, acrosome, axoneme, and the matrix and cristae of mitochondria. The second fraction consists of SDS-insoluble sperm head composed of the nucleus and perinuclear substance. The third fraction is constituted by the SDS-insoluble tail consisting of basal plate, connecting piece, outer dense fibres, fibrous sheath and outer mitochondrial membranes. Each of these three fractions can be resolved into multiple protein bands of SDS-PAGE eliminating the possibility of overlapping of polypeptides. The utilization of this fractionation technique has made possible the determination of the temporal sequence in the synthesis of sperm-cell proteins because the kinetics of mouse spermatogenesis are well known (see Chap. I). O'Brien and Bellvé (1980b), utilizing this procedure, have shown that proteins synthesized in the primary spermatocyte form a major portion of the protein components present in the fully differentiated sperm cell. But developing spermatid also synthesizes proteins de novo for incorporation into the sperm-cell components. The proteins of the tail components, especially of the outer dense fibres and fibrous sheath, are synthesized mainly in the haploid spermatids, concomitant with the morphogenesis of the SDS-resistant tail structures. The SDS-soluble fraction shows a heterogeneous group of proteins, some of which are formed during spermiogenesis. Those proteins synthesized during spermiogenesis may be the various enzymes of the acrosome and mitochondria, as the SDS-soluble fraction contains protein components of the acrosome and mitochondrial matrices. The synthesis of these and other stage-specific proteins will be discussed also in Chap. V.

e) Miscellaneous Components

Some miscellaneous components have been observed in the developing spermatids of some mammalian species. In other words, these do not form the characteristic feature of the developing spermatid of each species. Certain bodies appearing in the neck region of developing human spermatozoa were called lateral junction bodies and centriole-associated bodies by de Kretser (1969), and Holstein and Roosen-Runge (1981). These appear to correspond to chromatoid bodies that occur in the neck region of elongating spermatid, as discussed above. De Kretser has assumed that these structures may constitute precursor material for the formation of the dense components of the connecting piece, axonemal complex, and principal piece, since these decrease in size during spermiogenesis (Kerr and de Kretser 1981). The human spermatid also shows a complex of electron-dense granules next to the mitochondria in the cytoplasm (Holstein and Roosen-Runge 1981). The granular complexes are believed to be ribosome aggregates.

Rattner and Brinkley (1970) have reported the presence of a transitory tubular structure, termed the tubular complex, which appears at the proximal end of the principal piece in mid-spermiogenesis of the cotton top marmoset. This structure is formed by tubules 30–33 nm in diameter that are arranged both perpendicularly to and helically around the axial filament; its disappearance occurs concomitantly with mid-piece formation.

Wartenberg and Holstein (1975) have reported the presence of a spindle-shaped body (Fig. 49) which appears at the beginning of the late cap phase of the

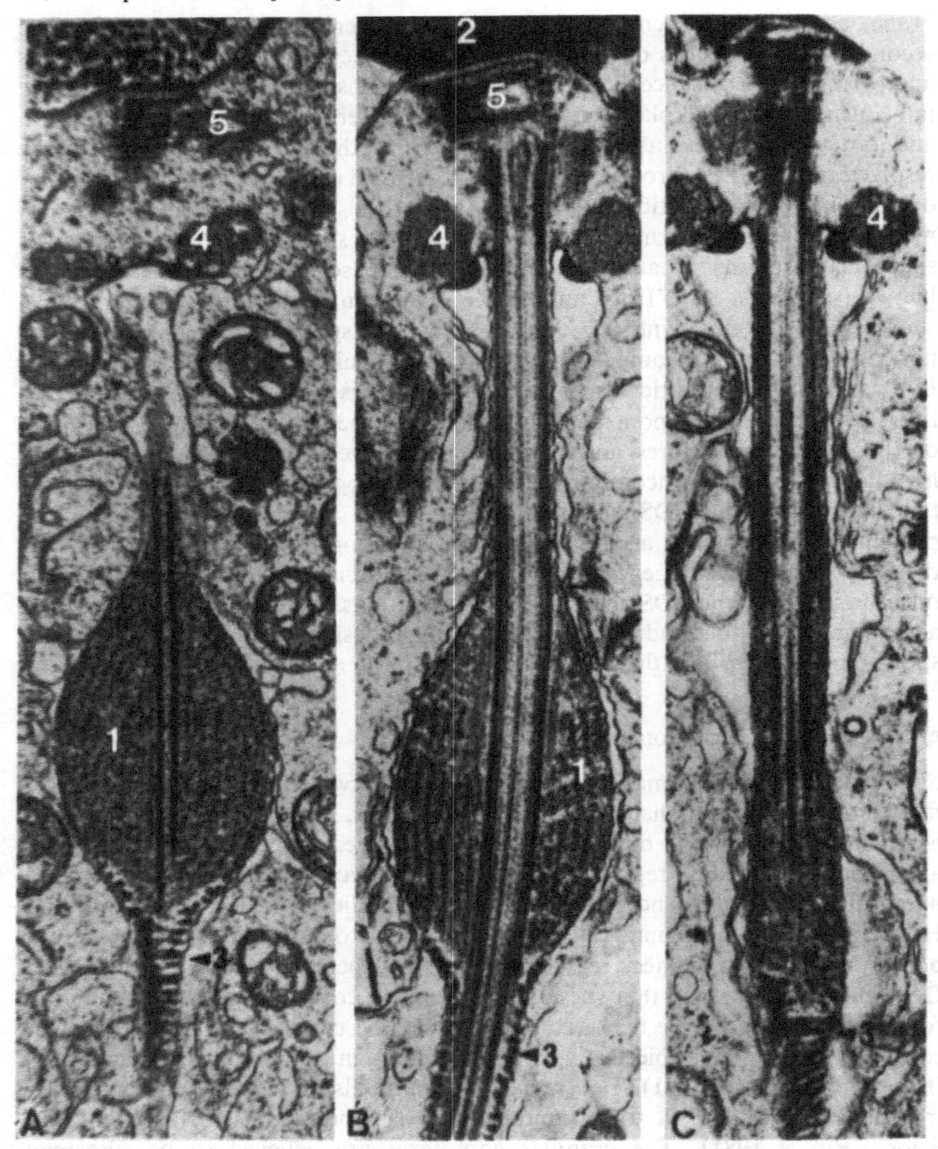

Fig. 49. A 47-year-old man; **B** 33-year-old man; **C** 37-year-old man. Spindle-shaped bodies (*1*) have reached their full size (**A** – **C**) containing a multilayered arrangement of microtubules. When nucleoplasm (*2*) has attained its greatest density (**C**), the spindle-shaped body diminishes, whilst the fibrous sheath fibres (*3*) increase in density; *4* annulus; *5* centrioles (From Wartenberg and Holstein 1975)

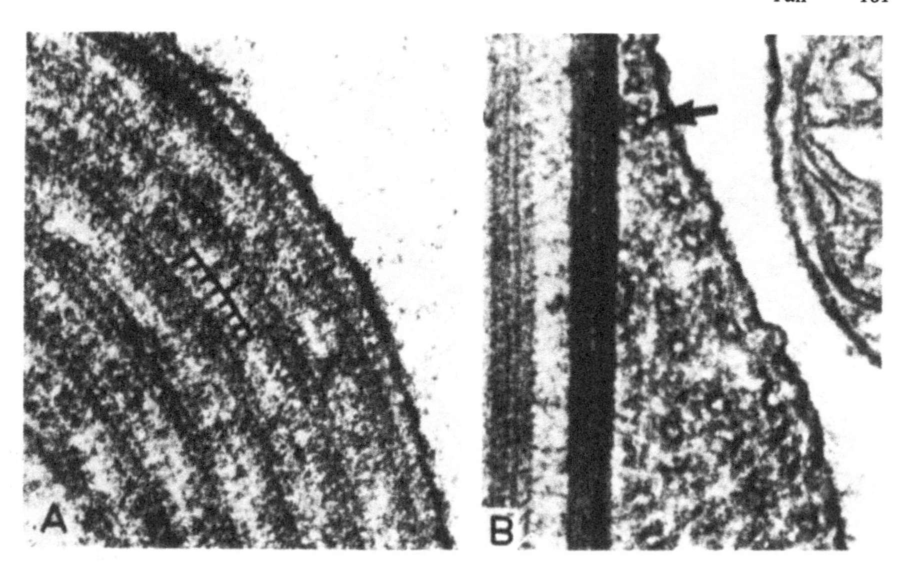

Fig. 50. A Cross-section through the tail of developing spermatids of a 47-year-old man. The microtubules display a central tubule (*double arrow*) accompanied by periodically arranged peripherally oriented subunits (*multiple arrow*); **B** Longitudinal section through the upper part of the spindle-shaped body. *Arrow* indicates cross-section through the central part of a microtubule, which is surrounded by a halo of tubule-like subunits (From Wartenberg and Holstein 1975)

normal differentiation of human spermatids and occupies part of the future mid-piece (Holstein and Roosen-Runge 1981). The spindle-shaped body of human spermatids also contains microtubules, which run circumferentially and form a single layer just beneath the plasmalemma (Fig. 50). Actually, its tubular structures are wound like yarn. It disappears when the fibrous sheath is completed and the annulus moves down, giving rise to the mid-piece (Holstein 1976, Holstein and Roosen-Runge 1981). The annulus is believed to form the fibrous sheath via the microtubular system of the spindle-shaped body. The continuities between the tubular structures of the spindle-shaped body and the ribs of the fibrous sheath of the principal piece suggest that the tubules may be involved in the formation of the ribs. The structure, similar to the spindle-shaped body, has not been observed in oligozoospermic men and in some cases of subfertility (Ross et al. 1971, 1973).

Another, hitherto neglected cytoplasmic component or structure appears transiently during the differentiation of human spermatids: flower-like structures consisting of a central, electron-dense granular core and surrounding vesicle clusters (Holstein and Schirren 1979, Holstein and Roosen-Runge 1981). Susi and Clermont (1970) also have described comparable structures in spermatids of rat (Fig. 51) and considered these to be derivatives of the chromatoid body. According to Holstein and Schirren (1979) these flower-like structures in humans develop during Stage II of spermatid development close to the nucleus and chromatoid body, undergo a series of characteristic changes, and are incorporated ultimately into the residual body where they disintegrate (Holstein and Schäfer 1978, Holstein and Roosen-Runge 1981).

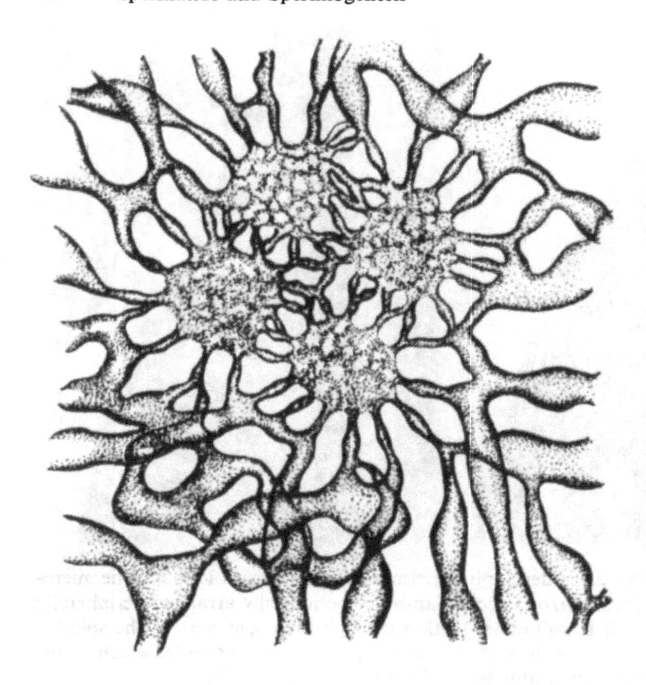

Fig. 51. A portion of the radial body and its connections with the endoplasmic reticulum, as seen in stereo pairs of thick sections of rat testes stained with Ur-Pb-Cu. The radial body is composed of several subunits, each composed of a diffuse centre surrounded by narrow fusiform or tubular collapsed cisternae. At the periphery of the radial body these narrow cisternae are continuous with distended cisternae of the rest of the network. Within the radial body, narrow cisternae surrounding the diffuse centres and common to the subunits frequently show a criss-crossing pattern (Redrawn from Clermont and Rambourg 1978)

f) Malformations of Spermatid Differentiation

The detailed morphological analysis of malformed spermatids, as reported already in various sections of this chapter, has revealed that in most cases the development of only one component of the cell, such as the acrosome, the nucleus or the tail, is disturbed (Holstein and Schirren 1979, Williamson et al. 1984). There are seen combined malformations of different elements of the cell, which may appear during its life. Such disturbances of the development of the spermatid may be the cause of severe malformations of the spermatozoa (Gould 1980, Williamson et al. 1984) resulting in their immotility and hence infertility. The detailed study of their genesis may help us to get an insight into normal spermiogenesis.

6. *Residual Cytoplasm and Droplets*

Corresponding to various morphological and biochemical alterations in the nucleus and the formation of the acrosomal cap and tail of the spermatozoon, as already discussed, there occur morphological, histochemical and biochemical changes in the cytoplasm and organelles of developing spermatid in mammals including primates (Breucker et al. 1985). Beckman et al. (1978) have demonstrated differences in content of total phospholipids, individual classes of phospholipids and triglycerides in spermatocytes, spermatids and late spermatids, besides fatty acids which have also been studied by Grogan et al. (1981). Kornblatt (1979) has studied the synthesis and turnover of sulphogalactoglycerolipid (SGG), a membrane lipid, during rat spermatogenesis. SGG is conserved in the mature spermatozoon. All these morphological and chemical changes have been included in the "maturation" phase by Leblond and Clermont (1952b).

a) Residual Cytoplasm and its Organelles

The Golgi complex, after separating from the acrosomal vacuole in marmoset and chimpanzee, begins to show more vacuoles and meanwhile the intensity of its colouration with Sudan black is increased further, suggesting the appearance of more lipids due to some degenerative changes (S. S. Guraya, unpublished observations). Similar morphological and histochemical changes have been described for the Golgi complex of developing spermatids in other eutherian mammals (Bloom and Nicander 1961, Guraya 1965, de Kretser 1969). Electron microscope studies have shown that as the Golgi apparatus is displaced to the post-nuclear cytoplasm, it develops a looser and more vesicular structure than before (Breucker et al. 1985). It also seems to diminish in volume by the transfer of vesicles to the surrounding cytoplasm (Clermont 1956). The vesicles are believed to mingle with similar elements of the endoplasmic reticulum and thus can no longer be identified as Golgi elements. Dictyosome-like structures described in the primary spermatocytes of guinea pig have been observed in the residual bodies, but the stacked configuration (or dictyosome form) is seldom seen in these stages of spermatid differentiation (Mollenhauer and Morre 1981). The association of ER with annulate lamellae has been observed in developing spermatids (Gardner 1966, Sandoz 1970). Rat spermatids also show clusters of radially disposed cisternae (Mollenhauer and Zebrun 1960), referred to as radial bodies (Fig. 51) by Clermont and Rambourg (1978). Their connection with the surrounding distended cisternae has been observed. Structures resembling the radial body are reported also to occur in the monkey by Dym and Pladellorens (1977), and as a "starlike" array of membranes in the mouse by Sandoz (1970), who observed its relationship with annulate lamellae.

Clermont and Rambourg (1978) have observed that during the acrosome phase the cytoplasm of maturing rat spermatids shows an abundant ER which expresses various modifications (Fig. 52): (1) along the inside and outside of the caudal tube the cisternae connected by delicate lateral anastomoses form long tubes or plates which lie adjacent and parallel to the microtubules; (2) along the flagellum the ER forms a fenestrated sleeve made up of a close network of tubular cisternae; (3) similar networks are organized as "fenestrated spherules" enclosing large vesicles seen throughout the cytoplasm; and (4) at a short distance from the flagellum, the ER cisternae are continuous with a stack of annulate lamellae (Figs. 51 and 52). Collapsed cisternae called the "radial body" appear as a compact group of collapsed cisternae, composed of several subunits, each with its own diffuse centre. The continued presence of a close morphological relationship between the ER and the Golgi apparatus, once the latter has detached from the acrosomic system, suggests a continuing functinal relationship (Clermont and Rambourg 1978). This suggestion is supported by the recent observations of Tang et al. (1982), as the structure of the peripheral region of the Golgi apparatus is not modified markedly and maintains a close association with the network of ER cisternae. NADPase and TPPase activities continue to be seen in

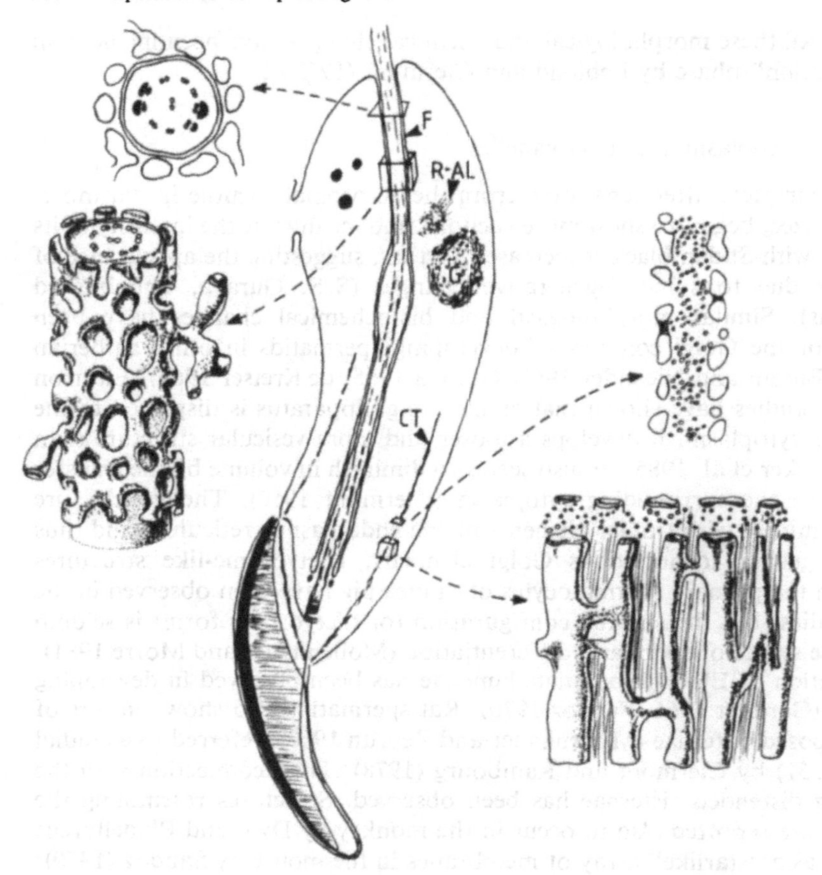

Fig. 52. A step-14 spermatid of rat, showing two regional specializations of the endoplasmic reticulum as revealed from stereo pairs of thick sections from Ur-Pb-Cu-stained testes. *On the left* a cross section of the flagellum (*F*) as seen in thin sections (*above*) and a three-dimensional reconstruction from thick sections (*below*) is depicted. In both drawings the axoneme surrounded by the invaginated plasma membrane of cell and the cisternal profiles of the "fenestrated sleeve" are illustrated. *On the right*, sections of a portion of the caudal tube (*CT*) composed of microtubules and the associated cisternae of endoplasmic reticulum is shown as seen in thin sections (*above*) and as reconstructed from thick sections (*below*). In the latter case the roughly tubular cisternae from a double curtain with the microtubules located in-between. The cisternae of the fenestrated sleeve along the flagellum and of the tubular cisternae along the caudal tube show numerous connections with the extensive network of cisternae distributed throughout the cytoplasm. *G* Golgi apparatus; *R-AL* radial body-annulate lamellae complex (Redrawn from Clermont and Rambourg 1978)

some elements of the Golgi apparatus that also incorporates [^{3}H]fucose, showing it to be still involved in glycoprotein synthesis, as also reported for the mouse spermatids following [^{3}H]galactose injection (Sandoz 1972, Sandoz and Roland 1976). But the actual role of the Golgi apparatus after the formation of the acrosome needs to be determined.

Baccetti et al. (1978) have studied the origin of the membrane investing the newly formed elongating organelles during spermiogenesis of man, rat, and bull.

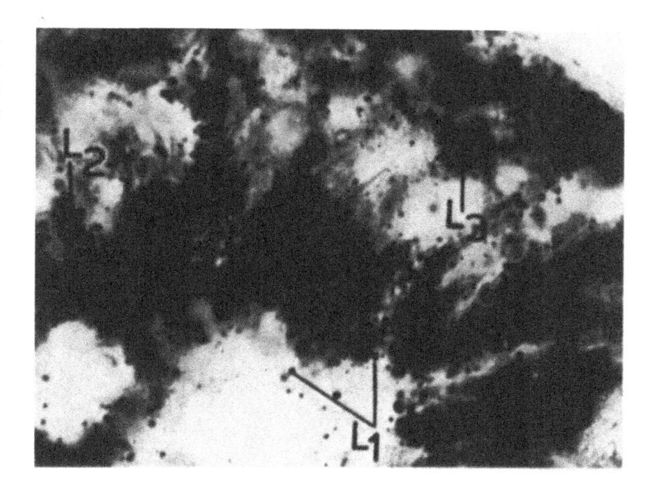

Fig. 53. Portion of the seminiferous tubule illustrating the accumulation of L_1 bodies in the cytoplasm of spermatids. Some of the L_1 bodies are being sloughed into the lumen of the tubule. Note also the L_2 bodies in the vicinity of accumulation of L_1 bodies. The L_3 globules lie in the peripheral portions of the tubule (From Guraya 1968)

This new surface is formed by several clusters of Golgi-originated vesicles which form a periaxonemal double cylinder; the latter finally fuses at its end with the old plasma membrane. Thus, the new periaxonemal plasma membrane is preformed in the spermatid body. The membrane surrounding the elongating head is, on the contrary, simply an extension of the old one.

There occur alterations in the mitochondria during spermiogenesis, as already discussed in detail. Sudanophilic lipid droplets consisting mainly of phospholipids and some triglycerides gradually accumulate in the cytoplasm of developing primate spermatids (Fig. 17), as described for other mammalian species (Guraya 1965, 1968, Dietert 1966). It will be very interesting to mention here that their amount is relatively greater in the later spermatids (Fig. 53).

During late spermiogenesis, the RNA-containing basophilic substance (or ultrastructural RNP particles and elements of ER) consisting of RNA and protein is transformed into large residual basophilic bodies of discrete nature, which are distributed sparsely throughout the cytoplasm (Fig. 17). Similar residual basophilic structures have been described for other mammalian species (Fig. 54) (Daoust and Clermont 1955, Smith and Lacy 1959, Dietert 1966, Guraya 1965, 1968). Late in spermiogenesis, the meiotic histone that was synthesized in the zygotene and pachytene spermatocytes and remained in the nucleus throughout early spermiogenesis is extruded into the spermatid cytoplasm. Actually, the fate of displaced histones is unknown except in biochemical terms (Bellvé, personal communication). Much of it may find its way, together with RNA, into the residual bodies as a result of the disengagement of the residual cytoplasm. Some may be eliminated with the cytoplasm from the head region of the spermatid; the characteristic evaginations of the spermatid cytoplasm, which normally act as anchoring locations for binding the heads of the spermatids to the surface of the Sertoli cell, perhaps provide additional means for the uptake and subsequent degradation of residual spermatid cytoplasm by the Sertoli cells (Romrell and Ross 1979, Russell 1979b, Vogl et al. 1983a, b, Russell and Peterson 1985, Tindall et al. 1985). Morphological aspects of spermatid residual bodies in the mouse have been revealed with scanning electron microscopy

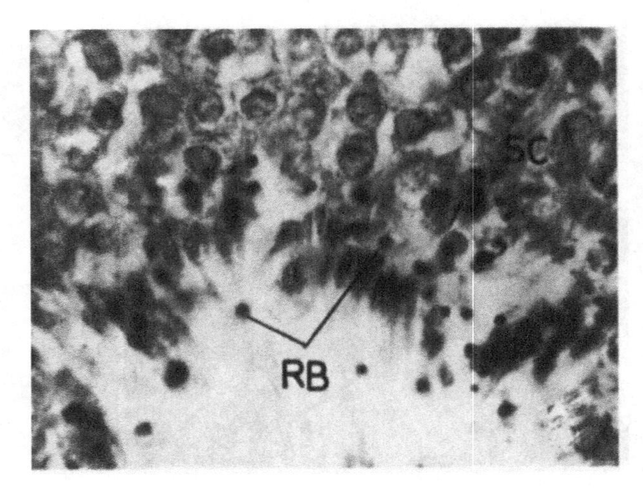

Fig. 54. Portion of the seminiferous tubule showing the RNA-containing residual bodies (*RB*) among the nearly mature spermatozoa. Note that some of the residual bodies (*RB*) lie in the lumen of the tubule; very few are seen in the cytoplasm of Sertoli cells (*SC*) (From Guraya 1968)

(Baradi and Rao 1983). Ultrastructurally, the residual bodies have a relatively dense cytoplasm which contains various organelles, such as mitochondria in groups, vesicles, fenestrated membranes, ribosomes in large complexes, lipid droplets, and remains of flower-like structures (in humans only) (Holstein and Roosen-Runge 1981, Breucker et al. 1985).

During late stages of spermiogenesis, some diffusely by distributed carbohydrates accumulate in the cytoplasm (Guraya 1965), which may be due to ultrastructural "post-acrosomic vesicles" containing glycoproteins that were omitted from the acrosome of the early elongating spermatids (Courtens 1978b).

Bloom and Nicander (1961), and Dietert (1966) have demonstrated the presence of some conspicuous ultrastructural modifications in the various organelles of residual cytoplasm. Clermont and Rambourg (1978) have observed the regression of ER during the last or maturation phase of spermiogenesis. The ER network fragments and then most of the cisternae disappear from the cytoplasm. Smith and Berlin (1977) have observed that breakdown of annulate lamellae in human spermatozoon residual bodies is indicated by a dilation of annulate lamellae cisternae to form vacuoles following the dissolution of pore complexes. The "radial body" is the last element of ER to be dissolved. These studies have demonstrated clearly that the ER undergoes extensive structural modifications during spermiogenesis (Fig. 52), indicating an active role of this organelle in the differentiation of spermatid into a spermatozoon. It becomes closely related with Golgi apparatus, caudal tube, flagellum and granules. The exact functional significance of these associations needs to be determined in future studies.

Corresponding to various cytoplasmic alterations, hydrolytic enzyme activities appear in the residual cytoplasm of the late spermatids (Niemi and Kormano 1965, Dietert 1966, Posalaki et al. 1968). The residual bodies in the seminiferous tubules of buffalo, goat, and ram contain nonacetylated mucopolysaccharides, glycogen, residual lipoproteins, acidic lipids or phospholipids (sphingolipids, choline-containing lipids, phosphoglycerides), unsaturated lipids and neutral lipids (triglycerides) (Bilaspuri and Guraya 1983c, d); some enzymes, such as acid phosphatase and esterases, are present (Bilaspuri and Guraya 1983a, e). All these changes could be considered as a sign of the gradual involution of the cytoplasm and organelles which will not be used for the construction of spermatozoon as these are sloughed off eventually (Figs. 17, 53 and 54). The residual bodies and spermatids in the rat testis contain antigotensin-converting enzyme activity (Vanha-Perttula et al. 1985), representing a new marker for the advanced differentiating spermatogenic cells.

This cast-off cytoplasm and organelles constitute conspicuous residual bodies in the wall of the seminiferous epithelium, which are believed to be phagocytosed by the Sertoli cells (Fig. 54) (Carr et al. 1968, Guraya 1968, Courtens 1984) so as to play some role in initiating the new wave of spermatogenesis (Lacy 1962). But this hypothesis has fallen into disfavour and is no longer considered valid. Johnson (1970) believes that the last stage of spermatogenesis is involved in the testicular-hypophyseal feedback mechanism in man. In this regard Johnson (1964) has raised the hypothesis that the cytoplasm split off from the spermatozoa, which is rich in newly formed fat globules (Fig. 53), contains a gonadotrophin-inhibiting substance (inhibin); the origin, characterization, purification, and biological significance of inhibin form the subject of several recent studies (see Chap. I). With the availability of isolation methods for the residual bodies (Nyquist et al. 1973), it will be possible to work out their exact chemistry and physiological significance. It will be interesting to mention here that Sertoli cells have been shown to be the source of inhibin (see Chap. I).

b) Spermiation

The molecular mechanisms involved in the release of spermatozoa from the seminiferous epithelium into the tubule lumen are poorly understood. This process of release of spermatozoa, termed "spermiation", starts with the breakdown of membrane junctional specializations between Sertoli cells and condensing spermatids (Ross 1976, Russell 1984, Russell and Peterson 1985). As the sperm loosens from the Sertoli cells, the spermatid cytoplasm becomes lobulated, but remains attached to the spermatid neck by a thin stalk which finally separates to release the spermatozoon (Fawcett and Phillips 1969b, Holstein and Roosen-Runge 1981). The Sertoli cells, therefore, are instrumental in separating the cytoplasmic bleb from the spermatozoon (Vogl et al. 1983a, b, Russell 1984, Russell and Peterson 1985). Electron microscope studies have demonstrated clearly that tubulobulbar complexes are formed by late spermatids and Sertoli cells in the rat testis (Russell 1979a, b, 1980b, Russell and Peterson 1985). These complexes consist of an evagination of one cell into an adjacent cell (2 – 4 μm in length) with the resulting structure having a tubular region at the evagination site with a bulbous dilation at the terminal end. The spermatid-Sertoli tubulobulbar complexes, which vary in numbers during the spermatogenic cycle, form devices for the normal elimination of cytoplasm from the head region of the late spermatids. Both the tubular and bulbous portions of the complex are regionally modified and show tight and gap junctions. The tubulobulbar complexes and cytoplasmic remnant residual body are phagocytized by the Sertoli cell and degraded through the activity of Sertoli lysosomes. The various studies have suggested the presence of two mechanism for the elimination of cytoplasm. One taking place via the formation of organelle-rich residual bodies, and the other taking place by way of numerous phagocytized waterish-appearing pockets of cytoplasm (tubulobulbar complexes). In the ground squirrel, numerous Sertoli cell processes develop in association with the residual cytoplasm of elongate spermatids (Vogl et al. 1983a). Following retraction into the epithelium, the residual cytoplasm condenses, the Sertoli cell processes disappear and, after sperm release, the residual bodies are interiorized and degraded by the Sertoli cells. But in other mammals Sertoli cell processes lack specialized contacts with germ-cells (Russell 1980b). Vogl et al. (1983a) have suggested that Sertoli cell microtubules possibly play a role in the retraction of residual cytoplasm from germ-cells. The disappearance of ectoplasmic specializations in Sertoli cells is related also to sperm release in the ground squirrel, where these structures are lost from around the spermatid heads in a sequence that correlates well with the retraction of the Sertoli cell cytoplasm from the germ-cells. Divergent views exist concerning the involvement of Sertoli cell microtubules in spermiation, as suggested by experiments with microtubule-disrupting agents (L.-M. Parvinen et al. 1978b, Aoki 1980, Russell et al. 1981). But Vogl et al. (1983b) have suggested that these probably are not involved directly with the mechanism of sperm release in the ground squirrel. Spermiation is believed to result in the physical segregation of two segments of the spermatid plasma membrane. Millette (1979a) has stated that the segregation of cell surface components into the residual body membrane is possibly nonrandom. For example, specific membrane receptors responsible for the adhesion of germ-cell to Sertoli cell could be sequestered on the residual body to allow sperm release. Some observations showing a greater density of concanavalin A sites on the surface of residual bodies in comparison with spermatids support the view that individual membrane constituents, in fact, are partitioned selectively during late spermiogenesis (Millette 1976). Further investigations are needed to determine the mechanism(s) of spermiation at the molecular and physiological levels.

c) Cytoplasmic Droplet

After the various residual organelles and excess cytoplasm are cast off, there are left behind some cytoplasm and organelles in the vicinity of the mid-piece, which constitute the cytoplasmic droplet (Fig. 17). Bloom and Nicander (1961), using the electron microscope, have described the formation of the cytoplasmic droplet and its components during the spermiogenesis of some mammals, as also described by Holstein and Roosen-Runge (1981) for humans. Its details, as structure, cytochemistry and possible functional significance, will be discussed in Chap. IX. It contains mitochondria, membranes, and numerous, often large vesicles with finely granular material in the developing human spermatid (Holstein and Roosen-Runge 1981).

d) Differentiations (or Specialization) of Membranes

Special differentiations and modifications of the spermatid membranes are known to occur during spermiogenesis (Courtens 1979a, Pelletier 1983, Pelletier and Friend 1983a). The Sertoli cell cytoplasm surrounding the sperm head shows considerable metabolic activity as determined by the presence of numerous vesicles and an accumulation of electron dense material (Horstmann 1961, Brökelmann 1963). This period correlates with the appearance of an intense nucleoside triphosphatase activity at the interface between the plasma membranes of the two cells, suggesting a transfer of materials from the Sertoli cell to the spermatid (Tice and Barnett 1963). The spermatid plasma membrane gradually becomes thicker and more electron-dense. The fusion of post-acrosomal vesicles with the plasma membrane of Stage 7–9 spermatids in the ram appears to add some substances (Courtens 1979a). By using electron microscopy of thin sections, freeze-fracturing, and filipin labelling, the membrane specializations and mapping of free cholesterol have been studied in the developing spermatid of guinea pig (Pelletier 1983, Pelletier and Friend 1983a). The development of the distinctive membrane specializations in various components closely correlates with the developmental steps of spermatid differentiation. Filipin-cholesterol complexes are abundant in the plasma membrane, less abundant in the acrosomal membrane, and least abundant in the nuclear membrane. This heterogeneous distribution of cholesterol-filipin complexes is believed to be indicative of lipidic contribution by the Sertoli cells for the control of spermiogenesis. The high levels of cholesterol demonstrated in the spermatid plasma membrane may decrease its permeability and thus prevent leakage of intracytoplasmic components outside the developing germ-cells. The lower levels of cholesterol observed in the more fluid acrosomal and nuclear membranes may be compatible with the shaping of the acrosome and the flattening of the nucleus. André (1963) has suggested that the surface coating of the late spermatid might be made of mucopolysaccharides, its role would be to reduce the adhesiveness of the cell surface, and thus to facilitate the release of spermatozoa into the lumen of the seminiferous tubule. The development, distribution, and function of antigens on the surface of spermatogenic cells are discussed in Chap. V.

Chapter V
Antigens During Spermatogenesis

A. Development and Distribution

The germinal epithelial cells in the mammalian testis undergo an elaborate and orderly sequence of differentiation from spermatogonia to testicular spermatozoa, accompanied by morphological and chemical changes as well as by a reduction of genome from diploid to haploid as already discussed in Chaps. II, III, and IV. The marked changes in the proteins during spermatogenesis are by no means limited to chromatin constituents, but occur in other parts of the germ-cell as well. For example, the polypeptide components of membranes in the differentiating male germ-cells show conspicuous alterations during specific stages of spermatogenesis (Millette 1979a, O'Rand and Romrell 1980, Balbontin and Bustos-Obregon 1982, Moore et al. 1985). Of the many peptides synthesized during spermatogenesis, not all are antigenic. In most cases, the functional or metabolic roles of these sperm-specific proteins has not been identified. Proteins that have been purified from testes include some of defined function, e.g. enzymes, and some that play an as yet unknown structural role, e.g. plasma membrane components. Many of the proteins that are formed during spermatogenesis represent potential autoantigens, as sexual maturation starts long after immunologic maturity. Besides the various testis-specific enzymes, plasma membrane constituents of developing spermatogenic cells could be antigenic. Several antigens have been identified on the surface of testicular cells (Radu and Voisin 1975, Koo et al. 1977a, b, O'Rand and Romrell 1977, Millette and Bellvé 1977, Romrell and O'Rand 1978, Tung et al. 1979, Bebe Han and Tung 1979, Millette 1979a, Balbontin and Bustos-Obregon 1982, Söderström et al. 1982, Goldberg 1983, Narayan et al. 1983, O'Brien and Millette 1984). The spermatogenic cell plasma membrane glycoproteins are identified by two-dimensional electrophoresis and lectin blotting (Millette and Scott 1984). The temporal expression of cell surface antigens in the mouse provides a good example; the germ-cells at advanced stages of spermatogenesis, namely pachytene primary spermatocytes, spermatids and also mature spermatozoa, contain antigenic membrane components which are undetectable before the pachytene stage of the first meiotic prophase, i.e. in spermatogonia and preleptotene, leptotene or zygotene primary spermatocytes (Millette and Bellvé 1977). O'Brien and Millette (1984) have made identification and immunochemical characterization of spermatogenic cell surface antigens that appear during the early meiotic prophase. Our knowledge of the molecular aspects of temporal development of these specific surface antigens or proteins is still limited. However, recent studies have shown that stage-specific synthesis of peptides is reflected in gene expression during rodent spermatogenesis (Silver and White 1982, Gold et al. 1983a, Stern et al. 1983a, b). Stage specificity is observed in mRNA populations both in terms of selective production and selective utilization of mRNA (Gold et al. 1983a).

Differentiation antigens have been defined experimentally as tissue-specific macromolecules, detectable usually on the cell surface, which are present at some stages of the tissue development (Boyse and Old 1969). Toullet et al. (1973) have described four autoantigens in guinea pig testicular cells and sperm, including a T antigen which is present on the cell surface as well as in the acrosomal membrane (Toullet and Voisin 1974). Radu and Voision (1975), who used IF study on testicular smears, have demonstrated the simultaneous development of acrosomal S, P, and T autoantigens in all stages of spermatids and perhaps even in late spermatocytes of guinea pig, which have been localized by fluorescent microscopy to the pro-acrosomic granules and all subsequent stages of acrosome development. Their simultaneous appearance suggests co-ordinate synthesis, although this has not been established. But several other autoantigens of mature guinea pig spermatozoa have been subjected to detailed biochemical characterization; their ontogeny remains to be determined (Jackson et al. 1975, 1976, Hagopian et al. 1976). In the mouse, Ia, H-2A variant, H-2D and H-2K antigens appear on leptotene-phase primary spermatocytes (Fellous et al. 1976a, Bhatnagar 1985). In contrast, the F_9 teratocarcinoma antigen, shared by spermatogenic cells has been observed on all developing male germ-cells including spermatogonia (Gachelin et al. 1976). Testicular cell differentiation antigen appears on primary spermatocytes (Koo et al. 1977a, b), but whether this is its first appearance during spermatogenesis is not clear.

Millette and Bellvé (1977) have provided evidence for the ordered temporal appearance of plasma membrane antigens specific to particular classes of mouse spermatogenic cells, such as pachytene spermatocytes, round spermatids, and residual bodies and mature spermatozoa. They have suggested that at late meiotic prophase, coincident with the production of pachytene spermatocytes, a variety of new components are inserted into the surface membranes of developing germ-cells. O'Brien and Millette (1984) have identified and characterized spermatogenic surface antigens that appear during early meiotic prophase. In the rabbit, sperm-specific plasma membrane isoantigenic components appear only after migration of spermatogonia to the luminal side of the barrier, i.e. pachytene primary spermatocytes (O'Rand and Romrell 1977). They are not present on type A, intermediate or type B spermatogonia. Millette and Bellvé (1980) after immunizing rabbits with type B spermatogonia, 82% to 88% pure, from the mouse testis, have observed antigenic components on membranes of spermatogenic cells at all stages of their differentiation, ranging from primitive type A spermatogonia to mature spermatozoa. Certain antigenic membrane components of early spermatogenic cells are partitioned selectively during spermiogenesis into that portion of the plasma membrane which will form a part of the residual body. Boitani et al. (1979) have demonstrated protein synthetic patterns in mouse spermatocytes and spermatids after culturing seminiferous tubules or isolated germ-cells. The middle to late pachytene spermatocytes and round spermatids synthesize soluble proteins that are resolved as about 250 radiolabelled spots, in comparison to only 100 spots from intermediate spermatids. The number of newly synthesized proteins is considerably decreased during spermiogenesis. Only few new molecular species are translated post-meiotically. Stern et al. (1983a) have observed that murine pachytene spermatocytes and round spermatids synthesize approximately equivalent numbers of polypeptides, with less synthesis in elongating spermatids and residual bodies. Gerton and Millette (1983) have studied in vitro synthesis of plasma membrane proteins by isolated mouse pachytene spermatocytes and round spermatids. Autoradiograms of membranes of these spermatogenic cells after incubation with [^{35}S]-Met, show overall similarity, but some differences also are observed. P^{151}/ 06.0, which is synthesized most prominently by cultured pachytene spermatocytes, represents the first surface protein whose synthetic pattern during spermatogenesis is analyzed biochemically. This is the major Con A-binding protein in mouse germ-cell membranes (Millette and Scott 1982) and could be of great significance in germ-cell-Sertoli interaction (Gerton and Millette 1983). Lectin binding sites on human spermatogenic cells have been studied by Mao-Chillee and Damjanov (1985).

Romrell and O'Rand (1978), using rabbit sperm isoantisera, have investigated the relative abundance, mobility, and ultrastructural localization of sperm surface isoantigens at different developmental stages. Their findings have confirmed the first appearance of surface isoantigens on pachytene spermatocytes. They have observed further that capping of surface isoantigens occurs on primary spermatocytes and early and mid-condensing spermatids when the number of isoantigens on the surface is small, but does not occur on late spermatids and residual bodies when the number of surface isoantigens is maximal. In the mouse, however, plasma membrane antigens, which first appear on primary pachytene spermatocytes, do not cap to a significant extent on the surfaces of any spermatogenic cells assayed (Millette 1976, 1979a). Experiments conducted with Con A and wheat germ agglutinin have revealed that late mouse spermatogenic cells do not cap readily (Millette 1979a). Patching of male germ-cells, however, has been detected in all experiments. Highest levels of autoantigens have been reported in the mid-pachytene spermatocytes and late spermatids of the rat testis (Söderström and Anderson 1981). They first appear in early pachytene spermatocytes and continue to be present in the spermatogenic cells up to late spermiogenesis. Millette and Moulding (1981b) have reported the complexity of the plasma membrane components of differentiating spermatogenic cells of the mouse testis. The pachytene primary spermatocytes and round spermatids show at least 25 to 30 polypeptides. None of these proteins is specific to either cell type. Stage-specific protein synthesis has been investigated also by Kramer and Erickson (1982) for the mouse spermatogenesis. Relatively, more active protein synthesis is observed in pachytene spermatocytes than either early or late spermatids. All three cell types show approximately 85% of the total extractable supernatant proteins. The fractions of early and late spermatids show 20 proteins which are not seen in other stages. Four proteins are confined to only pachytene spermatocytes, three to spermatocytes and early spermatids, ten to early spermatids, and only one protein to late spermatids. Stage specificity is observed in regard to solubilized particulate proteins, although 75% to 80% occur in all cell types.

Further biochemical and immunological studies are needed (1) to investigate the restrictive mechanisms underlying the lack of mobility by the plasma membrane of the spermatozoa and (2) to determine more precisely the generation of these mechanisms at the molecular level during spermatogenesis. These additional experiments investigating the lateral mobility of spermatogenic cell membrane components will be of great importance for answering questions in regard to when, where and how the developing germ-cell surface restricts the movement of its components to achieve the apparently rigid phenotype of the mature spermatozoan surface.

Tung and Fritz (1978) have reported the initial appearance on the plasma membranes of pachytene primary spermatocytes of germ-cell-specific antigenic determinants in the rat, which are retained during subsequent stages of spermatogenesis. In addition, evidence has been produced that other antigenic determinants present on pachytene spermatocytes are absent not only from germinal cells at earlier stages of development, but from maturing spermatids or from maturing spermatozoa obtained from the epididymis.

Tung et al. (1979) have described a group of differentiating antigens present on the cell surface of guinea pig testicular cells and spermatozoa. These first become detectable in late spermatids and continue to persist on both testicular and epididymal spermatozoa. They have designated these cell surface antigens collectively as the testicular cell-sperm differentiation antigens (TSDA), which differ from most of the previously described testicular cell (differentiation) antigens in being on the cell surface and found exclusively in spermatids and sperm. T antigen reported by Toullet and Voisin (1974) appears to be a component of TSDA.

The data obtained by Millette (1979a) have indicated that rabbit antimouse type B spermatogonia Ig G (ATBS) recognizes at least two classes of cell surface antigenic determinants. One class of antigens is shared by germ-cells and somatic cells. These are present on testicular, epididymal, and vas deferens spermatozoa and show a nonrandom topographic localization on these cells. The other surface antigens are present on Sertoli cells, on spermatogenic cells at early stages of differentiation, and on residual bodies. These are not seen on spermatozoa at any stage of epididymal maturation.

The origin and processing of these surface constituents have not been defined. However, Radu and Voisin (1975) have observed that the sperm antigens S, P, T develop in the Golgi granules of spermatids, and less certainly as diffuse low-concentrated substances in the Golgi region of spermatocytes II. In addition, specific antigens have been observed in the intracellular fractions of spermatogenic cells (Hagopian et al. 1975, Jackson et al. 1975). It is yet to be determined whether the surface antigens demonstrated by Tung and Fritz (1978) are related to the intracellular antigens. But the various membrane events are correlated with the appearance of the enzyme sulphotransferase that is capable of transferring sulphate from 3′-phosphoadenosine 5′-phosphosulphate to a glycerogalactolipid (Kornblatt et al. 1974). The product formed, a sulphoglycerogalactolipid, may become an integral component of the spermatocyte plasma membrane.

In addition to surface isoantigens discussed above, other sperm-specific components have been demonstrated to appear first during the primary spermatocyte stage: testis-specific form of cytochrome C, designated cytochrome C_t (Wheat et al. 1977, E. Goldberg et al. 1977b, E. Goldberg 1977), LDH-X (E. Goldberg 1983, Hintz and Goldberg 1977, Meistrich et al. 1977, Blanco 1980, Kido et al. 1983), and PGK-2 (E. Goldberg 1977). Other testicular enzymes showing presumptive stage-specific expression are carnitine acetyltransferase (Vernon et al. 1971), sorbitol dehydrogenase (Bishop 1968), and sperm hexokinase (Sosa et al. 1972). Acrosin and hyaluronidase to be discussed in Chap. VII are possibly testis-specific and autoantigenic (Mancini et al. 1964, E. Goldberg 1977). Synthesis of these new intracellular proteins coincides temporally with the appearance of new surface antigens. Recent studies have reported the ontogeny of LDH-X in spermatogenic cells (Hintz and Goldberg 1977, Meistrich et al. 1976b, 1977, 1978). The intracellular synthesis of LDH-X has been observed to be initiated during the late primary pachytene stage and continued during the first half of spermiogenesis. But no data in regard to its eventual expression on the cell surface are available. Similarly, cytochrome C_t is absent from the developing germ-cell until the development of pachytene primary spermatocytes (Wheat et al. 1977), although it has not been determined when is cytochrome C_t first synthesized. It is possible, therefore, that co-ordinate gene activation during the first meiotic prophase is responsible for the simultaneous development of LDH-X, cytochrome C_t, and yet uncharacterized cell surface markers (Wheat et al. 1977). LDH-X appears to be distributed throughout the cytoplasm of advanced spermatocytes and is not restricted to the plasma membrane (Hintz and Goldberg 1977). But in spermatozoa, LDH-X has been described to be present only on the surface (Erickson et al. 1975a, E. Goldberg 1979).

As a rule, the testis-specific isozymes of LDH are absent in the immature testis and develop during puberty, at about the same time as the pachytene primary spermatocytes (Blackshaw and Elkington 1970, R. B. Goldberg and Hawtrey 1967, E. Goldberg 1983). In the mouse, LDH-X is absent in the leptotene or early pachytene primary spermatocytes; it begins to be seen during the mid-pachytene stage which also marks the formation of the testis-specific cytochrome C_t (Wheat et al. 1977). In seasonally breeding wild animals the testis-specific forms of LDH are not seen in either immature or regressed testes, but are detectable at the time of sperm formation and storage (Blanco et al. 1969, Gutierrez et al. 1972, Holmes et al. 1973, Baldwin et al. 1974). In the adult mouse, LDH activity is confined to the germinal epithelium giving the strongest tetrazolium reaction in the spermatid, and the weakest in the peripheral region of the tubules (Hawtrey and Goldberg 1968). Within the germ-cells LDH-X is found largely in the mitochondria, but partly also in the cytosol. Because of this peculiar pattern of localization, it may be acting as a shuttle for the transfer of reducing equivalents from cytosol to the mitochondria, possibly in conjunction with the branched-chain amino acid aminotransferase (L-leucine: 2-oxoglutarate aminotransferase). The subcellular distribution of this transaminase appears to be closely similar to that of LDH-X (Montamat et al. 1978).

The form in which the testis-specific isozymes of LDH exist varies between species. The most common is LDH-X, which is a C_4-homotetramer consisting entirely of one type of subunit, namely C, and would therefore be more correctly called LDH-C_4. This isozyme has been obtained in crystalline form from mouse testes (E. Goldberg 1977, 1983) and its molecular features, immunolgical properties, and amino acid composition have been determined for the mouse and rat (Musick and Rossmann 1979a, b, Pan et al. 1983, Li et al. 1983). In addition to LDH-C_4, several heterotetramers have been reported which are association products of C with A, B and possibly some other subunits as well. The exact composition of all these subunit hybrids remains to be determined and their nomenclature agreed upon. For example, it has been suggested that the C_3 A tetramer, which is found in the testes of the marsupial *Schoinobates vulgaris* (greater glider of Eastern Australia) should be designated as LDH-X, and that the two other testicular isozymes, which are present in this animal and are believed to be C-B subunit hybrids, ought to be called X^{II} and X^{III} (Baldwin et al. 1974). Since in the normal mammalian testis the appearance of LDH-C_4 is related closely to the formation of pachytene primary spermatocytes, it is not surprising that damage to the germinal epithelium leading to either delay or complete arrest of spermatocyte formation postpones or completely prevents the appearance of this enzyme.

LDH-C_4 is so typical and so closely bound up with well-defined stages of spermatogenesis, that it has been proposed as a source of antigens for immunosuppression of fertility (E. Goldberg 1983). It also is one of the few testis-specific enzymes available in pure form. The availability of relatively large amounts (hundreds of milligrams) of mouse LDH-C_4 makes possible rigorous analyses of the effects of immunization on fertility (Wheat and Goldberg 1983, 1984, E. Goldberg 1983). When rabbits of either sex are injected with the purified mouse enzyme emulsified in Freund's adjuvant, specific antibodies are formed (E. Goldberg 1977, 1983, R. B. Goldberg and Wheat 1976). Subsequent studies with inbred strains of mice treated with murine LDH-X have shown that aside from variations in the magnitude of the response the kinetics of antibody production shows strain-dependent differences (Kille et al. 1978). Even less dependable than active immunization is passive immunization of female mice with rabbit antiserum specific for murine LDH-C_4; in vitro a moderate inhibitory effect of such an antiserum on fertilization has been observed, but this may have been due to interaction with spermatozoa rather than the eggs (Erickson et al. 1975b). The study on mice and rabbits has been extended to nonhuman primates which were immunized with mouse LDH-C_4 and then mated (E. Goldberg et al. 1983). In a series of breeding experiments, 22 out of 30 matings, or 74%, were infertile, as compared to 28% in control matings. E. Goldberg (1979, 1983) has discussed the various aspects of LDH-C_4 in relation to immunological approaches for controlling fertility (see also Wheat and Goldberg 1984). The F_9 antigen (Jacob 1977) is present on all classes of spermatogenic cells, including spermatogonia and stem cells. Schachner et al. (1975) have identified NS-4 antigen. The antigenic determinants reported by Tung and Fritz (1978) appear to be different from LDH-X, F_9 and NS-4.

By employing immunofluorescence complement-mediated cytoxicity, and quantitative assays, the presence of a variety of antigens has been demonstrated during spermatogenesis of the mouse (Millette 1979a). Plasma membrane components with defined temporal expression during spermatogensis include RSA-1 (O'Rand and Romrell 1981) and 1B3 (Gaunt 1982). These appear during late pachynema of the first meiotic prophase and in that regard resemble antigenic determinants demonstrated serologically in mouse and rat cells (Tung and Fritz 1978), using two-dimensional electrophoresis. These antigens have been observed to be expressed by all late pachytene primary spermatocytes, round spermatids, and residual bodies in a uniform diffuse fashion on the surface membranes. Mature spermatozoa also show these surface components which are confined to specific regions of their plasma membrane, as will also be discussed in Chap. XI. Earlier spermatogenic cells, such as spermatogonia, preleptotene, leptotene, zygotene and pachytene spermatocytes do not express demonstrable antibody binding sites. The appearance of new surface antigens during the primary spermatocyte stage of spermatogenesis is not restricted to sperm-specific antigens. In the mouse, Ia, H-2H and H-2K antigens appear on leptotene phase primary spermatocytes (Fellous et al. 1976a). Testicular cell differentiation antigen also appears on primary spermatocytes (Koo et al. 1977b), as already described, but whether this is its first appearance during spermatogenesis is not clear. The presence of presumptive stage-specific marker proteins for mouse pachytene spermatocytes (proteins Pa, Pb) and round spermatids (proteins RSa – d) has been shown (Millette and Moulding 1981a). Narayan et al. (1983) have investigated by two-dimensional polyacrylamide gel electrophoresis the total cellular polypeptide composition of separated human and mouse spermatogenic cells, which show extensive homology. Identified marker proteins specific to human spermatocytes include a group of polypeptides at p45/5.9 as well as a protein at p67/5.2. Proteins specific to mouse germ-cells include component p65/5.5. Further comparisons of this kind must be made in future studies to determine molecular features specific to individual mammalian species. Further studies are needed also to determine the cell surface markers specific to individual classes of human spermatogenic cells. The results of various studies have suggested that multiple spermatogenic cell proteins are formed and expressed with temporal specificity as differentiation and maturation of functions of the seminiferous epithelium advance. M. H. Johnson (1970) has suggested that later stages of spermatogenesis are more antigenic than earlier ones (W. L. Johnson and Hunter 1972). This suggestion is supported by the data of Romrell and O'Rand (1978) who have provided direct evidence that spermatids have more surface isoantigenic label than primary spermatocytes.

The initial formation of a specific germinal cell antigenic determinant on pachytene spermatocytes possibly indicates selective gene epxression which is being activated during the late prophase of meiosis. It is established that synthesis of RNA and protein is greatly enhanced during this time, as already discussed in detail in Chap. III, and that several new enzymes are being synthesized. The development of new plasma components correlates well with this general increase in metabolic activity.

B. Functions

The possible physiological function(s) of the different spermatogenic cell surface antigens that have been identified using serological techniques must remain speculative as yet, although it is implicit that these may represent gene products involved in regulating cell surface characteristics in differentiation and perhaps in functions that are organ- or tissue-specific (Boyse and Old 1969). Tung et al. (1979) have made several suggestions in regard to the functional significance of differentiation antigens which may function as receptors for enzyme or soluble inducers during capacitation and acrosomal reactions, or as molecules involved in the interaction between sperm and the zona pellucida or the egg vitelline during fertilization (Shapiro and Eddy 1980). Alternatively, these may be participating in cellular interactions during spermiogenesis. Further detailed investigations of the temporal expression of spermatogenic cell membrane antigens would be essential before it is feasible to formulate hypotheses relating these surface components to functional roles during sperm development. Actually, the detailed identification and biochemical analysis of these surface constituents of pachytene spermatocytes and successive classes of terminal cells, and other unique antigenic determinants on testicular somatic cells, would be very useful in the better understanding of mammalian spermatogenesis at the molecular level (Moore et al. 1985). Bebe Han and Tung (1979) have developed a quantitative assay for antibodies to surface antigens of guinea pig testicular cells and spermatozoa, as also described for the mouse testicular cells (Millette 1979a). To facilitate the biochemical characteristics of cell surface antigens, methods have been developed for the purification of plasma membran fractions from separated mouse spermatogenic cells (Millette 1979a). The availability of these purified cell membranes should (1) permit the application of immunoprecipitation techniques to concentrate particular antigens, and (2) allow the production of monoclonal antibodies directed against specific membrane determinants using hybridoma technology (Barnstable et al. 1978, E. Goldberg 1983, Bellvé and Moss 1983). These preparations should allow the eventual identification of individual cell surface antigens important for the regulation of mammalian spermatogenesis (Millett 1979a).

Now better methods have become available to obtain highly purified populations of spermatogenic cells. Meistrich et al. (1981) have developed a method for obtaining highly purified fractions of rat testicular cells. Shepherd et al. (1981) have reported a technique permitting the preparation of pachytene spermatocytes enriched to 75% purity from normal human testis. Further modification of this method, including a Leydig cell separation step, has now led to the isolation of both human pachytene spermatocytes and round spermatid populations in at least 90% purity (Narayan et al. 1983). These methods will facilitate biochemical analyses of human spermatogenic cells in order to compare and contrast the molecular features of human spermatogenesis with those of rodent spermatogenesis. The functional or metabolic sperm-specific roles of proteins need to be identified in future studies. Such identification would represent a major accomplishment in the investigation of autoantigenicity of testicular constituents by facilitating isolation methods. Narayan et al. (1983) have suggested that purified spermatogenic cell populations may be analyzed biochemically to determine constituents which are involved in the regulation of human sperm development. It is in this area that molecular and cellular biology can be advantageously applied in future studies. For example, recombinant DNA methodologies may provide sufficient quantities of regulatory factors, receptors and enzymes that are difficult to purify using classical techniques; monoclonal antibodies can now be directed against sperm and testis-specific antigens to help in understanding the developmental aspects (Bellvé and Moss 1983, Moore et al. 1985). The mechanisms of gene expression and regulation for key testicular hormones and proteins can be investigated with great sensitivity, using complementary DNA probes. These and other related investigations can benefit greatly from application of the established and emerging techniques.

Part Two Spermatozoa

Chapter VI
General Considerations

Mammalian spermatozoa are small and motile and show a general uniformity in their internal and external structure. The spermatozoon, which performs the function of carrying genetic material from the male to the oocyte, consists of two principal parts, such as the head and the tail (Fig. 55). The tail consists of four components, such as the neck, mid-piece, principal piece, and end-piece (Phillips 1975a). The cytoplasmic droplet is present in association with the mid-piece of immature spermatozoa (Fig. 17). Silver nitrate can differentiate many of the gross morphological features of spermatozoa, including the acrosome, subacrosomal region, perforatorium, post-acrosomal sheath, neck, dense outer fibres of the core of the mid-piece, annulus, principal piece, and end-piece (Elder and Hsu 1981). Silver-staining patterns of spermatozoa have revealed both species-specific and strain-specific differences, especially of the sperm head.

The basic structure of the spermatozoon involves packaging of genetic material in its head. It is very poorly endowed with cytoplasm; an energy-producing mid-piece wrapped with mitochondria, the energy source; and a motile tail. The form of the sperm nucleus is generally somewhat paddle-shaped and relatively symmetrical, exceptions to this arising only in some rodent families. Despite this underlying similarity, a relatively wide variety of minor characteristics is superimposed upon this basic structure of spermatozoon, as it is possible to observe a marked variation in structure and overall appearance of the membranes over the sperm head (D. S. Friend 1982, Holt 1984) and in the size of the nucleus itself when specimens are examined from divergent families.

The dimensional characteristics of ejaculated spermatozoa vary greatly in different species of mammals (Table 3A and B) (Sidhu and Guraya 1977, Cummins and Wooddall 1985). G. F. Friend (1936), while making an analysis of the sperm of British murid rodents, first suggested that it is possible to recognize any of the species from the sperm alone and decide the subfamily to which they belong. The recent transmission and scanning electronmicroscopic studies have revealed variations in sperm head structure of the Australian rodent genus, *Pseudomys* (Muridae) (Breed 1983) and Hydromyinae (Rodentia: Muridae) (Breed 1984). It is also possible to detect which representative in each subfamily is the most primitive one. Several studies have also been carried out on the sperm morphology of primates in order to use it for making further analysis of primate phylogeny (Gould 1980). Braden (1957) demonstrated the first genetically controlled differences in size and shape of the mouse spermatozoon head. Inherited variations in the dimensions of spermatozoa exist between inbred strains, which are not affected by the environment. Significant breed differences have been reported in

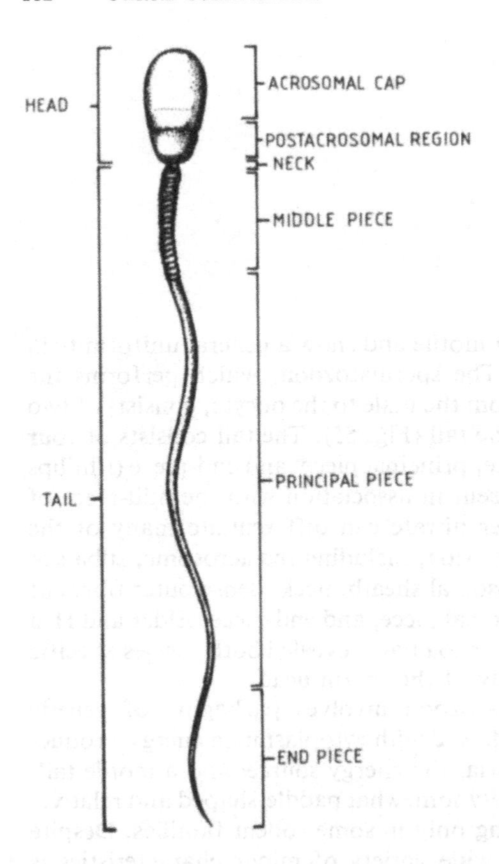

Fig. 55. Drawing of a spermatozoon as seen with the light microscope. It is depicted without the cell membrane (Redrawn from Fawcett 1975b)

the characteristics of goat (Das and Sidhu 1975) and buffalo spermatozoa (Mukherjee and Singh 1965).

During the past few years, the genetic control of the size and shape of the different parts or organelles of spermatozoa has been clearly demonstrated in various studies on different mammalian species (Beatty 1972). According to Beatty (1975) phenogenetics of spermatozoa explains phenotype in terms of genotype. Genetically, a spermatozoon has been subjected to a "diploid effect" mediated by the pre-meiotic diploid genotype and by the genotype of the soma, and also, potentially, by a "haploid effect" mediated by the post-meiotic haploid gene content, as already discussed in Chap. IV. "Additive" genetic variation is probably the main genetic determinant of the dimensional sperm phenotype.

The role of environmental factors cannot be overlooked in regard to the control of the size and shape of different components of spermatozoa, as the extremes of temperature (lower and higher) have been found to increase the percentages of various abnormalities in different components of buffalo spermatozoa (Guraya and Sidhu 1976, Sidhu and Guraya 1985a). The primary abnormalities in head and tail of buffalo spermatozoa are significantly more than the secondary (Saxena and Tripathi 1980). But it is still disputed whether the environmental factors bring about the spermatozoa abnormalities directly or indirect-

Table 3 A. Mean dimensional characteristics of spermatozoa in different species of Muridae[a, b] (From Sidhu and Guraya 1977)

Species	Head				Flagellum			
	Dimensions (microns)				Dimensions (microns)			
	Acrosomal length	Acrosome width	Acrosome area	Post-acrosome area	Mid-piece length	Mid-piece width	Mid-piece area	Mid-piece length
Millardia meltada	6.80	2.40	15.60	5.75	25.40	0.90	22.85	91.60
Bandicota bengalensis	5.10	3.00	15.30	6.60	22.20	0.85	18.85	56.20
Albino rat	5.80	2.70	14.54	5.15	23.60	0.80	18.85	70.00
Rattus norvegicus	7.70	1.50	11.60	3.825	55.00	0.60	33.00	70.00
Mus booduga	6.50	3.00	19.50	6.45	22.00	0.75	16.50	78.00
Rattus rattus	6.80	3.20	21.75	6.70	23.00	0.55	12.60	55.00

[a] Five samples in each case.
[b] Number of cells counted per five samples 500.

Table 3B. "F" values to test the significance of differences in spermatozoal dimensions between and within different species of Muridae (From Sidhu and Guraya 1977)

Sources of variation	Acrosome length	Acrosome width	Acrosome area	Post-acrosomal area	Mid-piece length	Mid-piece breadth	Mid-piece area	Main-piece length	Ratio of flagellum to head
Between species	24.87[a]	14.33[a]	10.16[a]	6.112[a]	129.55[a]	0.969[a]	19.53[a]	45.89[a]	11.113[a]
Within species	0.202[b]	0.090[b]	0.986[b]	0.123[b]	0.043[b]	0.292[a]	2.55[b]	0.701[b]	0.133[b]

[a] Significant at 1% and 5% levels.
[b] Insignificant at 1% and 5% levels.

ly. Hendrikse et al. (1976) have observed that with ageing of stallions the percentage of abnormal spermatozoa is increased slightly. Meanwhile, the percentage of head abnormalities and detached heads is decreased among the live spermatozoa, whereas the percentage of detached heads shows a marked increase among the dead spermatozoa.

During passage through the epididymis, the various components of spermatozoa undergo conspicuous alterations, these being morphological (including migration of the cytoplasmic droplet, changes in acrosome, membrane configurations in the neck region, condensation of mitochondrial cristae and midpiece surface membrane), chemical (including changes in acrosomal enzymes, differences in lipoprotein content between caput and cauda epididymal spermatozoa, changes in activity of glyceraldehyde-3-phosphate dehydrogenase, etc.), and physiological (including modifications in the cohesiveness between the outer acrosomal membrane and the underlying plasma membrane, changes in Na- and K-activated ATPase, stabilization of chromatin by disulphide linkages, variations in the susceptibility of spermatozoa to modification of iodoacetamide, carboxyphenol-maleimide, eosine stain, development of sperm motility). These alterations, which occur in response to changing external environment (Brooks 1984), are collectively referred to as maturation (Bedford 1979, Nicolson and Yanagimachi 1979, Harding et al. 1979, Egbunike 1980, Mann and Lutwak-Mann 1981, Moore 1981, Peterson 1982, Bhattacharyya and Bhattacharyya 1985). More precisely, the physiological, biochemical and biophysical changes include a loss in phospholipids, modification of the accessible proteins and membrane-bound enzymes, an increase in motility and cyclic AMP levels, and the appearance of high concentrations of carnitine. These changes, by which spermatozoa develop progressive motility and fertilizing ability, will be discussed in detail in subsequent chapters dealing with the structure, chemistry and function of different components of the spermatozoon. We still lack sufficient factual knowledge about the process of sperm maturation in the epididymis and about the special quality and role of the environment in the epididymis that supports maturation or sperm storage (Brooks 1984, Rajalakshmi 1985). But as a result of these biological and biochemical changes the spermatozoa progressively gain the ability to move and fertilize ova (Moore 1981, Hinrichsen and Blaquier 1980).

The mammalian spermatozoa must also reside for some defined period in the female reproductive tract before they can fertilize the egg, as reported by Austin (1951) and Chang (1951), employing delayed insemination experiments (see also Lambert et al. 1985). This physiological process is called the capacitation. Though the capacitation is achieved synergistically and efficiently in the female reproductive tract, it can also be accomplished outside the female reproductive tract in vitro in various well-defined media in serveral species of mammals. These in vitro studies have indicated that capacitation is biochemical in nature (O'Rand 1982, Bedford 1983, Farooqui 1983, Clegg 1983, Go and Wolf 1983, Tash and Means 1983, Meizel 1984, Nelson 1985, Guraya and Sidhu 1986). Actually, the capacitation should be considered as a two-way process undergone by the sperm-cell in the female reproductive tract as it achieves the competence to penetrate the investments of the ovum and enter vitellus. Regulatory substances adsorbed onto the sperm-cell surface as it passes from the testis down into the seminal plasma of the male reproductive tract are removed during incubation in female tract fluids (Fraser 1984). Meanwhile, changes in the sperm-cell surface render it more permeable to various constituents of the female fluids, both organic substances and inorganic ions that may function as co-factors mediating the motility activation shown by capacitated spermatozoa (Singh et al. 1978). The subtle alterations included under the fabric of capacitation are preliminary physiological events which prepare the spermatozoon for the reorganization, fusion, and vesiculation of the plasma and acrosomal membranes that acrosome undergoes (the acrosome reaction) (see Sect. C of Chap. VII) when the spermatozoon liberates a number of lytic enzymes that open the sperm's pathway through the extraneous layers of cells and membranes surrounding the egg. Without having undergone the acrosome reaction, the spermatozoon cannot pass through the zona pellucida. Capacitation, which forms the prerequisite for the calcium-dependent acrosome reaction, is believed to consist of reversible reaction of the sperm plasma membrane, the nature of which continues to remain poorly defined at the molecular level. The subcellular, biochemical, molecular and physiological changes observed in most of the in vitro studies facilitate the influx of Ca^{2+} needed for some acrosome reaction. Not much is known what happens in vivo. Controversies exist about the mechanism of capacitation. Several recent theories explaining molecular architectural alterations on sperm surfaces, intramembranous molecular alteration to facilitate Ca^{2+} influx and physiological changes in sperm energetics by alterations in O_2 uptake and glucose utilization manifested in the form of changes in pattern of flagellar beat (hyperactivated motility) during sperm capacitation have been set forth (see Meizel 1984, Courtens et al. 1984, Guraya and Sidhu 1986). Capacitation, therefore, involves certain changes in sperm metabolism (Boel 1985), as well as alterations in the distribution of intramembranous particles (proteins, phospholipids) and lectin binding sites over the sperm plasma membrane (changes of the surface coat); capacitation may also involve removal of seminal fluid components that suppress the metabolic activity and the ability of capacitated sperm to fertilize eggs. Some decapacitation factors present in epididymal fluid and seminal plasma are believed to be glycoproteins (Reyes et al. 1975, Sundhey et al. 1984), as judged by their deleterious effect on fertilization; the role of cholesterol in sperm decapacitation is also indicated (Meizel 1984). Extensive purification and careful analysis of homogeneous decapacitation factors are needed before the mechanism of such an inhibitory effect is determined more precisely. Both initiation of mobility and capacity to fertilize eggs are associated with morphological and molecular changes and alterations in enzyme activity of different components of the sperm, as will be discussed in subsequent chapters. Tremendous species variations are reported in regard to the time, site, and stimulus of capacitation and acrosome reaction in different species of mammals. Many significant advances in our understanding of the morphological, biochemical, and biophysical alterations in spermatozon during their period of maturation in the epididymis as well as during fertilization are the result of development and modifications of many techniques for collecting spermatozoa and fluids from various regions of the genital tract of different mammalian species (Pikó 1969, Bedford 1975, Voglmayr 1975, Yanagimachi 1978a, 1981, Brooks 1984, Nelson 1985).

Chapter VII
Head

A. Shape and Size

The sperm head consists mainly of the nucleus and acrosome (Figs. 55, 56 and 57). Its shape, size and structure vary greatly in different species of mammals including marsupials (Guraya 1965, 1971a, Phillips 1975a, Harding et al. 1979, Sarafis et al. 1981, Holstein and Roosen-Runge 1981, Heath et al. 1983, Gledhill 1986). The spermatozoa of various primate species show, more or less, the same basic configuration of the head, except for some minor variations in its shape and size (Gould 1980). The greatest variation in morphology of primate spermatozoa has been shown to involve the size and shape of the acrosomal membrane and the number, size, and organization of the mitochondria in the sperm mid-piece (Bedford 1967, Pedersen 1970a, Zamboni et al. 1971, Guraya 1971a, Bernstein and Teichman 1972, Lung and Bahr 1972, Semczuk 1977, Holstein and Roosen-Runge 1981). The sperm nucleus can be seen to extend as far as the apex of the head (Figs. 56 and 57). Scanning electron microscope provides the most ideal method available at present for the study of sperm morphology, especially for reconstructuring an impression of a three-dimensional structure from two-dimensional images (Gould 1980, Breed 1983, 1984). In general, the head of the primate spermatozoon is somewhat flattened and paddle-shaped, with evidence of an increasing posterior thickness with increasing evolutionary development towards the great apes. The shape and size of the sperm head in chimpanzee and man are apparently very similar (Figs. 17 and 56).

According to Zamboni et al. (1971), aside from patently abnormal forms studied in all samples of human spermatozoa, the shape of the heads of apparently normal human spermatozoa varies so conspicuously as to defy any attempt at detailed description. According to them, the problem is further complicated by the impossibility of comparing their observations with those made in studies in which the morphology of human spermatozoa was investigated utilizing parameters of standard semen analysis (MacLeod 1965, Amelar 1966, Freund 1968). This was due to the relatively low number of samples and spermatozoa studied by Zamboni et al. (1971) and the difficulty of obtaining overall impressions from samples investigated with the electron microscope. The well-marked heterogeneity of the sperm population of the human semen, which has been described and discussed in several studies (Zamboni et al. 1971, Semczuk 1977, Fredricsson et al. 1977, Gould 1980, Holstein and Roosen-Runge 1981) contrasts with the uniformity of shape, size and structure of spermatozoa of other subhuman primates. Variability of forms involves among others differ-

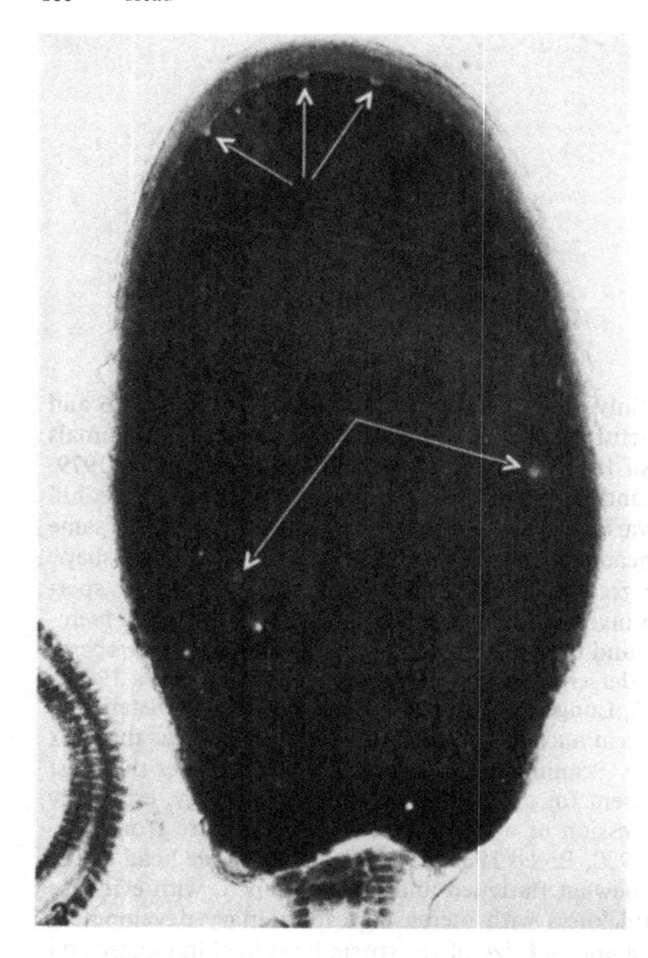

Fig. 56. Head of a *Macaca mulatta* spermatozoon sectioned along a dorsal or ventral plane. Nucleus, acrosome, and membranous envelopes are easily discernible. *Arrows* point to shallow indentations of the nuclear surface (From Zamboni et al. 1971)

ences in degree of condensation of the nuclear chromatin, in size and position of the cytoplasmic droplet, in the degree of differentiation of the mid-piece, and in the fine structures, such as the fibrils of the axial filament complex (Holstein and Schirren 1979, Holstein and Roosen-Runge 1981). Zamboni et al. (1971) believe that the polymorphism of human spermatozoa and, in particular, the variability in the configuration of the head, are the expression of a process of cellular differentiation much more complex in man than in other species of primates. The underlying cause could relate to social phenomena, such as the subtle effect of clothing on scrotal temperature, not seen in animals. The heterogeneity of human spermatozoa should be kept in mind whenever attempts are made to differentiate "normal" from abnormal sperm forms.

The "round-headed spermatozoon" is a specific defect in man affecting the whole sperm population. Its most common features include round head, absence of acrosome and post-acrosomal sheath, and infertility (Baccetti et al. 1977, Holstein and Schirren 1979, Gould 1980). An occurrence of round-headed sper-

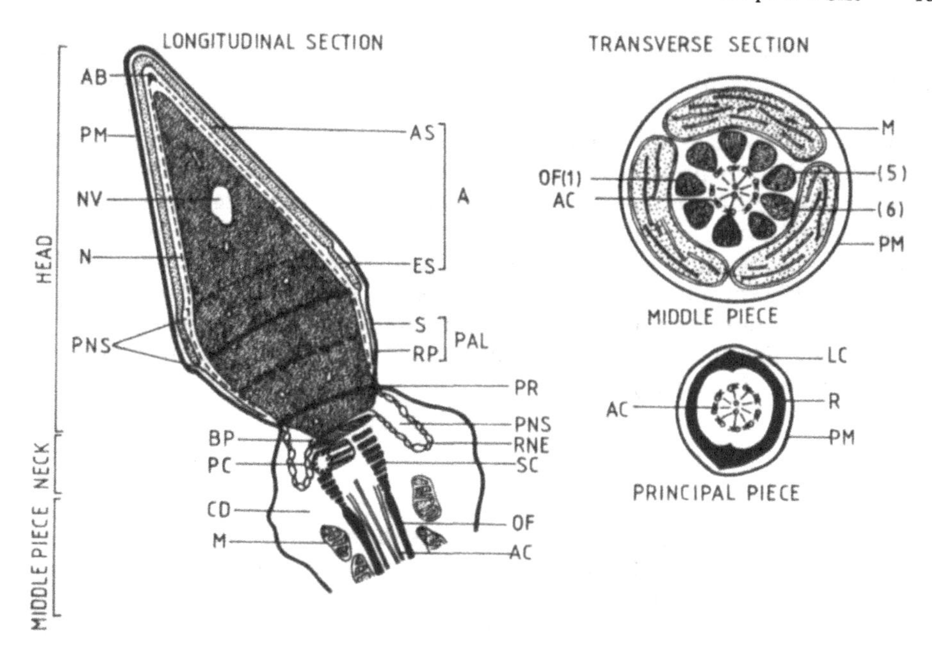

Fig. 57. Schematic representation of the ultrastructure of human spermatozoon (Composite diagram from data obtained by transmission and scanning electron microscopy and special techniques like freeze-etching and ultracytochemistry). *A* acrosome; *AB* apical body; *AC* axonemal complex; *AS* anterior segment; *BP* basal plate; *CD* cytoplasmic droplet; *ES* equatorial segment; *LC* longitudinal column; *M* mitochondria; *N* nucleus; *NV* nuclear vacuoles; *OF* oute fibers; *PAL* post-acrosomal lamina; *PC* proximal centriole; *PM* plasma membrane; *PNS* perinuclear substance; *PR* posterior ring; *R* rib; *RNE* redundant nuclear envelope; *RP* rows of particles; *S* striations; *SC* striated columns (Redrawn from Bustos-Obregon et al. 1975)

matozoa in two brothers (Kullander and Rausing 1975) has suggested that this defect may have a genetic origin. Escalier (1983) has made a quantitative ultrastructural study of human spermatozoa with large heads and multiple flagella.

The structural heterogeneity of spermatozoa within an ejaculate has also been reported for other mammalian species (Bielanski et al. 1982, Bongso 1983, Sidhu and Guraya 1985a). These structural variations are believed to originate during spermatogenesis (Holstein and Schirren 1979), maturation in the proximal tubules of the epididymides, the acrosome reaction just prior to fertilization and the degenerative changes associated with senescence. Jones (1975) has emphasized that spermatozoa in an ejaculate are at varying stages of their life history depending on their normality, stage of maturity, and degree of cytolysis, which should be taken into account in assessing semen quality.

Cooper and Bedford (1976) have described differences in the labelling pattern of surface iron colloid between the dorsal and ventral surfaces of the musk-shrew (*Suncus murinus*) sperm head. The dorsal aspect showed an aggregated pattern of colloid binding, whereas the ventral aspect yielded a more evenly distributed pattern. Its spermatozoa show several unusual features (Phillips and Bedford 1985). Koehler (1977) has demonstrated textural differences between the dorsal and ventral aspects of the sperm head, as well as between the acrosomal and post-acrosomal regions of the musk-shrew sperm. The sperm head in the mink (*Mustela vision*) shows six swellings on the dorsoventral aspects, two connected hump-like swellings at the anterior border of the equatorial segment of the acrosome and at the post-acrosomal sheath on each side (Kim et al. 1979). These swellings represent a species-specific structural feature which might be needed for recognition of ovum and attachment to it in fertilization. Harding et al. (1979) have discussed the unusual features of the sperm head in marsupials as also discussed for different species and genera of Muridae (Breed 1983, 1984). In spite of these morphological variations, the main structures present in the sperm head of mammals are the nucleus, the acrosome (or acrosomal cap), and membranous envelopes (Figs. 56 and 57).

B. Nucleus

The main characteristic of the sperm nucleus is the extremely condensed state of the nuclear material (Fig. 56) which results in a strong electron opacity and a very reduced size compared to somatic nuclei, as already discussed in Chap. IV. The nucleus, consisting of DNA and proteins, forms most of the head of the spermatozoon (Figs. 56 and 57). In general, the configuration of the nucleus is similar to that of the head. Therefore, it would seem reasonable to expect that the size of sperm nucleus is directly proportional to the amount of DNA in the haploid nuclei of the species. This expectation is not supported by the fact that different sperm types undergo different degrees of chromatin condensation. The nuclear volume of the human sperm (7 μm^3) is just about half that of bovine or rabbit sperm, in spite of similar DNA content (Koehler 1972).

The nucleus of spermatozoa stains homogeneously for DNA and basic proteins (Guraya and Sidhu 1975c, Bilaspuri and Guraya 1986c, d). However, it shows some vacuoles in the human and chimpanzee spermatozoa (Figs. 17, 56 and 57), which react negatively with various histochemical techniques, as already described. The sperm nuclei contain the paternal genetic material (DNA) consisting of a haploid set of somatic chromosomes and, in addition, one sex chromosome (X or Y).

1. Sex Chromosomes

Biologists have periodically attempted to fractionate spermatozoa into X- and Y-chromosome-bearing cells. Their separation is aimed at controlling the sex of the offspring (Gledhill 1983a, Corson et al. 1984). Previous attempts to separate them included head length measurements (Zeleny and Faust 1915), acid or alkaline douching (Unterberger 1930), electrophoresis of semen (Schroeder 1941), and density centrifugation (Lindahl 1958, Ericsson et al. 1973). However, all of these separations were challenged as ineffective (Cole et al. 1940, Rothschild 1960, van Duijn 1961, Nevo et al. 1961, Schilling and Thormäehlen 1975). It is still controversial whether spermatozoa having X or Y chromosome differ in their nuclear size and shape. DNA measurements on a population of X and

Y spermatozoa have revealed that X sperm contain 3.7% more DNA than the Y sperm and has 7% more surface area (Pearson et al. 1975). Gledhill (1984) has made cytometric analysis of shape and DNA content in mammalian sperm. This method could lead to sexing mammalian sperm.

The discovery that the human Y chromosome can be selectively demonstrated by staining it with fluorescent quinacrine dyes (Pearson et al. 1970, Lambovot-Manzur et al. 1972) has made it possible to distinguish the male-determining spermatozoa by the presence of a brightly fluorescent spot in the nucleus. This dye has the property of binding specifically to the DNA on the adenine-thymine-rich zones of the molecules (O. J. Miller et al. 1973). When stained with quinacrine, the Y chromosome of the sperm becomes fluorescent in the two distal thirds of the long arm. This region of the Y chromosome remains heteropycnotic in the interphasic nuclei, it being evident as a strongly fluorescent body, named Q or F body, or Y chromatin (Beatty 1975, Hegde et al. 1978). Sumner et al. (1971) combined fluorescent and Feulgen techniques to infer a 2.7% difference in DNA between X- and Y-bearing human spermatozoa.

It was to be hoped that selective staining of the Y chromosome with fluorescent quinacrine dyes would produce evidence for or against a consistent arrangement of the chromosomes within the nucleus of the spermatozoon. Unfortunately, it has not done so. The human sperm heads contain up to five fluorescent "F-bodies", the spermatozoa being symbolized as 0F, 1F, 2F, 3F, 4F, and 5F (Beatty 1975). As the bodies mark the presence of the Y chromosomes, the spermatozoa designated 0F, 1F, and 2F are believed to contain X, Y, and YY chromosomes, respectively. But Beatty (1975) has presented an analytical model that predicts with great accuracy the relative numbers of spermatozoa in the different F classes. On this model, none of the 2F spermatozoa need to be considered as YY. Therefore, a simple correspondence between F-bodies and Y spermatozoa seems unlikely and the true incidence of YY spermatozoa appears to be lower than currently thought. It is unlikely to exceed 1.6% and may be much lower, as suggested by the estimation of incidence of nondisjunction of the Y chromosome in secondary Y spermatocytes. A preliminary estimate of the true incidence of YY sperm is 0.4% or lower. According to Sumner and Robinson (1976), the measurements have suggested that many of the spermatozoa with two quinacrine-fluorescent spots are not YY-bearing as previously thought but might be incompletely condensed Y-bearing spermatozoa.

During spermatogenesis, the meiotic process results in the formation of haploid spermatids. In the heterogametic sex, the heterochromosomes X and Y segregate into these spermatids and thereafter into the spermatozoa, bearing the expected 1:1 ratio. In the past 15 years, the literature offers different X to Y ratios for spermatozoa in human semen. Using the quinacrine method, Barlow and Vosa (1970) found a frequency of 44% of sperms with Y chromatin (Y spermatozoa). By employing the same methods, the other workers observed a frequency of 33% (Diasio and Glass 1971), 41.65% (Evans 1972), 39% (Quinlivan and Sullivan 1974), and 46.2%. Kaiser et al. (1977) have studied the frequency of Y-chromatin-positive spermatozoa in relation to semen analysis. Broer et al. (1977), using fluorescent microscopic technique observed an average frequency of 48.2% ± 2.9% of sperms with Y chromatin. The ratio of X:Y spermatozoa has been related to the duration of the penetration tests, varying temperatures during penetration, varying acid values of semen and cervical secretion and also to the value of sperm motility in the semen analysis and cervical factors. None of the above variables had any measurable influence on the percentage distribution of X and Y spermatozoa during in vitro penetration. Jorge Sans et al. (1977) have studied the ratio of X to Y spermatozoa in normo and oligozoospermic human semen with the fluorescent technique using quinacrine. Out of a total of 44 samples (22 normo and 22 oligozoospermic samples) a ratio of 1:1 in X to Y spermatozoa is observed in both groups (Struck 1974). The frequency of X to Y spermatozoa in the ejaculate is not related to sperm count. It is also independent of morphology, motility, vitality, and seminal plasma pH (Kaiser et al. 1974).

From the distribution of chromosomes 1, 9 and Y in double-headed spermatozoa, the rate of nondisjunction has been estimated to vary between 2% and 5% (Pearson et al. 1975). Computer stimulation studies have indicated, because of compensating errors, that high nondisjunction rates are compatible entirely with an ability to discriminate between the DNA content of X and Y spermatozoa by Feulgen photometry. Hegde et al. (1978) have recently observed that the number of human spermatozoa with F bodies is increased by 2% – 9% after dithiothreitol treatment. According to Pawlowitzki and Bosse (1971), the Y chromosome is either localized within vacuoles or is attached to them.

The availability of the fluorochrome staining technique for the demonstration of the Y chromosome has stimulated renewed interest in the isolation of human X and Y spermatozoa with the goal of choosing the sex of the offspring, by artificial insemination in humans and farm animals (Roberts 1972, Ericsson et al. 1973). The fluorescent quinacrine dyes can yield valuable morphological data, but only for human spermatozoa (Hegde et al. 1978, Singer et al. 1986). Rabbit spermatozoa separate in two populations after nonequilibrium sedimentation (Stambaugh and Buckley 1971) or centrifugation into Ficoll gradients (Branham 1970, Shastry et al. 1977). These can be isolated on the basis of the ability of the Y sperms to swim in more superficial levels (Roberts 1972). Bovine sperms also appear to sediment into two classes (Schilling 1971). Bhattacharya (1976), using convection counter-streaming sedimentation, has separated the bull semen into lighter and heavier fractions, which have been used in the field tests. The proportions of X- and Y-chromsome-bearing spermatozoa have been determined in samples of bovine semen (Pinkel et al. 1983a). Garner et al. (1983) have made quantification of the X- and Y-chromosome-bearing spermatozoa of domestic animals by flow cytometry. Pinkel et al. (1982) have made separation of sperm bearing Y and 'O' chromosomes in the vole *Microtus oregoni*.

With the sephadex gel-filtration method after Steeno et al. (1975) X spermatozoa have been claimed to be separated successfully from the Y spermatozoa in semen samples from 81 donors with the possibility of practical use with the aid of artificial insemination to promote pregnancy (Adimoelja et al. 1977). A better quality of spermatozoa is obtained after this method of filtration, which will support the spermatozoa in having the potency of fertilization. There is indeed a decrease in concentration of spermatozoa after filtration, but the samples still fulfil the minimal concentration required in artificial insemination.

Human spermatozoa appear to fall into two categories in regard to their shape as examined with scanning electron microscopy (Baccetti and Afzelius 1976), as well as in regard to cell surface sialic acid content (Satoru et al. 1984, Kancko et al. 1984), but not with respect to gravity (Quinlivan and Sullivan 1974). Katz and Pedrotti (1977) have considered the possible geotactic distinctions between X-and Y-bearing human spermatozoa. Rohde et al. (1973) and Ericsson et al. (1973) have been able to obtain fractions of human spermatozoa which contain Y-bearing sperms up to 85%. Schilling and Thormäehlen (1976) have made attempts to separate X and Y spermatozoa in the rabbit and humans by density gradient centrifugation (Schilling and Thormäehlen 1977). Their results demonstrate that X- and Y-chromosome-bearing spermatozoa can to some extent be separated by centrifugation. The differences in their weight or size are, however, very small and highly variable, and therefore more than 70% of Y spermatozoa could not be obtained. Nasr et al. (1979) have reported a simple technique for collection and separation of Y spermatozoa of rat. David et al. (1977) have made a study of motility and percentage of Y- and YY-bearing spermatozoa in human semen samples after passage through bovine serum albumin. A selection of human spermatozoa has been made according to their relative motility and their interaction with zone-free hamster eggs (Forster et al. 1983). Goodall and Roberts (1976) have observed that Y-bearing spermatozoa, identified by the quinacrine staining technique, are significantly more motile than X-bearing spermatozoa. This difference is consistent with the current estimates of the difference in mean head DNA content. Broer et al. (1976) have observed the frequency of Y-chromatin-bearing spermatozoa in the intracervical and intra-uterine post-coital tests. Moruzzi (1979) has identified 24 mammalian species expected to have a maximal difference in bulk chromatin between X- and Y-bearing spermatozoa. The spermatozoa of these species may provide more favourable material to make correct separation of X- and Y-bearing spermatozoa in future studies. The results of various studies, as discussed here, indicate there still exists uncertainity about the precise separation of X- and Y-bearing spermatozoa.

2. *Nuclear Chromatin and Vacuoles*

a) Chromatin

Most noneutherian and eutherian mammalian spermatozoa have uniformly condensed chromatin, which is highly stable. The nucleus of spermatozoa in human and subhuman primates, studied with electron microscope, also appears as a homogeneous body consisting of highly compact, electron-opaque chromatin with occasional areas of decreased electron opacity (Fig. 56) (Zamboni et al. 1971, Evenson et al. 1978, Kumar 1985). These areas represent invaginations of the nuclear surface and are seen in the form of deep indentations of the anterior portion of the nuclear profile. A varying number of human spermatozoa, even within a single ejaculate, show a wide variation in the extent of condensation. Some are found to have a, more or less, granular chromatin (Schultz-Larsen 1958, Bedford 1967, Pedersen 1969, 1970a, Bedford et al. 1973,

Holstein and Schirren 1979, Holstein and Roosen-Runge 1981). This could be an expression of a varying degree of maturity, as a progressive condensation of chromatin takes place during spermiogenesis. This heterogeneity in nuclear ultrastructure in the human ejaculates suggests that the process of chromatin condensation does not proceed to completion in all members of the sperm population (Evenson et al. 1978). Its effects on the fertilizing capacity of spermatozoa are still not known in humans. The level of condensation and stabilization of the human spermatozoal chromatin is believed to be determined in part by the extent of available protein-bound thiol groups that have been oxidized to form disulphide cross-links (Bedford et al. 1973). The latter process is usually completed during passage of the spermatozoa through the epididymis. When exposed to agents that reduce disulphide bonds, these nuclei decondense more rapidly than those with dense homogeneous chromatin.

Recent studies with the freeze-cleaving technique have shown that the arrangement of nucleoprotein in the mammalian spermatozoa is in the form of layers of lamellae giving rise to a strong orientational dependence in birefringence (Friend and Fawcett 1974, Kaczmarski 1978). The ordered condensation of chromatin must have a major role in producing the lamellae, like DNA ordering observed in mature sperm head pieces (Lung 1972, Kaczmarski 1978). Veres and Očsenyi (1968) have shown the presence of a system of colloid parallel plates in the bull sperm head. Lung and Bahr (1972), using scanning electron microscope, have recently reported the fibrous nature of human sperm chromatin. The chromatin fibres form an irregular network lacking detectable architectural order. Fibres within the nuclear mass appear to coalesce. A treatment of the spermatozoa with 1.5% of the detergent Sarkosyl results in decapitation, removal of plasma and acrosomal membranes, and variable degrees of nuclear chromatin decondensation (Evenson et al. 1978). The subsequent addition of 0.02 M dithiothreitol (DTT) causes further chromatin decondensation, revealing the presence of meshwork of 2 to 3 nm fibres interconnecting cord-like and spherical chromatin bodies about 30 – 75 nm in diameter. The thinner fibres approximate the diameter of DNA molecules. Trypsin digestion in the presence of 0.02 M DTT results in the conversion of the chromatin bodies into fibrous structures that are continuous with the fibrous network interlacing the chromatin bodies. Wagner and Yun (1979) have observed that human sperm chromatin is characterized by fibres composed of discrete spherical organizational units unlike somatic nucleosomes. During the decondensation process fibres composed of joined spherical units of both 40 and 15 nm diameter are seen. Further study of sperm chromatin co-acervation as a mechanism of sperm nucleus condensation is needed. But the various alterations in the physical state of the chromatin are usually believed to be a stratagem of nature to protect the genome from damage on the perilous journey to the site of fertilization (Nicolle et al. 1980) and as a means of decreasing nuclear volume in order to streamline the spermatozoon and facilitate its motility.

b) Vacuoles

The nucleus of chimpanzee (Fig. 17) and human (Figs. 56 and 57) spermatozoa is characterized by the presence of vacuoles which are usually called nuclear vacuoles. These are unlike other vacuoles in not being limited by a membrane (Holstein and Roosen-Runge 1981). The nuclear vacuoles vary considerably in number, size, and location (Figs. 56 and 57) (Bedford 1967, de Kretser 1969, Pedersen 1970a, Zamboni et al. 1971, Fawcett 1975b, Holstein and Roosen-Runge 1981). The nuclear vacuoles are particularly seen in 90% of the normal human sperm after sodium dodecylsulphate dithiothreitol treatment (Bedford et al. 1973). These vacuoles are indicative of an incomplete condensation of the chromatin which shows differences between individual spermatozoa. Fawcett and Ito (1965) have described similar vacuoles in the nucleus of bat spermatozoa. The vacuoles generally are highly irregular in outline and, even in the absence of a limiting membrane, the demarcation between them and surrounding chromatin is always distinct. The vacuoles generally are occupied by a pleomorphic content consisting mostly of finely granular material and whorls of membranes. Evenson et al. (1978) have observed that decondensed spermatozoal nuclei contain non-membrane-bounded "vacuoles" which frequently contain 2 to 3 nm fibres and other thicker fibrillar structures. The vacuoles may be single or multiple and some large enough to distort the shape of the sperm head. The origin and functional significance of nuclear vacuoles remain unknown. These are believed to be formed as a result of incomplete chromatin condensation during spermiogenesis (see Chap. IV). Whether these anomalies of nuclear condensation have an effect upon fertilizing capacity is not known.

3. Nuclear Envelope

The nucleus is covered by a double-layered nuclear envelope which is exceptional in several respects. Its entire area under the acrosomal cap and in the post-acrosomal region is devoid of nuclear pores (Fig. 57) and the two membranes of the envelope are separated by only 7 to 10 nm (Fawcett and Chemes 1979, Holstein and Roosen-Runge 1981). Actually, the nuclear pores and perinuclear cisterns disappear during spermiogenesis, when the two laminae of the nuclear envelope adhere to each other, but several pores are conserved in the posterior area of the nucleus (Fig. 57). Caudal to the posterior ring to be described later, however, the two membranes diverge by the usual 40 to 60 nm distance. The two sheets of the nuclear membrane are adherent closely to the underlying condensed chromatin and, therefore, are difficult to resolve as separate units (Horstmann 1961, de Kretser 1969, Bedford and Nicander 1971, Zamboni et al. 1971, Holstein and Roosen-Runge 1981, Jamil 1982). However, in the posterior region of the nucleus these become separated from the chromatin by an irregular space. In the area of separation, the nuclear envelope is redundant, folded many times and characterized by numerous "pores" or "fenestrations" (Figs. 57 and 58) (Stackpole and Devorkin 1974, Fawcett and Chemes 1979, Holstein and Roosen-Runge 1981). The numerous pores of the redundant portion of the nuclear envelope are arranged in a hexagonal pattern, as revealed by freeze-etching. The nuclear envelope of the head and the redundant nuclear envelope are separated by a constriction (Stackpole and Devorkin 1974), which is apparently analogous to the striated band described by Pihlaja and Roth (1973) on the surface of bovine sperm and the posterior ring on human sperm (Koehler 1972).

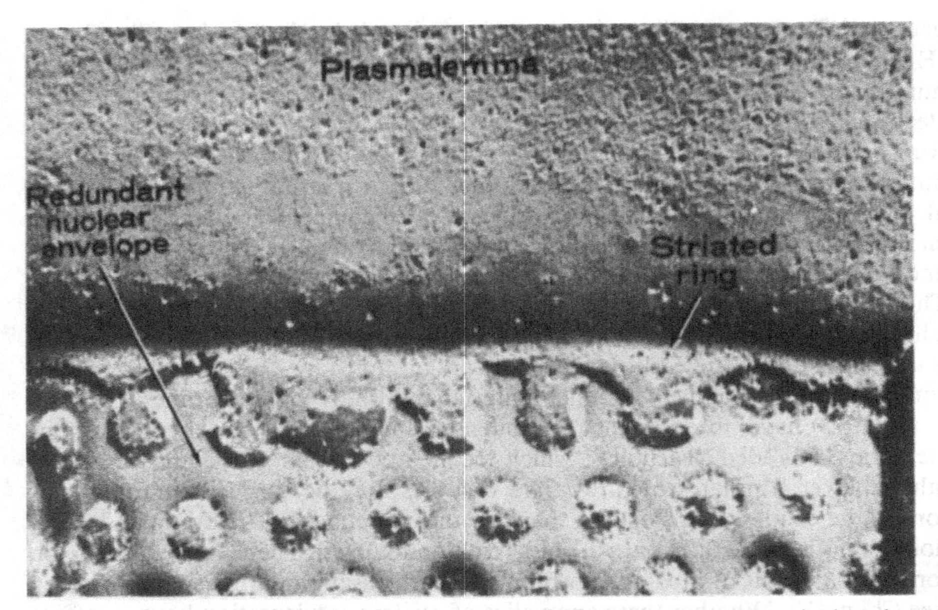

Fig. 58. Freeze-cleave preparation of the posterior region of a guinea-pig sperm head. The B face of the plasmalemma is shown and the posterior ring therefore apears as a ridge instead of a groove. Immediately behind the ring are very regularly arranged pores in the redundant portion of the nuclear envelope (From Fawcett 1975b)

The folded redundant nuclear envelope constitutes very conspicuous structures at the base of the head, which are called lamellar bodies (Pikó 1969). The space between the chromatin and the redundant portion of the nuclear membrane is occupied by flocular and/or filamentous material of relatively low electron opacity (Zamboni et al. 1971, Holstein and Roosen-Runge 1981). The presence of a redundant portion of the nuclear envelope separated from the chromatin at the nucleus and extending posteriorly towards the neck has been described in the spermatozoa of other mammalian species (Fawcett 1965, Nicander and Bane 1962, Fawcett and Ito 1965, Wimsatt et al. 1966). It is generally believed that the formation of the redundant portion of the nuclear envelope could result from the fact that the decrease of the volume of the nucleus during spermiogenesis is not accompanied by a corresponding decrease of the extension of the nuclear envelope. Zamboni et al. (1971) have stated that the development of the redundant portion of the nuclear envelope could be of functional significance because the posterior region of the nucleus, which is adjacent to the cytoplasm, is the only region of the spermatozoon where exchanges of material and information between the nucleus and cytoplasmic components can occur. This is supported not only by the presence of numerous pores in the redundant portion of the nuclear envelope (Fawcett and Chemes 1979), but also by the observation that the scrolls of the nuclear envelope are especially prominent in human spermatozoa, which are provided with large cytoplasmic droplets endowed with apparently active organelles and components (Zamboni et al. 1971, Holstein and Roosen-Runge 1981).

Bedford (1967) has observed a scroll of evaginated nuclear membrane in the neck region of spermatozoa from the bush-baby and the slow loris. The bush-baby, unlike the slow loris, exhibits an elaborate system of membranes in the neck region, which seem to be derived from the nuclear envelope. The excess nuclear envelope also forms one or two scrolls of concentric structures flanking the neck region (Fawcett and Ito 1965, Fawcett 1970). The nuclear pores may be concerned with the loss of nuclear contents. Fawcett and Ito (1965) have related the formation of the redundant nuclear membrane to the diminution of nuclear volume which occurs in spermatid of bat during spermiogenesis, and expressed surprise that it is not resorbed in the "mature" sperm. They raised the question as to whether the excess nuclear membrane might be reincorporated into the nuclear envelope after fertilization, as the sperm enlarges to form the male pronucleus. But Wimsatt et al. (1966) have stated that at least in the bat the redundant nuclear membrane is resorbed or otherwise disposed of before the sperm penetrates the egg.

On the posterior surface of the nucleus, the nuclear envelope again comes into close contact with the condensed chromatin. This portion of the nuclear envelope does not show the presence of pores and the membranes are in very close apposition (Fig. 59). It lines the implantation fossa, the site of attachment of the tail to the head. In the region of the implantation fossa a narrow interspace (6 to 7 nm) between the two nuclear membranes is transversed by regular periodic densities about 6 nm wide and 6 nm apart (Fig. 59A). Freeze-fracturing has revealed that this portion of the nuclear envelope is highly specialized within the plane of the membranes (Friend and Fawcett 1974). A particle-free, relatively smooth region of the membrane is seen in the central portion of the fossa but on either side of this featureless area there is present a very dense population of relatively large (15 nm) intermembranous particles spaced about 20 nm apart (Fig. 59B, C) which appear to correspond to the periodic densities seen traversing in the interspace between the membranes in thin sections (Fig. 59A). This region of the nuclear envelope is covered on its outer surface by a thick layer of very dense material, the basal plate at the base of the head, which lines the fossa and provides attachment for a large number of fine filaments that extend into it from the articular surface of the connecting piece.

4. Proteins and Nucleic Acids

a) Proteins

The transformation of spermatid chromatin from a nucleohistone to a nucleoprotamine complex occurs in a wide variety of species. Actually, the DNA of the sperm nucleus is conjugated with the protamines and histones during spermiogenesis, as already discussed in Chap. IV. The eutherian mammals show the basic keratin "stable protamine" type of sperm protein or arginine-rich histones (Reid 1977, Tanphaichitr et al. 1978, O'Brien and Bellvé 1980a). Formation of nucleoprotamine is considered responsible for the repression of gene transcription and for the extreme condensation of sperm chromatin (Pogany et al. 1981). Actually, in mammals, both gene repression and the initial condensation of

Fig. 59. A Electron micrograph of a thin section through the implanation fossa of a late Chinese hamster spermatid, showing the close apposition of the membranes of the nuclear envelope and the periodic densities that traverse the narrow cleft between them (*at arrows*); **B** A freeze-cleave preparation of rat sperm in which the fracture line has passed obliquely across the base of the flagellum. It illustrates a particle-poor area in the centre of the implantation fossa, surrounded by an area of closely packed particles; **C** Higher magnification of the intramembranous particles in the nuclear envelope lining the implantation fossa. Some of the particles appear to have a central hole or pore (*at arrows*) (From Fawcett 1975b)

chromatin occur prior to the first appearance of protamine, as discussed in Chap. IV. Nevertheless, the presence of cystein-rich protamine(s) does confer unusual sclerotic characteristics upon the nucleus of their spermatozoa (Bedford and Calvin 1974a).

Somatic histones have been reported in the sperms of some mammals (Uschewa et al. 1982). It is not clear whether mature sperms have been used in some of these studies. But Uschewa et al. (1982) have claimed that mature sperm nuclei contain tightly bound somatic histones. Kistler et al. (1973) reported the isolation from rat spermatozoa of a protein with high arginine and cyst(e)ine content, which is identical to the protamine described by Kumaroo et al. (1975). In addition, Kumaroo et al. (1975) have also isolated new protamine from rat sperm heads which has high arginine content but appears to be devoid of lysine and cyst(e)ine.

Biochemical analyses have shown that while the sperm nuclei of some species, e.g. rat and guinea-pig, contain only one basic protein, those of humans and mouse contain two or more (Bellvé 1979, O'Brien and Bellvé 1980a). Critical electrophoretic analysis of the heterogeneity of mouse sperm protein fraction has revealed two very close major bands and trace amounts of components of greater electrophoretic mobility (Calvin 1976), with the latter possibility being attributable to proteolysis of the sperm protein (Marushige and Marushige 1975a). O'Brien and Bellvé (1980a) have observed that besides the two subfractions of the SDS-insoluble sperm heads each containing one of the two mouse protamines, there are present acidic and moderately basic head fractions, each containing a limited number of distinct protein bands with molecular weights ranging from 14,000 to 76,000. These proteins are believed to be derived either from the nucleus or the associated perinuclear material. One- and two-dimensional electrophoresis on acetic acid- urea polyacrylamide gels indicates that the moderately basic fraction may contain minor components that resemble certain histones and/or spermatidal basic nuclear proteins.

Chang and Zirkin (1978) have studied proteolytic degradation of protamine during thiol-induced nuclear decondensation in rabbit sperm. Zirkin et al. (1980) recently have studied the involvement of an acrosin-like proteinase in the sulphydryl-induced degradation of rabbit sperm nuclear protamine. These studies have revealed that when rabbit acrosin serves as the exogenous proteinase, the peptide banding pattern seen is identical to the pattern characteristic of the nuclear-associated proteinase. These results have suggested that the proteinase associated with the rabbit sperm nuclei and involved in sperm nuclear decondensation in vitro appears to be acrosin-like.

Rodman et al. (1979) have observed that mouse sperm protein is immunogenetic in a heterologous species. Its antigenic sites can be observed in spermatozoa and spermatids of all stages, but not in primary spermatocytes. These antigenic sites are masked at about Stage XV of spermatogenesis and may be unmasked by treatment with a reducing agent. These results have indicated that one or more components of mouse sperm protein are assembled at the beginning of spermiogenesis and undergo an alteration in the final intratesticular stage of spermatid differentiation. That alteration has been presumed to be the formation of disulphide linkages between the cysteine residues.

Protamine(s) synthesis in mammalian species occurs during the terminal (or condensing) stage of spermiogenesis after the complete repression of gene transcription. It is retained within the nucleus during the remainder of spermiogenesis and sperm maturation, as already discussed in Chap. IV. Phosphorylated protamine derivatives have been reported in the testis chromatin of serveral mammals (Marushige and Marushige 1975b, 1978). Their protamine(s) vary in the degree of phosphorylation. These phosphorylated protamine molecules are associated preferentially with condensed chromatin that is derived from late spermatids (Marushige and Marushige 1978). By the time spermatozoa enter the epididymis and protamines are completely dephosphorylated, their cysteine residues develop intermolecular disulphide bonds which effectively stabilize the genome of spermatozoa (Marushige and Marushige 1974, 1975b). Chulavatnatol et al. (1982) have made a comparison of phosphorylated proteins in intact rat spermatozoa from caput and cauda epididymis.

The protamines are more heterogenous in humans than in other animals (Tanphaichitr et al. 1978). There are three main fractions, HP_1, HP_2, HP_3 and two minor fractions, HPS_1, HPS_2 (Puwaravutipanich and Panyim 1975). Analysis of the amino acid compositions has revealed high arginine content (Pongswasdi and Svasti 1976). Two other testis-specific histones TH_1 and TH_2B have also been described (Branson et al. 1975, Kistler et al. 1975b, Shires et al. 1975) with amino acid compositions similar to H_1 and H_2B. But TH_1 and TH_2B of rat and mouse have a higher acidic: basic amino acid ratio, resulting in a slower electrophoretic mobility than H_1 and H_2B in acid-urea gels. TH_2B has also been shown to contain one cysteine. This amino acid has been found only for one or two residues in H_2, but at a considerably higher level in mammalian protamines (Shires et al. 1976).

In rats, the testis-specific histones first appear in the primary spermatocytes and their amounts increase when the germ-cells reach the spermatid stages (Shires et al. 1976, Kumaroo et al. 1975, Mills et al. 1977). These results have suggested that testis-specific histones may be intermediates during the sequential transition of histones to protamines. Comparison of the electrophoretic patterns of histones from human testis, testicular sperm and ejaculated sperm implies that the histones may be removed in the order H_2A and H_1 before H_3 H_4, and H_2B and TH_2B (Tanphaichitr et al. 1978). TH_2B, which is the major histone fraction in ejaculated sperm, has no longer a strong affinity to DNA. Marushige and Marushige (1983) have studied proteolysis of somatic type histones in transforming rat spermatid chromatin.

Basic protein with a high content of arginine and containing cyst(e)ine has been isolated from bovine (bull) and porcine sperm heads (Brill-Petersen and Westenbrink 1963, Coelingh et al. 1969, Tobita et al. 1982). Basic nuclear proteins of buffalo spermatozoa are rich in arginine and poor in lysine as determined with cytochemical techniques (Sidhu and Guraya 1985a). The best-known basic nuclear proteins of a mature sperm are the histone of the bull (Bril-Petersen and Westenbrink 1963) which contains 47 amino acid residues, 24 of which are arginine residues and 6 half-cystines (Coelingh et al. 1972). The half-cystine residues are responsible for the cross-linking of the deoxyribonucleoproteins, which gives to the mature sperm nucleus a chemical resistance. Partial sequences

Table 4. Amino acid composition of mammalian spermatozoan protamines[a] (From Bellvé 1979)

Amino acid	Bull	Stallion	Ram	Boar	Rat	Mouse		Human	
						1	2	1	2
Arginine	24	27	27	24	31	33	35	22	24
Lysine					2	3	3		2
Histidine	1		1	1			12	1	8
Cysteine	6	6	6	8	9	9	6	5	4
Serine	2	4	2	3	4	3	3	5	4
Threonine	3	1	3	1	1	1		1	2
Tyrosine	2	2	2	2	3	4		4	1
Alanine	1	1	1	2	1	1		2	
Glycine	2		1				2		2
Valine	2	3	3	2					
Leucine	1	1	1				1		1
Isoleucine	1			1		1	1		1
Proline				2				2	
Phenylalanine	1			1		1			
Methionine								1	
Glutamic acid	1	3	1					4	2
Total residues	47	48	49	46	52	55	53	47	51
%Basic residues	53	56	57	54	64	65	79	50	67
Amino acid diversity	13	9	12	10	8	8	8	10	11

[a] Compositions given as numbers of residues per mole protein

also are available for the analogous protein of the ram, boar, stallion (Monfoort et al. 1973) and the rat (Kistler et al. 1976). Bellvé (1979) has discussed the details of amino acid composition of mammalian spermatozoon protamines (Table 4). Aside from human and mouse protamine 2, all the other polypeptides isolated have alanine at the amino terminus. The coexistence of protamines 1 and 2 in human and mouse spermatozoa represents a major shift from the single component that is present in other mammalian species. Reid (1977) has observed differences in protamines of sperm heads in inbred and outbred mice.

The differential cleavages of disulphied cross-links of protamines are demonstrated in boar sperm nuclei (Tobita et al. 1983). It has been demonstrated in different species that the lowest zinc level in sperm nucleoplasm corresponds to the higher rate of disulphide formation (Baccetti et al. 1976a). The persisting high level of zinc is believed to interfere with the formation of disulphide bonds in the nucleus whose chromatin remains uncompacted in the round-headed human spermatozoa (Baccetti et al. 1977). A low phosphorus concentration in their zinc-rich nuclei is another characteristic of poorly compacted chromatin. Persistent lysine in the same nuclei indicates that the somatic histones have not yet been subsituted by those rich in arginine and cystine, typical of the mature mammalian spermatozoa, as already discussed. Huang and Nieschlag (1984) have observed alteration of free sulphydryl content of rat sperm heads by supression of intra-testicular testosterone.

Cornelish et al. (1973) have demonstrated relatively large species variations in the amino acid composition and carboxyl-terminal structure of some basic chromosomal proteins of mammalian spermatozoa. Basic protein synthesized during the spermatogenesis of mouse has equal molar content of lysine and arginine (Lam and Bruce 1971). Gouranton et al. (1979) have made an autoradiographic study of the apparent lysine incorporation by the nuclei of spermatozoa in the dog epididymis. All these observations provide evidence for a complicated series of changes in basic chromosomal proteins during spermatogenesis in different mammalian species, as already discussed in detail in Chap. IV.

A variety of functions has been assigned to basic nuclear proteins in spermatozoa, as already discussed in Chap. IV. Species uniqueness of the sperm basic nuclear proteins of mammals contrasts with the highly conserved function of those proteins, to mediate extreme condensation of sperm chromatin. Despite their molecular uniqueness, the sperm basic nuclear proteins of all eutherian mammals share certain characteristics: low molecular weight, high percentage of arginine and cystine residues and, in the final step of chromatin condensation, extensive disulphide cross-linking. Pruslin et al. (1980) have studied interspecies immunologic cross-reactivity of mammalian sperm basic proteins. They have suggested that immunologically cross-reactive sites represent the homologous regions of the various species of sperm basic nuclear proteins that serve the function of chromatin condensation. Immunologic reactivity is shown by sperm made to swell by treatment with 2-mercaptoethanol. Unswollen sperm are not reactive, showing that reduction of disulphide bonds is necessary to render the antigenic sites of sperm basic nuclear proteins available for immunologic recognition. The formation of disulphide bonds within that protein appears to be a molecular mechanism of antigen sequestration in vivo. However, Kvist et al. (1980) have suggested that the unique occurrence of disulphide-stabilized structures in eutherian spermatozoa may serve to protect the spermatozoa from structural degradation by its own proteolytic activity during the relatively slow passage through the eutherian egg investments. A characterization of spermatozoa proteins and the attainment of their detailed map are necessary as first steps towards understanding the roles of the different sperm proteins, as well as towards following changes in existent proteins and the apparence of new ones during cell processes. The biochemical study of spermatozoa proteins has been greatly simplified by an initial fractionation of spermatozoa structures through the use of combined chemical, enzymatic and physical methods, followed by analytical techniques (Bellvé et al. 1975, Mujica et al. 1978, O'Brien and Bellvé 1980a, b). Mujica et al. (1978) have made a detailed study of electrophoretic patterns of total nuclear and flagellar proteins from ejaculated human spermatozoa, as also studied by O'Brien and Bellvé (1980a).

b) DNA

The function of the spermatozoon is to deliver the DNA to the egg. Sperm nucleus contains only half the amount of chromatin and thus its DNA content is haploid, that is half that of the diploid somatic cells (J. R. Clarke et al. 1980). On the assumption that all mature spermatozoa have haploid nuclei, one would expect the amount of DNA to be exactly the same throughout; in fact, however, while the majority are haploid, spermatozoa with diploid or polyploid DNA also occur in semen (Mann and Lutwak-Mann 1981). The DNA of the sperm nucleus is comprised mainly of four deoxyribonucleotides: adenylic, guanylic, cytidylic and thymidylic acids. But small quantities of certain other bases, especially methyl cytosine, are present. The content of DNA per sperm nucleus and its composition with respect to the ratio between the four mononucleotides are species-variable but they are characteristically constant within a species, at least in normal sperm (Mann 1964, 1975, Mann and Lutwak-Mann 1981, Alonso-Lancho et al. 1981).

As the mammalian spermatozoa pass from the epididymis into the vas deferens and ampulla, the nucleus of spermatozoa undergoes some chemical changes. Various studies have indicated significant changes in the deoxyribonucleoprotein complex of the sperm nucleus during epididymal passage (Bedford 1975). The most obvious change is the progressive loss of "Feulgen stainability", the ampullary spermatozoa yielding significantly lower values than those obtained from the epididymis, probably owing to a higher proportion of older, i.e. "aged" spermatozoa (Esnault et al. 1974, Mann 1975, Mann and Lutwak-Mann 1981). J. R. Clarke et al. (1980) reported some changes in the quantity of Feulgen DNA in vole spermatozoa recovered from the testis, epididymis, and vas deferens. In the water buffalo, the heads of ejaculated spermatozoa stain less intensely with the Feulgen technique than those of testicular spermatozoa (Guraya and Sidhu 1975c). In spite of the diminished response of the ampullary spermatozoa to the Feulgen reagent, their true DNA content, as determined by direct ultraviolet absorption, remains unaltered (Esnault et al. 1974). In the absence of quantitative change in total DNA, the reduction in Feulgen reactivity has been interpreted to indicate significant qualitative change in the molecular relationship of the deoxyribonucleoprotein complex during epididymal maturation (Bedford 1975, J. R. Clarke et al. 1980). Mann and Lutwak-Mann (1981) have stated that, all in all, it would appear that the changes in Feulgen stainability, which have been reported in relation to spermateleosis in the testis, sperm maturation in the epididymis, capacitation in the female genital tract, or sperm ageing under storage conditions in vitro, do not indicate real differences in DNA content as such, but must be considered as a rather crude indicator of chemical events that have occurred in the state of the deoxyribonucleoprotein complex as a whole. Now it is well established that the content of DNA in the spermatozoa does not change, either during the passage in the male and female genital tract or under storage conditions. But DNA content varies more widely in abnormal spermatozoa (Ebert et al. 1982). The use of proper methods is important for assessing the changes in DNA content more precisely.

Feulgen stainability cannot be considered as a reliable criterion for determining the content of DNA of spermatozoa, as the ability to stain with the fuchsin-sulphurous acid reagent depends not only on DNA content but also on the composition of the basic nuclear proteins (protamines and histones), as well as other factors such as the state of DNA-protein bonds (Mann 1975, Mann and Lutwak-Mann 1981). The results obtained with quantitative radioautography have further revealed that there is a progressive decrease in the ability of the deoxyribonucleoprotein complex of spermatozoa to bind [^{3}H]actinomycin D as spermatozoa mature in the testis and then pass from the testis into the ejaculate (Brachet and Hulin 1969, Darzynkiewicz et al. 1969, Gledhill 1971).

Sidhu and Guraya (1979b) have demonstrated significant variations ($p<0.01$) in the DNA content of buffalo semen in different seasons of the year, which may have some bearing on the fertility. However, Greesh et al. (1977) reported non-significant differences in DNA content of buffalo spermatozoa in summer and winter seasons. The discrepancy in the results of these two studies may be due to some differences in the materials and methods used. Amir et al. (1977) have observed that oral treatment of bulls with ethylene dibromide causes a temporary reduction of the DNA and protein content and the head area of epididymal and ejaculated spermatozoa. Both relative DNA-stain content and a measure of cell flatness can now be determined simultaneously for each cell at the high rates possible with flow cytometry (Dean et al. 1978, Otto et al. 1979, Gledhill 1983a). As in animals, the DNA content of normal human spermatozoa also appears to be fairly constant, and varies within the range of 2.2 – 2.6 mg/10^9 cells (Meyhöfer 1963). However, in subfertile and infertile men, the DNA content of spermatozoa appears to vary considerably (Leuchtenberger 1960, Meyhöfer 1964, Joel 1966, Mann and Lutwak-Mann 1981).

Bedford et al. (1972) have concluded that the nucleus released at spermiation is structurally immature and that during epididymal passage the nuclear chromatin becomes stabilized by the establishment of disulphide bonds between the nucleoprotein chains in the eutheria, though not in marsupial spermatozoa, in which there occurs a reorientation of the nucleus almost parallel to the axis of the tail (Temple-Smith and Bedford 1976, Harding et al. 1979). During this stabilization, many of the free SH groups in immature spermatozoa become oxidized to form disulphide bonds (Pellicciari et al. 1983). These alterations are accompanied by a progressive decrease in the binding of tritiated actinomycin-D to DNA as its binding sites are obliterated by cross-linking (Darzynkiewicz et al. 1969). The stability and resistance of eutherian sperm nuclei to disruption have been attributed to the extensive cross-linking by the disulphide bonds between the chromosomal proteins bound to DNA. During the passage of the spermatozoa through the epididymis the proteins rich in arginine and cystine show qualitative and quantitative variations (Esnault et al. 1974, Calvin 1975, Marushige and Marushige 1975b, Mann and Lutwak-Mann 1981).

Numerous methods have been used to promote the release of chromatin and access to the nuclear proteins (Mann and Lutwak-Mann 1981, Carranco et al. 1983). R. J. Young and Cooper (1983) have studied the dissociation of intermolecular linkages of the sperm head and tail by primary amines, aldehydes, sulphydryl reagents and detergents. The harshness of these methods produced not only chromatin release but decondensation of the chromosomal materials as well, precluding any meaningful investigation of the ordering of the DNA molecules of the chromosomal material. A correlative electron microscopic and biochemical study has been made of the organization of human sperm chromatin decondensed with sarkosyl and dithiothreitol (Sobhon et al. 1982). Treatment with dithiothreitol, which specifically reduces disulphide bonds, results in a reactivation of the DNA-Feulgen material of the spermatozoa (Esnault et al. 1974).

A similar reactivation of the DNA-Feulgen material can be seen also after the sperms have been recovered from the female genital tract. The possible degree of reactivation of the Feulgen reaction can be used as a measure of the degree of the maturation of spermatozoa in domestic animals. Nicolle et al. (1980) have made a cytophotometric study of ejaculated ram spermatozoa (controls) and of spermatozoa after their incubation in the utero (2 – 2 h) for DNA and nucleoproteins. The nuclear DNA measured with UV light does not show any significant change, but the Feulgen DNA is increased in sperm-incubated samples. Most of the incubated samples (19/21) show the highest extinction values when stained with Fast green for arginine-rich basic proteins. The nuclear area of in utero-incubated sperm is increased in comparison to the control cells. Differences in staining ability have been attributed to modifications in chromatin structures. Pinkel et al. (1983b) have observed radiation-induced DNA content variability in mouse sperm.

Sipski and Wagner (1977), using physico-chemical and theoretical techniques, have revealed the unique ordered conformation of the DNA molecules in the chromosomal fibres of equine sperm nuclei. DNA molecules, maintained in a modified B-form secondary structure, are compacted by coiling into a superhelical tertiary structure. These ordered molecules are aligned parallel to each other in planes, each plane slightly rotated as these stack upon each other forming a quarternary cholesteric liquid crystal arrangement with a right-hand screw direction and a pitch of 176 nm. The human sperm chromatin shows a nucleosomal structure (Wagner and Yun 1981). As yet little is known of the mechanism involved in sperm nuclear swelling and DNA synthesis in vivo. However, Zirkin and Chang (1977) have demonstrated an involvement of endogenous proteolytic activity in thiol-induced release of DNA template restrictions in rabbit sperm nuclei. Perreault and Zirkin (1982) have also suggested the role of sperm-associated proteinase in sperm nuclear decondensation in vivo.

Swollen and disrupted human sperm nuclei have been shown to contain particulate complexes of high-molecular-weight RNA and DNA polymerase (Witkin et al. 1975, Bendich et al. 1976, Witkin and Bendich 1977). When the complex is disrupted, DNA polymerase activity can be demonstrated in the presence of synthetic deoxy- or ribo-nucleotide polymers. Philippe and Chevaillier (1976) have made a characterization of a DNA polymerase activity in mouse sperm nuclei. Richards and Witkin (1978) have isolated and partly characterized a DNA polymerase from bull sperm heads. Its apparent molecular weight and synthetic template utilization resemble that of DNA polymerase Y. It probably is located in the nucleus. The DNA polymerase extracted from human sperm heads has an apparent molecular weight of 79,000 – 89,000, and its activity is inhibited by concentrations of potassium phosphate exceeding 10 mM (Witkin 1980). Presumably, DNA polymerase-like acrosin is present within the sperm heads in an inactive or bound form and becomes active only after fertilization, while the sperm nucleus undergoes decondensation. But the mechanism of the in vivo decondensation of the sperm nucleus needs to be determined.

c) RNA

Divergent views have been expressed about the presence and synthesis of RNA in the mature spermatozoa. Markewitz et al. (1967) have shown the absence of RNA synthesis in the shed human spermatozoa, as also supported by the observations of Moore (1972), who has not observed any RNA polymerase activity in sperm collected from the epididymides or vasa deferentia. The biochemical studies have not revealed the presence of RNA in the mature spermatozoa (Mann and Lutwak-Mann 1981). But several other biochemical studies have demonstrated the presence and synthesis of small amounts of RNA in mammalian spermatozoa (Abraham and Bhargava 1963, Betlach and Erickson 1973, 1976, MacLaughlin and Turner 1973). This RNA is synthesized mainly on the mitochondrial, rather than the nuclear, DNA template and the minute amount of protein synthesized by mammalian spermatozoa is of mitochondrial origin, as will be discussed in Chap. X. There is no conclusive proof that RNA polymerase exists in an active form within the nucleus of a normal mature and intact spermatozoon, or that RNA is generated by such a nucleus to any appreciable extent. All transcription appears to stop at the time when the sperm nucleus reaches full maturity, that is after the nuclear protein has been dephosphorylated and the DNA has become fully condensed (Chap. IV). After the protamine messenger RNA (mRNA) has reached the cytoplasm of developing spermatid, first the amino acid residues are translocated by diribosomes, and then the newly synthesized protamine is phosphorylated by ATP under the influence of protamine kinase before being transferred from the cytoplasm to the cell nucleus where it replaces the histone in chromatin and binds to DNA. Once histone displacement nears completion, dephosphorylation starts, and so does the concomitant DNA condensation, as a necessary step in the sperm maturation process; a nucleoprotamine model setting out the advantages of protamine structure as a DNA-condensing agent has been proposed (Warrant and Kim 1978).

C. Acrosome

During the past few years, structural and biochemical studies on the spermatozoa have shown a definite change-over from experiments carried out with whole sperm cells to observations made on their organelles, separated mechanically or by chemical means, thus providing isolated components of sperm (Bellvé and O'Brien 1979, Calvin 1979a, R. J. Young and Cooper 1979, Olson 1979, Pallini and Bacci 1979). This unique type of experimental analysis is reflected clearly in the great amount of recent work carried out on the structure and chemistry of the acrosome, which plays some important roles during fertilization. The functions of acrosome still need to be defined more precisely at the molecular level. Acrosome status has been evaluated in human ejaculated spermatozoa with monoclonal antibodies (Wolf et al. 1985).

1. Structure

The acrosome of sperm is the organelle which facilitates its penetration into the egg. The term "acrosome" is used here to denote the cap-like membrane-bound structure (Figs. 56, 57, 60 and 63). The name acrosome in eutherians will, therefore, include the structure often designated as "head cap" and "galea capitis". There is no structure conforming to the traditional descriptions of the galea capitis and this term has now been abandoned. Various electron microscope studies have clearly shown that all these are but parts of a single unit surrounded by a limiting membrane of its own, usually called as the "acrosomal membrane" (Fig. 59), which is a Golgi membrane, having its usual, ultrastructural appearance full of hexagonally packed 9-nm particles (Phillips 1975a, b, Oko et al. 1976).

Since the acrosomal membrane is continuous it can be referred to as the "acrosomal sac" or "acrosomal cap" (Fig. 61) and its contents as the "acrosomal material". The outer acrosomal membrane lying immediately beneath the plasma membrane is continuous at the posterior margin of the acrosomal cap with the inner acrosomal membrane which, in turn, is closely applied to the nuclear envelope (Fig. 57). Two acrosomal membranes lie parallel throughout most of their extent and enclose a narrow cavity occupied by the acrosomal content. In the human and chimpanzee spermatozoa the acrosome remains relatively small and does not extend anteriorly much beyond the leading edge of the nucleus (Figs. 17, 23 and 56). In the spermatozoa of bush-baby, slow loris, marmoset, and various species of monkeys investigated so far, the acrosome is relatively thickest around the anterior margin of the sperm head and thinnest from the equator to the posterior third of the nucleus where it terminates (Franklin 1968, Bedford and Nicander 1971, Zamboni et al. 1971, Rattner and Brinkley 1970, 1971, Guraya 1971a, Kumar 1985). The thin portion of the acrosome is usually referred to as the equatorial segment. In the spermatozoa of chimpanzee (Figs. 17 and 23) and man (Figs. 56 and 57) the thickness of the acrosome is, more or less, uniform throughout and only exceptionally increases around the anterior margin of the nucleus (Holstein and Roosen-Runge 1981). In the equatorial region of human spermatozoa the acrosome becomes abruptly thinner, and in many spermatozoa its posterior margin attains an expanded, bulbous form (Zamboni et al. 1971).

In many other species of mammals, a conspicuous thickening of the acrosomal cap extends anteriorly well beyond the nucleus (Fig. 60A). This part, designated the apical segment of the acrosome, often shows a shape specific to the species (Phillips 1975a, Oko et al. 1976, Berrios et al. 1978). For example, the spermatozoa of the guinea-pig, chinchilla, and ground-squirrel have a very large apical segment of the acrosome (Fawcett 1965, 1970, Guraya 1971a, Phillips 1975a). In the plains mouse (*Pseudomys australis*) – sperm head studied with transmission and scanning electron microscopy – the acrosomal material covers the dorsal hook and appears to constitute most of the ventral hooks which contain nuclear material only at the base (see details in Breed 1983, 1984 for species variations in hooks). The acrosome of Asiatic musk shrew (*Suncus marinus*) is noteworthy for its size (J. A. Green and Dryden 1976). It is an elongate, flattened structure 21.5 μm long, 17.5 μm at its greatest width, and 1 μm thick. The acrosome in marsupials shows no differentiation into the regions, such as apical, principal and equatorial segments (Harding et al. 1979), which have been recognized in eutherian sperm (Fawcett 1975b). In marsupials, the acrosomal vesicle is confined to the upper surface of the nuclear plate.

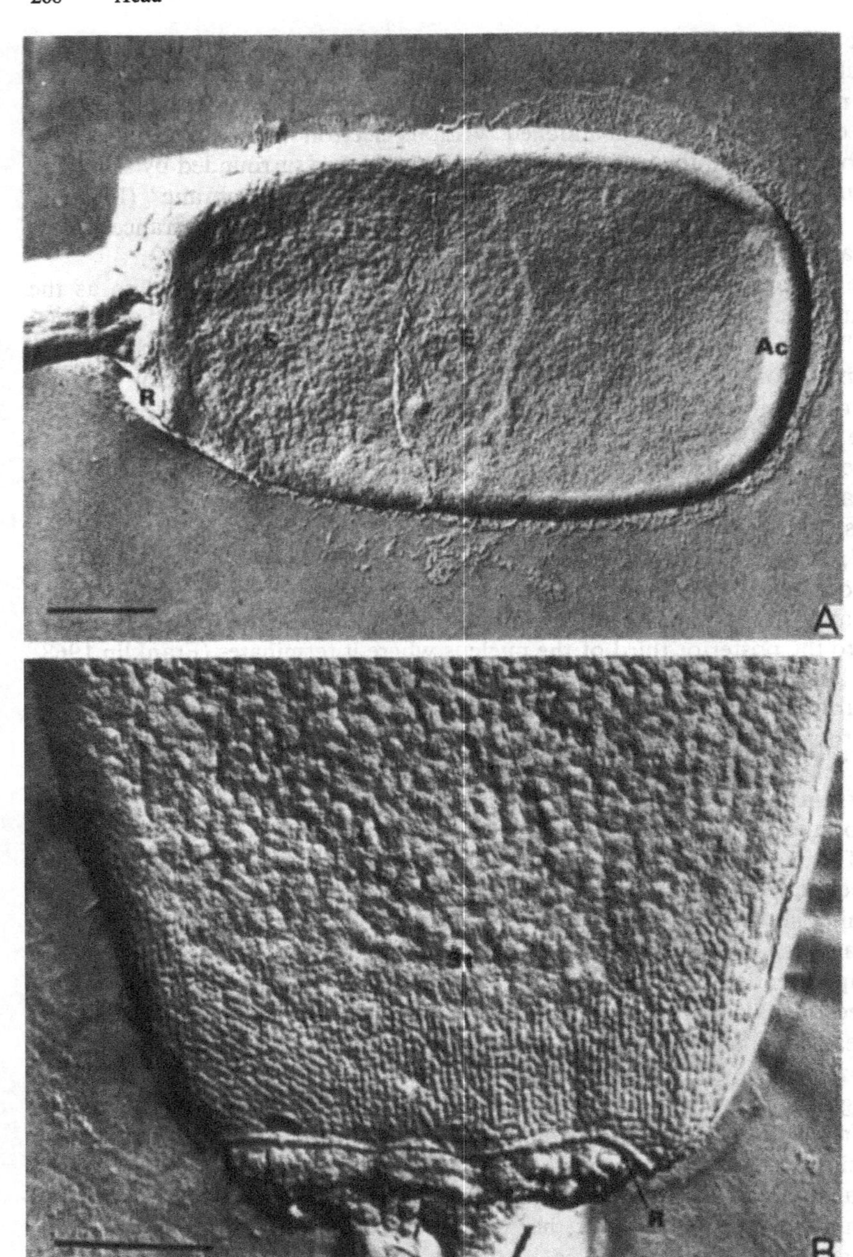

Fig. 60. A Surface replica of untreated buffalo sperm showing normal surface landmarks, acrosomal bulge (*Ac*), equatorial segment (*E*), post-nuclear sheath (*S*), and posterior ring (*R*); **B** Surface replica of untreated buffalo sperm showing post-nuclear sheath at higher magnification. The closely spaced basal striations (*St*) are evident just above the posterior ring (*R*). Circular profiles of nuclear pores associated with redundant nuclear envelope are seen below the ring (From Koehler 1973a)

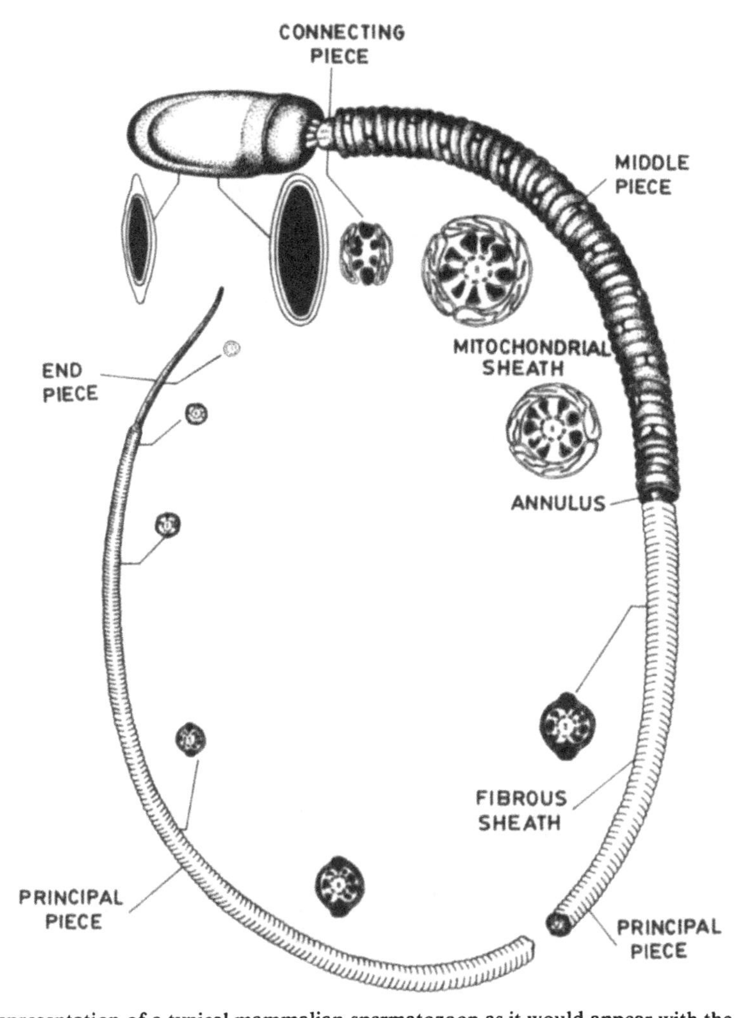

Fig. 61. Schematic representation of a typical mammalian spermatozoon as it would appear with the cell membrane removed to reveal the underlying structural components. An acrosomal cap covers the anterior two-thirds of the nucleus. The connecting piece is inserted in an implantation fossa in the posterior aspect of the nucleus. The internal structure of the sperm flagellum is best understood from the study of cross-sections at various levels. Running through the axis of the sperm tail for its length is the axoneme, a longitudinal bundle of microtubules similar to that found in cilia and flagella in general. Outside of the axoneme is a row of nine longitudinally oriented outer dense fibres that are not found in other flagella. Three segments of the sperm tail are defined by the nature of the sheaths that envelope the core complex of microtubules and dense fibres. The middle piece is characterized by a sheath of circumferentially oriented mitochondria. A dense annulus marks the caudal end of the middle piece. In the long principal piece, the core complex is enclosed in a fibrous sheath of circumferential dense fibres. The end-piece is the portion beyond the termination of the fibrous sheath, consisting only of the axoneme. The plasma membrane invests all of the structures shown (Redrawn from Fawcett 1975b)

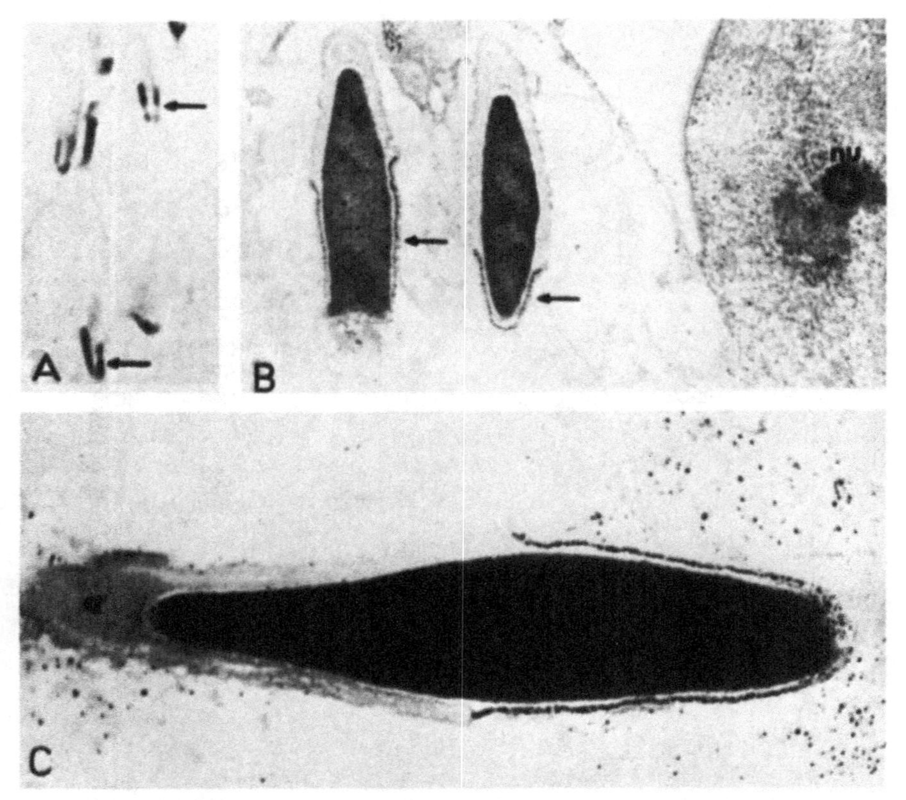

Fig. 62. A Thick section of a block treated with the silver method under the light microscope. The selective staining of the post-acrosomal lamina can be observed (*arrows*); **B** Electron microphotograph of a sample after the silver method. The post-acrosomal lamina appears positive (*arrows*) as well as the nucleolus of a young spermatid (*nu*). Silver method and uranyl staining; **C** Late spermatid showing the silver grains in the post-acrosomal region. Few silver granules can be seen in the nucleus, but the acrosomal region is free of granules. Silver method and uranyl staining (From Krimer and Esponda 1978)

At present, there is no satisfactory explanation for the marked specific differences in the shape and volume of acrosome in different species of mammals (Guraya 1971a, Fawcett 1975b), as well as in other different groups of invertebrates, protochordates, and vertebrates (Baccetti 1979). Actually, the acrosomal complex shows a great diversity in the animal kingdom (Adiyodi and Adiyodi 1983). The dimensions of the acrosome in the spermatozoa of the same species (e.g. rabbit) are altered during their passage through the head of the epididymis (Bedford 1975). Discrete maturation changes involving the fine structure, shape, and size of the acrosome have been reported for several species of mammals including marsupials (Fawcett and Hollenberg 1963, Bedford 1975, Berrios et al. 1978, Harding et al. 1979).

In the equatorial segement (Figs. 59, 62A) the inner and outer membranes come to lie more closely in parallel configuration (Bedford et al. 1979). This region is delineated in the light microscope as a crescent extending across the centre of the sperm head (Hancock 1957). From a structural viewpoint, the identity of the equatorial segment as a region distinct from the remainder of the acrosome is supported by its differential resistance in the face of mild disruptive agents (Wooding 1973, 1975, Russo and Metz 1974, Srivastava et al. 1974), and in degenerating spermatozoa by the presence of numerous bridges linking the inner and outer acrosomal membranes (Russell et al. 1980b) and by its structure in surface replicas (Phillips 1977b). It also remains intact during the acrosome reaction and after loss of the reacted elements and content of the acrosome, which must precede penetration of the zona pellucida (Bedford et al. 1979). The equatorial segment forms the characteristic feature of acrosome in all the eutherians investigated so far (Phillips 1977b, Bedford et al. 1979, Holstein and Roosen-Runge 1981). It has not been identified in the acrosome of marsupials (Temple-Smith and Bedford 1976, Harding et al. 1979).

The equatorial segment in eutherians varies somewhat in length (Fawcett 1970, Oko et al. 1976), being usually long in boar spermatozoa (Nicander and Bane 1962). In the human spermatozoa, linear differentiation within this segment gives its limiting a pentalaminar appearance (Pedersen 1974b). In guinea-pig spermatozoa a differentiation of the content of the equatorial segment results in a conspicuous palisade formation of parallel ridges (Friend and Fawcett 1974, Koehler 1973b). Numerous bridges linking the inner and outer acrosomal membranes of the equatorial segment of the acrosome in the boar spermatozoa are belived to play some role in maintaining the close spacing and parallel arrangement of the membranes in this portion of the acrosome. The presence of the equatorial segment of the acrosome in eutherian mammals suggests that it may have a role specific to their fertilization. Its exact functional significance is still controversial. However, Bedford et al. (1979) have suggested that the equatorial segment of the acrosome in eutherian mammals functions to preserve a discrete region of the "labile" plasmalemma overlying it, through which fusion with the egg is achieved. They have also discussed the need for this feature and possible reasons for its restriction to their spermatozoa. From the results obtained, Bedford et al. (1979) have also denied the role of the plasmalemma covering the post-acrosomal region in the intial fusion between the gametes.

A mixture of 10% phosphotungistic acid in 10% chromic acid selectively and differentially stain the acrosome and plasma membranes of ejaculated porcine sperm (James et al. 1974, Clegg et al. 1975). The acrosomal membranes in rodents have been investigated. Cleavage faces of the acrosomal membranes are characterized by a regular arrangement of tightly packed particles as well as other types of particles in the mouse spermatozoa (Stackpole and Devorkin 1974). Phillips (1975b) has observed in the surface of acrosomal membranes an ultrastructural appearance of hexagonally packed 9-nm particles, but Koehler (1972) in man observed particles of 20 nm in diameter. According to Pedersen, (1972a) the acrosomal membranes in the posterior area (equatorial segment) is pentalaminar. Phillips (1977b) in rhesus and other mammals has revealed in this region a system of regular rows of hexagonally packed particles, with a centre-to-centre spacing of 17 nm, while in the post-acrosomal region the material is arranged in parallel ridges with a centre-to-centre spacing of 15 nm; this region also contains actin in rabbit and mouse (Welch and O'Rand 1985a, b). In surface replicas of spermatozoa in bull, rabbit, hamster, guinea-pig, degu (*Octodon degus*), and rhesus monkey, the surface of the outer acrosomal membrane in the equatorial region is composed of regular rows of hexagonally packed particles with centre-to-centre spacing of 17 nm (Oko et al. 1976). These particles are limited to the equatorial segment. But the majority of the surface of the outer acrosomal membrane in mouse and rat is composed of similar 17-nm spaced particles, suggesting that a major portion of the acrosomal surface in mouse and rat is analogous to the equatorial segment of other species of mammals. Olson and Winfrey (1985) have investigated the structure of membrane domains and matrix components of bovine acrosome. Anionic sites have been detected on the cytoplasmic surface of the guinea-pig acrosome membrane (Enders and Friends 1985).

The acrosome membranes seem to undergo some changes during the maturation of spermatozoa (Bedford 1975, 1979, Jones 1975). Change in the acrosome ranges from a major reorganization expressed in the guinea-pig and chinchilla to the more modest change observed in the rabbit, elephant, hyrax, and some primates (Bedford 1979). It also occurs in some rodents, e.g. in *Mystromys albicauda*, but not in the ground-squirrel *Citellus lateralis*, nor in the muskshrew *Suncus murinus*, both of which have a prominent acrosome. Nor does it occur in man. Large particles are concentrated over the thickened apical portion of the acrosome and within the nuclear envelope directly beneath the acrosome. The tightly packed paracrystalline arrays of particles distinct from typical intramembranous particles seem to indicate that the acrosomal membrane is organized in a manner different from other sperm membranes. Similar arrangements of particles have also been described in freeze-cleaved acrosomal membranes of guinea-pig, rat, rabbit, bull, and human sperm (Koehler 1970a, 1972, 1975b, Plattner 1971, Oko et al. 1976). The freeze-etching results have indicated that the inner acrosomal membrane, that is the portion of the membrane adjacent to the nuclear envelope, is composed of or contains a highly organized paracrystalline component. Koehler (1975b) has suggested that the two separate compartments, that is intraacrosomal and membranous, may serve to segregate enzyme activities needed at different times during fertilization. Phillips (1972a) has demonstrated a 4.2-nm periodicity in rat acrosomes in the cortical region, the area just below the plasma membrane on the convex surface of the acrosome.

In spite of these various observations, still little is known about the precise structure and function of acrosomal membranes and their alterations during capacitation and acrosome reaction (Olson and Winfrey 1985). However, very recently the possibility of isolating the outer acrosomal membrane from boar spermatozoa (Töpfer-Petersen and Schill 1981, 1983) tempted Töpfer-Petersen and co-workers to start their studies on the mammalian acrosome with the characterization of this membrane (Töpfer-Petersen et al. 1983a, b, 1984). By electron microscopical examination of boar spermatozoa, using the lectin-peroxidase and lectin-ferritin technique, it could be shown that in addition to the plasma membrane both acrosomal membranes – inner and outer acrosomal membrane – contain binding sites for concanavalin A (Con A) and R. Communis-agglutinin-120 (RCA-120). For the outer acrosomal membrane these observations could be confirmed by the carbohydrate analysis of the isolated membrane fraction by which mannose, galactose, N-acetylglucosamine, N-acetylgalactosamine and neuraminic acid in the molar ratio of about 3:6:4:5:3 and traces of fucose could be determined. RCA-120 binds preferentially to the exposed terminal galactose residues of carbohydrate side-chains. These galactose residues could be easily radiolabelled with tritium, thus facilitating the detection of RCA-120 receptor proteins which possess binding sites for Con A, too. Four further Con A receptor proteins with the apparent molecular weights of 120,000, 110,000, 88,000, and 66,000 could be identified and partially isolated by lectin-affinity chromatography, HPLC combined with a specifically developed enzyme-linked lectin-assay (ELLA) and SDS-PAGE. Most of the smaller proteins (MW 20,000) contained no carbohydrates. The biological roles of these proteins and glycoproteins of outer acrosomal membrane need to be determined in future investigations with immunological techniques. The results of immunological studies have suggested that antigenic structures of these large glycoproteins are protected within the membrane unit. There is also now a possibility for the isolation of the inner acrosomal nuclear membrane complex for biochemical studies (Rahi et al. 1983).

According to Koehler (1975b), the acrosomes of spermatozoa from several mammalian species show a coarsely granular content as viewed through irregular fenestrations in the surface membranes of replicas. These fenestrations may occur either through trauma or degeneration, or are possibly related to the process of capacitations (M.C. Chang and Hunter 1975, R.C. Jones 1975). The acrosomal contents, as seen in surface replicas, appear similar to the image obtained in cross-fractured acrosomes in freeze-etching preparations. Cryofixation procedures have revealed a "cobblestone"-like fracturing pattern in the acrosome of bovine and rabbit spermatozoa (Plattner 1971, Fléchon 1974b). A substructure of the acrosome has also been claimed in bovine (Wooding 1973, Olsen and Winfrey 1985) and porcine spermatozoa (D.J. Moore et al. 1974). According to Kaczmarski (1978), the contents of the acrosome show electron-dense areas or zones of paracrystalline structures.

The finger-like projections associated with the posterior border of the acrosome in the guinea-pig (Friend and Fawcett 1974, Koehler 1974, 1975b) have not been observed in the musk shrew sperm (Koehler 1977) nor is there any ornamentation of the post-acrosomal sheath as in the "picket fence" arrangement seen in the rabbit spermatozoa (Koehler 1970a, 1975b). The very densely packed band of intramembranous particles seen just superior to the posterior ring is, however, quite distinctive in the musk shrew sperm. Koehler (1972) believes that the acrosomal membrane particle arrays might represent membrane-associated enzymes involved in fertilization and present in the acrosome, as will be discussed below.

Several "acrosomal" enzymes have also been detected in the overlying plasma membrane (Bernstein and Teichman 1968, Gordon and Barnett 1967, Gould and Bernstein 1973) that does not possess such particles (Stackpole and Devorkin 1974).

Tests for several phosphatases have been applied to the acrosomal membrane, but only an ATPase active at pH 9 in the presence of calcium ions has been observed in both epididymal and ejaculated sperms of many rodents and man (Gordon and Barnett 1967, Gordon 1973). Working and Meizel (1981) have provided evidence for the function of ATPase in the maintenance of the acidic pH of the hamster sperm acrosome. In the rabbit acrosome two phosphatases have been reported; both hydrolases, and one of them is stimulated by Zn^{2+} (Gonzales and Meizel 1973). Mathur (1971) has reported the presence of glucose-6-phosphatase, ATPase, and monoamine oxidase in the acrosomal region of mouse sperm. The precise localization of these enzymes could not be determined. Immunochemical studies at the fine structural level may be of great utility to determine the intra-acrosomal localization of various acrosomal enzymes. Some work has already been carried out on the immunological localization of various acrosomal enzymes (McRorie and Williams 1974, Zaneveld 1975, D. B. Morton 1976).

In the early electron microscope studies, the contents of acrosome were described as homogeneous. But later studies have revealed distinct areas of differing density in the acrosome of guinea-pig (Fawcett and Hollenberg 1963), chinchilla (Fawcett and Phillips 1969b), and other mammals (Kaczmarski 1978). There is often a pale outer zone around an inner zone of greater density (Fig. 56). In the rabbit spermatozoa, discrete dense bodies with ill-defined boundaries are also reported in this region of the acrosome. In electron micrographs, these show a fine periodic structure interpreted as crystal lattice (Fléchon 1973, Kaczmarski 1978). A highly ordered substructure with a periodicity of 4.2 nm has also been reported in a superficial zone on the convex portion of the curved acrosome of the rat spermatozoa (Phillips 1972a). A similar periodicity is also observed in the acrosome of the human spermatozoa (Pedersen 1972c). Our knowledge is still very meagre about the distribution and molecular arrangement of the various chemical components within the acrosome (Fléchon 1974a, 1976, Kaczmarski 1978, Berrios et al. 1978, Baccetti 1979). It is not known whether the different enzymatic components are uniformly distributed or aggregated in the cyrstalline areas and zones of differing density within the acrosome, which have been clearly demonstrated with the electron microscope. The electron-dense areas or zones of paracrystalline structures have been suggested to reflect the compact spatial arrangement of acrosomal enzymes (Kaczmarski 1978). Stambaugh and Smith (1978) have used immunofluorescence and electron microscope to reveal the presence of tubulin and microtubule-like structures in rabbit and rhesus monkey sperm acrosomes. The distribution of the tubulin within the acrosome appears similar to the distribution of proteolytic activity, suggesting the association of proteinase with these structures. Electron microscopic study of rhesus monkey sperm incubated in vitro has revealed a progressive condensation of the acrosomal matrix into microtubule-like structures.

2. *Chemical Components and Their Significance*

The PAS-positive acrosomal substance in the spermatozoa of human and subhuman primates, which completely fills the membranous acrosomal cap, is generally homogenous (Figs. 17, 23 and 56) (Horstmann 1961, de Kretser 1969, Zamboni et al. 1971, Guraya 1971a). Various cytochemical and biochemical studies on the spermatozoa of different species of mammals have shown that it consists of carbohydrates (neutral polysaccharides), proteins, lipids and various hydrolases (Table 5).

a) Carbohydrates, Proteins and Lipids

The acrosomal mucopolysaccharides contain galactose, mannose, fucose, and hexosamine (Table 5). Kopecny (1976) has made a study of labelling of acrosomes in mouse spermatozoa by L-[1-^{3}H]fucose. The highest concentration of grains on the dorsal surface correspond to the thickest part (or principal segment) of the acrosome. Acrosomes of rabbit spermatozoa are also labelled by [^{3}H]-fucose introduced into the cells during spermatogenesis (Kopecny et al. 1980). The labelling has been analyzed simultaneously by autoradiography and biochemically. In compact intact acrosomes, the labelling is confined strictly to the acrosomal region of the sperm head. In swollen and detached acrosomes, the autoradiographic grains are associated mostly with the acrosomal cap. In acrosomal extracts, a considerable share of radioactivity coincides with gel filtration fractions showing esterase activity (N-α-benzoyl-L-arginine ethyl ester splitting) akin to that shown by acrosin, to be discussed later. In a recent paper, Kopecny et al. (1984a) have discussed the cytochemical significance of acrosome labelling by radioactive sugars.

The various carbohydrates demonstrated combine with protein to form an alcohol-precipitable fraction consisting of a carbohydrate-protein complex, later named "acrosomin" (Clermont and Leblond 1955). But according to other workers it contains a lipoglycoprotein complex associated with a wide range of enzymes which act on egg envelopes (Srivastava et al. 1965, Hartree and Srivastava 1965). The glycoprotein fraction contains glutamic acid as the predominant amino acid. Besides the various sugars described above, sialic acid is also believed to be present. But since sialic acid is known to be present on the plasma membrane, dissolved by the detergent, the contribution of the acrosomal material itself remains to be determined.

Biochemical studies demonstrated the acrosomal lipids, which had a high content of phospholipid of a composition similar to that of whole spermatozoa, i.e. consisting mainly of plasmalogen and lecithin (Piko 1969). But the present author has not observed any lipid content in the acrosome of marmoset, chimpanzee, guinea-pig, rabbit, opossum, goat, sheep, and buffalo (Guraya 1965, 1971a, Guraya and Sidhu 1975b), which could be demonstrated with modern histochemical techniques for lipids. The lipid component demonstrated biochemically (Srivastava et al. 1965, Hartree and Srivastava 1965) may, therefore, be derived from the outer and inner membranes of the acrosome (Fig. 57), which have already been described. In other words, the lipids may be due to contamination from nonacrosomal parts of the sperm cell. Mann (1964) has discussed the evidence, indicating the presence of large molecular carbohydrate-protein-lipid conjugates in the acrosome. The polysaccharides themselves could be of several kinds, as already described. The latter view is corroborated by the apparent antigenic complexity of the acrosome (Hathaway and Hartree 1963, Voisin et al. 1974).

Table 5. Acrosomal hydrolases of mammalian spermatozoa

Enzymes	Properties				Physiological significance
	Km	Ki	MW	Subunit (if any)	
A. *Acrosome-associated enzymes*					
1. Hyaluronidase	–	–	60,000	4–5	Dispersion of cumulus cophorus
2. Proacrosin-acrosin					
Proacrosin	–	–	55,000 53,000	Dimer	Penetration of zona pellucida
Acrosin					
α-acrosin	0.05 mM (BAEE)	4.4 mM (Benzamidine)	49,00	–	
β-acrosin	0.05 mM (BAEE)	4.0 mM (Benzamidine)	34,000	–	
γ-acrosin	–	–	35,000	–	
3. Aryl sulphatase	–	–	–	–	May play an adjunct role in penetration of the cumulus cell layer
4. Nonspecific esterase	–	–	–	Multiple forms	–
5. Acid-alkaline phosphatases	0.05–0.18 mM (p-nitrophenyl-phosphate)	–	47,000 125,000	Multiple forms	May involve in acrosome reaction indirectly by influxing Ca^{+2} into the cytosol around the acrosome

B. *Acrosomal or/and head-associated enzymes*					
6. β-glucuronidase	–	–	112,000	1–2	Influences the capacitation of hamster epididymal spermatozoa
7. Arylamidase	–	–	112,000	1–2	–
8. β-N-acetylglucosaminidase	–	–	140,000	–	–
9. β-N-acetyl galactosaminidase	–	–	140,000	–	–
10. α-Mannosidase	–	–	–	20,000, 30,000 } monomers	–
11. Phospholipase	–	–	–	–	May help in hydrolysing phospholipids for endogenous respiration of spermatozoa
12. β-Aspartyl-N-acetyl-glucosamine-amido hydrolase	–	–	–	–	–
13. Corona-penetrating enzyme	–	–	–	–	Dispersion of the cells of corona radiata
14. Neuraminidase	–	–	–	–	–
15. Acid proteinase	–	–	40,000	–	–
16. Collegenase-like peptidase	–	–	110,000	–	–
17. α-glucosidase	–	–	–	–	–
18. Cathepsins	–	–	–	–	–

Key to abbreviations: –, Not known or have not been investigated; BAEE, N-α-benzoyl-L-arginine ethyl ester

b) Enzymes

A large number of enzymes, especially hydrolases, have been identified in sperm extracts (Table 5). Although biological functions for all of these enzymes have not yet been established (see Nelson 1985), several have been located specifically in or around the acrosomal region in the sperm head. Some fertility disorders have been related to alterations in some enzymic acrosomal constituents (Schill 1974, Singer et al. 1980b, c, Mohsenian et al. 1982). Several excellent reviews are available which detail the procedures and results obtained from the biochemical point of view (McRorie and Williams 1974, D.B. Morton 1976, Mann and Lutwak-Mann 1981).

The acrosome contains two most important enzymes such as hyaluronidase and zona lysin or acrosin, which function during sperm entry into the ovum (Florke et al. 1983, C.R. Brown 1983, Mack et al. 1983, Meizel 1984, Nelson 1985). During the past few years, several other lysosomal enzymes have also been obtained from the mammalian acrosome. These include acid phosphatase, β-glucuronidase, arylamidase, arylsulphatase A and B, β-acetylglucosaminidase, phospholipase A, nonspecific esterases and β-aspartyl-N-acetylglucosamine amido hydrolase (Table 5) (Mann and Lutwak-Mann 1981, Sidhu and Guraya 1985a). Isozymes of several acrosomal enzymes including acrosin and acid phosphatase have also been reported (Sidhu and Guraya 1978a). Bhalla et al. (1973) have demonstrated the presence of β-aspartyl N-acetyl glucosamine amido hydrolase in the acrosomal extracts of ejaculates from squirrel monkey, boar, ram, and humans, indicating its origin from the acrosome. The greater amount of enzymes per ejaculate was present in the acrosomal extracts from ram and boar. The relative distribution of various enzymes within the acrosome is still not precisely known (Koehler 1978). Evidence has accumulated that some of these acrosomal enzymes, particularly a protease, are involved in fertilization (Yanagimachi 1978a). The acrosomal enzymes probably aid the spermatozoon in contacting and penetrating the oocyte investments. Meizel (1984) has discussed the importance of hydrolytic enzymes in relation to acrosome reaction of mammalian spermatozoa.

The presence of trypsin-like proteinase activity has been demonstrated cytochemically and biochemically in the acrosome of spermatozoa in different mammals (McRorie and Williams 1974, D.B. Morton 1976, Hartree 1977, Koehler 1978, Baccetti 1979, Parrish and Polakoski 1979, Primakoff et al. 1980, C.R. Brown 1982, Florke et al. 1983). This proteinase has attracted interest following the discovery that trypsin inhibitors impede the penetration of spermatozoa through the zona pellucida. This enzyme was obtained in purified form from the acrosomes of rabbit and boar spermatozoa and was designated as acrosin (Polakoski et al. 1972). Based on this evidence, the commitee on Enzyme Nomenclature of the International Union of Biochemistry accepted the term acrosin as the official name of this proteinase and numbered it EC 3.4.21.10 (Parrish and Polakoski 1979). Acrosin activity is believed to the partially regulated by natural protein proteinase inhibitors which are present in seminal plasma, as will be discussed later on.

Hyaluronidase. Hyaluronidase is the first emzyme to be released during the acrosome reaction (Peter et al. 1985, Zao et al. 1985). It emerges from the acrosome of capacitated spermatozoon through the holes formed by the membrane fusion during the early stages of acrosomal reaction (Rogers and Morton 1973, Rogers and Yanagimachi 1975). Hyaluronidase is reponsible for the penetration of the spermatozoa through the cumulus oophorus (Lewin et al. 1982). This enzyme causes the dispersion of the cells of cumulus oophorus, and the spermatozoon appears to employ this enzyme to lyse a passage through this investment. Indeed, inhibitors of hyaluronidase specially prevent the penetration of spermatozoa through the follicle cell layer (Zaneveld 1982). Hyaluronidase is also found to dissolve a component in the hamster zona pellucida (Talbot 1984).

The structural localization of hyaluronidase within the acrosome is controversial. Pikó (1969) has stated that since hyaluronidase is the first to function during sperm entry, it might be expected to occupy the apical region (or the leading edge) of the acrosome (Fléchon 1976). Mancini et al. (1964) using a fluorescent labelled antibody to hyaluronidase, observed that the activity is associated with the acrosome of bull sperm and, furthermore, that it is present in the perinuclear (Golgi) region of developing spermatids. Fléchon and Dubois (1975), utilizing the fluorescent antibody method, have also observed that hyaluronidase activity is associated with the acrosome of rabbit spermatozoa. The various immunohistochemical, histochemical, and biochemical studies have indicated the localization of hyaluronidase in the outer acrosomal membrane (Fléchon 1975, 1976, Fléchon and Dubois 1975, C.R. Brown 1975). Gould and Bernstein (1973) in the bull, and D.B. Morton (1975b) in the ram, have localized it immunocytochemically in the central acrosomal zone, far from the inner acrosomal membrane. D.B. Morton (1975b) has observed that spermatozoa, which had acrosomes removed, still retain about 20% of the activity and stain with antibody, suggesting that a fraction of the hyaluronidase may remain bound to the inner acrosomal membrane after acrosome loss. Kopecny and Fléchon (1981), after labelling the glycoproteins of the acrosomes during spermiogenesis in the rabbit and guinea-pig, have demonstrated the label on the expanded acrosomal material and on the acrosomal vesicles, but not on the inner acrosomal membrane after acrosome disruption or induction of acrosome reaction in vitro, confirming its localization in the apical region of the acrosome. Waibel et al. (1984) have detected cytochemically the hyaluronidase activity in single human and mouse sperm by an improved substrate-film technique.

Talbot and Franklin (1974a, b) have investigated the release of hyaluronidase from guinea-pig and hamster spermatozoa undergoing the acrosome reaction (see also Peter et al. 1985, Zao et al. 1985). A significant amount of hyaluronidase activity is observed even before ultrastructural indications of acrosome reaction are seen, indicating that a substantial fraction (50%) of hyaluronidase is released independently of the acrosome reaction. Metz (1978) has suggested that a major portion of hyaluronidase in the rabbit sperm is present in association with the plasma membrane. Rogers and Yanagimachi (1975) have observed a very good correlation between the release of hyaluronidase and acrosome loss in guinea-pig spermatozoa that have been synchronized by Ca^{2+} to undergo the acrosome reaction within a few minutes. In this way, it is possible to correlate hyaluronidase release with the morphological acrosome reaction in an optimal way and at the same time account for extracellular enzyme contribution from moribund spermatozoa. These results have revealed that the amount of extracellular hyaluronidase is a direct reflection of the percentage of dead spermatozoa in the population, producing a background activity which is greatly increased on the addition of Ca^{2+} to trigger the acrosome reaction. The role of a K^+ ion influx and Na^+K^+-ATPase activity in the hamster sperm acrosome reaction has been suggested (Mrsny and Meizel 1981). Holzmann et al. (1978) have not found any strict correlation between hyaluronidase activity and sperm morphology. There is also a lack of correlation between acrosomal pathologies of human spermatozoa and hyaluronidase activity in sperm (Singer et al. 1983). Koehler (1978) has suggested that in the absence of other definitive experimental proof, it must be assumed that claims of surface localization of hyaluronidase are based on absorbed enzyme released from other spermatozoa in the population, as also supported by the observations of Gould and Bernstein (1973). All the studies made so far have indicated that most of the hyaluronidase activity occurs in the acrosomal matrix and is released during the acrosome reaction. It would be rewarding to combine a fine-structural localization study with the Yanagimachi and Usui (1974) synchronized acrosome reaction experiment. In this way, it might be possible to determine more precisely the localization of hyaluronidase with cell surface and inner acrosomal membrane.

Hyaluronidase has been reported in the sperm acrosome of various mammalian species investigated with biochemical techniques (D. B. Morton 1975a, Mann and Lutwak-Mann 1981, Singer et al. 1982, Bhattacharyya and Bhattacharyya 1985). But conspicuous species variations exist in regard to the hyaluronidase content of their spermatozoa. Its amount decreases repectively from ram, rabbit, bull, human, boar, rat, and stallion sperm down to rooster sperm, which do not possess this enzyme (Zaneveld and Schumacher 1972, Zaneveld et al. 1973b). Their results have also indicated that the sperm hyaluronidase is identical to testicular hyaluronidase but differs from lysosomal hyaluronidase of other organs. Hyaluronidase has been observed to undergo biochemical changes during sperm maturation (Bhattacharyya and Bhattacharyya 1985).

There are some doubts concerning the molecular weight of hyaluronidase. Molecular weight of purified hyaluronidase has been estimated by various methods. Ultracentrifuge has shown molecular weigths ranging between 11,000 and 55,000 (Khortin et al. 1973), gel chromatography between 61,000 and 126,000 (D. B. Morton 1973) (although higher molecular weights reported may be due to dimerization), and dodecyl sulphate-gel electrophoresis between 60,000 and 65,000 (D. B. Morton 1975b, Yang and Srivastava 1975). The consensus is in the order of MW 60,000. Khortin et al. (1973) have reported that testicular hyaluronidase consists of four identical subunits of 14,000 MW each. The optimum pH is 3.8 for bull, 4.3 for ram (Yang 1972), and is different from lysosomal hyaluronidase present in organs other than the testis (Zaneveld et al. 1973b). Multiple forms of hyaluronidase activity have been reported, which could be separated by chromatography on DEAE-52 (D. B. Morton 1975a, b). Acrosomal hyaluronidase has a requirement for cations; K and Na have a greater effect than Ca, Mg or Mn.

Yang and Srivastava (1974) have reported the separation and properties of hyaluronidase from ram sperm acrosomes. They have been able to separate acrosin from the hyaluronidase. Good techniques for their separation are now available, i.e. isoelectric focussing, gel chromatography, CM-cellulose, hydroxy-apatite ion exchangers, as well as affinity chromatography with ligands for acrosin, such as soyabean trypsin inhibitor, p-amino-benzamidine (McRorie and Williams 1974, D.B. Morton 1976). There are indications that sperm acrosomal proteinase and hyalurindase show a variety of biochemical and immunochemical differences as compared to the respective enzymes prepared from sources other than sperm or testicle (Zaneveld et al. 1973a, b). Buffalo resembles bull closely in regard to hyaluronidase content of its spermatozoa (Sidhu and Guraya 1985a). The enzymatic acitivity of bull acrosomal and testicular hyaluronidase is inhibited by antiserum against bull sperm acrosomal extract, using various tests (Zaneveld et al. 1971). Inhibition of sperm hyaluronidase by antibody has been shown to be highly species-specific and capable of inhibiting fertilization in vitro (Metz 1973). This specificity may be exploited in future to develop methods for control of fertility by enzyme inhibitors or isoimmunization (Goldberg 1979, Zaneveld 1982). During the past decade, two types of antihyaluronidases have been evaluated for their contraceptive activity: hyaluronidase antibodies and the chemical inhibitors (Zaneveld 1982). Antibodies prepared against sperm hyaluronidase inhibit the lytic activity of this enzyme towards synthetic substances and the follicle cell layer, and can prevent in vitro fertilization of spermatozoa. The chemical inhibitors prevent feritilization. But no effect on sperm mobility or forward progression is observed at the concentration of inhibitors used.

Acrosin. By far the most extensively studied and best characterized sperm protease is acrosin (zona-penetrating enzyme, or zona pellucida proteolytic enzyme, or trypsin-like enzyme; see M.C. Chang and Hunter 1975). This enzyme is a unique sperm proteinase which is needed for mammalian fertilization. Studies on acrosin have not only led to a better understanding of fertilization at the molecular level, but have also indicated that specific inhibition of this enzyme has contraceptive potential (Zaneveld 1982). Consequently, many studies have been carried out on numerous species of mammalian sperm (McRorie and Williams 1974, Zaneveld et al. 1975, Hartree 1977, Parrish and Polakoski 1979). However, the nature of this system has only recently been recognized and partially demonstrated, as will be discussed here.

The localization of acrosin within the acrosome is controversial. It appears to be localized closer to the inner acrosomal membrane than hyaluronidase, and is apparently not bound to the outer membrane of the acrosome, because after membrane isolation more than 70% of the activity remains associated with the sperm pellet with which the inner membrane is still associated (McRorie and Williams 1974). However, Faltas et al. (1975), using immunohistochemical techniques, suggested that acrosin in the rabbit sperm is localized on the inner surface of the outer acrosomal membrane, as also suggested by other workers (S.W. Brown and Hartree 1974, Srivastava 1973). Proacrosin has also been localized on the inner acrosomal membrane of spermatozoa in rabbits and hamsters (Bradford et al. 1981). Removal of the outer membranes of the sperm head by artificial means has provided some evidence for the presence of acrosin on the inner acrosomal membrane and/or equatorial segment (D. B. Morton 1975b, Zahler and Doak 1975, C. R. Brown and Hartree 1976a, b). Stambaugh et al. (1975) observed in the rabbit a complex distribution consisting of six linear loops of evenly spaced granules, running diagonally across the flat side of the sperm head in a criss-cross pattern with the two most anterior loops. An immunoperoxidase staining procedure that readily demonstrates acrosin in the rostral portion of the acrosome fails to detect acrosin in the equatorial segment of spermatozoa in bull, boar, rabbit, and humans (Garner et al. 1977). Fléchon et al. (1977) have made immunocytochemical localization of acrosin in the anterior segments of acrosomes of ram, boar, and bull spermatozoa (see also Hundeau et al. 1984), as also made by Castellani-Ceresa et al. (1983) for the boar spermatozoa. Berruti and Martegani (1982) produced cytochemical demonstration that acrosin is unavailable in intact ejaculated boar and bull spermatozoa. They have used dansylalanyllysyl-chloromethyl ketone as a fluorescent probe for localization of acrosin activity in boar and human spermatozoa (Berruti and Martegani 1984). In general, the results of various studies, as discussed by D. B. Morton (1976), are not consistent among various species, and in some experiments even suggest higher levels of activity associated with the sperm tails than with heads. There is also good evidence for a more surface orientation of acrosin (Garner et al. 1975). Immunochemical evidence has indicated that the acrosin is a peripheral membrane-bound protein, possibly associated with the inner acrosomal membrane (Garner et al. 1977). Castellani-Ceresa et al. (1982) have made immunoelectron localization of acrosin in the acrosomes of mammalian spermatozoa. L. A. Johnson et al. (1983a, b) have localized immunocytochemically acrosin on both acrosomal membranes and acrosomal matrix of porcine spermatozoa. These observations are of special interest, because results of recent studies have suggested that the enzyme properties of membrane-bound acrosin differ significantly from the properties of the soluble enzyme (C. R. Brown and Hartree 1976a, b, Parrish and Polakoski 1979). D. P. L. Green and Hockaday (1978) have made a histochemical study of acrosin in guinea-pig sperm after the acrosome reaction. It is distributed unevenly over the outer surface of the newly exposed inner acrosomal membrane but does not extend to the equatorial segment. These findings have suggested that acrosin is a candidate for the role of zona lysin.

Acrosin is extracted differently from the inner and outer membrane of acrosome. According to Sidhu and Guraya (1985a), most of the acrosin activity in the buffalo spermatozoa can be isolated from acrosome by hyamine treatment alone. The remaining activity can be extracted only after combined treatment of hyamine and triton, showing that this fraction of acrosin is possibly bound to the inner acrosomal membrane and equatorial segment. Bhattacharyya and Zaneveld (1978) have suggested that both acrosin and acrosin inhibitor (acrostatin) are present in two localizations on the human spermatozoon: (1) associated with the outer acrosomal membrane either within the acrosomal matrix or on the surface; and (2) closely adherent to the inner acrosomal membrane. Initially, spontaneous release only of the acrosin and acrostatin that are associated with the outer acrosomal membrane occurs. The results of this study also demonstrate that the quantity of surface acrosin forms only a small percentage of total amount of acrosin present (probably about 10%, based on the 10-h incubation value). After 10 h of incubation, the acrosin associated with the inner acrosomal membrane, which is probably primarily in the proacrosin form, will also begin to disperse and zymogen activation will take place. This occurs more readily than does acrostatin liberation, accounting for the increased amount of free acrosin after that time. Gaddum-Rosse and Blandau (1977) have suggested that proteinases may be lost from guinea-pig spermatozoa as a result of the acrosomal reaction in vitro. From the discussion of results of ultrastructural, cytochemical and biochemical observations, it can be suggested that the localization of acrosin needs to be determined more precisely. Its identification with the acrosomal matrix, equatorial segment, or specific acrosomal membrane awaits a definite study, perhaps based on specific antibody-labelling procedures (Koehler 1978). It will be interesting to determine how the acrosin is stored in the acrosomal components. Stambaugh and Smith (1978) and Stambaugh (1978), using immunofluorescence and electron microscopy, have observed tubulin in the rabbit and rhesus acrosome and shown that microlubules polymerize after in vitro incubation in a silver-proteinate solution. Acrosin and tubulin show the same distribution in the acrosome, suggesting the possibility that tubulin is the binding structure for the proteinase.

The trypsin-like proteinase of the acrosome was designated as acrosin because of its substrate specificity and other properties (Polakoski et al. 1972). The free enzyme appears to exist as a monomer dimer mixer of varying proportions, with a monomeric molecular weight of about 28,000 (Fink et al. 1972, Polakoski et al. 1972, 1973a, b). The properties of solubilized acrosin from various species have been reported by serveral workers (H. Fritz et al. 1972, Multamaki and Niemi 1972, Schiessler et al. 1972, Stambaugh and Buckley 1972, Zaneveld et al. 1972a, b, Elce and McIntyre 1982). Acrosin seems to be somewhat different in rabbits and boars; in the first species, it is a dimer of two units of 27,300 MW; in the second, the molecular weight is 30,000 without dimeric form. The optimum pH in both cases is about 8. The human acrosin (Zaneveld et al. 1972a) appears to resemble that of the boar. For bull the situation seems to be similar to that of rabbit, but needs to be clarified (McRorie and Williams 1974). The bovine acrosin has a molecular weight of 39,000, whereas the human acrosin 49,000 and traces of a 38,000 MW component (Elce and McIntyre 1982). Garner and Easton (1977) have not observed any difference between acrosin of various mammals from an immunological point of view. Some differences in response to inhibitors are known and have been discussed by McRorie and Williams (1974). Dudkiewicz (1982) studied immunological relationships between rabbit acrosin and pancreatic trypsin. Rabbit acrosin has been isolated from arcosomal extracts by immunoaffinity chromatography (Dudkiewicz 1983a).

The extensive biochemical characterization of acrosin from several mammalian species has also been reviewed by various other workers (McRorie and Williams 1974, R.G. Harrison 1975, D.B. Morton 1976, Primakoff et al. 1980, Parrish and Polakoski 1979, Kaneko and Moriwaki 1981, Mann and Lutwak-Mann 1981, C.R. Brown 1982, Elce and McIntyre 1982, Siegel and Polakoski 1985). Wendt et al. (1976) have isolated acrosin in highly purified form from boar spermatozoa. The main form obtained has a molecular weight near 38,000, is a glycoprotein and has a substrate-splitting specificity similar to pancreatic trypsin. Dudkiewicz and Powledge (1982) have studied the effects of sonication and detergents on the extractability of acrosin and sperm neuraminidase from boar and rabbit spermatozoa. C.R. Brown (1982) has studied the purification of mouse sperm acrosin, its activation from proacrosin and effect on homologous egg investments. Gelatin-substrate film technique has been used for detection of acrosin in single mammalian sperm (Ficsor et al. 1983).

Isoenzymes of acrosin have been reported for the spermatozoa of some mammalian species (Sidhu and Guraya 1978a, Berruti 1980, 1981, Mann and Lutwak-Mann 1981). Recently, three forms of acrosin with calculated approximate molecular weights of 36,000, 34,000 and 25,000 have been reported for the bovine spermatozoa (Berruti 1981). The amino acid composition of the 36,000 MW form shows a relative similarity with the molecular content of the other known acrosin forms, human and rabbit acrosomal proteins. But the isozymes of acrosin are believed to represent activation stages of proacrosin. Zelezna and Cechova (1982) isolated two active forms of acrosin from boar ejaculated sperm. R.A. Anderson et al. (1981a) have characterized a high-molecular-weight form of human acrosin and compared it with human pancreatic trypsin.

Parrish and Polakoski (1977) have suggested that polyamines can serve as in vivo modulators of the proteolytic activity of acrosin and the activation of proacrosin; high levels a calcium-dependent modulator protein have been demonstrated in the sperm head (H.P. Jones et al. 1978). Fink and Wendt (1978) have reported the presence of a trypsin-like protease (acrosin) in acidic extracts of mouse spermatozoa by an electrophoretic method for separating the enzyme from a putative inhibitor, as also described by C.R. Brown and Hartree (1976a, b). Bhattacharyya et al. (1979) have recently demonstrated neutral proteinase, acrosin, in the mouse spermatozoa, which is largely (70% – 80%) present in the zymogen (proacrosin) form. Acid extraction yields higher amounts of acrosin than detergent extraction. Synthetic inhibitor studies have indicated that mouse acrosin has a serine and histidine at its active site and hydrolyses the peptide bonds of lysine and arginine but not of phenylalanine. N-terminal amino acid sequence of boar acrosin is determined (Fock-Nuzel et al. 1980). Muller-Esterl et al. (1980) have observed that boar acrosin (a glycoprotein) tends to aggregate in the absence of detergents and lipids. Molecular weights ranging from 44,000 up to 237,000 are found, corresponding to acrosin monomer up to hexamer. Involvement of the active site of the serine proteinase in the formation of oligomers is shown. Only monomeric acrosome has full activity, while a marked decrease in specific activity is observed upon aggregation. This indicates that acrosin has hydrophobic binding sites modulating the proteinase activity. Muller-Esterl and Schill (1982) studied liberation of acrosin from the acrosome and activity of the free proteinase in the presence of Nonoxinol-9. Kaneko and Moriwaki (1981) have made some studies on the purification and characterization of boar acrosin. The active sites of acrosin and trypsin are similar to each other. Kaneko et al. (1983) related acrosin to sperm motility. Purification of mouse sperm acrosin has also been studied (C.R. Brown 1983).

Acrosin is completely and irreversibly inhibited by various chemical inhibitors as well as by a number naturally occurring proteins (see R.G. Harrison 1975,

Muller-Esterl and Schill 1982, Long and Williamson 1983). Schill and Fritz (1976) have studied the enhancement of sperm acrosin activity by glycerol pretreatment. Zaneveld et al. (1973c) have reported that detergent extracts of hamster epididymal sperm hydrolysed the acrosin substrate α-N-Benzoyl-L-arginine ethyl ester. Acrosin is involved in the dissolution of zona pellucida during fertilization (S.W. Brown and Hartree 1974, D.B. Morton 1975a, McRorie et al. 1976, Yanagimachi 1981). C.R. Brown (1982) studied the effects of ram sperm acrosin on the investments of sheep, pig, mouse, and gerbil eggs (see also Urch et al. 1985a, b). The actual biological substrate of acrosin is still unknown, although it is very likely an essential structural component of the zona pellucida (Hedrick et al. 1985). It will be interesting to mention here that acrosin is capable of acting on and degrading sialic acid-containing glycoproteins present on cell surfaces (Uhlenbruck et al. 1972). Similar compounds have been demonstrated in the zona pellucida (Guraya 1974, 1985, Gwatkin 1979). Bhattacharyya and Zaneveld (1978) believe that the acrosin associated with the outer acrosomal membrane is probably not important for sperm penetration through the zona pellucida of the eggs, since this membrane is lost before the spermatozoon enters this layer (Yanagimachi 1978a). It may play a role in the acrosome reaction, which involves the vesiculation and disappearance of the outer acrosomal membrane and plasma membrane of the spermatozoon (Yanagimachi 1978a, Bedford et al. 1979).

Pro-acrosin. The presence of pro-acrosin, an inactive zymogen precursor of acrosin was first demonstrated in rabbit testis and cauda epididymal sperm (Mukerji and Meizel 1979, Syner et al. 1979, Kennedy et al. 1981, Goodpasture et al. 1981a). It appears to be localized on the inner acrosomal membrane of spermatozoa in rabbits and hamsters (Bradford et al. 1981). Its localization is studied for ram also (Hundeau et al. 1984). In an in vitro study proacrosin is shown to be associated with anionic (porcine) phospholipid membranes through apparent electrostatic charge interactions (Straus and Polakoski 1982a). But its exact mode of activation is not clear. Polakoski (1974) has produced evidence for the presence of acrosin in boar ejaculated sperm. Similarly proacrosin has also been reported in ejeaculated buffalo spermatozoa (Sidhu and Guraya 1982) as well as during the differentiation of spermatozoa in the rabbit, boar, bull, and ram (Phi-Van et al. 1983). According to Wendt et al. (1976) active acrosin is produced from an inactive zymogen, proacrosin, with a molecular weight near 70,000, by initial proteolysis via several active forms. Meizel and Mukerji (1976) have made biochemical characterization of proacrosin and acrosin from hamster cauda epididymal spermatozoa. According to Bhattacharyya et al. (1979) the acrosin in the mouse spermatozoa is largely (70% – 80%) present in the proacrosin form. But the total amount of acrosin (nonzymogen acrosin + proacrosin) in untreated guinea-pig spermatozoa is three to five times that in mouse and human spermatozoa (Goodpasture et al. 1981a). A reduction of 50% in the total amount of sperm acrosin occurs after induction of the acrosome reaction in vitro. Whether spermatozoa are untreated, capacitated or acrosome-reacted, 90% of the acrosin present in the spermatozoa is in proacrosin form. The total amount of acrosin released after induction of the acrosome reaction is never more than 55% of the original amount present, even if more than 80% of the spermatozoa have undergone acrosome reaction. No, or very little, proacrosin activation occurs during in vitro capacitation of guinea-pig spermatozoa.

Mukerji and Meizel (1979) have made a study of purification, molecular weight estimation, and amino acid and carbohydrate composition of rabbit testis proacrosin. Tobias and Schumacher (1977) have demonstrated two proacrosins (P_1 and P_2) in extracts of human spermatozoa, which are inactive precursors of acrosin. Similar observations have also been made in boar spermatozoa by Polakoski and Parrish (1977). Schleuning et al. (1977) have studied multiple forms of boar acrosin and their relationship to proenzyme activation. The direct estimates of the proacrosin content of human spermatozoa have shown that about 90% of the acrosin is in this zymogen form (Polakoski et al. 1977). The consistent proacrosin is isolated from spermatozoa in the presence of benzamidine required to prevent premature conversion of proacrosin to acrosin (Polakoski and Parrish 1977). Recently, Siegel and Polakoski (1985), have made an evaluation of the human sperm proacrosin-acrosin using gelatin-sodium dodecyl sulphate-polyacrylamide gel electrophoresis. Syner et al. (1979) have made a radioimmunoassay study of immunologic changes associated with the conversion of rabbit testicular proacrosin to acrosin.

Meizel and Mukerji (1975) have suggested that the proacrosin becomes activated during capacitation, possibly by changes in intracellular cation concentration and or in pH. The proacrosin is activated during the capacitation process when the "acrostatin" previously bound to the enzyme is removed (Bhattacharyya and Zaneveld 1978). This sequence of events has been reported in the boar system (Polakoski and Parrish 1977, Parrish and Polakoski 1979). Parrish and Polakoski (1979) have discussed the molecular properties and possible regulation of this important enzyme system. Proacrosin, which is stable at pH 3.0, is rapidly activated to acrosin at pHs 6.0–9.0. The conversion of highly purified boar proacrosin into acrosin follows a classically sigmoid-shaped autoactivation curve and the following sequence is observed.

Proacrosin → m α-Acrosin → m β-Acrosin → m γ-Acrosin

The approximate molecular weights of the acrosin were 49,000, 34,000, and 25,000, respectively. The activation is concentration-dependent in the presence of 0.05 M calcium chloride and is proteolytically stimulated by m α-acrosin, m β-acrosin, or trypsin, but not by chymotrypsin (Polakoski and Parrish 1977). Schleuning et al. (1976) observed that porcine pancreatic and urinary Kallekrein, but not plasmin, thrombin, or urokinase, accelerated the conversion process. These results indicate that the conversion of proacrosin into acrosin can occur via a biomolecular process and probably involves the hydrolysis of an arginine or lysine peptide bond. However, two other nonacrosin proteinases present in sperm extracts have been demonstrated capable of initiating the conversion of proacrosin into acrosin. The first is acrolysin (McRorie et al. 1976) and the second is an enyzme active at pH 5.5 (Polakoski and Zaneveld 1976). There is no compelling evidence that acrolysin (thermolysin-like enzyme) is present in the acrosome (Shapiro and Eddy 1980). Many proteolytic enzymes can activate proacrosin in vitro. A uterine fluid, glycosaminoglycan, can also activate proacrosin in vitro, perhaps by inducing a change in structure that allows a few molecules to be activated; the process then proceeds autocatalytically (Wincek et al. 1979).

R. A. P. Harrison and Brown (1979) have shown that the zymogen form of acrosin in testicular, epididymal, and ejaculated spermatozoa from ram show autoactivation and there is no evidence for the involvement of any auxiliary enzyme for activation. Thus, activation of only a few molecules of proacrosin need take place, for the acrosin thus formed could activate the remainder. According to Polakoski et al. (1979), most of the proacrosin (70% – 80%) is converted into acrosin in boar sperm incubated in vivo for 120 min while less than 3% of the proacrosin is converted into acrosin in the control sperm. Electrophoretic comparison of the enzymatic activities in the sperm extracts to homogenous mα-acrosin and mβ-acrosin has suggested that the in vivo conversion sequence is: proacrosin → mα-acrosin → mβ-acrosin. Kennedy et al. (1982) have provided evidence for an intrazymogen mechanism in the conversion of proacrosin into m-acrosin (Kennedy and Polakoski 1981). An improved method for the preparation and purification of boar m-acrosin has been developed by Kennedy et al. (1981). This simple, two-step procedure includes selective and quantitative conversion of proacrosin into m-acrosin at pH 8, using excess leupeptin, a highly effective acrosin inhibitor, dissociation of a tightly acrosin-bound protein by denaturation in 6 M guanidine hydrochloride at pH 3, and gel filtration over Sephadex G-100 superfine resin. Thus, activation of only a few molecules of proacrosin need take place, for the acrosin thus formed could activate the remainder. R. A. P. Harrison (1982) studied the interaction of ram proacrosin and acrosin forms with antiserum raised against ram mβ-acrosin.

The newly formed acrosin appears to play a role in stimulating the acrosomal reaction (Meizel and Lui 1976, Bhattacharyya and Zaneveld 1976), which can be inhibited by inhibitors in vitro (Meizel 1981). A related effect of sperm acrosin on its own physiology is in allowing the enzyme to be released from the sperm head; acrosin slowly diffuses away from the acrosome-reacted sperm and this release is prevented by acrosin inhibitors to be discussed later. C. R. Brown and Harrison (1978) have studied the proacrosin in spermatozoa from ram, bull, and boar and found that the activation of proacrosin to acrosin takes place by disruption of acrosomes. The acrosin activity of canine spermatozoa has been used as an index of cellular damage (Froman et al. 1984). Parrish et al. (1978) have observed that aqueous dispersions of synthetic phospholipids in the form of anionic, single bilayer vesicles stimulate the appearance of boar sperm acrosin esterase activity from its zymogen precursor. The liposomes that were used resulted in an approximately 100-fold increase in the rate of conversion of proacrosin into acrosin. Association of proacrosin with phospholid membranes and comparisons of membrane-associated and solubilized enzyme have been studied by Straus and Polakoski (1982a, b).

Goodpasture et al. (1981a) have studied acrosin, proacrosin, and acrosin inhibitor of guinea-pig spermatozoa capacitated and acrosome-reacted in vitro. Perreault et al. (1982) studied the effect of trypsin inhibitors on acrosome reaction of guinea-pig spermatozoa. Wincek et al. (1979) have observed that a uterine glycosaminoglycan stimulates the conversion of sperm proacrosin to acrosin. An acidic pH of 5 or less may serve to inhibit the activation or autoactivation of the acrosomal zymogen proacrosin to acrosin. Stambaugh and Mastroianni (1980) have observed the stimulation of the rhesus monkey proacrosin activation by oviduct fluid which appears to contain pronase-resistant low-molecular-weight stimulators. The stage of fertilization at which conversion of proacrosin to acrosin occurs is not known. A reasonable suggestion would be that the activation of proacrosin represents a part or perhaps an immediate consequence of the acrosome reaction, and is triggered by some factor present in the female tract secretions. A soluble protein fraction containing acrosomal enzymes is released during the acrosome reaction of guinea-pig spermatozoa (Primakoff et al. 1980). This fraction may be useful as a starting material for the purification of acrosomal proteins, as well as for the identification of functions in fertilization of various acrosomal proteins. Srivastava and Ninjoor (1982) studied isolation of rabbit testicular cathepsin and its role in the activation of proacrosin.

Zahler and Polakoski (1977) observed that Benzamidine at concentrations greater than 10 mM inhibits the conversion of proacrosin to acrosin (Sturzebecher 1981). These results indicate an autocatalytic conversion of proacrosin to acrosin. ρ-aminobenzamidine (Schleuning et al. 1976) and ρ-nitrophenyl-$\acute{p}$-guanidinobenzoate (Borhan et al. 1976) have also been shown to block the conversion of proacrosin into acrosin. This inhibition is not irreversible, for after dialysis to remove the inhibitor the proacrosin can still be converted into acrosin. Calcium (Polakoski and Zaneveld 1976) has been observed to retard the conversion of proacrosin into acrosin, but it has not been established whether this inhibition rests at the level of the proacrosin or the acrosin. Low-molecular-weight polyamines inhibit the conversion of proacrosin into acrosin (Parrish and Polakoski 1977). There is an apparent correlation between the size of the polyamine and the effectiveness of the inhibition of conversion, i.e. spermine > spermidine > cadaverine > putrescine. Interestingly, the polyamines stimulate the activity of boar acrosin but have no effect on human acrosin (Parrish et al. 1978), indicating that the polyamine inhibition of proacrosin conversion probably results from the polyamine binding to proacrosin with the resulting complex being less susceptible to proteolytic conversion.

Acrosin inhibitors. Acrosin appears to be involved in several aspects of the fertilization process. Inhibitors of this enzyme, which have been reported to prevent the sperm acrosome reaction, sperm binding to the zona pellucida, and sperm passage through the zona pellucida and the vitelline membrane (Green 1978a, b, c, Zaneveld 1982, Ichikawa et al. 1984, Van der Ven et al. 1985), have been extensively studied during recent years. Schumacher and Zaneveld (1972) have observed that the human cervical mucus contains proteinase inhibitors, which can inhibit the rabbit and human sperm acrosomal proteinases. Zaneveld et al. (1973c) have provided evidence for an acrosin-acrosin inhibitor complex in sperm acrosomal extracts of several mammalian species including humans and monkey, as also confirmed by Bhattacharyya and Zaneveld (1978), and Kennedy et al. (1982); its amounts vary among different species (Goodpasture et al. 1981a). The acrosin inhibitor, for which the name "acrostatin" has also been proposed, might be either a protein or one of the steroid sulphates, such as cholesteryl or desmosteryl sulphate, which normally occur in spermatozoa (Bhattacharyya and Zaneveld 1978, Burck and Zimmerman 1980). Whether acrosin and acrostatin are separately compartmentalized within the acrosome or somehow complexed together remains to be determined more precisely. However, the acrosin inhibitor present within the spermatozoa is not identical with the acrosin-inhibiting proteins of the seminal plasma (Zahler and Polakoski 1977, C.R. Brown and Harrison 1978, Wincek et al. 1979, Kotonski et al. 1980).

Cechova and Fritz (1977) have made a chracterization of the proteinase inhibitors from bull seminal plasma and spermatozoa, as also recently reported by Cechova et al. (1979a), who have observed that the inhibitor is a basic polypeptide. It is not strictly specific in its effect as it also inhibits trypsin and to a lesser degree chymotrypsin in addition to bull and boar acrosin. Cechova et al. (1979b) have isolated three natural proteinase inhibitors with low isoelectric points from bull seminal plasma, which are designated as B_1, B_2, and A. Isoinhibitors B_1 and B_2 have identical amino acid composition. Isoinhibitor A contains six amino acid residues less than isoinhibitors B_1 and B_2. Their heterogeneity has been attributed to the sugar component. All of them inhibit acrosin, trypsin, and chymotrypsin.

Anderson et al. (1981a) have observed inhibition of human acrosin by fructose and other monosaccharides. Structure-activity relationships have also been determined after inhibition of human acrosin by monosaccharides and related compounds (Anderson et al. 1981b). Schirren et al. (1977) have studied acrosin and trypsin inhibitor activity in the human ejaculate following testosterone treatment in cases of autoagglutination of spermatozoa, as also studied by Kotonski et al. (1980) in the boar semen; sperm acrosin has also been studied in patients with unexplained infertility (Mohsenian et al. 1982). Bhattacharyya et al. (1979) have observed that an inhibitor of acrosin is associated with mouse spermatozoa, capable of preventing the activity of at least 60% of all available acrosin. A proacrosin conversion inhibitor present in boar spermatozoa has been purified and initially characterized (Kennedy et al. 1982). This purified proacrosin conversion inhibitor does not inhibit acrosin. Thus it acts to prevent proacrosin conversion by selectively inhibiting the zymogen's self-catalysed conversion mechanism. A new acrosin inhibitor has been obtained from boar spermatozoa (Tschesche et al. 1982).

The acrosin (or proteinase) inhibitor is added to the acrosin while the spermatozoa are passing through the male ductular system (Dietl et al. 1976). This suggestion is supported by the observations of Veselsky and Cechova (1980) who have demonstrated the distribution of acrosin inhibitors (BUSI I and BUSI II) in the tissues and fluids of bull seminal vesicles and ampullae, and on the acrosomes of ejaculated and ampullar spermatozoa. BUSI II is also observed in the epididymal fluid and on the acrosomes of epididymal spermatozoa. Antisera to both inhibitors cross-react with boar seminal vesicle fluid and ram seminal plasma. There is no cross-reaction with the components of blood serum. Thus the acrosin is blocked by complexing with an acrosin inhibitor, which is likely to be removed in the female genital tract before fertilization. Wendt et al. (1976) have isolated an inhibitor in pure form with high activity to acrosin from boar seminal plasma, which is a glycoprotein and occurs in several isolated forms. It is structurally homologous to the pancreatic secretory trypsin inhibitor. Gecse et al. (1979) studied bradykinase and protease inhibitors in seminal plasma of fertile and infertile men. The naturally occurring inhibitor is produced by the epithelium cells of the male genital tract and is absorbed to the spermatozoa during ejaculation and later on removed during sperm migration in the female genital tract. It is bound to the outer acrosomal membranes or plasma membrane as found by immunofluorescent techniques (Dietl et al. 1976), and is obviously for penetrating the acrosomal coatings. The removal of acrosin inhibitor during capacitation is believed to activate the acrosin on the surface of the spermatozoon, which then proceeds to digest the outer sperm membranes. Recent studies have indeed indicated that proteinase inhibitors can prevent the acrosome reaction (Meizel and Lui 1976, Wolf 1977, Perreault et al. 1982), or can inhibit intra-acrosomal acrosin in boar (Muller-Esterl et al. 1983), or can interfere with the process of sperm-egg fusion (Wolf 1977, Zaneveld 1982, Beyler and Zaneveld 1982, Ichikawa et al. 1984). Reddy et al. (1982) have investigated the properties of a highly purified antifertility factor from human seminal plasma. Electron microscopic study of sperm in guinea-pig undergoing ionophore-induced acrosome reaction in the presence of the acrosin inhibitors benzamidine, p-amino-benzamidine, and phenylmethyl sulphonyl fluoride show that the acrosome reaction proceeds normally but the dispersal of the acrosomal content is inhibited, suggesting that acrosin activity is important in solubilizing acrosin (Green 1978a, b, c). Bhattacharyya et al. (1979) have also observed that acrosin is essential for fertilization because natural and synthetic inhibitors of mouse acrosin prevent the union of gametes. In contrast to the naturally occurring inhibitors, certain synthetic agents can bind irreversibly or pseudo-irreversibly to proteinases, and should have more immediate clinical potential as contraceptives (Bhattacharyya et al. 1980). A number of synthetic acrosin inhibitors have been tested for their ability to prevent conception in vivo (Sturzebecher and Markwardt 1980, Burck and Zimmerman 1980, Borlin et al. 1981, Schill et al. 1981). The relative inhibitory activity of synthetic agents towards acrosin runs parallel to their antifertility activity. Gossypol, a known antispermatogenic agent, is also found to be a potent inhibitor of human sperm acrosomal proteinase (Johnsen et al. 1982a, Tso and Lee 1982). It is found to effectively inhibit the highly purified sperm proacrosin-acrosin proteinase enzyme system by irreversibly preventing the autoproteolytic conversion of proacrosin to acrosin and reversibly inhibiting acrosin activity (Kennedy et al. 1983). These results have indicated the possibility of testing low levels of gossypol for their contraceptive action when placed vaginally. Antifertility effects of tetradecyl sodium sulphate in rabbits have been reported (Zimmerman et al. 1983). Connors et al. (1974) have recently reported the kinetics of rabbit acrosomal proteinase by 1-chloro-3-tosylamido-7-amino-2 heptanone. Most of the acrosin present on ejaculated spermatozoa is in a zymogen form, called proacrosin, as already discussed. Proacrosin conversion to acrosin is probably a necessary step for the fertilization process. In some species of mammals, acrosin activity and proacrosin conversion can be inhibited by zinc ions, which also inhibit the acrosome reaction of hamster spermatozoa (Zaneveld 1982). Steven et al. (1982), after studying inhibition of human and bovine sperm acrosin by divalent metal ions, have suggested the possible role of zinc as a regulator of acrosin activity. Johnsen et al. (1982b) observed the inhibition of the gelatinolytic and esterolytic activity of human sperm acrosin by zinc.

Inhibition of sperm acrosin by antibody appears to be species-specific and capable of inhibiting fertilization (E. Goldberg 1979, Zaneveld 1982, Beyler and Zaneveld 1982, Dudkiewicz 1983b, Srivastava et al. 1984). There is a strong possibility of exploiting this immunological approach for control of fertility (E. Goldberg 1979, D.B. Morton and McAnulty 1979, Syner et al. 1979, Zaneveld 1982). But antibody studies have been complicated because of the difficulty in preparing enough pure acrosin as antigen, so that high antibody titres can be produced (Zaneveld 1982). Srivastava et al. (1984) observed that immunized rabbits failed to impregnate females as long as the antibody titre remained high. As the titre decreased, these animals became fertile and the females delivered normal litters.

According to currently available views, the main function attributed to the protein inhibitors is to inactivate free acrosin which, by exuding from ageing or otherwise damaged sperm-cells, could impair penetration of the zona pellucida by healthy spermatozoa. The proteinase inhibitors, which occur in the seminal fluid coat the sperm surface and become firmly attached to it; their removal appears to coincide with capacitation. Human urinary trypsin inhibitor inhibits the human acrosomal proteinase acrosin (Sumi and Toki 1980). A major inhibitor of acrosin in rhesus monkey and rabbit oviduct fluid is the secretory immunoglobulin A (Go et al. 1982). But Bhattacharyya and Zaneveld (1978), after investigating the release of acrosin and acrosin inhibitor from human spermatozoa, have made the hypothesis that both are associated with the outer acrosomal membrane and inner sperm head membranes. Natural proteinase inhibitors have also been isolated and characterized from boar spermatozoal acidic extracts (Schmitt and Polakoski 1980). The fact that p-nitrophenyl-p-guanidinobenzoate (NPGB) binds to about one-half of the available acrosin suggests that at least this portion is in the free forms, since NPGB would not bind readily to an already existing enzyme-inhibitor complex. This particular substance is a most effective acrosin inhibitor, notoriously difficult to dislodge from the sperm surface; it also appears to be an effective contraceptive, when tested in the mouse either in vitro or after vaginal application (Bhattacharyya and Zaneveld 1978).

Corona radiata-penetrating enzyme (C.P.E.). Another enzyme released during acrosomal reaction is called "corona radiata-penetrating enzyme" (C.P.E.). This is a hydrolase discovered and investigated in the laboratory of Williams (Zaneveld and Williams 1970, Tillman 1972). C.P.E. dissolves the intercellular material of the corona radiata, active at pH 7.7, and inhibited decapacitation factors (McRorie and Williams 1974). However, Hartree (1971) and Bedford (1975) doubt the existence of such enzyme in spermatozoa. Further studies are needed to determine its structural localization within the acrosome, as well as its chemical properties and its natural substrate in the reproductive process.

Leucine-aminopeptidase. The acrosome of buffalo spermatozoa contains leucine-aminopeptidase that selectively hydrolyses L-leucine-β-naphthylamine at pH 7.0 (Sidhu and Guraya 1985a). A similar enzyme has also been reported in bull spermatozoa (Meizel and Cotham 1972). Aminopeptidase with broad substrate specificity has been isolated, purified and characterized from boar spermatozoa (Sidhu 1980). There is no correlation between leucine aminopeptidase and acrosin activity in spermatozoa (Meizel and Cotham 1972). The physiological role of leucine-aminopeptidase needs to be determined. Like acrosin, it may also be involved in the dissolution of zona material during fertilization.

Arylsulphatases. Arylsulphatase A and B are present in almost similar amounts in the acrosome of buffalo spermatozoa (Sidhu and Guraya 1985a). Their 50% activity appears to be bound to the outer acrosomal membrane and equatorial segment, as observed by their behaviour during extraction procedure. Both arylsulphatases have also been reported in the rat and rabbit spermatozoa (Allison and Hartree 1970, Seiguer and Castro 1972, Yang and Srivastava 1973, 1974). Arylsulphatase, like hyaluronidase, is relatively easy to extract from sperm heads. During treatment of ram or rabbit spermatozoa with $MgCl_2$, followed by detergents, the bulk of arylsulphatase activity together with hyaluronidase passes into the $MgCl_2$ extract, while the more firmly bound enzymes, such as neuraminidase, are solubilized by the detergents (Srivastava et al. 1974). The extraction of rabbit spermatozoa with $MgCl_2$, followed by chromatography on a DEAE-Cellulose column, yielded preparations containing arylsulphatase A and B (Yang and Srivastava 1974). Arylsulphastase A of the rabbit testis has been purified 900-fold by using multiple chromatography and shown to be a glycoprotein which is capable of hydrolysing arylsulphate ester linkages (Dodgson et al. 1956). The role of arylsulphatases in normal fertilization is still to be determined. These may play an adjunct role in penetration of the cumulus cell layer (Allison and Hartree 1970), as their substrates such as sulphated polysaccharides have been shown histochemically between corona cells and in the zona pellucida (Guraya 1974, 1985). The purified arylsulphatase is effective in dispersing cumulus cells in rabbit ova (Farooqui and Srivastava 1979a).

Neuraminidase. Srivastava and Hussain (1977) have studied the purification and properties of rabbit spermatozoal acrosomal neuraminidase which is exclusively bound to the inner acrosomal membrane (Srivastava et al. 1970). Two form of neuraminidase have been reported, one is identical to that present in other mammalian tissue, and the other, known as neuraminidase-like factor (NLF), has somewhat different substrate specificity (Srivastava et al. 1970, 1974). Neuraminidase and NLF have been speculated to be involved in zona reaction of the egg during fertilization because pre-treatment of ova with neuraminidase greatly decrease both in vivo and in vitro fertilization by inhibition of sperm penetration through zona.

Miscellaneous enzymes. In addition to acrosomal enzymes discussed above, the activities of several other hydrolases have been shown to be associated at least with the head of mammalian spermatozoa. Their physiological significance is obscure. These enzymes include acid and alkaline phosphatases, β-glucuronidase, α-glucosidase, α-mannosidase, several exopeptidases, β-N-acetylglucosaminidase, phospholipase A, nonspecific esterases, β-aspartyl-N-acetylglucosamine-amino-hydrolase, and acid proteinase (Couchie and Mann 1957, Dott and Dingle 1968, Allison and Hartree 1970, Meizel and Cotham 1972, Bernstein and Teichman 1973, Bryan and Umaithan 1973, Bhalla et al. 1973, Polakoski et al. 1973b, Mann and Lutwak-Mann 1981). The collagenase, like peptidase, is peculiar to mammalian sperm. It has been isolated from the acrosomes of human, rat, and bovine spermatozoa (Koren and Milkovic 1973). Its localization in the acrosome is not quite evident, the pH optimum is 7.5, molecular weight is about 110,000.

The physiological role of β-N-acetylglucosaminidase and β-N-acetyl-galactosaminidase, which have been demonstrated in the spermatozoa of buffalo (Sidhu and Guraya 1985a), and ram and rat (Allison and Hartree 1970), is still obscure. Similarly, the role of acid and alkaline phosphatases and esterases, which have also been demonstrated in the acrosome of mammalian spermatozoa, is not known more precisely.

c) Hydrolytic Enzymes and Acrosome Reaction

Very divergent views have been expressed about the role of hydrolytic enzymes and other factors in sperm capacitation and acrosome reaction (Meizel 1984, Nelson 1985, Guraya and Sidhu 1986). Gordon (1966) suggested that phosphatases of acrosome are involved in sperm capacitation and acrosome reaction. These may also be involved in the transport of Ca^{2+} from external fluids of the female genital tract (Restall and Wales 1968) into the cytosol around the acrosome. The outer membrane of the acrosome in the human sperm selectively binds Ca^{2+} (Roomans 1975). Calcium ions are well known to play both a critical and multifaceted role in the process of mammalian fertilization (Shapiro and Eddy 1980, Yanagimachi 1981, Shapiro 1984, Hyne 1984, Courtens et al. 1984, Nelson 1985). These are required for the acrosome reaction in spermatozoa from all species studied up to date, suggesting that calcium provides the primary signal (Yanagimachi and Usui 1974, Talbot et al. 1976, Triana et al. 1980, Shapiro and Eddy 1980, Byrd 1981, Tash and Means 1983, Meizel 1984, Nelson 1985, Guraya and Sidhu 1986). Rogers and Yanagimachi (1976) have observed that the Mg^{2+}/Ca^{2+} ratio in the medium influences the occurrence of the acrosome reaction in guinea-pig spermatozoa in vitro. They have suggested the possibility that the Mg^{2+}/Ca^{2+} ratio in vivo is involved in regulating the occurrence of the acrosome reaction. Yanagimachi (1978b) has observed that sperm penetration of zonae-free guinea-pig eggs occurs only in the presence of Ca^{2+}. Ca^{2+} is also required for mouse sperm to bind to the zona pellucida of mouse eggs (Saling et al. 1979, Yanagimachi 1978a, Shapiro and Eddy 1980). Increased concentration of Ca^{2+} in the compartment between the plasmalemma and acrosome may promote fusion of these membranes and, secondly, activate the acrosomal proteinase which starts the acrosome reaction (Gordon 1966, Meizel 1984). Triana et al. (1980) have observed that release of acrosomal hyaluronidase follows increased permeability to Ca^{2+} in the presumptive capacitation sequence for spermatozoa of the bovine and other mammalian species. Actually, there occurs an enhanced influx of Ca^{2+} (Tash and Means 1983, Meizel 1984, Nelson 1985). Green (1978a, b, c, 1982) has recently made several observations on the physiology of acrosome reaction of the guinea-pig sperm. He has suggested that the immediate cause of the acrosome reaction is an increase in the cytoplasmic free Ca^{2+} concentrations (Green 1982). Jamil and White (1981) induced acrosome reaction in sperm with ionophore A23187 and Ca^{2+}. Calmodulin, which is present in abundance in spermatozoa (H.P. Jones et al. 1980, Tash and Means 1983, Nelson 1985), is believed to play a key role in organizing events associated with calcium uptake and function (Peterson et al. 1983d). Lenz and Cormier (1982) have studied the effects of anti-calmodulin drugs on the guinea-pig acrosome reaction, as also studied by Peterson et al. (1983d) for boar spermatozoa. Hyne et al. (1984) have shown sodium requirement for capacitation and membrane fusion during the guinea-pig sperm acrosome reaction.

Besides the Ca^{2+}, the pH of the culture medium also influences the capacitation and acrosome reaction (M.L. Lee and Storey 1985, Sidhu et al. 1986c). Added biological factors apparently are needed to capacitate cells at lower pH values. A correlation of increased intracrosomal pH with the hamster sperm acrosome reaction has suggested that intracrosomal pH change may be one of the events necessary for acrosomal reaction (Working and Meizel 1983, Meizel 1984). Meizel (1984) has suggested that the function of the influx of K^+ just before the acrosome reaction in the hamster is possibly to stimulate directly or indirectly, the H^+-efflux needed for the increase in intraacrosomal pH occurring during capacitation. Dithiothreitol, a disulphide-reducing agent, inhibits capacitation, acrosomal reaction and interaction with eggs by guinea-pig spermatozoa (Yanagimachi et al. 1983).

Szöllösi and Hunter (1978) have described the sequence of membranous events that occur during the acrosome reaction in spermatozoa of the domestic pig. The outer acrosomal membrane undergoes several changes during capacitation and the acrosome reaction, as will be discussed in Chap. XI. According to Talbot and Franklin (1976) the acrosome becomes swollen, crenulated and fragmented during acrosome reaction, and many other workers describe fusion, vesiculation (fenestration), and loss of the outer acrosomal membrane and the overlying sperm plasma membrane, and the release of acrosomal matrix or enzymes (Roomans and Afzelius 1975, Parimakoff et al. 1980, Meizel 1984, Nelson 1985). This process, an exocytotic event, is necessary for the penetration of the zona pellucida and the fusion of the sperm and egg membranes (reviewed by Yanagimachi 1981). The mechanisms by which sperm undergoing the acrosome reaction initiate and control these organized membrane events are poorly understood. Freeze fracture study has revealed disintegration of the geometric pattern of the intramembrane particles (D.S. Friend and Rudolf 1974). Meizel (1984) has suggested that a cleaning of intramembrane particles may be a common event of both the mammalian sperm acrosome reaction and somatic cell exocytosis; such sites become cholesterol-poor prior to sperm acrosome reaction and exocytosis. The membrane fusion during acrosome reaction appears to involve proteolytic degradation (caused by trypsin-like enzyme, i.e. acrosin) of membrane proteins leading to increased freedom of movement by these proteins (Dravland et al. 1984, Huang et al. 1985). This suggestion is supported by the fact that acrosome reaction is inhibited by the synthetic trypsin inhibitors (Meizel 1984). However, there is also some disagreement about the role of trypsin-like activity in membrane events of the acrosome reaction, as discussed by Meizel (1984), who has proposed four ways of acrosin action in inducing acrosome reaction: (1) acrosin may activate putative zymogen form of phospholipase and/or acrosin, which is involved in acrosome reaction, as will be discussed later on; (2) acrosin hydrolyses sperm membrane proteins, thereby promoting membrane fusion, (3) acrosin may stimulate sperm adenylate cyclase, thus increasing intracellular cAMP which is believed to act as the primary signal for capacitation/acrosome reaction (Tash and Means 1983); and (4) acrosin may also modify the sperm surface, thus increasing Ca^{2+} influx resulting in membrane vesiculation. The results of various studies, as discussed by Dravland et al. (1984), have also suggested that the sperm trypsin like enzyme has a direct or indirect role in the membrane events of acrosome reaction. The increased trypsin-like activity (possibly acrosin), phospholipase-A_2 and Na^+, K^+-ATPase activities, and decreased Mg^{2+}-ATPase (H^+-pump) activity are involved in the membrane events of the golden hamster acrosome reaction or in events of hamster capacitation, which are needed for the acrosome reaction (Meizel 1984). The precise nature and localization of Ca^{2+}-ATPase in mammalian sperm needs to be determined in future studies. The distribution and functional significance of ATPases will be discussed in relation to sperm plasma membrane in Chap. XI. Their relative significance in acrosome reaction is controversial (Meizel 1984). Further studies are needed to define the regulation and roles of various enzymes of the sperm head in the acrosome reaction in vivo at the molecular level (Nelson 1985). However, the exogenous acrosome reaction of hamster spermatozoa is enhanced by the proteolytic enzymes, such as Kallikrein, trypsin and chymotrypsin (Shinohara et al. 1985). This enhancement of the acrosome reaction by exogenous proteinases is believed to be due in part to accelerated removal or alteration of the sperm coat (glyco-protein) by the enzyme prior to the acrosome reaction (Shinohara et al. 1985). The exogenous proteinases may also function synergistically with endogenous (acrosomal) proteinases (and other enzymes) in bringing about changes in membrane proteins and dispersing the acrosome matrix during the course of the acrosome reaction. The results of this study have indicated that participation of exogenous hydrolysing enzymes present in the lumen of specific regions of the female genital tract in capacitating spermatozoa is a certain possibility. The presence of lysosomal hydrolases in the female genital tract (oviduct and uterus) during the oestrous cycle of the hamster has been demonstrated (Rahi and Srivastava 1984). But sites and effects of stages of oestrous cycle on acrosome reaction in vivo need to be determined for different species of mammals (Herz et al. 1985). However, Sidhu et al. (1986a, e) have shown that capacitation-inducing ability of the oestrous uterine fluid is present in albumin-like fraction which is nondialysable heat-labile. The uterine factors also cause acrosome reaction in buffalo spermatozoa in vitro.

Some aspects of acrosome reaction have also been reported in other studies (Srivastava et al. 1974, Churg et al. 1974, Metz 1978, Cornett and Meizel 1978, Aitken et al. 1983a, Singleton and Killian 1983, Plachot et al. 1984). The kinetics of the normal acrosome reaction of mouse sperm have been studied (Dudenhausen and Talbot 1982, Töpfer-Petersen et al. 1985). Mouse sperm capacitation has been assessed by kinetics and morphology of fertilization in vitro (Fraser 1983a). The effects of drugs, ions, and other factors on capacitation and acrosome reaction form the subject of many recent in vitro studies (Meizel 1984,

Kyono et al. 1984, Koichi et al. 1984, Courtens et al. 1984, Nelson 1985, Shinohara et al. 1985, Chen 1985, Fleming and Armstrong 1985, Fleming and Kuehl 1985, Go and Wolf 1985, Kyono et al. 1985, M. L. Lee and Storey 1985, Tso 1985, Shi and Friends 1985, Sidhu et al. 1986a – e, C. N. Lee et al. 1986) and very divergent views have been expressed in regard to modes of their action at the molecular level. Most of these reagents or factors may modify the surface components or structural components of the sperm plasma membrane, making them more permeable to Ca^{2+} needed for acrosome reaction. The primary role of capacitation appears to be to regulate the timing of Ca^{2+} influx. The onset of acrosome reaction, which is an endogenous phenomenon, appears to be coordinated by complex events occurring during capacitation (Bedford 1983). The acrosome reaction in hamster spermatozoa can be accelerated by lysolecithin (Ohju and Yanagimachi 1982), chondroitin sulphates (Lenz et al. 1983, Ax et al. 1985), cis-unsaturated fatty acids (Meizel and Turner 1983b), and serotonin or its agonist 5-methoxytryptamine (Meizel and Turner 1983a). Glycosaminoglycans can induce an acrosome reaction in ejaculated rabbit, hamster, human, and bovine spermatozoa (Lenz and Ax 1982, Lenz et al. 1983, Parrish et al. 1985, C. N. Lee et al. 1983, 1986, Tesarik 1985, Meizel and Turner 1986). Mammalian zona pellucida glycoprotein can also stimulate the acrosome reaction in vitro (Bleil and Wassarman 1983, Cherr et al. 1986). These studies have indicated that glycosaminoglycans (at least one of which, hyaluronic acid, is present in the cumulus oophorus and/or zona) can rapidly stimulate the acrosome reaction in motile previously capacitated spermatozoa. The role of prostaglandins (PGs) in the acrosome reaction and fertilization has been suggested for the mouse and guinea-pig (Nuzzo et al. 1983, Meizel and Turner 1986). Meizel and Turner (1984) have suggested that arachidonic acid metabolites (e.g. PGE_2) produced by the sperm and by the female reproductive tract are important for the mammalian sperm acrosome reaction. It is controversial whether prostaglandins are effective Ca^{2+} ionophores. Fraser and Quinn (1981) have observed that addition of glucose in the medium is obligatory to initiate both the acrosome reaction and the whiplash motility associated with fertilizing ability of mouse spermatozoa. The changes in motility also accompany the acrosome reaction in hyperactivated hamster spermatozoa (Suarez et al. 1984). However, Rogers (1979) observed that glucose retards acrosome reaction in various mammalian spermatozoa. K^+ ions are observed to modulate expression of mouse sperm fertilizing ability, acrosome reaction, and hyperactivated motility in vitro (Fraser 1983b). Actually, very divergent views exist about the control of changes in metabolism, respiration and motility during capacitation, and acrosome reaction. Bradley et al. (1984) have studied the modulation of sperm metabolism and motility by factors associated with eggs.

From the results of various in vitro studies it can be suggested that during capacitation and acrosome reaction occurring during the passage of spermatozoa through the female reproductive tract, there occur a series of endogenous physiological alterations as a result of effects of various factors, which facilitate multiple fusion events of the plasma membrane with the underlying outer acrosomal membrane. Actually, the molecular and physiological changes that occur during capacitation prepare the sperm for acrosome reaction. This vesiculation process, termed acrosome reaction, is initiated by the influx of Ca^{2+} down the electrochemical gradient. This erosion of the plasma and acrosomal membranes can also be provoked more specifically by various factors (R. A. P. Harrison 1983). How calcium brings about vesiculation of membranes is also a much debated question. Based on numerous in vitro studies, various recent concepts have been set forth involving Ca^{2+} in acrosome reaction by either promoting electrostatic interaction between opposing membranes or by destabilizing the opposed membranes by inducing domains of crystalline acidic phospholipids which promote point fusions (Guraya and Sidhu 1986). The calcium is also implicated in the activation of proacrosin/acrosin, phospholipases and ATPases. These enzymes are believed to play a role in acrosome reaction (see Shinohara et al. 1985). Some recent studies have suggested that cAMP may provide the primary signal for the acrosome reaction (Tash and Means 1983). Acrosin might stimulate sperm adenylate cyclase, thus increasing intracellular cAMP (Meizel 1984).

Of considerable interst in connection with the physiological alterations defined as capacitation is the induction of these processes and of the acrosome reaction by adrenal gland extracts (Bavister et al. 1976). Epinephrine and α and β-adrenergic agonists in the presence of a protein-free ultrafiltrate of bovine adrenal cortex stimulate the activated flagellar motion, known as whiplash, and the acrosome reaction in hamster sperm. It is believed that epinephrine stimualtes both α- and β-adrenergic receptors. However, α-adrenergic agonists are more potent than β-adrenergic agonists. Activation of α-adrenergic receptors leads to calcium influx, which is required for the acrosome reaction to occur and it might lead to increase in cGMP levels intracellularly (Cornett and Meizel 1978, 1979, Meizel and Working 1980, Meizel et al. 1985). Meizel et al. (1985) have suggested that biogenic amines from follicular fluid may help to stimulate capaciation. The serotonin and its agonist show a more direct effect on the hamster sperm acrosome reaction than other biogenic amines, and the effect is receptor-mediated (Meizel and Turner 1983a). According to Bize and Santander (1985), the epinephrine decreases the K^+ requirement of hamster sperm capacitation. Furosemide blocks the effect of epinephrine.

The acrosome reaction is a prerequisite for fertilization (Plachot et al. 1984), allowing the release of lytic enzymes necessary for penetration of the outer investments of the ovum and possibly the exposure of receptor sites for the sperm-egg interaction (Bleil and Wassarman 1983, Monroy and Rosati 1983, R. A. P. Harrison 1983, Shapiro 1984, Nelson 1985). The mechanism, biochemistry, significance and historical perspective of sperm capacitation are discussed in several recent reviews (O'Rand 1982, Bedford 1983, Farooqui 1983, Clegg 1983, Go and Wolf 1983, M. C. Chang 1984, Meizel 1984, Nelson 1985, Guraya and Sidhu 1986). Takei et al. (1984) have studied phospholipids in guinea-pig spermatozoa before and after capacitation in vitro. The results of various studies discussed in these reviews have indicated that there occur changes in lipids of sperm head during capacitation and acrosome reaction (see also Fleming and Yanagimachi 1984, Yanagimachi and Suzuki 1985, Go and Wolf 1985), which lead to alterations of permeability in the sperm head membranes. Meizel (1984) has discussed the roles of phospholipase-A_2 and phospholipase-C in these lipid changes (see also Kyono et al. 1984), leading to increased membrane fluidity (possibly caused by cis-unsaturated fatty acids) that may be important to the acrosome reaction; both the enzymes are associated with the sperm head membranes. It will be interesting to mention here that there is observed an increased incorporation of [1-^{14}C]-acetate and [U-^{14}C]-glucose into the phospholipids during capacitation of buffalo spermatozoa (Sidhu et al. 1986b). The phospholipids as substrate of sperm-associated phospholipase could give rise to lysophospholipids which facilitate membrane fusion during acrosome reaction. Acrosin may activate inactivated phospholipase A_2 which induces membrane vesiculation. According to Srivastava et al. (1982), phospholipase-C has the advantage that it does not form more damaging lysophosphatides as phospholipase A. In addition to its sterol-accepting activity, albumin could also be involved in cellular phospholipid modulation, i.e. accepting fatty acids formed during phospholipase action on phospholipids and preventing end-product inhibition of enzyme. The effects of components of media used for the study of sperm capacitation and acrosome reaction in vitro are also important. Llanos and Meizel (1983) observed an increase in phospholipid methylation during capacitation of golden hamster sperm in vitro, suggesting it some role in capacitation. Physiological changes accompanying sperm capacitation are expressed in sperm motility, respiration, and substrate utilization, as already stated in Chap. VI. The optimum energy source for capacitation may be different for different species, but various energy sources used affect metabolism, which in turn affect capacitation (Guraya and Sidhu 1986). Actually, the molecular mechanisms involved in capacitation and acrosome reaction are still very controversial (Meizel and Turner 1986). These are endogenous molecular events at the membrane level which can be modulated by external environment. In future, various extrinsic and intrinsic molecular membrane probes will certainly resolve the controversies involved in the elucidation of the mechansims of capacitation and acrosome reaction.

In conclusion, it can be stated that there occur the following important changes during acrosome reaction of mammalian sperm: (1) Membrane changes (both surface and intramembranal) during capacitation facilitate Ca^{2+} influx, possibly causing some alterations in membrane ATPases. (2) increased Ca^{2+} ions stimulate vesiculation by either neutralizing negative charges of the opposing membranes, or by causing formation of crystalline domains of acidic phospholipids in the membrane; Ca^{2+} also inhibits Mg^{2+} ATPase (proton pump) and thus facilitates water influx into the acrosome, inducing approximation of the opposing membranes and leading to vesiculation; and Ca^{2+} stimulates acrosin involved in acrosome reaction. (3) Inhibition of Mg^{2+} ATPase by Ca^{2+} also increases acrosome pH; alkaline acrosomal pH induces activation of proacrosin to acrosin which may cause membrane protein hydrolysis leading to vesiculation; acrosin and alkaline pH also stimulates acrosomal phospholipase A_2 which acts on phospholipids and produces fatty acids and an important fusogen, lysophosphatide responsible for membrane vesiculation; the end-product inhibition of phospholipase may be prevented by removal of fatty acids and lysophosphatides by albumin in vitro or by some other protein in vivo; and albumin may also remove sperm membrane cholesterol, thus destabilizing the membranes for vesiculation. The important areas for further detailed study are those of extracellular protein; integral protein interaction, lipid bilayer composition and fluditity, Ca^{2+} binding and action within the cell during capacitation, and acrosome reaction. These alterations need to be investigated in the acrosomal region as well as in regions not involved in acrosome reaction, both in vivo and in vitro conditions. Courtens et al. (1984) have observed that bulbo-urethral gland secretion plays an active role in the induction of the acrosome reaction in the spermatozoa of goat in the presence of milk, and seminal vesicle secretion. Srivastava et al. (1982) believe that contaminating enzyme, such as proteinases, phospholipase C, and cathepsin D, may also promote acrosome reaction in rabbit. The regulation of the effects of calcium ions, activators of many enzymes including the various flagellar ATPases and cyclic nucleotide metabolism, is subject to modulation by calmodulin (Meizel 1984). The sperm-cells concentration of the calcium regulatory protein exceeds that of the cells of many other mammalian tissues 20-fold (Dedman and Means 1980).

d) Release of Acrosomal Enzymes Under Experimental Conditions

One of the characteristics of ejaculated spermatozoa is their susceptibility to cold shock which occurs when semen above 0 °C is rapidly cooled. This causes an irreversible loss of viability which is accompanied by an increased permeability of the cell and presumably changes to the cell lipoprotein membrane (Guraya and Sidhu 1975c, Pinatel et al. 1980, Menger and Menger 1981b, Mann and Lutwak-Mann 1981). The release of GOT and GPT into the extracellular media has been taken as an index of the measure of injury to spermatozoa (Graham and Pace 1967). Zavos et al. (1980) studied mobility and enzyme activity of human spermatozoa stored for 24 h at 5 °C and −196 °C. The storage at 4 °C causes morphological change of the acrosome in the motile bovine spermatozoa (Aaseth and Saack 1985). Menger and Menger (1981b) have made comparative light and electron microscopic studies of aminotransferase release and fertility of stallion semen before and after freezing.

Sidhu and Guraya (1978b) have studied the effect of cold shock on various enzyme releases in buffalo spermatozoa. Significant amounts of hyaluronidase, arylsulphatase A and arysulphatase B, which have been shown to be associated with acrosome, are liberated. Their presence in the extracellular media after cold shock has been attributed to detached acrosomes (Guraya and Sidhu 1975c); acrosome damage detachment with cold shock has also been reported in other farm animals (Hancock 1952, Iype et al. 1963, Quinn et al. 1969, Chinnaiya and Ganguli 1980, Patil et al. 1981). Quinn et al. (1969) have shown the release of acrosomal material consisting of protein, nondialysable phosphorus and polysaccharides into extracellular media with cold shock and freezing. Moroz et al. (1977) have observed that hyaluronidase activity of spermatozoa during freezing is significantly reduced, which is one of the causes of the low fertility of the frozen-thawed semen. Storage of whole human semen at −20 °C has deleterious effects on proacrosin and acrosin inhibitor, but not on nonzymogen acrosin (Goodpasture et al. 1981b). R. A. P. Harrison and Fléchon (1980) made immunocytochemical detection of acrosomal damage following cold shock. There was seen loss of acrosin from the acrosomal region of ram and bull spermatozoa.

The semen extender used by Antonyuk (1976) in experiments on boar sperm showed hyaluronidase, acid cathepsins, β-D-galactosidase and β-glucosidase, revealing their release; acid phosphatase was not released. Temperature effect on sperm (e.g. deep-freezing in liquid nitrogen and subsequent thawing) activated hyaluronidase, acid cathepsins, β-D-glucosidase and β-D-galactosidase significantly, but exerted an inhibitory effect upon acid phosphatase. These results have suggested that significant changes in sperm biological properties caused by temperature effect are due to breaks in the functional status of cell hydrolytic mechanisms and obviously to changes in the permeability of membrane lysosomal structures. Sidhu and Guraya (1979c) have studied the behaviour of various enzymes of buffalo spermatozoa in different modified dilutors, which are released progressively with increasing times of storage. Supplementation of diluted samples with chloroquine-diphosphate (a membrane stabilizer) controls the release of most of the enzymes. Kato et al. (1979) have studied the effect of catalase in diluent on survival and acrosome system of boar spermatozoa stored at 4 °C. The addition of catalase improves sperm survival. Reduced glutathione (GSH) or sodium laurylsulphate (SLS) prolong sperm life when these are used singly, but the beneficial effect of these agents is reduced more rapidly than that of catalase alone. Both GSH and SLS have no beneficial effect when these are combined with catalase. Ethylenediamineteracetate is not beneficial to sperm survival, regardless of the presence or absence of catalase.

e) Acrosome as a Lysosome

The presence of acid hydrolases in the acrosome has contributed to formulate the concept that the acrosome could be analogous to a lysosome (Zamboni et al. 1971, Fawcett 1975b, Nelson 1985). The various studies have provided support for the lysosomal origin of the acrosome. The acrosome is regarded as a specialized lysosome which evolved to facilitate fertilization. But the acrosome and lysosomes differ in regard to their mode of action. Lysosomes bring about the digestion of exogenous or endogenous intracellular substrates (De Duve 1963). The acrosomal enzymes are believed to digest the extracellular substrates, such as the intercellular matrix of the cumulus oophorus and the zona pellucida (McRorie and Williams 1974, M. C. Chang and Hunter 1975, D. B. Morton 1975a, Yanagimachi 1978a, Metz 1978, Nelson 1985). Acrosome and lysosome also differ, as lysosomes contain mostly (with some exceptions) acid hydrolases, but acrosome contains enzymes active at alkaline pH values also. The acrosomal and lysosomal acid hydrolases differ significantly in physico-chemical properties, pH optima, K_m, and heat stabilities (Majumder and Tarkington 1974). These authors have also suggested, besides the facilitating of penetration of the sperm through the investments of the egg, that acrosomal enzymes may also function to release specific glycosyl side-chain enzymes that can change the ovum glycoproteins, allowing interaction between specific sperm-cell surface receptors and thereby facilitating the fertilization process. Hyaluronidase from bull testis and bull sperm acrosomes seem to be immunologically similar, but these differ from lysosomal hyaluronidase from other tissues (Zaneveld et al. 1973a, b, c).

Like lysosome, the acrosome of spermatozoa is derived from the Golgi apparatus during spermatogenesis. Also the acrosome stains similarly to lysosome with acridine orange. It will be interesting to mention here that the respective contribution of the acrosomal vesicle (or vacuole) and granule to the acrosomal cap vary greatly in the developing spermatozoa of different mammalian species including primates (Guraya 1971a); in the chimpanzee and humans the acrosomal vesicle makes relatively very little contribution to the acrosomal material which is mainly formed by spreading of material from the acrosomal granule (Figs. 17 and 23). According to Guraya (1971a), these differences must be related, somehow, to the thickness and chemical composition of egg coverings, which are considered to be dissolved by the acrosomal enzymes of penetrating spermatozoa.

The mature acrosome does not show other cytoplasmic structures or organelles of any kind. There also does not occur a new synthesis of hyaluronidase, a major acrosomal chemical component (Pikó 1969). However, the maturation of the acrosome continues during sperm descent through different segments of epididymis, as judged from some morphological alterations (or modifications) in its substance and enclosing membranes (Bedford 1972, Fléchon 1974a, Bustos-Obregon 1975, Mann und Lutwak-Mann 1981). The functional significance of these changes in the composition and structure of acrosome is obscure, but these have been usually related somehow to the known progressive increase in fertilizing capacity (or maturity) of the spermatozoa as they pass through the epididymal tract (Fléchon 1974b, Jones 1975).

D. Subacrosomal Space

The inner surface of the acrosome is separated from the outer leaflet of the nuclear envelope by a space which is usually called subacrosomal space, containing granular or amorphous material of low electron opacity in primates (Fig. 57) (Bedford 1967, Pedersen 1970a, Bedford and Nicander 1971, Zamboni et al. 1971) and rodents (Breed 1984). In spermatozoa of rats and mice, this space is more capacious and occupied by a moderately dense, resistant material which is often called the perforatorium, of a pyramidal shape (Clermont et al. 1955, Bane and Nicander 1963, Hancock 1966, Baccetti et al. 1980). In Rodentia the material occupying the subacrosomal space becomes more and more abundant, as in the guinea-pig and the chinchilla, until it reaches an enormous size in the myomorpha rodents (Fawcett and Phillips 1970, Fawcett 1970). Although this subacrosomal structure has been designated as perforatorium in rodents, it is not a fibrous and stiff structure (Breed 1984). Fawcett (1975b) has stated that this is an unfortunate term because it implies a mechanical function in sperm penetration, which has not been revealed. Bedford (1967) has denied the presence of subacrosomal space in the human spermatozoa. But according to Pedersen (1970a) and Zamboni et al. (1971), a distinct subacrosomal space is also present in human spermatozoa (Fig. 57), which contains actin and myosin as demonstrated by immunofluorescent staining using specific antibodies (Campanella et al. 1979). In both monkey and human sperm, the subacrosomal space is open posteriorly and it communicates freely with the space separating the post-acrosomal cap from the outer sheet of the nuclear envelope (Zamboni et al. 1971).

Considerable importance has been attributed to the subacrosomal space, mainly because the earlier workers believed that the anterior region of this space is identical to, or was occupied by, a structure called perforatorium (Blom and Birch-Andersen 1965, Pikó 1969, Baccetti 1979, Baccetti et al. 1980), or apical body (Fig. 57) (Hadek 1963a, b, Bedford 1964). This hypothesis has been supported by the observations of Gordon (1969), who has found that the most anterior region of the subacrosomal space of guinea-pig and human spermatozoa contains a distinct body stainable with phosphotungstic acid. However, Zamboni et al. (1971) have denied the presence of any specialized structure in the subacrosomal space of both monkey and human spermatozoa, which contains only granular material of low electron density (see Pedersen 1970a). Similarly, Bedford (1967), and Bedford and Nicander (1971) have not described any specialized structure in the subacrosomal space of spermatozoa in different subhuman primates. These observations on the primate spermatozoa are in agreement with those of Fawcett (1965), and Fawcett and Ito (1965), who have also denied the presence of a perforatorium or any other specialized structure in the anterior portion of the mammalian sperm head (Baccetti 1979). The subacrosomal space is much less prominent in the spermatozoa of cat and horse, but in many other species it has been shown to be occupied by amorphous electron-dense material (Nicander and Bane 1966).

In agreement with Bedford (1967), Zamboni et al. (1971) have also rejected the view that the anterior region of the sperm head in primates is provided with an apical filament, a structure reported to be present in rabbit (Dickmann 1964) sheep (Dziuk and Dickmann 1965), and pig spermatozoa (Dickmann and Dziuk 1964), progressing through the zona pellucida. Zamboni et al. (1971) believe that the subacrosomal space merely represents a cytoplasmic region which has been trapped between the nucleus and acrosome during the process of sperm differentiation and development. This cytoplasmic area is occupied exclusively by a ground substance whose electron density and staining affinity may show regional differences due to some physico-chemical changes during sperm maturation. Jones (1970, 1971) described a connection between the perforatorium (or apical body), subacrosomal substances and the material present on the nuclear envelope under the post-acrosomal lamina in ram and boar spermatozoa. Continuous perinuclear material has also been shown incidentally in sperm of golden hamster, rabbit, and tupaia (Franklin et al. 1970, Gordon 1972a, b, Bedford 1974).

Courtens et al. (1976) have restricted the terms perforatorium and subacrosomal substance to their classical meanings (Nicander and Bane 1966), and used the general term "perinuclear substance" which is present as a thin, continuous layer of a proteinaceous substance lacking sugar residues (Courtens and Loir 1975b, Courtens et al. 1976), located between the nucleus on one side and the acrosome and post-acrosomal lamina on the other in boar, bull, ram, and rabbit spermatozoa, as also studied with freeze-fracturing technique (Fléchon 1974b). The lysine accumulates in this substance during spermiogenesis, as a result of protein turnover in the nucleus (Courtens et al. 1976), and at the same time -S-S-cross (Calvin and Bedford 1970, Bedford and Calvin 1974a). Olson et al. (1976b) have observed that the perinuclear substance in the rat consists of a single polypeptide, of 13,000 MW, with a content of 6.5% cystein. It has been demonstrated by immunofluorescent staining that the whole perinuclear space in human sperm contains actin and myosin (Baccetti 1979, Campanella et al. 1979, Baccetti et al. 1980). All these observations have indicated that in the typical mammalian sperm the acrosome is placed in a matrix containing actin and myosin and also proteins that can be stabilized by disulphide bonds. Baccetti (1979) has concluded that the subacrosomal material of mammals starts from a very involuted situation where a true perforatorium is absent and is represented by a thin layer of actin and myosin in the superficial region of the sperm head cytoplasm. The perforatorium-like organelle of rodents sperm is a newly acquired structure, consisting only of a cystein-rich cytoskeletal protein specific of their sperm. The perforatorium and equatorial blisters (rabbit) are part of the perinuclear substance, which first appears when the acrosomal vesicle apposes on the nucleus of young spermatids (Baccetti et al. 1980).

Morphological and cytochemical modifications of the prinuclear substance occur during spermiogenesis and epididymal maturation while disulphide bonds, already present in the spermatid perinuclear substance, are increased (Calvin and Bedford 1971). Such cross-linked proteins are relatively resistant to pronase extraction on thin sections of rabbit spermatozoa (Fléchon 1974a). Bedford and Nicander (1971) have also reported the existence of changes in the disposition of the subacrosomal material during the last stages of spermiogenesis and during epididymal passage. However, the rostral concentration of subacrosomal material persists apparently unchanged during the dispersion of the rabbit nucleus within the egg (Bedford 1970).

The various observations, as discussed above, have shown that the perinuclear substance appears as a structural and biochemical entity, suggesting that it plays a functional role. This role may be important for fertilization, as the perinuclear substance is the underlying substrate of the acrosome and post-acrosomal lamina and these sperm components are related to egg penetration and attachment (Yanagimachi 1978a). The perinuclear substance may act as a cement between the nucleus and the overlaying elements of the sperm head (Fawcett 1970, Fléchon 1973, Franklin 1974). Kaczmarski (1978) has suggested that the perinuclear substance most likely functions in cementing the acrosome to the anterior part of the nucleus and in stiffening the apical portion of the head. In the post-acrosomal region, the post-acrosomal lamina and the perinuclear substance are spatially linked, since in mature rabbit spermatozoa their mutual opposition is not disturbed by the presence of clear spaces over the nucleus, which have been reported to contain phospholipid material (Teichman et al. 1974) and disappear in uterine spermatozoa; meanwhile the perinculear substance returns close to the nuclear surface. The cement hypothesis in the acrosomal region is supported by the persistence of the inner acrosomal membrane on spermatozoa deprived of the anterior segment (C. R. Brown et al. 1975) and the process of ovum penetration (Franklin et al. 1970, Yanagimachi and Noda 1970a, Courtens and Loir 1975b). The scleroprotein nature of the perinuclear substance may also give rigidity to the perforatorium, thus facilitating penetration of the zona pellucida (Yanagimachi and Noda 1970a). The physiological significance of actin and myosin present in the subacrosomal region of human sperm needs to be determined (Campanella et al. 1979). The various proteins present in the perinuclear substance may also be playing some role in the release of enzymes at the time of acrosomal reaction. According to Baccetti (1979), actin is very abundant around the nucleus in rat testicular material, but absent in epididymal spermatozoa (Baccetti et al. 1980). The 13,000 MW protein appears only in the sperm from the epididymis. From these observations he has suggested that the subacrosomal region changes role during sperm maturation. Actin appears to play only a morphogenetic role during the moulding of the head region, and the newly appeared 13,000 MW protein endows the region with a skeletal role. It is not known whether actin is conserved in an undetectable form or is eliminated. However, Talbot and Kleve (1977) have shown the presence of actin by immunological methods along the concave margin of the acrosome in the hamster.

E. Post-nuclear Cap

1. Structure

The spermatozoa of marmoset and chimpanzee treated with the silver nitrate technique of Ayoma, show a strongly argentophil layer covering the posterior half of the nucleus in the testicular spermatozoa of marmoset and chimpanzee (S.S. Guraya, unpublished observations). A similar layer, usually referred to as the "post-nuclear cap" (Fawcett 1970, Pedersen 1972a) or the post-nuclear sheath (Koehler 1970a, b) or more precisely post-acrosomal dense lamina or sheath (Figs. 57 and 60) (Fawcett and Phillips 1969a, Phillips 1977b, Krimer and Esponda 1978, Bae and Kim 1981, Czaker 1985c), has also been demonstrated in the sperms of a variety of mammalian species (Bishop and Walton 1960, Oko et al. 1976, Krimer and Esponda 1978). The post-nuclear cap forms the post-acrosomal region of the sperm head which lies behind the posterior margin of the acrosome (Fig. 57). The presence of the post-nuclear cap has been clearly demonstrated in the mouse, human, monkey, and bovine, including water buffalo, spermatozoa by the use of the scanning electron microscope, freeze-etching, and replica methods (Fig. 60) (Arthur 1968, Koehler 1970b, 1973a, Pedersen 1972a, Stackpole and Devorkin 1974, Oko et al. 1976, Czaker 1985c, Holstein and Roosen-Runge 1981). Krimer and Esponda (1978), using light and electron microscopy, have observed preferential staining of the post-acrosomal lamina in the mouse spermatids after silver impregnation (Fig. 62) as also demonstrated by Czaker (1985c), who has provided more details.

In electron microscopical studies of thin sections of spermatozoa in primates and other mammalian species the post-nuclear cap, situated caudal to the acrosomal cap, appears as a thin intermediate layer uniting the sperm plasma membrane with the outer leaf of the nuclear envelope (Fawcett and Ito 1965, Bedford 1967, Pedersen 1970a, Fawcett 1970, Zamboni et al. 1971, Holstein and Roosen-Runge 1981). The thin dense layer is called the post-acrosomal dense lamina of the post-acrosomal sheath (Fawcett and Ito 1965, Fawcett 1970, Bae and Kim 1981). It runs parallel to the plasma membrane at a distance of 15 to 20 nm and regular periodic densities about 12 nm apart project from its outer aspect to the inner surface of the plasma membrane (Fawcett and Ito 1965, Fawcett 1970). Actually, the post-nuclear cap shows, in the different species, a longitudinal or circumferential parallel striation (Koehler 1966, 1970a, b, 1973a, Pedersen 1972a) due to an arrangement of ridge-like elevations of the plasmalemma, which connect the lamina to the plasmalemma. In the centre there is a parallel monolayer of microtubules (Plattner 1971). In its fine structure the post-acrosomal sheath has been described as fibrous (Randall and Friedlander 1950, Pikó and Tyler 1964), microtubular (Blom and Birch-Andersen 1965), and membranous (Fawcett and Ito 1965). This layer is always tightly attached to the overlying plasma membrane and to a lesser extent to the nuclear membrane, showing a possible structural role in binding these membranes together. On the basis of this close structural relationship, Zamboni et al. (1971) have suggested that the fine morphology of the post-nuclear cap (or post-acrosomal cap) is comparable to that of septate desmosome.

According to Pedersen (1972a), the post-acrosomal sheath of human spermatozoa consists of two morphologically distinct areas. The anterior one shows a structural pattern consisting of evenly spaced, circumferentially oriented densities extending to the inner aspect of the cell membrane. The posterior one consists of obliquely oriented cord-like structures extending from the posterior margin of the anterior area to the striated band, which is a circumferential close union between the cell membrane and the nuclear envelope at the posterior margin of the post-acrosomal sheath (Pedersen 1974a, b). The substructural pattern seen in the anterior area of the human sperm is present throughout the post-acrosomal sheath of *Macaca arctoides,* but here the linear densities are oriented longitudinally instead of circumferentially.

The post-nuclear cap in the mouse spermatozoa consists of layers of material displaying regular parallel striations as revealed by freeze-etching (Stackpole and Devorkin 1974); its dense material, lying between the plasma membrane and the nuclear envelope, is not membrane-bound. Phillips (1977b), using surface replicas, has observed that the post-acrosomal region in several species of mammals is composed of material arranged in parallel ridges with centre-to-centre spacing of 15 nm. Koehler (1973a), using replica, freeze-etching, and thin section preparations, has also considered the post-nuclear sheath of water buffalo spermatozoa to be a complex composed of matrix material, formed elements termed "striations", and a caudal posterior ring (Fig. 60). The latter completely encircles the sperm and is noted to be a specific feature of mammalian spermatozoa by a number of investigators (Koehler 1973a). It is possibly a differentiation of the plasma membrane (Pikó 1969), is believed to form ringlike seal at the caudal border of the post-nuclear sheath between the plasma and nuclear membranes, and it has been speculated that it plays a role in compartmentalizing the sperm cytoplasm into head and tail regions (Koehler 1973a).

The posterior ring is full of densely packed 20 nm particles (Koehler 1972). Posterior to the ring, another striated area exists in human and rat sperms (Woolley 1970, Pedersen 1972a, c). Here, next to the inner surface of the plasmalemma, which apparently has five layers (Pedersen 1972c), there is a redundant nuclear envelope with pores (Pedersen 1972a), which will be discussed in relation to the neck region where it is usually present (Wooding and O'Donnell 1971) and is united to the plasma membrane by homogeneous material (Pedersen 1972c). Anteriorly, the post-nuclear cap is limited by the posterior edge of the acrosomal sac, which often shows a local thickening in this area (Bedford 1967, Pikó 1969, Zamboni et al. 1971).

A narrow cleft between the post-acrosomal sheath and the nucleus is closed behind by the posterior ring, a narrow circumferential band of fusion of the plasmalemma to the underlying nuclear envelope (Woolley 1970, Pedersen 1972a, D. S. Friend and Fawcett 1974). The origin, formation, and chemical composition of the post-nuclear or post-acrosomal sheath are still not fully known. During spermiogenesis of marmoset and chimpanzee there appears an argentophil nuclear ring in spermatids at the posterior border of the acrosome and extends progressively in a caudal direction (S. S. Guraya, unpublished observations). It is very similar to that described by previous workers (Hancock 1957, Hancock and Tevan 1957). According to Burgos and Fawcett (1955), the nuclear ring probably arises through a local differentiation of the cytoplasm of the sperm plasma membrane.

2. Chemistry

The evidence so far available indicates that the post-nuclear cap may contain phospholipids and proteins rich in basic amino acids, especially arginine (Pikó 1969) Guraya and Sidhu (1975b), using cytochemical techniques, have demonstrated the presence of phospholipids in the region of the post-nuclear cap of buffalo spermatozoon, which are combined with protein to form lipoprotein. Cummins and Teichman (1974) have demonstrated the presence of melachite green-stainable phospholipid in the post-acrosomal region (or postnuclear cap) of rabbit spermatzoa, which consists predominantly of choline plasmalogen and accumulates within the spermatozoa during their final stages of maturation in the epididymis. It is believed to be involved in capacitation. The post-nuclear cap of testicular spermatozoa in buffalo, goat, and ram also contains sudanophilic, unsaturated lipids consisting of phospholipids and lipoproteins (G. S. Bilaspuri and S. S. Guraya, unpublished observations). The proteins with disulphide bonds are also present. The presence of basic proteins rich in arginine and lysine, reported previously in the post-nuclear cap (Pikó 1969, Koehler 1975a, Courtens et al. 1976), could not be confirmed in these studies. The functional significance of phospholipids demonstrated in the post-nuclear cap remains obscure. But recent studies have also revealed the presence of plasmalogen (phosphatidyl choline) in the post-acrosomal segment of the sperm head, which accumulates during the final stages of sperm maturation (Cummins et al. 1974, Teichman et al. 1974).

Wislocki (1949) reported a faint acid phosphatase reaction in the post-nuclear cap area of deer sperm. Teichman and Bernstein (1971) have also reported the presence of acid phosphatase in the post-acrosomal segment of rabbit and bull sperm heads. In rabbit spermatozoa the activity is localized predominantly in two regions, the equatorial segment of the acrosome and the post-acrosomal segment of the sperm head; in bull spermatozoa it is confined chiefly to the anterior third of the post-acrosomal segment. In the mouse, a positive Gomori reaction for acid phosphatase is given by the post-acrosomal dense lamina, the subacrosomal space, and the space between the plasma membrane and the outer acrosomal membrane (Poirier 1975). Guraya and Sidhu (1975a) have demonstrated the presence of various hydrolytic enzymes, such as alkaline and acid phosphatases, 5′-nucleotidase, and ATPase in the region of the post-nuclear cap of ejaculated buffalo spermatozoa. Mathur (1971) has reported the presence of cholinesterase and monoamine oxidase in the post-nuclear cap of mouse sperm. Several enzymes, such as acid phosphatase, alkaline phosphatase, thiamine pyrophosphatase, 5′-nucleotidase, Mg^{2+}-ATPase, Ca^{2+}-ATPase, alkaline pyrophosphatase, lipase, choline esterase, and monoamine oxidase, have been observed to be present in the post-nuclear cap of testicular spermatozoa of buffalo, goat, and ram (G.S. Bilaspuri and S.S. Guraya, unpublished observations). A preferential localization of various elements (Ca, K, S, P) has been demonstrated in the head portion of human spermatozoa that corresponds to the post-acrosomal cap (Fain-Maurel et al. 1984). Courtens (1979b) has studied the appearance of new electrical properties in the plasma membrane of ram spermatids. Their en bloc staining with electronegative colloidal Fe at pH 9 causes particle precipitation only on the portions of plasma membrane covering both the post-acrosomal lamina and the perinuclear ring. This area of the plasma membrane implicated in fusion with the oocyte is, therefore, characterized by unique electrical properties appearing in the testis.

3. Function

In regard to the function of the post-nuclear cap, Fawcett and Ito (1965) believe that it could be purely structural device for maintaining firm cohesion between the plasma membrane and the nuclear envelope in this area. Presumably, it plays a mechanical protective role in maintaining the head shape. This structure is, in fact, very resistant to any extractive agent (Wooding 1973, Koehler 1973a). However, Zamboni et al. (1971) have introduced the hypothesis that the post-nuclear cap plays a role in establishing and maintaining adhesion between spermatozoa and the ovum at the time of gamete conjugation. Stefanini et al. (1969), and Yanagimachi and Noda (1970a, b, c) have also made a similar suggestion. They have found that some changes in the physiological properties of the post-nuclear cap or region constitute a necessary preliminary to the membrane fusion between sperm and egg-cells. The presence of various hydrolytic enzymes in the post-nuclear cap, as discussed above, may be ultimately related to sperm penetration or gamete membrane fusion mechanism active during fertilization process. Clarke and Yanagimachi (1978) have detected an actin-like substance in the post-acrosomal region of mammalian (human, rabbit, dog, bull, rat, mouse, guinea-pig and golden hamster) spermatozoa by indirect immunofluorescence. These actin-like contractile proteins are believed to play some important role in sperm function and sperm-egg interaction during fertilization. But in more recent studies, the function of establishing and maintaining adhesion between sperm and egg during fertilization has been assigned to the equatorial segment of the acrosome (Bedford et al. 1979). The functional significance of the post-nuclear cap still needs to be determined more precisely.

Chapter VIII
Neck

The neck is a short, slightly constricted segment of the sperm located between the base of the head and the first gyre of the mitochondrial helix of the mid-piece (Figs. 55 and 61) (Fawcett and Ito 1965, Zamboni et al. 1971, Fawcett 1975b, Phillips 1975a, Holstein and Roosen-Runge 1981, Sato and Oura 1985). The neck differs clearly from both the head and the tail in certain morphological features of the plasmalemma, a sharp demarcation of its upper limit by the posterior ring, and lack of continuity between the segmental columns and the outer dense fibres of the tail (Figs. 57 and 63) (Kaczmarski 1978).

The nucleus in spermatozoa of marsupials, except the koala, is flattened in what has come to be described as the dorsoventral plane, and the neck is inserted into the central region of the long ventral axis, while the acrosome lies on the anterior dorsal surface (Harding et al. 1979). In contrast, the neck in eutherian sperm is inserted into the short caudal surface of the nucleus (Fawcett 1975b). This difference in the location of neck insertion is the result of events occurring during spermiogenesis in marsupials, namely, the plane in which nuclear flattening takes place with respect to the long axis of the flagellum.

Very divergent opinions have been expressed about the relations of various components in the neck of the mammalian spermatozoon which, according to Fawcett and Ito (1965), are complex and difficult to visualize in three dimensions from the study of thin sections. Much of this controversy pertains to the presence or absence of distal centriole. Hughes (1976) has observed the relatively undegenerated proximal and distal centrioles within a striated neck cylinder of spermatozoa in the marsupial devil (*S. harrisii*). All the previous workers have agreed in regard to the presence of proximal centriole, connecting piece consisting of striated columns, and basal plate placed in the implantation fossa (Figs. 57 and 63) (Fawcett and Ito 1965, de Kretser 1969, Bedford 1967, Pedersen 1970a, 1974a, b, Zamboni et al. 1971, Fawcett 1975b, Phillips 1975a, Sato and Oura 1985, Holstein and Roosen-Runge 1981).

A. Basal Plate and Connecting Piece

Immediately behind the sperm head is the connecting piece (Fig. 57). This complex structure has a dense, convex articular region called the capitulum which conforms to the concavity of the basal plate lining the implantation fossa of the nucleus (Fig. 63) (Fawcett 1975b, Phillips 1975a, Bae and Kim 1981, Holstein and Roosen-Runge 1981). Fine filaments traversing the narrow, electron-lucent

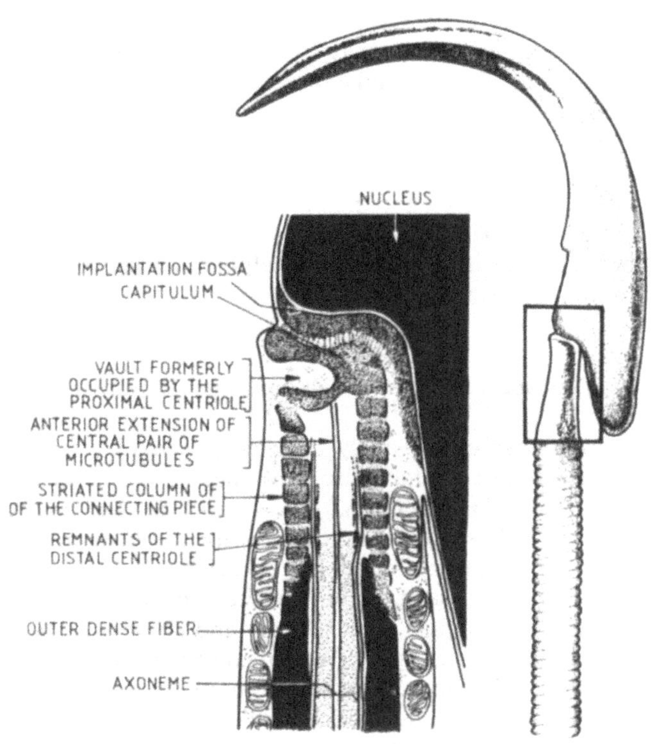

Fig. 63. Rat sperm head. *At the left* is a diagram of the structures of the neck region as they would appear at higher magnification in an electron micrograph (for orientation, see the area enclosed in the *rectangle on the right-hand figure*). The relationship of the connecting piece to the implantation fossa and to the outer fibres of the tail are shown, and the absence, in this species, of both centrioles. The main features of the diagram apply to all mammalian sperm, but in most species a proximal centriole is retained in a niche beneath the capitulum (Redrawn from Fawcett 1975b)

space between the capitulum and the basal plate appear to be mainly responsible for attachment of the head to the tail.

As revealed by freeze-etching (Stackpole and Devorkin 1974), the basal plate, the point of articulation between the head and the tail of the mouse sperm, closely resembles a gap junction due to the tight apposition of the inner and outer leaflets of the nuclear envelope and the packing of intramembranous particles within the inner leaflet (Goodenough and Revel 1970, Raviola and Gibila 1973). Since the basal plate must maintain the integrity of this attachment despite active motion of the large tail, the intramembranous particles may contribute a structural rigidity to this region.

Extending backwards from the capitulum are nine segmented columns, $1-2$ µm in length (Figs. 57 and 63), which at their caudal end overlap the tapering anterior ends of the nine dense fibres of the flagellum to which these are firmly attached. These columns forming the connecting piece seem to be continuous with the outer dense fibres, but the two have different origins, as already discussed in Chap. IV. These have fused secondarily along an oblique line (Fawcett

1975b, Phillips 1975a). As observed in longitudinal sections, the columns of the connecting piece appear to consist of dense segments alternating with narrower light bands (Figs. 57 and 63). A very thin intermediate line bisects the light bands. Each dark segment show 11 fine transverse striations. The neck of the mink sperm appears to show a dorsoventrally continuous but laterally separated capitulum which is followed by two major and five minor columns, forming at first a striated ring and then joining with the dense fibres of the axial fibre bundle (Kim et al. 1979). According to Bedford and Calvin (1974a), the mammalian connecting piece is stabilized by disulphide bonds, formed during sperm maturation in the epididymis.

B. Centrioles and Their Relationship with Other Components

A transverse or obliquely oriented proximal centriole is generally placed in a niche or vault in the dense substance of the connecting piece just beneath the capitulum (Figs. 57 and 63) (Fawcett 1975b, Phillips 1975a, Bae and Kim 1981). In both the bush-baby and the slow loris, the connecting piece and proximal centriole are commonly offset in a somewhat eccentric position, with respect to the implantation fossa at the base of the nucleus (Bedford 1967). A major difference in the fine structure of bush-baby and slow loris sperm lies in the neck region, which in the latter species often exhibits a modest scroll of evaginated nuclear membrane. In the neck region of monkey sperm, the connecting piece/centriole complex abuts against a thick basal plate set symmetrically in a relatively shallow implantation fossa (Bedford 1967).

The head and neck regions of the human spermatozoa are apposed over an almost flat source, and thus the true implantation fossa described for subhuman primates and other mammalian species does not show any appreciable development (Nicander and Bane 1962, Blom and Birch-Andersen 1965, Fawcett and Ito 1965, Bedford 1967, de Kretser 1969, Pedersen 1970a, 1974a, b, Zamboni et al. 1971). However, according to Pedersen (1970a) the nucelar membrane at the posterior end of the head in man is modified corresponding to the implantation fossa (Holstein and Roosen-Runge 1981). When viewed from the dorsoventral aspect, the proximal centriole of the human spermtozoon appears to be set at right angles to the long axis of the sperm and not at 45° as in many other mammals (Pedersen 1970a, 1974a, b). Bedford (1967) has attributed this difference to the absence of an implantation fossa in the human sperm. De Kretser (1969) has called the proximal centriole of other workers the transverse centriole in the human sperm, in order to distinguish it from the longitudinal or distal centriole.

Electron microscope studies by Fawcett (1965), and Fawcett and Ito (1965) on the spermatozoa of guinea-pig and bat respectively showed that the connecting piece is derived from the transformation of the distal centriole and represents the basal body and centre of flagellar activity. This conjecture has been recently abandoned following the observation that the connecting piece is not a modified distal centriole but a structure which forms around both distal and proximal centrioles (Fawcett and Phillips 1969a, 1970, Fawcett 1972). On the basis of this finding and in the apparent absence of any typical centriolar structure connected with the contractile elements of the flagellum (Fawcett and Ito 1965), it has been concluded that the sperms lack a basal body and that their flagella can beat even in the absence of a kinetic centre (Fawcett and Phillips 1969a, Fawcett 1972). However, in an excellent study of the fine structural organization of the various components of the neck of mature spermatozoa of rabbit, monkey, and man, Zamboni et al. (1971) have observed that the striated columns of connecting piece implant on the proximal centriole, that the distal centriole does not disappear but persists albeit in a modified form, and that the central tubules of axoneme of the flagellum terminate at the lower vault of the proximal centriole. Of the two centrioles, the most plausible candidate for the role of the basal body of the flagellum and centre of sperm motility is the proximal centriole, as supported by the apparent continuity of this centriole with all the contractile elements of the flagellum, and indirectly by the consideration that the distal centriole cannot be a basal body in that its lumen is traversed throughout by a pair of the tubules. The orientation of the proximal centriole at an angle to the flagellum, a unique situation since basal bodies are normally oriented on the same axes of cilia and flagella, may be related to the particular type of motility of the spermatozoon.

C. Other Elements

The other elements of the neck are two or three mitochondria (Fawcett and Ito 1965, Zamboni et al. 1971). These generally establish close relationship with either end of the proximal centriole by wrapping around the lateral surface of the latter. These mitochondria are continuous with the uppermost mitochondria of the mid-piece helix. No mitochondria are seen in the upper part of the neck region in the goat spermatozoon, but there is present membranous scroll material (Bae and Kim 1981). In the spermatozoa of the bush-baby there occurs an intricate system of membranes which extend caudally from the head-neck junction to surround the anterior part of the mitochondrial sheath (Bedford 1967). Bedford (1967) has stated that is seems unlikely that the whole complex shroud of membranes originates from extensions of the nuclear membrane. The membrane complex has not been seen in the neck of spermatozoa in other primates, including man.

Chapter IX
Cytoplasmic Droplet

A. Structure

Retzius (1909) described the cytoplasmic droplet in the spermatozoa of various mammals including primates. The cytoplasmic droplet also forms the conspicuous feature of spermatozoa in the marmoset, chimpanzee, and buffalo (Fig. 17) (Guraya 1965). The cytoplasmic droplet in the spermatozoa of marmoset is a pear-shaped structure which surrounds the neck region from the base of the head to the anterior part of the mitochondrial sheath (S. S. Guraya, unpublished ob-

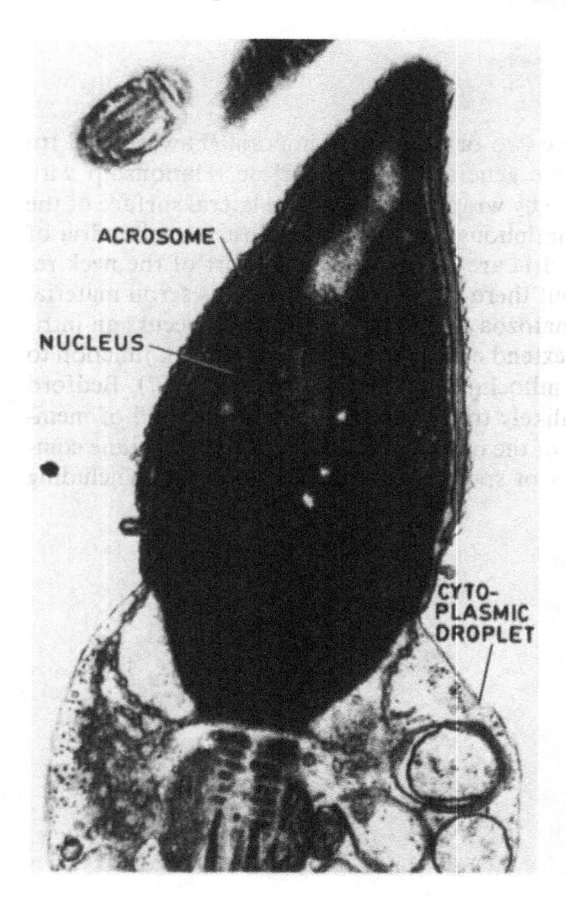

Fig. 64. Head and part of the cytoplasmic droplet of a human spermatozoon. The coarse aggregation of the chromatin is reminiscent of that of spermatid nuclei. At the point where the nucleus becomes surrounded by the droplet, the two sheets of the nuclear membrane detach from the chromatin and form a complex system of folds provided with numerous pores and extending deeply into the droplet cytoplasm. The folded membranes are probably involved in nucleo-cytoplasmic exchanges (From Zamboni et al. 1971)

servations). In the chimpanzee, it is an elongated structure which completely surrounds the mid-piece (Fig. 17). Actually, it forms a tube-like structure around the mid-piece. The droplet is limited by the cell membrane covering the whole spermatozoon (Fig. 64) (Bedford 1967, Zamboni et al. 1971). The contents of the droplet border on the mitochondrial sheath and the centriolar structures of the neck region (Fig. 64). Bloom and Nicander (1961) have stated that the remarkably large size of rat droplets has its counterpart in the very long mid-piece of rat spermatozoa. The cytoplasmic droplet of human spermatozoa is also of large size (Zamboni et al. 1971, Holstein and Roosen-Runge 1981).

The cytoplasmic droplets in the marmoset and chimpanzee show the presence of some deeply sudanophilic lipid granules (Fig. 17) which consist of phospholipids and some lipoproteins. The cytoplasm of the droplets also shows some diffuse lipoproteins and very little RNA. Similar cytoplasmic components have also been described in the droplet of spermatozoa in ruminants (Guraya 1963, 1965).

The ultrastructure of cytoplasmic droplet in the spermatozoa of mammals including primates has been either described or simply illustrated in electron micrographs (Nicander and Bane 1962, Kojima and Tshikawa 1963, Fawcett and Ito 1965, Wimsatt et al. 1966, Bedford 1967, Zamboni et al. 1971, Dott and Dingle 1968, Arthur 1968, de Kretser 1969, Bedford and Nicander 1971, Hruban et al. 1971, Temple-Smith and Bedford 1976, Harding et al. 1979, Holstein and Roosen-Runge 1981). A close examination of various electron micrographs given in these papers shows that the cytoplasmic droplets of spermatozoa in different mammalian species contain variable amounts of membranous structures (fine, curved tubules or lamellae, arch-like double membranes, and many small and some large vesicles of endoplasmic reticulum) and a ground cytoplasm having different electron density (Figs. 64 and 65). Even the distribution and form of these elements vary greatly in the cytoplasmic droplets of different mammalian species investigated with electron microscope (Figs. 64 and 65). These are generally restricted to the peripheral regions of the cytoplasmic droplets.

The cytoplasmic droplets of human spermatozoa show a very complex structural organization (Zamboni et al. 1971), suggesting the presence of some diverse cytoplasmic structures in it (Fig. 64). The cytoplasmic droplet in the spermatozoa of the flying squirrel shows a complex and organized structure of membrane system (Hruban et al. 1971), as also observed for the possum (Fig. 65) (Temple-Smith and Bedford 1976), and other marsupials (Harding et al. 1979).

The phospholipids and lipoproteins demonstrated histochemically in the cytoplasmic droplet (Guraya 1963, 1965) may be constituting the various ultrastructural cytoplasmic components (Figs. 64 and 65). The various elements seen in the cytoplasmic droplets represent the residual cytoplasmic components of the spermatid, which have not been sloughed off. The ultrastructural smooth-surfaced tubular and vesicular elements are believed to be derivatives of the endoplasmic reticulum or possibly of the Golgi complex (Bloom and Nicander 1961, Fawcett and Ito 1965). An other structure to be considered as a possible source of the tubular elements in the droplet is the manchette characterized by fine tubules (as already discussed) which disappear just before the development of the mitochondrial sheath.

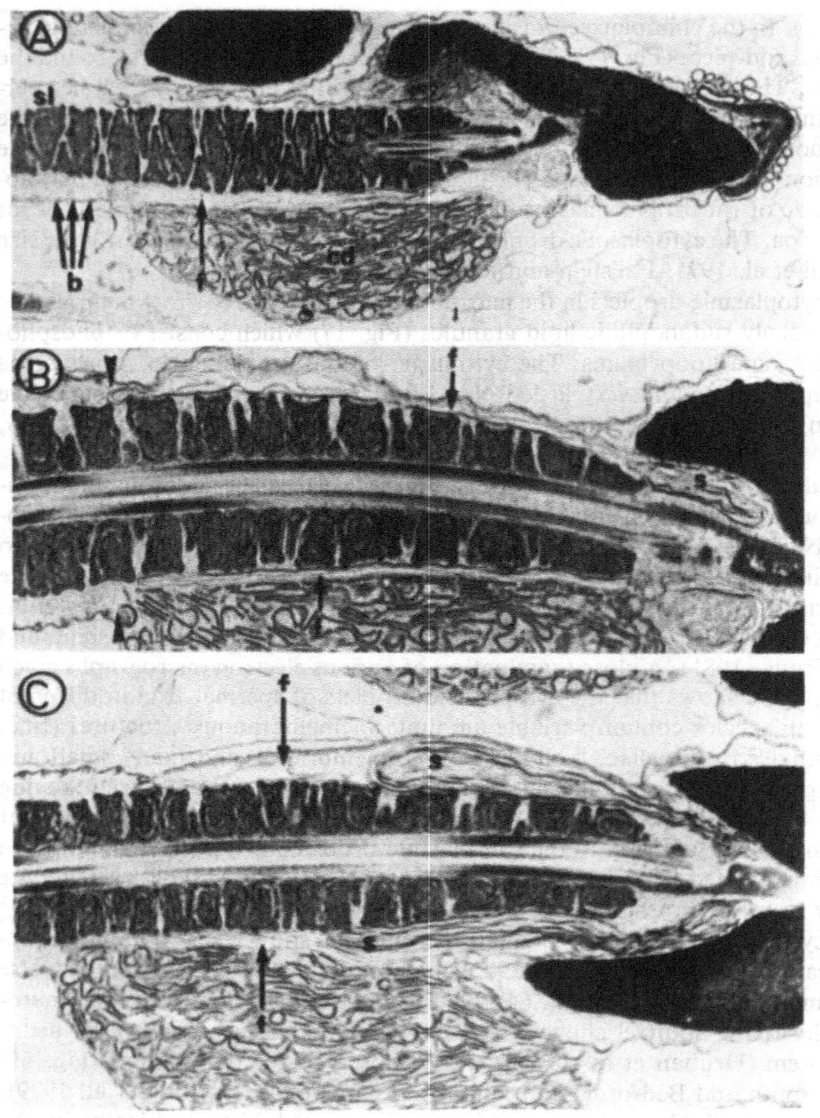

Fig. 65. A Longitudinal section of a spermatozoon from region 2 of the epididymis. The mid-piece already shows the beginning of its differentiation into anterior and posterior segments. No organized membrane lamellae have yet appeared in the anterior segment of the mid-piece, but a fissure (*f*), which extends to the distal edge of the droplet, has formed around it. The thick peri-mitochondrial cytoplasmic sleeve (*sl*) has appeared in the posterior segment and, coincidentally, the outer layer of cytoplasm has disappeared from the droplet (*cd*), which now consists only of an eccentrically placed mass of flattened vesicles. Fibrous bands (*b*) have formed beneath the plasma membrane of the posterior segment, but the flask-shaped membrane invaginations have not yet developed between them. **B** Spermatozoon from region 3 of the epididymis showing, at higher magnification, a later stage in segmentation of the mid-piece. The membrane-lined fissure (*f*) extends between the droplet and the mitochondria to the interface (*arrow heads*) between the anterior and posterior segments. Scrolls of membrane(s) have begun to appear around the neck. **C** Spermatozoon from region 4 of the epididymis showing more profuse stacks of parallel membrane(s) infiltrating the fissure (*f*) in the anterior segment of the mid-piece (From Temple-Smith and Bedford 1976)

B. Chemistry

Histochemical tests have demonstrated the presence of phospholipids, lipoproteins, RNA, etc. in the cytoplasmic droplets (Guraya 1963, 1965), which show various enzymes, including β-N-acetyl-glucosaminidase, β-glucuronidase, acid proteinase, arylsulphatase, ribonuclease, deoxyribonuclease, 5′-nucleotidase, hyaluronidase, several glycosidases, esterases, alkaline and acid phosphatases, aspartate aminotransferase (glutamate-oxaloacetate transaminase), antigotensin-converting enzyme, choline esterase, hexokinase, glucose phosphate isomerase, aldolase, lactate dehydrogenase, sorbitol dehydrogenase, cAMP-dependent protein kinase, and calmodulin binding proteins (Dott and Dingle 1968, R. A. P. Harrison and White 1969, Garbers et al. 1970, Bavdek and Glover 1970, Moniem and Glover 1972a, b, Dott 1973, Chakraborty and Nelson 1976, Weitze 1976, Farooqui and Srivastava 1979b, Mann and Lutwak-Mann 1981, Tash and Means 1983, Vanha-Perttula et al. 1985). The nonlysosomal enzymes described are possibly incorporated into the droplets at the time of generation and sequestration from the Golgi apparatus in the testis. On the other hand, it is conceivable that some of them are taken up by the droplet from other sperm organelles, especially the mid-piece, during the course of the sperm-ripening process in the epididymis, when droplet changes position from the neck region to the posterior part of the mid-piece of mammalian spermatozoa, or simply during the isolation procedure of the droplets from ejaculated semen.

C. Function

Now the question arises: what is the possible physiological significance of the cytoplasmic droplet in relation to the sperm metabolism. Very divergent views have been expressed in this regard. Various workers believe that the cytoplasmic droplet is the last remnant of unsequestered cytoplasm and is related to the residual body, and thus not performing any specific function in sperm physiology (Zamboni et al. 1971, Phillips 1975a). With maturation of the epididymal sperm this element, like the residual body, is lost, suggesting the cellular immaturity of spermatozoa (Bedford 1975, 1979, Phillips 1975a, Kazuyoshi and Kubota 1980, Mann and Lutwak-Mann 1981). There is at present no explanation of the functional significance, if any, of the movement and loss of the cytoplasmic droplet. But alterations occur in the fine structure, shape, and position of the cytoplasmic droplets during the passage of spermatozoa through the epididymis. Regional alterations in the ionic composition of the epididymal fluids (Howards et al. 1979) are believed to be responsible for the distal movement and ultimate loss of the droplet (Bedford 1975, 1979). The immaturity of spermatozoa is also reflected in certain chemical and metabolic properties (Bedford 1975), best illustrated by the behaviour of sperm phospholipids, as will be discussed in Chap. X.

Most of the human ejaculated sperms has a voluminous droplet located around the head of mid-piece (Fig. 64). The presence of droplets in a significant percentage of human spermatozoa appears to be indicative of pathology of sperm maturation and sometimes of infertility (Bedford 1975). Moniem and Glover 1972a, b) have stated that the alkaline phosphatase in spermatozoa, concentrated as it is in the cytoplasmic droplet or in the mid-piece, plays a part in the function of the cell by dephosphorylation and transport of phosphate groups between epididymal plasma and the sperm-cell. It is possible that some of the phosphate groups might be derived from the phospholipids demonstrated histochemically in the cytoplasmic droplet (Guraya 1963, 1965). Guraya (1963, 1965) has suggested that the droplet may have a nutritive role in the economy of the spermatozoon; it may be the source of a metabolizable endogenous substrate (phospholipids).

Clegg et al. (1975) have studied the phospholipid composition of porcine semen cytoplasmic droplets, which show large amounts of membrane within them, and this is reflected in a high content of 28 μg lipid phosphorus per 10^9 droplets. Phosphatidylinositol constituted 6% of the droplet phospholipids. Roberts et al. (1976) have developed a method for the isolation and characterization of the cytoplasmic droplet in the rat, which is believed to be important in sperm maturation in the epididymis possibly through its role in inositol synthesis and metabolism (Eisenberg and Bolden 1964). The suggestion that the cytoplasmic droplet plays a significant role in the physiology of epididymal spermatozoa is strongly supported by the studies of Van Rensburg et al. (1966), who have found that the live sperm from different parts of the caput epididymis showed considerable activity by violent lashing of the tails in all spermatozoa with a cytoplasmic droplet. This activity ceases as soon as the droplet was discharged, suggesting that it is of great significance in sperm activity. Mann (1975) has suggested that lysosomal enzymes of the droplet may perhaps prepare the spermatozoon for the final stage of its maturation. Further studies are needed to determine the functional significance of the cytoplasm droplet. The cytoplasmic droplets are rich in cAMP-dependent protein kinase, phosphoproteins, calmodulin and calmodulin-binding proteins, some of which are present on the external surface of the cytoplasmic droplets. Therefore, Tash and Means (1983) have suggested that the functions of these components in sperm function must be studied in ejaculated sperm freed from cytoplasmic droplet contamination.

Chapter X
Tail

The motor apparatus of the spermatozoon tail is the axoneme or axial filament complex which consists of the usual central pair or axial fibrils (or microtubules) surrounded by an inner row of nine evenly spaced doublet microtubules, each with two rows of arms that project towards the adjacent doublet tubule, one row of radial spokes that radiate inwards towards the central pair of microtubules, and an outer ring of nine coarse longitudinal fibres (Figs. 45, 57 and 61) (Fawcett 1975b, Harding et al. 1979, Linck 1979, Satir 1979, Amelar et al. 1980, Bae and Kim 1981, Holstein and Roosen-Runge 1981). Actually, all the structural components of the flagellum, which include the connecting piece, 9+2 axoneme, fibrous sheath, and outer dense fibres, are structurally interlocked into one functional unit. There occur some variations in the size of coarse filaments (usually nos. 1, 5, 9). A central sheath, made up of projections has been described surrounding the two central tubules (Fig. 66). It is connected by nine spokes, radial links to the nine doublets, which are also connected to each other by the interdoublet links (Fig. 66).

Relatively very little information is avialable about the comparative ultrastructure, chemistry and interrelationships of various components of tail flagellum in the spermatozoa of different mammalian species; however, some studies in this regard have been carried out on the spermatozoa of rat (Linck 1979). Therefore, comparative studies will be very useful in revealing the nature of axial fibrils and interrelations (or connections) existing between them; especially regarding the dynein and spokes. The details of ultrastructural characteristics and macromolecular organization of eukaryotic cilia and flagella in relation to

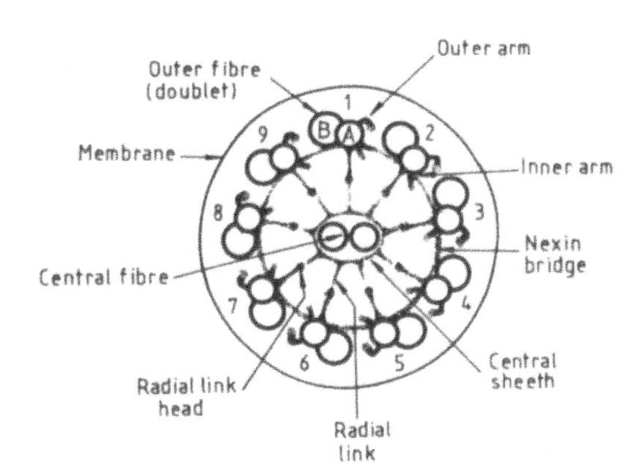

Fig. 66. Current interpretation of the organization of the axoneme of cilia and flagella (Redrawn from Fawcett 1975b)

their motility have been discussed in recent reviews (Baccetti and Afzelius 1976, Woolley 1979, Satir 1979, Gibbons 1979, Linck 1979, Amelar et al. 1980), and thus will be omitted here. Only important features in this regard will be described in relation to the tail of mammalian sperm.

A. Axoneme

The occurrence of the 9+2 pattern in cilia and flagella, with only rare exceptions, forms their universal feature in plants and animals (Linck 1979, Amelar et al. 1980). Recent investigations have increased our knowledge about the structural and chemical nature of the microtubular components of the axoneme, their mode of assembly, their energy source and the manner in which these interact to produce bending (Olson and Linck 1977, Woolley 1979, Satir 1974, 1979, Gibbons 1979, Linck 1979, Amelar et al. 1980). Most of these previous studies have been carried out on the cilia and flagella of spermatozoa in invertebrates. But the basic mechanism are similar in all motile cell processes including those of mammalian spermatozoa, which also possess a 9+2 axoneme. Fawcett and Phillips (1970) have clearly shown the origin of the satellite fibrils from the sides of outer dense fibres, which have a clearly defined cortical layer 20 to 30 nm. Olson and Linck (1977) have observed that in the dissociated flagella of rat spermatozoa each axonemal doublet microtubule remains adjoined to its associated outer dense fibre, and the central pair remains as an intact complex (Linck 1979).

1. Peripheral Fibres (or Doublet Microtubules)

The nine peripheral doublets of a typical axoneme are equidistantly spaced (Figs. 45, 57 and 66). According to Warner and Satir (1973), and Warner (1974), who measured cilia and sperm flagella, these are separated from one another by a distance of 18 nm occupied by the dynein arms and interdoublet links (Fig. 66). The whole axoneme has a diameter of 0.25 μm. The two central tubules are separated from each other by 9 nm. A number can be assigned to each doublet microtubule. Number 1 lies in the plane perpendicular to that of the central tubules; increasing numbers are in the direction of the arms. When the axoneme is viewed from base to tip, the arms project clockwise (Baccetti and Afzelius 1976, Linck 1979). Flagella extracted in the presence of ATP show a specific pattern of axonemal disintegration: doublet microtubules-dense fibres numbered 4, 5, 6 and 7 are extruded in a proximal direction from the fibrous sheath, while doublets numbered 8, 9, 1, 2 and 3 remain associated as a half cylinder together with the central pair complex in the fibrous sheath. Of the extruded fibres, doublets numbered 5 and 6 always remain adjoined to each other, suggesting the possibility for a structural linkage similar to that found in typical 9+2 flagella. It is commonly called the 5–6 bridge. Linck (1979) has recently discussed the structural and chemical composition of the "arms" constituting the 5–6 bridge, which cannot be assumed to be the same as the other eight sets of dynein arms. The 5–6 bridge is apparently not always well preserved and has not been observed in many species.

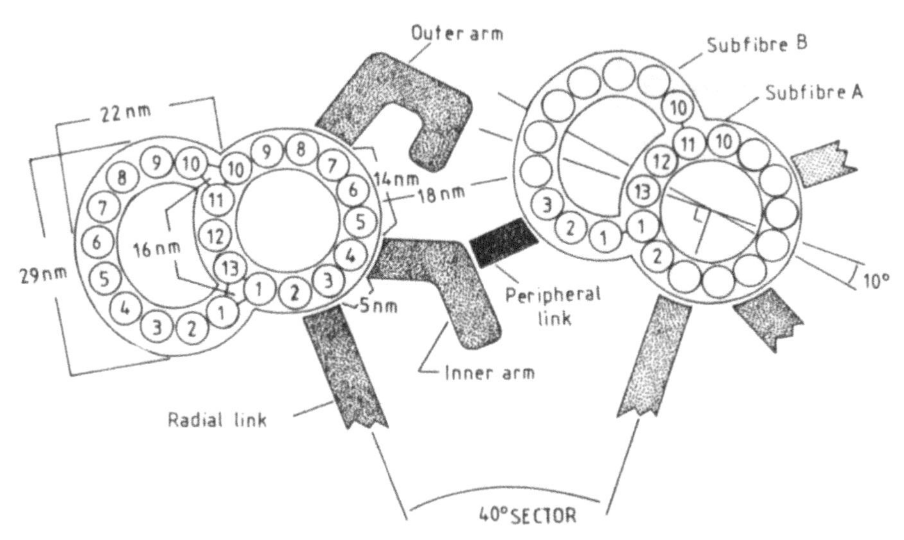

Fig. 67. Protofilament organization in the doublet microtubules. Thirteen protofibrils make up the wall of *subfibre A,* and ten are found in the wall of the subfibre B. *Subfibre B* shares a 3 to 5 protofibril segment of the wall subfibre A. A proposed scheme of numbering of the protofilaments is presented. It is not yet clear whether protofilaments 1 and 10 of subfibre B are bounded to 1 or 2 protofilaments on the wall of A. Alternative possibilities are illustrated on the two doublets shown (Redrawn from Fawcett 1975b)

The details of the construction of the doublet and singlet microtubules of the axoneme are now well elucidated. Each axonemal doublet consists of two microtubules or subfibres, called the A and B subfibres (Fig. 66). The A subfibre is a complete microtubule which appears circular in cross-section (Fig. 66) and is about 25 nm in diameter, with a wall 5 nm thick. The B subfibre is C-shaped in cross-section with its ends attached to the wall of the A subfibre (Fig. 66). It is, therefore, an incomplete slightly smaller (20 to 30 nm) cylinder. In the axonemal complex of mink sperm the A subfibre is larger than the central fibre, while the B subfibre is the smaller (Kim et al. 1979). Both subfibres of each doublet microtubule are made up of parallel, monolayered protofilaments (Linck 1979). The cylindrical wall of the A subfibre consists of 13 protofilaments 4 nm in diameter. Each protofilament is made up of 8 nm protein dimers designated tubulin (Figs. 67 and 68). These dimers are placed end-to-end (Amos and Klug 1974, Linck 1979). It is believed that the dimers of adjacent protofilaments are in staggered array. The B subfibre, which is shorter, also consists of dimeric subunits of tubulin in about 10 to 11 protofilaments (Fig. 67) (Warner 1972, Warner and Satir 1973, Tilney et al. 1973); Linck (1979) demonstrated 11 protofilaments for the B subfibre. It sits alongside the A subfibre like an intersecting, incomplete cylinder. Previously, different protofilament numbers (between 10 and 13) have been mentioned for different kinds of axonemes (Baccetti and Afzelius 1976), but the numbers 13 and 10 (Fig. 67) accepted by Warner (1974) and by Amos and Klug (1974) appear to be definitive and universal for doublet microtubules. Four or five protofilaments are shared by the A and B subfibres.

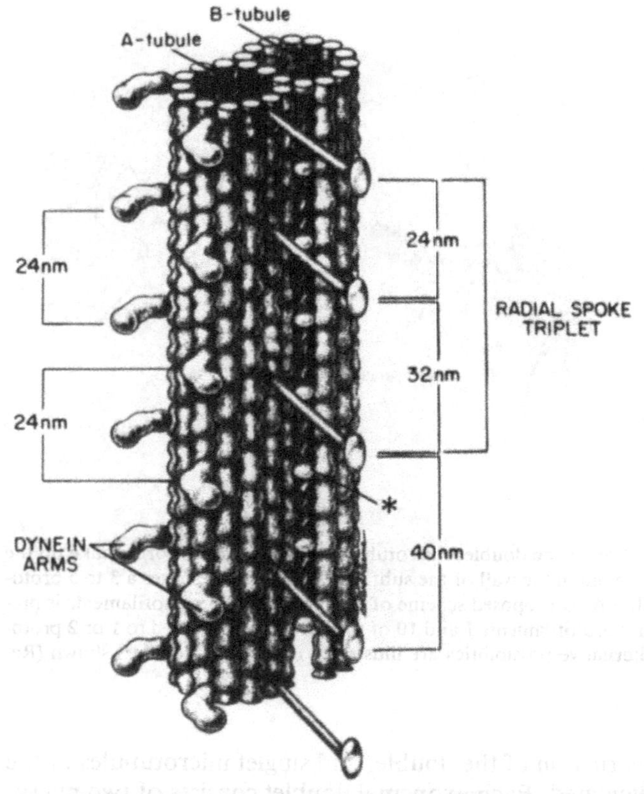

Fig. 68. A three-dimensional model of the doublet microtubule complex based on data from electron microscope images, optical diffraction analysis and computer reconstruction of flagellar axonemes (Amos et al. 1976, Olson and Linck 1977). The doublet tubule is oriented with the proximal or basal end down. Certain structural features of the model have been established: (1) The two rows of dynein arms are arranged each with axial repeats of 24 nm, and the arms are slightly staggered such that the outer arms are approximately 9 nm nearer the flagellar tip relative to the inner arms. The dynein arms appear to be composed of three morphological subunits as demonstrated by Warner and Mitchell (1978). (2) Members of the radial spoke triplets have alternate spacings of 24 and 32 nm, with 40 nm separating adjacent triplets. (3) The polarity of the spoke triplets is that determined by Olson and Linck (1977) and Warner and Satir (1974), i.e. the 32-nm spacing is proximal to the 24-nm spacing. Other details of the model are shown for the sake of illustration only; the exact attachment sites of the arms and spokes on the surface lattice of the tubule are not known, nor is the axial position (i.e. phase relationship) of the spokes relative to the inner dynein arms; periodic subunits (*asterisk*) are shown connecting the inner wall of the B tubule to the A tubule (Linck 1976) (From Linck 1979)

Two proteins constitute the tubulins (Mohri 1968) and are generally referred to as α and β tubulin. The protofilaments of the A subfibre, when negatively stained, are seen to consist of a linear chain 4-nm subunits which form the smallest protofilament component described thus far. The 4-nm subunit represents the globular (Amos and Klug 1974), monomeric form of the microtubular proteins (Warner 1974, Amos and Klug 1974). Tubulin and other microtubule-associated proteins of high molecular weight, MAP_1 and MAP_2, have been localized immunocytochemically in the human spermatozoa (Maunoury and Hill (1980).

Tubulin is present in the axonemal region, the centriolar adjunct, and the equatorial part of the acrosome; in contrast, MAP_1 and MAP_2 are present in the postnuclear region and in the fibrous sheath of the principal piece of the flagellum.

The two fractions of tubulin, α and β, can be distinguished electrophoretically. Dimers of doublet microtubules are made up of one subunit of each. Some studies have indicated additional heterogeneity of tubulins as evidenced from the variable solubility properties of the α- and β-subunits (Behnke and Forer 1967, Stephens 1974, Linck 1973). Heating doublets at 37 °C causes selective solubilization of the subfibres. Doublet microtubule preparations have been shown to contain at least nine minor protein components accounting for 25% – 35% of the total protein (Linck and Amos 1974). Some of these proteins may be concerned with the attachment of the B subfibre of the doublet to the A subfibre, or the attachment of the arms, radial spokes, or nexin (interdoublet) links, as discussed in detail by Linck (1979). Baccetti and Afzelius (1976) have also discussed in detail the comparative molecular composition of the A and B subfibres. Their α and β tubulins are different, but related proteins. Each one has been largely unchanged in the course of evolution; however, there do exist differences in their molecular weights (Witman et al. 1972a, b). According to Bryan (1974), the two tubulins have identical molecular weights. These also possess identical antigenic properties (Fulton et al. 1971). Morisawa and Mohri (1972, 1974) have recently observed that sea-urchin sperm microtubules bind zinc and iron.

The A subfibre in cross section has been observed to form the site of attachment for the two diverging arms that project towards the adjacent doublet (Figs. 66, 67 and 68). They show a clockwise directionality when viewed in the axoneme from base to tip. Both the arms differ in their length and morphology (Figs. 66, 67 and 68) as well as in chemistry (Gibbons 1977). The arms are made up mainly of dynein, which is a protein with ATPase activity (Gibbons 1979, Satir 1979, Amelar et al. 1980, Nelson 1985); the chemistry of the dyneins has been discussed by Gibbons (1977) in detail. The ATPase protein dynein consists of two components which sediment at 14S and 30S, respectively. The former is composed of globular particles with 8×15 nm with a molecular weight of 600,000. The latter is a linear polymer of the former, varying from 440 to 550 nm in length. Both fractions have similar enzymatic properties. Warner and co-workers (Warner and Mitchell 1978, Warner et al. 1977) have recently shown that in negatively stained preparations the dynein arm appears to be composed of three morphological subunits with a molecular weight of approximately 350,000. These findings are also in agreement with biochemical data of Gibbons et al. (1976). Johnson (1984) has used scanning transmission electron microscopy in the study of dynein arms at the molecular level.

The ATPase activity of dynein is associated with the flagellar movement. ATPase activity has been demonstrated around all the axonemal elements of the sperm tail of bull and invertebrates (Baccetti 1975). Adenosine triphosphatase and adenosine diphosphatase localization deposits have been observed within axial and peripheral filaments of the axoneme of the mature spermatid tail (Barham et al. 1976). Histochemical ATPase stainings performed recently by Baccetti et al. (1981) have indicated that this enzyme is localized mainly in the

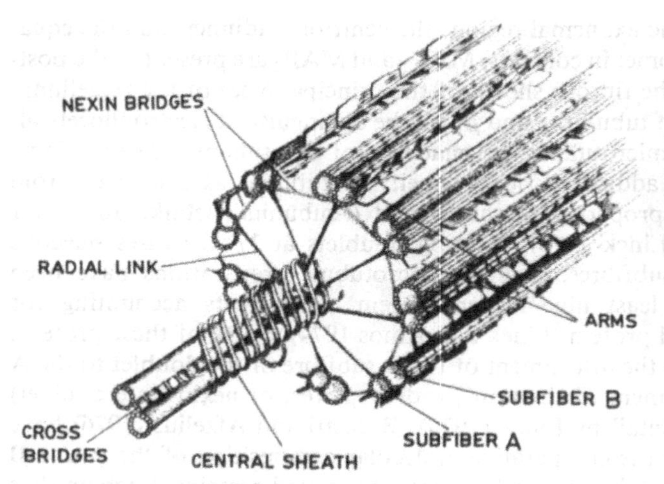

Fig. 69. Highly schematic three dimensional reconstruction of the axoneme and its associated structures. There is no basis for depicting the arms as rectangular, it is simply intended to indicate that they are periodic and not continuous along the doublet. The details of the attachments of the nexin bridges and radial links remain to be worked out (Redrawn from Fawcett 1975b)

doublet arms, and to a lesser extent in the central structures of tail in human sperm. It will be interesting to mention here that in certain men whose semen showed good sperm density but completely lacked motility, the dynein arms of the axoneme were missing (Afzelius et al. 1975, Pedersen and Rebbe 1975, Nelson 1985). Baccetti et al. (1981) have made a study of human dynein and sperm pathology. Such results provide good illustrative examples of the newer investigative approach to problems relating to male fertility and of the advantages to be gained from close integration of morphological and biochemical methods of sperm study (Nelson 1985). Williams et al. (1984) have reported ultrastructural sperm tail defects associated with sperm immotility.

Two slender nexin links are also attached to each subfibre A. These connect it to the adjacent doublets (Stephens 1974). Two types of links, namely γ links and peripheral links, have been described in cilia and flagella (Linck 1979). The protein of the links is called nexin and is 2% of the total axoneme protein with a molecular weight of 150,000 – 165,000 (Stephens 1974). A radial spoke attached to each subfibre A joins it to a helical sheath around the central pair of microtubules (Fig. 69). The arms of the doublets are seen to space at regular intervals of 24 nm along the subfibre A (Linck and Amos 1974).

2. *Radial Spokes*

The radial spokes, as investigated by negative staining of dissociated axonemes, are placed in groups or triplets (Fig. 69) (Warner and Satir 1974, Linck 1979); Olson and Linck (1977) have also observed that the radial spokes in the axoneme of rat sperm are arranged along the doublet tubule in sets of three; the spoke sets show a centre-to-centre spacing of 96 nm. Within a set, two spokes (Nos. 1 and 2) are spaced 32 nm centre-to-centre, while spokes 2 and 3 are spaced 24 nm centre-to-centre. The polarity of the spoke triplets is such that the 32-nm spoke space is directed towards the anterior of the flagellum. An additional axonemal structure associated with the doublet tubule is located between spokes 1 and 2 of rat spermatozoa; this globular component lies close to the tubule surface and successive globules are spaced 96 nm centre-to-centre. The diameter of the globule is approximately 17 nm. The globule appears most likely to represent a hitherto unsolved structural component of the radial spoke triplet; however, whether it is specific to rat spermatozoa or is a universal structure of the flagellar axoneme remains to be determined. Fine filamentous linkages extending between adjacent outer dense fibre-doublet tubule complexes are found to approach and join the doublet tubule at the level of the globule or spoke number and may be phosphoproteins (Tash and Means 1983); successive filamentous linkages are spaced at 96-nm intervals. Olson and Linck (1977), and Linck (1979) have discussed the relationship of these components to the circumferential nexins described in other species. A single nexin fibre may run continuously from one doublet to the next with structural contacts to both A and B tubules of each doublet tubule (Fig. 69). Nexin fibres are believed to perform two functions in the axoneme: (1) a purely structural role to maintain the ninefold symmetric cylinder of doublet microtubules in the "resting" state of the axoneme and (2) an active role as elastic elements to regulate the amount of shear displacement between adjacent doublet tubules during active sliding. Our knowledge is still meagre about the function of radial spokes.

3. *Central Tubules*

The central tubules in thin section show a circular shape and a diameter of 20 to 30 nm (Figs. 45 and 66). The tubule wall thickness varies between 5.5 and 8.5 nm (Warner 1974), depending on the staining used. Each number of the central pair of microtubules in the axoneme with 9+2 consists of 13 parallel protofilaments, as described for the subunit A of the doublets (Amelar et al. 1980). The central tubule's lumen usually appears translucent but may show some structural components. The central tubules are composed of tubulin, like the doublets. But there is evidence of some difference (Stephens 1974). The two central microtubules are attached to one another along their length by regularly arranged bridges 13.5 nm apart (Pedersen 1970b, 1972b).

4. *Central Sheath*

Many studies have revealed a central sheath associated with the central pair microtubules (Figs. 66 and 69), and this sheath has been variously described as a helical filament wrapped about the central pair (Fig. 69) or as sets (or rows) of regularly spaced projections arising from the central tubules (Warner 1974, Olson and Linck 1977, Linck 1979). Although the interpretation of the central sheath structure has varied widely, one consistent finding has been the description of a regular periodicity of about 16 nm asscociated with sheath. The projections measure 18 nm long. With certain exceptions, each microtubule of the central pair possesses two rows of projections which subtend at an angle of 120°. The tips of the projections in each row are in close apposition to the tips of the projections emanating from neighbouring tubule; however, whether the tips of one row of projections are in morphological contact with the projections on the adjacent tubule is not clear. This description is also characteristic of rat spermatozoa, as reported by Olson and Linck (1977) who, using negatively stained preparations of central pair microtubule complexes from rat sperm, have observed that these retain periodic "sheath" components. Optical transforms of negatively stained central pairs reveal a regular set of layer lines indexing on a fundamental 32-nm repeat. This 32-nm reflection arises from only one member of the central pair. In negatively stained images and optically filtered reconstructions, the 32-nm periodicity originates from a set of "barb-shaped" structures which arise from the lateral edge of one central pair microtubule. The barb-shaped projections are pointed towards the proximal end of the flagellum and are attached along the tubule nearly in register with alternating "bridges" and "projections" both of which are spaced axially at 16 nm along the central pair complex. Filtering of images where the central pair retains its structural integrity with a doublet microtubule has provided evidence of structural interactions between the 32-nm periodic sheath components and the radial spokes.

5. Structural and Chemical Interactions Between Components of the Axoneme

The principal force-generating mechanism of the 9 + 2 axoneme is believed to involve cyclic interactions between dynein arms and the adjacent doublet microtubules, resulting in relative sliding movements between the nine outer doublet microtubules (Nelson 1975, 1985, Satir 1979, Woolley 1979, Gibbons 1979, Linck 1979, Amelar et al. 1980). But the mechanisms responsible for converting interdoublet sliding into bending movements as well as those responsible for determining polarity need to be determined more precisely. However, results of recent studies as integrated by Woolley (1979), Satir (1979), and Gibbons (1979) have thrown some further light on these mechanisms at the molecular level. In earlier ultrastructural investigations it was shown that the ciliary or flagellar beat is perpendicular to the coaxial plane of the central pair microtubules, involving the central pair in control of beat polarity (Olson and Linck 1977). More recent studies in this regard have further revealed that the structural interactions between the radial spokes of the doublet microtubules and the sheath components of the central pair may be participating in the interconversion of the relative sliding movements of doublet microtubules into bending couples (Warner and Satir 1974, Woolley 1979, Satir 1979, Gibbons 1979). The detergent-extracted sperm models have produced direct evidence for this view (Summers and Gibbons 1971, 1973, Gibbons 1979). Trypsin treatment in these models destroys the linkages between the doublets and central pair tubules, as well as the circumferential nexin linkages between adjacent doublet tubules. During ATP-dependent reaction, flagella show no bend formation and have only doublet microtubule sliding. Olson and Linck (1977) have observed that the tips of the barbs are in close contact with the radial spoke heads, suggesting that these may interact with the spokes in some manner. Warner and Satir (1974) have shown sliding to take place between the central sheath projections and the outer doublet radial spokes. They have suggested that this spoke-sheath interaction may convert outer doublet sliding into bending waves. Olson and Linck (1977) have recently suggested that a second kind of interaction, i.e. between radial spokes and the central pair barbs, may also influence the propagation of bending waves and/or the directionality of beat, as also discussed by Linck (1979). The nature of biochemical interactions occurring between the radial spoke and the central sheath should be determined in future studies to support these suggestions.

Spermatozoa of a number of species are known to contain a myosin like protein, an actin like protein, and even a tropomyosin like protein (Tilney 1975, Talbot and Kleve 1978, Clarke and Yanagimachi 1978, Campanella et al. 1979, Nelson 1985, Welch and O'Rand 1985b). The functional and spatial relationships of these proteins to each other and to flagellar structural units remain to be clarified. However, immunocytochemical evidence has suggested that the outer fibres are composed of proteins with actin and myosin-like antigenic properties (Plowmann and Nelson 1962, Nelson 1967, 1975, 1985) but there is, as yet, no more compelling evidence for their contractility. However, Young and Nelson (1968) have suggested that sperm tail actin may be involved in flagellar beat generation. An actin-like material has been obtained from human spermatozoa by Clarke et al. (1982), who concur with the suggestion of Young and Nelson (1968). Determination of the function of sperm actin needs further studies (Nelson 1985). The finding that the outer fibres arise as outgrowths from the wall of the axonemal doublets (Fawcett and Phillips 1970, Nelson 1975) provides additional circumstantial evidence that these are of similar composition and may also be composed of contractile protein. Functional significance of spermosin and flactin (flagellar myosin and actin) in relation to sperm motility will be discussed also in Chap. XII. Most of the phosphoproteins are localized in the sperm flagellum (Chulavatnatol et al. 1982). But it needs to be determined whether any of these phosphoproteins are subunits of dyein. However, Tash and Means (1983) have suggested that both sperm dyein ATPase and the membrane ($Ca^{2+} + Mg^{2+}$) ATPase are phosphoproteins, as it appears that other Ca^{2+}-calmodulin-dependent enzymes can be phosphorylated in a cAMP-dependent manner at or near the calmodulin binding site. The cAMP or Ca^{2+}-dependence of these phosphorylations needs to be determined precisely.

The axonemal complex described above is embedded in a matrix constituting half of the total flagellar protein, and in several cases is high in glycogen content (Baccetti and Afzelius 1976). In the sperm, LDH usually follows the distribution of glycogen. Among the enzyme activities of the matrix special attention has been focussed on the protein kinases of bovine spermatozoa. A cyclic AMP-dependent protein kinase activity has been reported in sperm which is 100-fold higher than in other bovine tissues. This protein kinase forms one of the major components of the sperm cytosol (Hoskins et al. 1972). It occurs in three forms which show different molecular weigths of 120,000, 78,000, and 56,000, respectively. Each of these is composed of a catalytic subunit of 30,000 MW and of cyclic AMP-binding proteins with molecular weights of 78,000, 35,000, 40,000, and 17,000 to 18,000, respectively. (Garbers et al. 1973a). These enzymes are believed to be involved in the regulation of the metabolism and motility, which are affected by cyclic AMP. Moreover, in vitro capacitation causes a modification of sperm adenyl cyclases (B. E. Morton and Albagli 1973). The caffeine, a cyclic AMP phosphodiesterase inhibitor, stimulates sperm motility and influences sperm metabolism (Garbers et al. 1971b), as will be discussed in detail in Chapt. XII. The physiological protein substrate of sperm cytosol protein kinases needs to be determined.

B. Mid-piece and Sperm Metabolism

Anatomically, the mitochondrial sheath and the outer ring of coarse fibres characterize the mammalian sperm mid-piece (Fawcett 1975b, Phillips 1975a). It is that part of the flagellum which lies between the neck and annulus (Figs. 55 and 61) and forms the most important site for various metabolic activities of the sperm. Its mitochondrial sheath is believed to be the source of energy for sperm motility. The size and length of mid-piece vary among different species of mammals. The mid-piece is of medium length in the mink sperm, when compared with other mammalian species (Kim et al. 1979).

1. Dense Fibres

a) Morphology

The axoneme of the mammalian sperm is surrounded by nine outer dense fibres which are also called the coarse or accessory fibres (Figs. 45A, 57, 61 and 70). These run for the major part of its length, thus constituting a 9+9+2 cross-sectional pattern (Figs. 45A, 57, 61 and 70) (Baccetti 1982). Both the coarse fibres and the bundles of satellite fibrils are prominent in the proximal part of the mid-piece of the European common shrew (Plöen et al. 1979).

The dense fibres No. 1, 5, and 6 in the spermatozoa of mammals are generally larger in size than the others. But in the mink sperm, the dense fibres numbered 9, 1, 5, and 6 are larger than the rest (Kim et al. 1979). In the Korean native goat, dense fibres numbered 9, 1, 5, and 7 are larger in diameter than the rest of the dense fibres (Bae and Kim 1981). The cross-sectional configuration of dense fibres in mammalian sperm appears to be species-specific. Marked interspecific differences in the development of the outer fibres have also been observed. The outer fibres in some species are conspicuously thick and run through the entire length of the principal piece (Hughes 1976), whereas in others these are relatively slender and end about half way along the principal piece (Phillips 1975a). Near its termination each fibre is apparently fixed to the wall of the corresponding doublet (Fig. 61). Each dense fibre consists of a thick medulla, and a thin cortex of lower density. The cortical layer is continous over the abaxial surface of the fibre but it is generally absent in the side nearer to the axoneme. The dense outer fibres in the shrew spermatozoa show bilateral rather than radial symmetry (Green and Dryden 1976); the bilateral symmetry being achieved from an odd number of fibres as a result of pronounced lobation of one of the fibres. It is stained deeply with phosphotungstic acid (Gordon and Benesch 1968).

An oblique striation or distinct periodicity consisting of globular subunits is present in the surface replicas of isolated dense fibres. The long repeat distance is about 50 nm in bull (Baccetti et al. 1973), although shorter periods, about 20 nm,

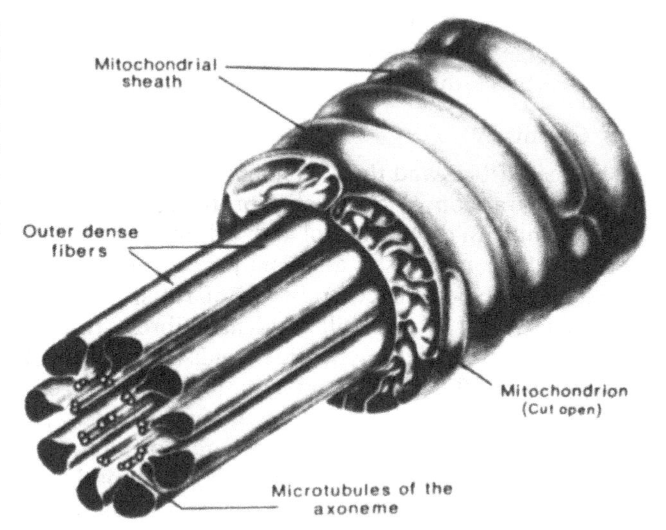

Fig. 70. An extended diagram of a segment of the middle piece of a typical mammalian spermatozoon, showing the close proximity of the mitochondrial sheath to the outer dense fibres, and the relationship of the latter to the doublets of the axoneme (From Fawcett 1975b)

also occur in longitudinal sections of the medulla (Woolley 1971, Pedersen 1972c). In the main period, nine subbands can be distinguished, grouped in two bands, each about 20 – 30 nm thick (Baccetti et al. 1973). Woolley (1971), and Phillips and Olson (1975) described oblique striations of the cortex of dense fibres in specimens of rat spermatozoa subjected to dehydration and desiccation. Olson and Sammons (1980) observed that surface replicas and positively or negatively stained whole mount preparations reveal a cross-banding pattern with a 40-nm major period repeat obliquely over the outer dense fibre surface in the rat sperm. These periodicities are confined to the outer dense fibre cortex. In their freeze-etch study of dense fibres in rat spermatozoa Espevik and Elgsaeter (1978) confirmed the presence of oblique striations on their cortex. The dense fibres show a main periodicity of approximately 40 nm, and a secondary one of about 20 nm. The dense fibre cortex appears to consist of globular substructure units with a diameter of about 19 nm. According to Olson and Sammons (1980), the dense fibre cortex consists of a single lamina of globular particles 6 – 7 nm in diameter, whereas the medulla appears electron-dense with no obvious substructure. The flagellar matrix lying between the dense fibres of the mid-piece and proximal portion of the principal piece shows small punctate or angular profiles, which in longitudinal sections appear as linear profiles running parallel to the axoneme and other fibres (Fawcett and Ito 1965). For their staining properties, the angular profiles resemble the cortex of the dense fibres. It is believed that these originate by exfoliation from the free edge of the cortex on the sides of the dense fibres. These structures are called satellite fibrils (Fawcett and Phillips 1970). These show a greater development in the bandicoot (Cleland and Rothschild 1959), marsupial devil (Hughes 1976), and ground squirrel (Fawcett 1970, Fawcett and Phillips 1970), which possess very thick sperm tails.

Paddock and Woolley (1980) observed that when dense fibres are released from rat, hamster, and rabbit spermatozoa by ultrasonic treatment, their fragments spontaneously adopt helical forms. The pitch of the helicas is positively correlated with fibre thickness. These results suggest that either the axonemal complex is maintained under tension in vivo, or the tendency to coiling is suppressed by readily dissociated components of the dense fibres.

b) Chemistry

The dense fibres and their satellite fibrils resist the action of chemical agents used in the extraction of contractile proteins, but these can be put into solution by treatment with dithiothreitol and sodium dodecyl sulphate (Fawcett 1975b). Vera et al. (1984) have used a simple procedure to isolate the fibrillar complex for studying polypeptide composition of rat sperm outer dense fibres. The electrophoretic study of solubilized dense fibres from rat and bull sperm tail has revealed four major polypeptide bands (Price 1973, Baccetti et al. 1973, 1976a, b, Calvin et al. 1975, Olson and Sammons 1980). In each case, one high-molecular-weight component and several low-molecular-weight polypeptides are being identified. In the rat these correspond to polypeptides of MW 40,000, 25,000, 12,000, and 11,000. The 25,000 MW polypeptide accounts for 58% of the total (Price 1974). But similar studies carried out on the dense fibres of bull sperm tail have demonstrated only three bands, which correspond to molecular weights of 55,000, 30,000, and 15,000; of these the 30,000 MW polypeptide is most abundant (Baccetti et al. 1973). Amino acid analysis carried out in these studies has shown a high content (20%) of cysteine; proline (8%) is also present. Baccetti et al. (1976b) observed that purified mammalian dense fibres consist of two groups of disulphide-cross-linked polypeptide chains. The high-molecular-weight chains (42,000 – 72,000) are rich in aspartic acid, glutamic acid, and leucine; the low-molecular-weight chains (29,000 – 31,000) are rich in cysteine and proline. According to recent studies of Olson (1979), the outer dense-fibre fraction from the rat sperm contains a four polypeptide component. The three low-molecular-weight polypeptide bands have a high cysteine content (10% – 12%), while the high-molecular-weight component (87,000) contains 3% cysteine. The low-molecular-weight components are located in the medulla.

Zinc, which is needed for spermatogenesis in mammals (Gunn and Gould 1970, Underwood 1977), is incorporated into the tails of late spermatids (Miller et al. 1961), where it is localized in the dense outer fibres (Baccetti et al. 1974, 1976a). Various studies have revealed that most of the zinc in mature mammalian sperm is associated with dense fibres, which contain cysteine-rich proteins (Calvin et al. 1975, Calvin 1975, Baccetti et al. 1973, 1974, 1976a, b) whose sulphydryl groups are probably responsible for the retention of zinc by the rat sperm tail (Calvin et al. 1973, Baccetti et al. 1976a). Calvin (1979b) produced electrophoretic evidence for the identity of the major zinc-binding polypeptides in the rat sperm tail. Selective binding of [^{65}Zn] by some, but not all, of the cysteine-rich polypeptides in the sperm tail extracts may be of physiological significance, particularly since those macromolecules that demonstrated the greatest affinity for zinc are believed to be the major constituents of dense fibres. Calvin (1981), after studying comparative labelling of rat epididymal spermatozoa by intratesticularly administered [$^{65}ZnCl_2$] and [^{35}S]cysteine, suggested that primary sites of labelling are in the sperm tail. The oxidation to disulphides can be catalysed by traces of copper present in the fibres (Baccetti et al. 1973).

The occurrence of low-sulphur and high-sulphur polypeptides, the zinc-binding properties, and the analogous localization in wave-generating flagella prompted Baccetti et al. (1976b) to distinguish the keratin-like proteins of sperm dense fibres with the new name of parergins. The infrared spectrum of whole fibres is similar to that of keratin. Protofibril-like structures, 2 nm thick, are detected in native fibres and become more evident after proteolysis or "renaturation" from guanidine-HCl solutions. The cross-striation of accessory fibres originates from the lateral packing of protofibril-like units. An appreciable content of bound trigylceride (103 μg/mlcl protein) in the dense fibres of mammalian sperm is an astonishing observation (Baccetti et al. 1973, Price 1973). Very divergent opinions exist about their carbohydrate content (Baccetti et al. 1973, Price 1974) and ATPase activity (Baccetti et al. 1973, Price 1974).

c) Function

Differing opinions have been expressed about the fucntional significance of dense fibres (Fawcett 1975b, Baccetti and Afzelius 1976). The early workers considered them as contractile, accessory motor elements which evolved to overcome the greater resistance to locomotion in the female genital tract (Fawcett and Ito 1965, Fawcett 1970). It has been observed that the mammalian spermatozoa with relatively thin dense fibres appear fairly flexible, whereas the sperm with large fibres appear very stiff when beating (Phillips 1972a, Phillips and Olson 1975). These observations indicate that the dense fibres are stiff, rather than the contractile elements. The mitochondria placed close to the dense fibres provide the necessary energy for the contraction of the latter. Their contractile nature was suggested by the results of immunohistochemical studies, which indicated that the dense fibres have both ATPase activity (Nelson 1985, Baccetti and Afzelius 1976) and antigenic similarities to actomyosin (Nelson 1985). Spermosin, a myosin-like ATPase protein, has been claimed to be localized in the core and an actin-like protein, termed spactin or flactin, at the periphery (Baccetti and Afzelius 1976). But recent chemical studies on the isolated dense fibres do not point to a contractile function (Baccetti et al. 1973, Price 1973, 1974) as there is little chemical resemblance between the dense fibres and any known contractile function. These workers denied the occurrence of the ATPase activity in the dense fibres. The activity localized in their cortical region appears to be derived from the surrounding matrix, and is detectable only histochemically. More recent data support the hypothesis of an extractable ATPase contained in the matrix of the bull fibres (Young and Smithwick 1975b), which can be lost during sperm fractionation.

The dense fibres are now believed to have significant passive elastic properties and may function to stiffen or provide elastic recoil for the sperm tail (Fawcett 1975b, Kaczmarski 1978). Their proteins stabilized by abundant disulphide cross-linking support this suggestion. Comparative cinematographic observations of sperm motility indicate that the sperm tail with large outer dense fibres shows bending waves of lower amplitude than those having smaller outer fibres. Thus, these appear to be stiffer (Phillips 1970a). Phillips and Olson (1975) suggested that the differently shaped beats (due to fibres) in different species appear to be suited to specific conditions in the female reproductive system. Kaczmarski (1978) suggested that dense fibres add to the elasticity of the tail which permits the mammalian spermatozoon to change its position in an environment of high viscosity. Their elasticity appears to be derived from specific proteins (panergins) which make up these fibres.

Baccetti et al. (1973, 1974) observed that the dense fibres consist of proteins with properties similar to metalloproteins and with a high zinc content. W. A. Anderson and Personne (1969) demonstrated calcium ions in the dense fibres. These observations indicated that the dense fibres may function as an intrinsic control mechanism for sperm movement by regulating the concentrations of divalent cations. This is supported by the fact that zinc ions may cause conformational changes in ciliary protein molecules (Morisawa and Mohri 1974). According to Baccetti and Afzelius (1976), the other role of dense fibres may be to remove toxic divalent ions. Lindholmer (1974) believes that a zinc-albumin complex coats mammalian spermatozoa and protects the cells against toxic substances in the seminal fluid. Future studies should determine whether there is any functional relationship between these two zinc-containing sperm components.

2. *Mitochondria*

a) Morphology

The arrangement and size of mitochondria in the mid-piece differ significantly in various mammalian species (Fawcett 1958, 1975b, Fawcett and Ito 1965, Hrudka 1968a, Pedersen 1970a, Green and Dryden 1976, Phillips 1975a, 1977a, Harding et al. 1979, Friend and Heuser 1981, Holstein and Roosen-Runge 1981). The mid-piece of musk shrew spermatozoa is long and possesses more than 100 mitochondrial gyres (Green and Dryden 1976). In general, the mitochondria of the mid-piece are arranged end-to-end to constitute a helix around the longitudinal fibrous elements of the tail (Figs. 61 and 70). The end-on junctions of the mitochondria are generally seen at random along the course of the helix, but in some mammals are spaced remarkably regularly. In some species mitochondria are arranged differently in the neck region than along the major portion of the mid-piece. The pattern and number of mitochondrial gyres are, however, almost identical among cells of one species. The mid-piece of mammalian sperm appears to consist of mitochondria of unequal length. This can be seen in replicas of monkey, rat, golden hamster, degu and guinea-pig (Phillips 1977a). In bull, rabbit, and mouse, demarcations between mitochondria in a row cannot be seen in replicas but in grazing sections the mitochondria do not seem equal in length.

Each turn of the mitochondrial sheath in the sperm of the bat (*Myotis*) shows two mitochondria of identical size (Fawcett and Ito 1965). Their ends always meet on the plane passing through the central pair of microtubules in the axoneme. Thus the end-to-end junctions of mitochondria in successive turns of the sheath are aligned in register along the dorsal and ventral aspects of the mid-piece for its whole length. Two mitochondria are seen in each ring of the mitochondrial sheath of some rodents. But their end-to-end contacts in successive turns are offset by 90° so that in surface replicas the junctions in every other turn

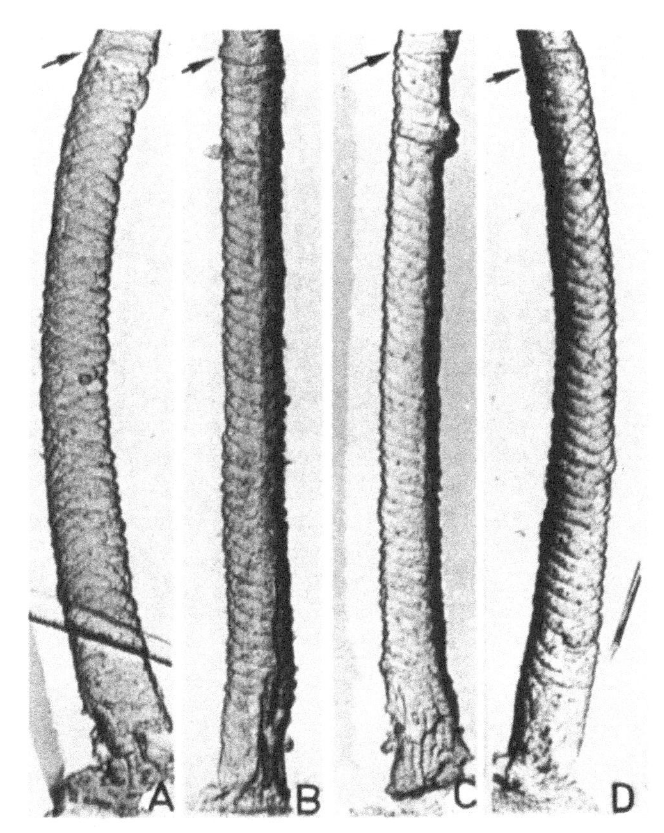

Fig. 71 A – D. Mid-piece of rabbit spermatozoa. In the neck region mitochondria are arranged parallel to the flagellum. Posterior to the neck, strands of mitochondria form a quadruple or quintuple helix of about 41 gyres. The helix can best be followed in **A**, **B**, and **D** where the platinum coated both sides of the mid-piece; *Arrows* annulus (From Phillips 1977a)

are aligned. A similar staggered arrangement is also observed in successive rings of four mitochondria in some marsupials (Phillips 1970a, b). This represents a rotation of 45 ° from row to row. Mitochondria must be similar in size in order to maintain the precise alignment of the junctions over the entire length of the mid-piece in these species. It is not known how this uniformity is obtained during spermatid differentiation or what advantage would be obtained by such a high degree of order and symmetry in the mitochondrial sheath. By employing scanning electron microscopy, Lung and Bahr (1972) found that ,in contrast to the smooth main-piece, the mid-piece showed a "bumpy" impression in relief with the mitochondria disposed helically.

Phillips (1977a) observed considerable interspecies variation in the manner in which mitochondria are arranged (Figs. 71, 72 and 73). Rabbit sperm possesses coils of mitochondria in a quadruple or quintuple helix (Fig. 71) The mitochondrial helix is triple or quadruple in bull sperm (Fig. 72), double or triple in mouse sperm (Fig. 73), and single or double in rhesus monkey sperm. The other faces of

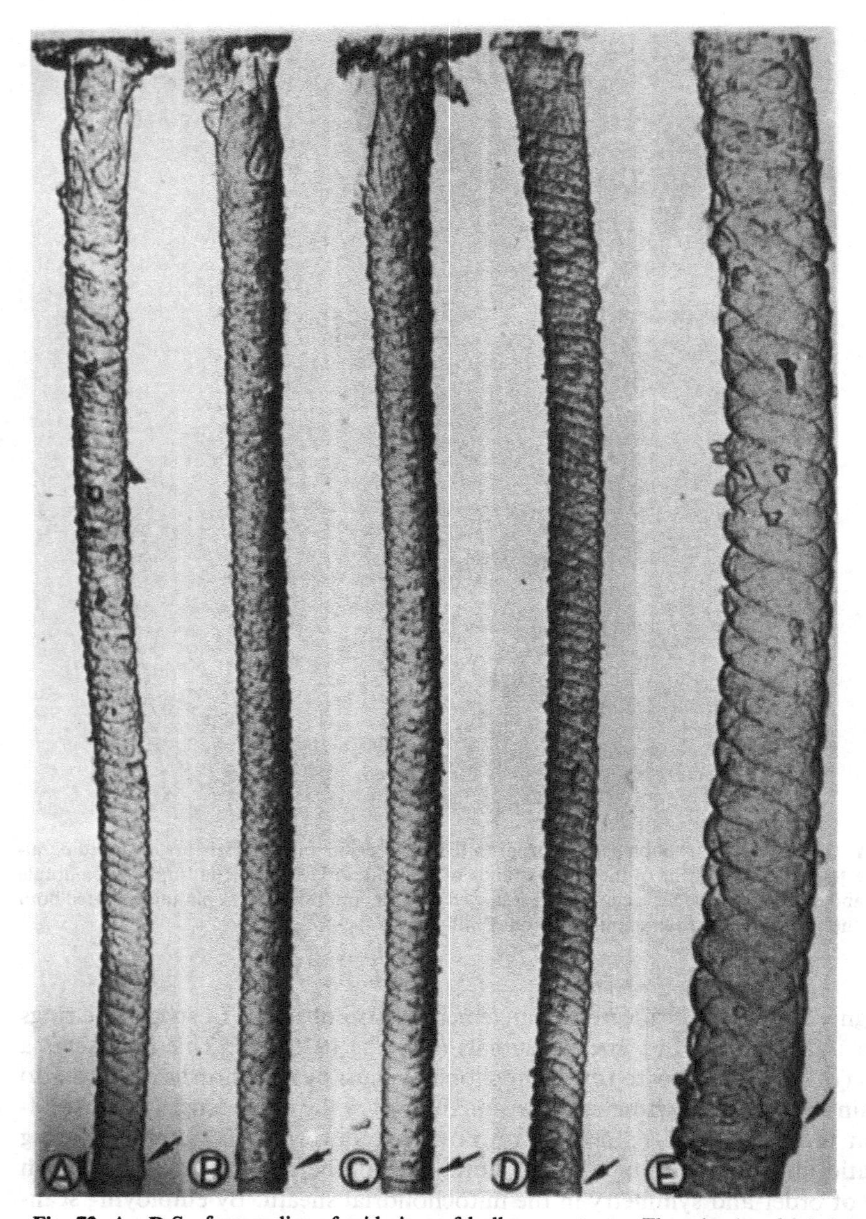

Fig. 72. A – D Surface replica of mid-piece of bull spermatozoa. The mitochondrial sheath of bull spermatozoa is composed of about 64 gyres of closely apposed mitochondria which spiral around the dense fibres in a triple or quadruple helix. In the neck region, however, mitochondria are disposed parallel to the flagellum; *Arrows* annulus; **E** This replica was shadowed at an angle, such that both sides of the mid-piece were coated. Mitochondrial strands can be followed along the entire sheath, revealing a triple helix and in some regions a quadruple helix; *Arrows* annulus (From Phillips 1977a)

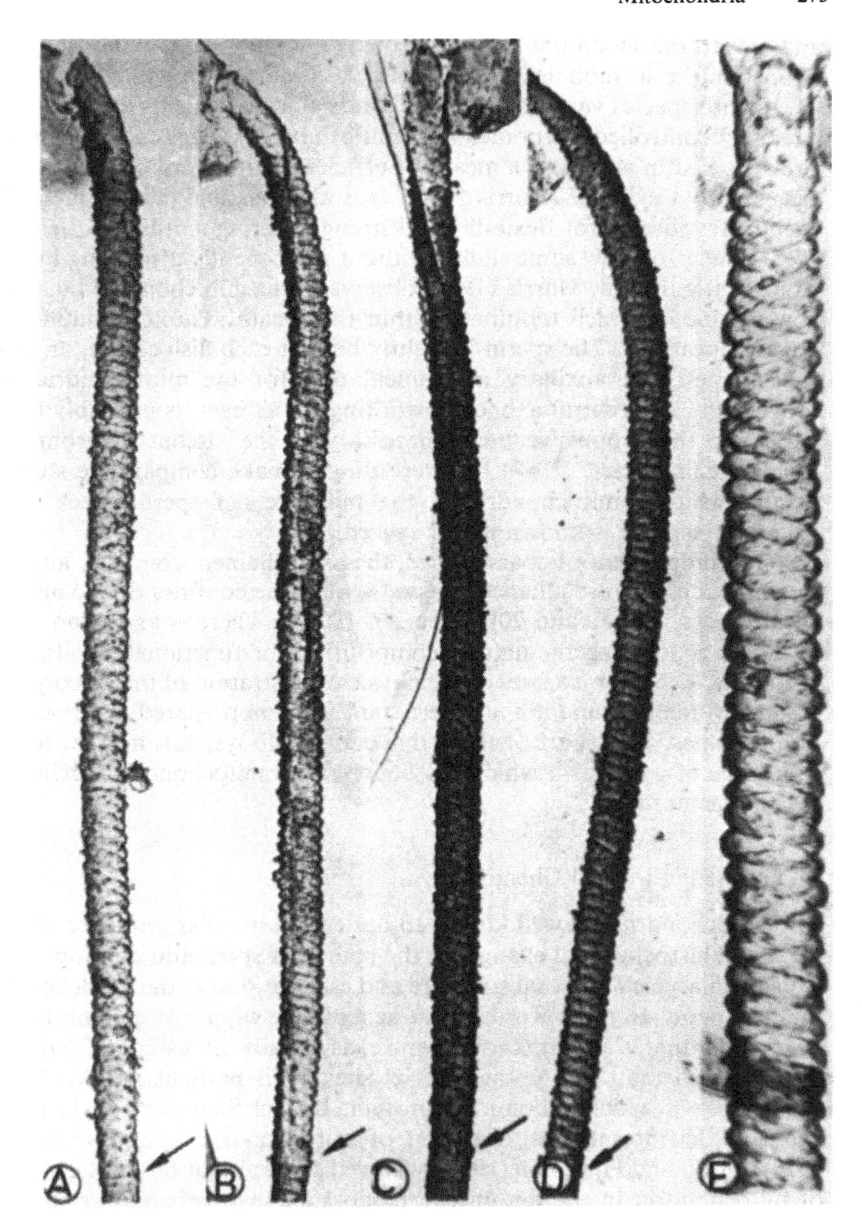

Fig. 73 A – E. Replicas of mid-piece of mouse spermatozoa. Mouse spermatozoa are composed of about 89 gyres arranged in a double helix or in some regions a triple helix; *Arrows* annulus (From Phillips 1977a)

individual mitochondria appear rectangular in Chinese hamster, diamond-shaped in rhesus monkey, S- or V-shaped in rat, and irregular in guinea-pig and degu. Interspecies variation in mitochondrial arrangements appears to be a very precisely controlled phenomenon. Phillips (1977a) suggested that mitochondria are arranged in a helix as a means of efficiently fitting mitochondria of unequal length into a cylinder (Harris 1976). It is also possible that the helical arrangement is favourable for flexibility and strength during motility. A helical arrangement may also allow some sliding without permanently displacing the mitochondrial configuration. Harris (1976) observed that mitochondria lie along helical rows, some of which terminate within the sheath. The terminations represent edge dislocations. The sperm is slightly bent at each dislocation, an observation that suggests an auxillary mechanical role for the mitochondrial sheath in movement. The central axoneme, working on its own, is probably incapable of supplying the propulsive thrust necessary in the viscous environment of the female genital tract. It will be interesting to make comparative studies on the arrangement of mitochondria in the mid-piece of spermatozoa in different primates, as little is known in this regard.

The complement of coarse fibres, the axial filament complex, and the matrix in which these lie are all encompassed within the confines of the mitochondrial sheath (Figs. 57, 61 and 70) (Pedersen 1970a). There is as yet no information available concerning the nature, composition, or functional significance of the matrix (i.e. whether it serves as the physical substratum of the glycolytic enzymes system normally found in the supernatant fraction prepared from other cell-type homogenates, or as part of the active contractile system), nor for the relatively thin layer of cytoplasm which lies between the mitochondrial surfaces and the plasma membrane.

b) Ultrastructure and Chemistry

The mitochondria are well known to undergo a peculiar sequence of ultrastructural and histochemical changes in the course of spermatid development that results in an atypical internal structure and composition in the mitochondria of the mature sperm. In other words, these accumulate whorls of membranes or lipids, or granular material of unknown nature, as already discussed in Chap. IV. Mann (1975) stated that the "mitochondrial sheath" is particularly rich in phospholipid, most of which is bound to protein. In fact there are considerable differences in the internal conformation of mitochondria among species (Fawcett 1970). Olson and Hamilton (1976) observed some alterations in the ultrastructure of mitochondria in the mid-piece of opossum spermatozoa during epididymal transit. Hughes (1976) also observed that the mitochondria of the mid-piece in the epididymal spermatozoa of the marsupial devil have a distinctive ultrastructure with concentrically arranged cristae, as also reported by Harding et al. (1979) in different marsupials.

Little is known about the fate of intramitochondrial lipids or whorls of membranes of spermatozoa within the genital ducts (Harding et al. 1979). However, Guraya (1971b, 1973) suggested that these endogenous lipids might be the source of oxidative energy utilized for sperm motility and other metabolic activities including maturation; phospholipids are also most essential for the proper functioning of mitochondria. This suggestion is supported by the fact that the endogenous respiration of the spermatozoa is believed to depend mainly on the oxidation of intracellular lipids (Mann and Lutwak-Mann 1981, Peterson 1982). In man, just as in the bull or ram, ejaculated spermatozoa carry no appreciable reserve glycogen (Mann and Rootenberg 1966). Moreover, these are unable to metabolize glycogen by means of glycogenolysis, since these do not posses the necessary enzyme phosphoglucomutase (Mann and Lutwak-Mann 1981). The virtual absence of glycogen has an obvious bearing on sperm survival in vivo. This shows that the spermatozoa of mammalian species, such as the bull, ram, or man, cannot depend for survival on their intracellular carbohydrate reserve. These do, in fact, utilize for their endogenous metabolism another intracellular reserve material, namely phospholipids (Hartree and Mann 1961, Poulos et al. 1973b, Mann and Lutwak-Mann 1981, Peterson 1982). Mitochondrial oxidative metabolism is capable of maintaining spermatozoal motility in the absence of exogenous substrates, further supporting the suggestion that endogenous oxidizable reserves, most probably lipids, are present in spermatozoa. The accumulation of lipids or ultrastructural whorls of membranes in the mitochondria during spermiogenesis, as demonstrated with correlative histochemical and electron microscopical studies (Guraya 1971b, 1973, Harding et al. 1979), must be of great significance in this regard.

Actually, little is known about the biosynthesis and breakdown of mitochondrial phospholipids (Borst 1969, Mann and Lutwak-Mann 1981, Peterson 1982). It is, therefore, suggested that further correlative histochemical, biochemical, and ultrastructural studies should be carried out on the mitochondria during spermiogenesis as well as during the passage of spermatozoa through the male and female ductular system. Olson and Hamilton (1976) and Temple-Smith and Bedford (1976) demonstrated ultrastructural changes in the mitochondria of the mid-piece in the wooly opossum and brush-tailed opossum spermatozoa during epididymal transit (Hughes 1976, Harding et al. 1979), which may provide a basis for correlating changes in sperm energy metabolism reported for the sperm of many species during epididymal transit (Voglmayr 1975, Mann and Lutwak-Mann 1981). Such studies will greatly increase our knowledge about the physicochemical and metabolic adaptations of mitochondria of the mid-piece in relation to storage and utilization of various chemical substances at different stages of sperm formation, maturation, and transport in mammals including primates. It has been suggested that the mid-piece mitochondria are capable of consuming components of their own cristae as an energy source for sperm motility (Baccetti and Afzelius 1976).

The mitochondria of the mid-piece constitute the energy mechanism for the mammalian spermatozoon and have been compared to a combustion engine (Mann and Lutwak-Mann 1981). No specific changes of mitochondrial ultrastructure could be revealed in human spermatozoa with decreased motility. However, in some cases of severe asthenospermia or necrospermia, defects are observed in the internal structure and arrangement of mitochondria (Amelar et al. 1980). There may also be structural derangement of the mitochondria in spermatozoa preserved by freezing, that may be responsible for the impaired motility.

Nelson (1967) stated that "the chemistry and physics of the mitochondria are as distinct from the filament complex of the flagella, and perhaps as well as distinct from somatic mitochondria". There is an ultrastructural difference between the outer and inner membrane of mid-piece mitochondria (Elfvin 1968). The outer membrane becomes usually thick (10 nm) and is five-layered rather than three-layered. Highly ordered particulate arrays on the mitochondrial outer membrane are among the more remarkable features of rapidly frozen guinea-pig spermatozoa (Friend and Heuser 1981). The organizational disparity between the convex and concave surfaces of the mitochondria not only affords evidence of a new mitochondrial, substructure, but represents a type of topographic heterogeneity rarely found except within specialized areas of the plasma membrane. According to Bedford and Calvin (1974a, b), the outer mitochondrial membrane of mammalian sperm is stabilized by intermolecular disulphide bonds (together with nucleus, coarse fibres, and other components), possibly for the strengthening of a structure which is unusually exposed to deformations due to the flagellar beats. Selenium is bound to a structural protein of mitochondria in the bull sperm and is believed to stabilize the outer mitochondrial membrane (Pallini and Bacci 1979). Calvin and Cooper (1979) also demonstrated a specific selenopolypeptide associated with the keratinous outer membrane of rat sperm mitochondria, which is in some way essential for the assembly of the mitochondrial sheath. Wallace et al. (1983) found progressive defects in the mitochondria of mouse sperm during the course of three generations of selenium deficiency. Recent studies have provided evidence that selenium is associated with the cysteine-rich structural protein of the mitochondrial capsules (Calvin et al. 1981). Pond et al. (1983) have studied the incorporation of selenium-75 into semen and reproductive tissues of bulls and rams. The stabilization of the outer mitochondrial membrane my be related to the higher resistance of sperm mitochondria to hypotonic conditions and to their relative impermeability to cytochrome C, as demonstrated in rabbit sperm by Keyhani and Storey (1973). Pallini et al. (1979) demonstrated a peculiar cysteine-rich polypeptide, which is related to some unusual properties of mammalian sperm mitochondria (e.g. maintenance of the crescent shape of the outer membrane). It consists of three polypeptide chains (MW 20,000, 29,000 and 31,000) cross-linked by disulphide bridges. The 20,000 MW polypeptide contains more than 20 prolines and about 18 cysteines per 100 residues. The well-documented unusual shape and resistance to swelling of mammalian sperm mitochondria have been attributed to the stabilizing effect of disulphide-hardened structural proteins, probably localized in the outer membrane.

Very divergent views have been expressed about the presence of nucleic acids and their role in protein synthesis in the sperm mitochondria. Spermatozoa contain mitochondrial DNA (mt DNA) (Bairati et al. 1980) as well as nuclear DNA. MtDNA is active in transcription and translation. In some studies it has been shown that mature mammalian spermatozoa actually synthesize small amounts of RNA and that this RNA shows biochemical characteristics very similar to that synthesized by the mitochondria (Premkumar and Bhargava 1972, 1973). MacLaughlin and Turner (1973) demonstrated that spermatozoa from the cauda epididymis in the rat and hamster are capable of synthesizing RNA of high molecular weight. Fuster et al. (1977) observed that washed mature spermatozoa from bulls incorporate ribonucleoside triphosphates into RNA using an endogenous template. This RNA synthesis was inhibited by ethidium bromide, rifampicin, acriflavine, actinomycin D, and caffeine, but not by α-amanitine or fifamycin SV. More than half of the total RNA polymerase activity of spermatozoa is associated with the tail fraction. Hecht and Williams (1978) provided evidence that separated sperm heads and tails are both capable of transcribing RNA from endogenous templates. The nuclear and mitochondrial fractions of spermatozoa contain different RNA polymerases (Mann and Lutwak-Mann 1981). Ejaculated bovine spermatozoa not only synthesize RNA but also proteins in their mitochondria; protein synthesis is also reported in the mitochondria of mouse spermatozoa (Bragg and Handel 1979). From the results of recent studies, it can be stated that both a mitochondrial DNA-determined transcription and a translation occur in this cell type, again at a stage when nuclear RNA synthesis and cytoplasmic protein synthesis have stopped. It will be interesting to determine the nature of proteins synthesized at this stage. Extracts of bull and ram sperm tails and mitochondria prepared by dithiothreitol and acetyltrimethyl ammonium bromide treatment showed DNase II, degrading only native double-stranded DNA at acid pH. At least three distinguishable DNase II activities are found, as judged by their sensitivity to thiol compounds and various anions and cations. This DNase II activity appears to be associated with the mitochondria. The existence of a mitochondrial DNA polymerase in bovine spermatozoa and of DNase II in the mitochondrial fraction of ram spermatozoa (Fisher and Bartoov 1980) both suggest that sperm mtDNA might be engaged in replication (Mann and Lutwak-Mann 1981). Bartoov et al. (1981) produced evidence for the replication of mtDNA in ram spermatozoa.

Ezzatollah and Storey (1973) found that the mitochondria of the mid-piece can maintain their energy conservation capacity and morphological integrity in hypotonically treated rabbit spermatozoa. A peculiar metabolism of pyruvate (Van Dop et al. 1977) connected with an intramitochondrial localization of LDH isozyme X (Montamat and Blanco 1976, Baccetti et al. 1975) has also been demonstrated. Calvin and Tubbs (1978) studied the mitochondrial transport processes and oxidation of NADH by hypotonically treated boar spermatozoa. According to Gerez de Burgos et al. (1978) a branched chain α-hydroxy acid/amino acid shuttle for the transfer of reducing equivalents from cytosol to mitochondria may be functional in mouse spermatozoa.

Hrudka (1968b) observed a striking pleomorphism of the inner compartment in bovine sperm mitochondria. Two kinds of organelles in a population of isolated spermatid mitochondria are observed (Bartoov and Messer 1976). A staining reaction with Altmann's aniline-acid fuchsin reagent gives a positive reaction in about half of the mitochondria of the bat sperm (Wimsatt et al. 1966). This has been interpreted as an expression of different physiological state of the mitochondria. A pattern in the distribution of cytochemical activity of nitro blue tetrazolium (NBT)-reductase has indicated that not all sperm mitochondria are in the same activity phase at the same time (Hrudka 1978). These observations point to either (1) heterogeneity or (2) cyclic changes in the sperm mitochondrial populations. Hrudka (1978) described and applied a technique for studying the fine morphology, changes in the conformation, and other morphologico-functional correlations on isolated sperm mitochondria. This study has also revealed that the 90 mitochondria of bovine sperm break into two groups that are synchronized in their activity: one in the expanded state and the other in the condensed state at any given time. The condensation appears to be a fast, active phase, which is associated with increased formazan production.

In spite of the fact that the bovine mitochondria are structurally modified, their biochemical properties resemble those of mitochondria from other tissues (Mohri et al. 1965). Isolated mid-pieces catalyse the aerobic oxidation of the Krebs cycle intermediates and have normal P:O ratios and respiratory control indices. Even if endogenous glycogen is absent, intracellular phospholipids are areobically oxidized, as well as fructose, polyols, and lactic acid arising from the seminal plasma (Mann 1975, Mann and Lutwak-Mann 1981, Peterson 1982). The characteristic lactate dehydrogenase is inhibited by low concentrations of pyruvate and shows a high affinity towards lactate, like the heart-type enzyme (Battellino et al. 1968). It is localized in the mitochondria (Machado de Domenech et al. 1972, Storey and Kayne 1977, 1978) in order to produce pyruvate in the cellular compartment where the compound is further oxidized. Hoppe (1976) demonstrated a glucose requirement of mouse sperm for capacitation in vitro, and it is oxidized primarily by the glycolytic pathway. Rogers et al. (1979) also studied the glucose effect on respiration in relation to a possible mechanism for capacitation in guinea-pig spermatozoa.

c) Lipids of Spermatozoa and Their Significance

There are very conspicuous species differences in lipid composition and phospholipid-bound fatty acids and aldehydes of mammalian spermatozoa. Mann and Lutwak-Mann (1981), and Sidhu and Guraya (1985a) have discussed in detail their variable amounts and functional significance in spermatozoa of different species of mammals. The contents of total lipids and cholesterol may vary in human sperm normally and in pathology (Stanislavov et al. 1978). By far the greatest proportion of the lipids present in mammalian spermatozoa consists of phospholipids. The major phospholipids present in the mammalian spermatozoa are sphingomyelin, phosphatidylcholine (lecithin), choline plasmalogen, ethanolamine, plasmalogen, phosphatidylethanolamine, phosphatidylserine gangliosides (acidic glycolipids characterized by the presence of sialic acid), and cardiolipin. Significant differences are apparent between species in the concentrations of some phospholipids (for example, the plasmalogens and phosphatidylethanolamine) (Mann and Lutwak-Mann 1981). It is now widely accepted that plasmalogen, either as phosphatidalcholine or phosphatidalethanolamine forms one of the major phospholipids. In ram spermatozoa, the plasmalogen is present mostly in the form of phosphatidalcholine and the main fatty aldehyde is palmitaldehyde. Among the lysophosphatides, only lysolecithin has been observed in mammalian spermatozoa. Buffalo spermatozoa contain not only lysolecithin but also lysocephalin and lysophosphatidylserine (Sidhu and Guraya 1985a). Lysolecithin and lysocephalins show relatively higher concentrations than phosphatidylserine. In most of the mammalian species studied so far, cholesterol forms the major neutral lipid followed by triglycerides which, however, form the major components of neutral lipids in buffalo (Sidhu and Guraya 1985a). Diglycerides are also reported in the neutral lipids of spermatozoa in some mammals. Polyunsaturated fatty acids (including docosapentaenoic or docosahexaenoic acid) are found in abundance in the sperm of various species (Ahluwalia and Holman 1969, L. A. Johnson et al. 1969, Neill and Masters 1973, Mann and Lutwak-Mann 1981). In the ejaculated spermatozoa of cattle and ram, docosahexaenoic acid (22:6) is the principal acid of phospholipids, and myristic acid (14:0) the major component of digylcerides (Mann and Lutwak-Mann 1981). It will be interesting to mention here that the content of total phospholipid and plasmalogen, as well as the ratio between lipid-bound unsaturated and saturated fatty acids or aldehyde, may vary in spermatozoa depending upon breed characteristics, breeding season, nutrition of the animal, percentage of abnormal sperm cells in semen, and frequency of ejaculation. The physiological role of significant ($p < 0.01$) seasonal variations in total lipids, phospholipids, cholesterol, and free fatty acids reported for buffalo spermatozoa (Sidhu and Guraya 1979b) also needs to be determined in relation to their motility, metabolism, and fertilizing ability. Changes in lipid and fatty acid composition of sperm occur during migration through the reproductive tract (M. H. Johnson 1970, Mann and Lutwak-Mann 1981). Evans et al. (1980) studied diacyl, alkenyl, and alkyl ether phospholipids in ejaculated, in utero, and in vitro-incubated porcine spermatozoa.

Guraya and Sidhu (1975b, 1976) made a correlative cytochemical and biochemical study of phospholipids of ejaculated buffalo spermatozoa, which consist of phosphatidylcholine, lysolecithin phosphatidyl ethanolamine, phosphatidyl inositol, phosphatidyl glycerol, and monogalactosyl diglyceride. These lipids are combined with protein to form a lipoprotein complex in the plasma membrane, post-nuclear cap, and mid-piece; the latter appears to form the major source of lipids in the sperm. According to Sharma and Venkitasubramanian (1977), phospholipids constituted 56.6%, cholesterol 4.16%, and glycerides 39.3% of total lipids in buffalo spermatozoa. Phosphatidyl choline and phosphatidyl inositol are the main fractions in spermatozoa phospholipids. Free cholesterol is higher in concentration than esterified cholesterol. Similar observations were also made by other workers (Jain and Anand 1975, 1976). Total phospholipid content of porcine ejaculated spermatozoa is 65.7 μg lipid $P/10^9$ sperm, of which 41% is alkyl ether and 23% is alkenyl ether glycerophospholipid (Evans et al. 1980). All of the other phospholipids are choline and ethanolamine glycerophospholipids. In order of decreasing amount (percent of total phospholipid), the phospholipids are choline and ethanolamineglycerophospholipids (49.9% and 28.2%), sphingolipid (10.6%), cardiolipin (5.5%), phosphatidylinositol (2.3%), phosphatidic acid (1.5%), phosphatidylserine (1.2%), and phosphatidylglycerol (0.8%). P-containing sphinolipid separated into two components during TLC. Sphingosine is the only long-chain base identified in either band.

In ram spermatozoa, the neutral lipid fraction, with a content of 158 μg fatty acids/10^9 spermatozoa consists almost entirely of saturated fatty acids. The phospholipid-bound fatty acids of mammalian spermatozoa contain high levels of polyunsaturated fatty acids, ranging from 70% by weight of the total fatty acid fraction in boar to approximately 40% in both humans and rabbit. The major saturated fatty acid found in mammalian spermatozoa is palmitic acid (Scott et al. 1967, Poulos et al. 1973b, Mann and Lutwak-Mann 1981). But in buffalo, stearic acid is the most abundant fatty acid followed by palmitic and myristic acids (Sidhu and Guraya 1985). Actually, the lipids and lipid-bound fatty acids have been studied in more detail for buffalo spermatozoa than for any other mammalian species. Evans et al. (1980) observed that major fatty acids with lower Rf in the ejaculated porcine spermatozoa are 16:0 (56%), 20:0 (23%), and 18:0 (11%) plus smaller amounts of 14:0, 18:1, and 22:0, while those in the band with higher Rf are 14:0 (30%), 16:0 (45%), and 18:1 (12%) plus smaller amounts of 18:1 and 20:0. Choline is the only water-soluble base present in the lower Rf sphingomyelin, while ethanolamine is prevalent in the higher Rf component. Palmitaldehyde forms the predominant aldehyde in the lipids of mammalian spermatozoa (Sidhu and Guraya 1985a). The spermatozoa of most of the mammalian species studied show more saturated than unsaturated aldehydes in their lipids. Although our knowledge about the lipid composition of spermatozoa has greatly increased in the past, the physiological meaning of different lipids is poorly understood. The roles of phospholipids are generally related to (1) chemical changes associated with sperm maturation in the epididymis, (2) preservation of sperm membrane integrity, and (3) metabolism of endogenously respiring spermatozoa. Mann and Lutwak-Mann (1981) have pointed out that technical reasons appear to be the cause of uncertainties in regard to the precise roles of lipids. Fatty aldehydes are yet to be studied for their physiological roles in the spermatozoa.

Various studies on sperm lipid biochemistry have revealed that there is much less phospholipid in ejaculated than in testicular ram spermatozoa and that both in the ram and in some other mammals the phospholipid content progressively decreases as the spermatozoa pass from the caput to the cauda epididymidis. The percentage composition of phospholipid-bound fatty acids varies in testicular, cauda epididymal, and ejaculated ram and bull spermatozoa (Scott et al. 1967, Poulos et al. 1973b, Brooks 1979b, Mann and Lutwak-Mann 1981). Palmitic acid constitutes 32% to 53% of the total phospholipid-linked fatty acids of the testicular spermatozoa, whereas in cauda epididymal and ejaculated spermatozoa it is only 19% to 36%. However, the proportion of myristic (14:0), linoleic (18:2), linolenic (18:3), and particularly docosahexaenoic acid (22:6) is higher in cauda epididymal or ejaculated than in testicular spermatozoa. The change in the percentage of palmitic acid has been related to an absolute loss of this fatty acid, whereas the increase in the percentage of docosahexaenoic acid does not reflect the absolute increase in the amount, but is mainly due to the loss of palmitic acid (Scott et al. 1967). L. A. Johnson et al. (1972) reported differences in the fatty acid pattern between spermatozoa obtained from the boar caput and cauda epididymidis. Voglmayr (1973) suggested that spermatozoa may control their own degradation of lipids by releasing well-controlled amounts of arachidonic acid from which the epididymis synthesizes prostaglandin $F_{2\alpha}$; phosphatidylethanolamine is the source of arachidonic acid in the rat spermatozoa during their passage through the male reproductive tract, but in other mammals arachidonic acid is present in all major phosphoglycerides of their ejaculated spermatozoa (Sidhu and Guraya 1985a). The antilipolytic prostaglandin may play a key role in a control mechanism for selective degradation of lipids in spermatozoa during their period of maturation in the epididymis, but this needs to be confirmed by further studies.

Various studies of sperm maturation have revealed the changing pattern of motility of spermatozoa as these pass through the epididymis (Bedford 1975, Voglmayr 1975, Mohri and Yanagimachi 1980, Peterson 1982). Mann (1968) stated that even though the motility of spermatozoa in the mammalian epididymis is low, these have to rely on the supply of at least some energy in order to sustain their metabolism and to survive. Bedford (1975, 1979), Voglmayr (1975), Brooks (1979b) and Peterson (1982) reviewed the various observations in regard to the metabolism of maturing spermatozoa in the epididymis. Testicular and mature spermatozoa differ in O_2 uptake and in energy substrate utilization. Species differences in this regard are also reported. Certain energy-producing and synthetic systems of spermatozoa undergo both qualitative and quantitative changes during epididymal passage (Brooks 1979b, Mann and Lutwak-Mann 1981), but much more work is needed to reveal their metabolic changes more precisely.

Fructose is not available in the epididymis since the bulk of that sugar is secreted in other accessory glands, such as the seminal vesicles, which are located much lower down the male reproductive tract (Mann and Lutwak-Mann 1981, Peterson 1982). The epididymal spermatozoa have to depend, therefore, mainly on their intracellular reserve of phospholipids stored mainly in the mitochondria of the mid-piece for their energy requirement and maturation changes. Poulos et al. (1973a), after studying the changes in the phospholipid composition of bovine spermatozoa during their passage through the male reproductive tract, also suggested that during maturation of spermatozoa a number of phospholipid molecules, via their long chain fatty acid residues, serve as a source of nutrients for the maturing spermatozoa. The changes in phospholipid content are mainly due to a marked decrease in the concentration of phosphatidalserine, phosphatidalethanolamine, ethanolamine, plasmalogen, and cardiolipin. However, the concentration of choline plasmalogen, the principal phospholipid in spermatozoa, is not significantly different among the various cell types. The maturation of spermatozoa in the epididymis is accompanied by changes in lipid metabolism (Bedford 1975, Voglmayr 1975, Mann 1975, Mann and Lutwak-Mann 1981). Adams and Johnson (1977) have observed that spermatozoa from the caput epididymidis of the rat (*Rattus norvegicus*) show a significantly greater ($P<0.05$) content of phospholipids, cholesterol ester, and free fatty acid than those from the cauda epididymidis. Spermatozoa from the corpus epididymidis have a significantly greater ($P<0.05$) content of monoglyceride than those from the caput epididymidis, and a greater content of phospholipid, cholesterol, free fatty acids, and monoglyceride than those from the cauda epididymidis. The lipid changes occur in boar spermatozoa during epididymal maturation (Evans and Setchell 1979a, b). Selivonchick et al. (1980) investigated the structure and metabolism of phospholipids in bovine epididymal spermatozoa. The results of this study have suggested that other lipids provide stable structural components of sperm membrane, while diacyl analogues undergo degradation and resynthesis.

Some oxidation of fatty-acid chains without destruction of the whole phospholipid molecule has also been suggested by Evans and Setchell (1978a, b) while investigating the effect of rete testis fluid on the metabolism of testicular spermatozoa. The source of phospholipids from which spermatozoa can draw energy for survival in the male ductular system needs to be determined more precisely, but it is suggested that these phospholipids may be provided by the mitochondria and protoplasmic droplet (Voglmayr 1975), which have already been shown to accumulate such lipids (see Chap. IV). At least some of the observed alterations in phospholipid concentration can be attributed to the loss of the cytoplasmic droplet as the spermatozoa pass through the epididymis. This also probably explains why the concentration of total phospholipid, phosphatidylcholine, phosphatidylethanolamine, phosphatidylinositol, and choline plasmalogen in ram spermatozoa taken from the caput is higher than in those collected from the cauda epididymis (Scott et al. 1967, Quinn and White 1967). Phosphatidylinositol plays a significant role in the overall lipid metabolism of spermatozoa in the epididymis as it is readily metabolized by spermatozoa obtained from any part of the ram genital tract. This probably explains the marked decrease of phosphatidylinositol in spermatozoa during their passage through the epididymis (Scott et al. 1967, Scott and Dawson 1968). This phospholipid has also been observed to stimulate the motility, respiration, and glycolysis of bull spermatozoa (Voglmayr 1973).

The decrease in the concentration of selected phospholipids together with a concomitant loss in acyl esters and a reduction in the chain length of the phospholipid-linked fatty acids support the suggestion that lipids (phospholipids) serve as substrates for spermatozoa in the epididymis of ram and bull, which have been extensively studied for this purpose (Mann and Lutwak-Mann 1981). The seminal lipids act as energy substrate for the spermatozoa (Abdel-Aziz et al. 1983). The endogenous lipid is not metabolized by washed sperm suspensions replenished with either glycolysable sugar or other extracellular oxidizable substances. Actually, spermatozoa preserve their intracellular lipid reserve intact in the presence of exogenous substrates, and for as long as their supply lasts. In the ram spermatozoa, deprived of exogenous oxidizable substrate (fructose and lactic acid in particular), the fatty acids of the phospholipids (chiefly plasmalogen) function as substrate for endogenous respiration (Mann and Lutwak-Mann 1981). Prasad and Singh (1980) studied utilization of endogenous lipids by ejaculated buffalo spermatozoa in vitro. There was no significant change in the values of lipid fractions (total lipids, phospholipids, and cholesterol) of washed spermatozoa after up to 4 h incubation in the presence or absence of glucose, suggesting that buffalo spermatozoa do not utilize even the extracellular lipids present in seminal plasma during incubation of whole semen and resemble bovine sperms in this regard, which do not have the ability to oxidize added phospholipids (Bomstein and Steberl 1957, Scott and Dawson 1968). Incubation of washed porcine spermatozoa in Ca^{2+}-free Ringer-fructose at 37 °C for 2 h has produced no significant change in the level of any of the phospholipids demonstrated in the ejaculated spermatozoa (Evans et al. 1980). However, incubation of washed sperm in the uterus for 2 h, in the presence of oviducal secretions, produced an increase in phosphatidylcholine from 7.2 to 10.2 μg lipid $P/10^9$ sperm. Further studies are needed to determine more precisely the utilization of lipids by spermatozoa of different species of mammals for energy production and other functions under different physiological situations. The mechanism suggested for the oxidative metabolism of endogenous lipids in spermatozoa involves the splitting of fatty acid residue from the phospholipid and oxidation as two-carbon acyl fragments, with acetylcoenzyme A functioning as the acetyl carrier. The enzymes of lipid metabolism including lipase, β-hydroxybutyrate dehydrogenase and glyceraldehyde-3-phosphate dehydrogenase, have been demonstrated both histochemically and biochemically in mammalian spermatozoa (Atreja and Anand 1985, Sidhu and Guraya 1985a). Lipase may help in hydrolysing lipids to be used endogenously. The presence of other enzymes suggests the utilization of fatty acids through β-oxidation (β-hydroxybutyrate dehydrogenase) and biosynthesis of lipids (glyceraldehyde-3-phosphate dehydrogenase). The lipid synthesis has been suggested for mammalian spermatozoa (see details in Sidhu and Guraya 1985a, Sidhu et al. 1986b). Sidhu et al. (1986b) have observed in vitro incorporation of [I-^{14}C] acetate and [U-^{14}C] glucose into lipids of buffalo spermatozoa. The physiological significance of these observations must be determined in future studies. A study has been made of phospholipase A_2, PG synthesis and PG metabolism during sperm maturation and transport (Ellis and Cosentino 1982, see also Atreja and Anand 1985). The surface-active phospholipase A_2 has been demonstrated in mouse spermatozoa (Thakkar et al. 1983). Srivastava et al. (1982) have observed hydrolysis of p-nitrophenylphosphorylcholine by alkaline phosphatase and phospholipase C from rabbit sperm-acrosome. The origin of phospholipases that initiate the degradation of phospholipids of spermatozoa remains to be determined more precisely (see Voglmayr 1975). Some of this activity has been suggested to originate in the sperm-cell itself. Two characteristic features of ejaculated boar sperm lipids are the absence of lysophosphatidylcholine and phosphatidylinositol (Clegg et al. 1975). Lack of lysophosphatidylcholine has provided evidence that boar sperm do not contain a phospholipase capable of attacking the choline phosphatides.

Of considerable interest is also the observation that the concentration of non-esterified cholesterol in both the spermatozoa and epididymal plasma decreases from the caput to the cauda (Mann and Lutwak-Mann 1981). Quinn and White (1967) suggested that the decrease in the phospholipid and cholesterol content of spermatozoa may be responsible for their increased susceptibility to "cold shock". White and Darin-Bennett (1976) observed a substantial loss of total phospholipid when ram, bull, and boar spermatozoa are cold-shocked, and an even greater loss after freezing and thawing, as also demonstrated for the buffalo spermatozoa (Sarmah et al. 1984). The spermatozoa of ram, bull, and boar have a high ratio of unsaturated fatty acids to saturated, and are much more susceptible to cold shock than those of rabbit, human, dog, and fowl, which have a lower ratio. The cholesterol content of ram and bull sperm (about 300 $\mu g/10^9$ sperm), which are particularly susceptible to cold shock, is lower than that of rabbit and human sperm (about 550 $\mu g/10^9$ sperm) which are not so susceptible. This suggests that cholesterol is also important in forming a reasonably impermeable and cohesive membrane structure. The liposomes of egg phosphatidyl choline protect sperm from cold shock and during cooling and storage at 4 °C (Parks et al. 1981). Addition of cholesterol to liposomes reduces the effectiveness of the protection provided by egg phosphatidyl choline. This study suggested that the protection provided to sperm immediately post-dilution is not the result of an altered sperm cholesterol/phospholipid ratio. The phospholipid protects ram spermatozoa from cold shock at a plasma membrane site (Quinn et al. 1980).

Various studies have shown that during aerobic incubation the sperm phospholipids undergo peroxidation, leading to the formation of fatty acid peroxides. Jones and Mann (1977a, b) suggested that lipid peroxides or their degradation products, whether introduced exogenously or derived from the peroxidation of endogenous phospholipids in semen, constitute a potential hazard to the structural and functional integrity of spermatozoa stored in vitro, particularly under conditions where oxygen is present (see also Mann 1978, Jones et al. 1979, Mann et al. 1980, Mann and Lutwak-Mann 1981, Alvarez and Storey 1982, 1984, 1985, Dawra et al. 1983, Sidhu and Guraya 1985b). These studies also indicated that the endogenous lipid peroxidation, which can be followed by the thiobarbituric acid reaction, is accompanied by loss of phospholipid, chiefly plasmalogen, and a rapid decline of motility and metabolism as a result of loss of cytosolic enzymes and essential substrates, such as adenine and pyridine nucleotides. Actually, intracellular enzymes and substrates escape from the spermatozoa through the damaged plasma membrane. Alvarez and Storey (1984) have studied the damage to the plasma membrane of rabbit epididymal spermatozoa during spontaneous lipid peroxidation by means of trypan blue uptake and expression of activity of the intracellular enzymes, lactate dehydrogenase and pyruvate kinase. Both the dye uptake and the expression of enzyme activity indicate cell damage from lipid peroxidation as loss of integrity of the plasma membrane. The process of lipid peroxidation is considered one of "autointoxication" of spermatozoa, leading to immobilization and profound and irreversible changes in fructolysis and respiration (Mann et al. 1980, Mann and Lutwak-Mann 1981).

Peroxidized lipids are also toxic to human spermatozoa (Amelar et al. 1980). The study of the rate of endogenous lipid peroxidation by the thiobarbituric reaction can provide a useful basis for biochemical appraisal of human sperm quality (Mann and Lutwak-Mann 1981). Fatty acid peroxides are highly spermicidal towards human spermatozoa. Jones et al. (1979) observed that an antiperoxidant factor present in human seminal plasma effectively counteracts the toxic effect of exogenous peroxidized fatty acids upon human spermatozoa, but is unable to restore motility lost by lipid peroxide action. This antiperoxidant factor is heat-stable, resistant to the proteolytic activity of seminal plasma, and is probably of a protein nature, although its exact chemical nature has yet to be determined. Johnsen and Eliasson (1978) observed that washed human spermatozoa produce significant amounts of lipid peroxides in the presence of Cu^{3+} and Fe^{2+}. Depletion of the Zn^{2+} content of spermatozoa by albumin, histidine, and EDTA strongly stimulates metal-catalysed production of lipid peroxides. This has shown that Zn^{2+} is important for the integrity of sperm membrane. The presence of superoxide dismutase in spermatozoa, either intracellularly or extracellularly, does not inhibit ascorborate/Fe^{2+} catalysed lipid-peroxidation reactions.

Alvarez and Storey (1982) have observed that rabbit spermatozoa undergo spontaneous lipid peroxidation in the absence of inducers with concomitant loss of motility. However, the rate of lipid peroxidation, as measured by production of malondialdehyde (MDA), depends on the ionic composition of the medium, but the production of MDA gives a linear correlation with motility loss, independent of the rate of production. Holland et al. (1982) suggested that apparently superoxide dismutase plays a major role in protecting rabbit epididymal sperm against damage from lipid peroxidation. Alvarez and Storey (1983a, b) observed that superoxide dismutase postpones but cannot prevent sperm "death". These authors also observed the inhibition of lipid peroxidation in rabbit spermatozoa by taurine, hypotaurine, epinephrine, and albumin which protect spermatozoa against loss of motility (Alvarez and Storey 1983c, d). The conditioning effect of seminal plasma on the lipid peroxide potential of washed human spermatozoa is shown by Johnsen et al. (1982a). The experimentally induced lipid peroxidation, that leads to irreversible damage of sperm structure and function, has suggested that possible analogous reactions may also occur as a part of the natural structural decay, decreasing motility, and declining metabolic activity in spermatozoa which are affected by ageing either within the reproductive tract in vivo or during storage in vitro (Mann and Lutwak-Mann 1981). The acrosomal region of the sperm plasma membrane shows a conspicuous structural damage due to lipid peroxidation in vitro, which resembles the structural changes induced by cold shock. The agents, which protect spermatozoa from the adverse effects of lipid peroxides, can also protect them partially from damage by cold shock. The oxygen damage originating from lipid peroxidation constitutes a major cause of the gradual decrease in motility and rate of oxygen consumption seen in endogenously respiring sperm suspensions. Mann and Lutwak-Mann (1981) have discussed the chemical aspects of oxygen damage to spermatozoa (see also Alvarez and Storey 1985). A relationship between lipid peroxidation within the male reproductive tract and the formation and secretion of prostaglandins is also discussed (Mann and Lutwak-Mann 1981). The rate of spontaneous lipid peroxidation in rabbit and mouse epididymal spermatozoa depends on temperature and oxygen concentration (Alvarez and Storey 1985).

The amount of phospholipid present in mammalian spermatozoa decreases during passage from the testis through the reproductive tract and there is evidence which suggests that the phospholipids may, via their long-chain fatty acid residues, serve as an important nutrient source for the sperm during maturation, as already discussed. The total phospholipid composition of the spermatozoa does not give any indication of the amount and nature of lipids present in their different components (Clegg et al. 1975, Guraya and Sidhu 1975b, 1976, Sidhu and Guraya 1979b). An analysis of isolated sperm components, which can be obtained by mechanical or chemical dissection (Millette et al. 1973, Gall et al. 1974, Mann 1978, Mann and Lutwak-Mann 1981), may be required before any definite conclusion can be reached in regard to the lipid composition of the midpiece and other components. However, studies with filipin, polymyxin B, and other cytochemical probes have revealed a heterogenous distribution of membrane cholesterol (Friend and Elias 1978, Elias et al. 1979, Friend 1980) and anionic lipids (Bearer and Friend 1980, 1981) in the guinea-pig spermatozoon. Filipin labelling has also been used as an indicator of membrane cholesterol distribution in developing spermatids of the guinea-pig (Pelletier and Friend 1983). Filipin-sterol complexes are abundant in the plasma membrane, less abundant in the acrosomal membrane, and still less abundant in the nuclear membrane. Further correlative biochemical and cytochemical studies will also be very useful in understanding the exact biochemical roles of lipids in different components of the sperm. At present, these are believed to be involved in determining membrane structure and permeability, in influencing or regulating enzyme reactions and also serving as an endogenous source of energy (Scott et al. 1963, 1967, M. H. Johnson 1970, Scott 1973, Poulos et al. 1973b, 1974, Mann and Lutwak-Mann 1981). Triglycerides and phospholipids are studied in relation to fertility (Kulka et al. 1984). The significance of bound fatty acids in human sperm has been studied in relation to fertility disorders (Nissen et al. 1981). The role of the sperm plasmalogen may become more evident with further study of the recently identified plasmalogen (phosphatidylcholine) occupying the post-acrosomal segment of the sperm head and accumulated by rabbit spermatozoa during the final stages of maturation (Cummins et al. 1974, Teichman et al. 1974).

In order to localize various biochemical effects at the membrane level, Clegg et al. (1975) developed procedures to isolate the plasma membrane and acrosome membrane from porcine sperm. These morphological and biochemical findings provided evidence of unique patterns of change in membrane lipids associated with spermatogenesis, sperm maturation, and capacitation in the pig. An early change involves the appearance of the phosphotungstic acid-chromic acid-staining component of the plasma membrane; a striking late change in the female tract is the appearance of phosphatidylinositol. Pelletier and Friend (1983a), using electron microscopy of thin sections, freeze-fracturing, and filipin labelling as an indicator of membrane cholesterol distribution, have demonstrated the development of membrane differentiations in the guinea-pig spermatid during spermiogenesis. A gradient is visible in the distribution of the cholesterol-filipin complexes, decreasing from the plasma membrane to the nuclear membrane. Even within the same membrane, cholesterol distribution varies from one pole of the cell to the other. Voglmayr (1975) suggested that the increase in the amount of unsaturated fatty acids in the phospholipid fraction of spermatozoa as these pass through the epididymis, together with a decrease in selected phospholipids and cholesterol, would greatly affect the characteristics of the plasma membrane. This is supported by the fact that the membrane structure is very labile and hence various kinds of lipids may influence the characteristics of spermatozoa lipoprotein membranes, as there is a close relation between the lipid composition and permeability of membranes (Van Deenen et al. 1972, Pelletier and Friend 1983a).

3. Metabolic Pathways and Enzymes

The mid-piece of mammalian spermatozoa shows well-organized distribution of mitochondria, as already discussed. Therefore, it forms the most important site for various metabolic activities of spermatozoa. Besides lipids, the mitochondria of the mid-piece abound in other components, including a variety of enzymes that play an active role in sperm metabolism (Mann and Lutwak-Mann 1981, Peterson 1982). Basically, spermatozoa possess two enzyme systems that are essential for activity. The first system is associated with the mid-piece and tail, and includes the enzymes involved in the membrane transport of compounds, glycolysis, citric acid cycle, oxidative phosphorylation, and other metabolic activities. These processes serve primarily to generate the energy needed for the sliding of the axonemal microtubules, that is, for sperm motility, which will be discussed in Chap. XII. The second enzyme system is associated with the acrosome, and includes various enzymes which are involved in fertilization, as already discussed in Chap. VII (see also Nelson 1985). Glycolysis and respiration constitute the main sources of energy for motility and metabolism in mammalian spermatozoa (Fig. 74), but the extent of involvement of the pentose phosphate cycle energy production is still not known precisely (Sidhu and Guraya 1979a, 1985a, Guerin et al. 1980, Mann and Lutwak-Mann 1981, Gandhi and Anand 1982, Bhela et al. 1982, Peterson 1982). Both glycolysis and respiration are bound up with the activity of spermatozoa and their rates largely depend on the concentration and motility of spermatozoa in semen. However, it is quite difficult to determine the relative contributions of glycolysis and respiration to the overall energy demands of the spermatozoa, as such a calculation depends on very precise knowledge of metabolic rates and parameters of hydrodynamic work, that are rather difficult to measure.

Spermatozoa undergo metabolic changes during their transit through the ductular system (Brooks 1979b, Mann and Lutwak-Mann 1981, Fisher-Fischbein et al. 1986). Our knowledge is still meagre in this regard in spite of the fact that this subject has been of great interest in the last decade. Panse et al. (1983) have observed shift of glycolysis as a marker of sperm maturation in the epididymis. No profound differences are observed in glycerol 3-phosphate oxidation between epididymal and testicular spermatozoa whether from ram or rat (Ford 1981). The glycerol 3-phosphate consumed by ram spermatozoa can be accounted for by the accumulation of dihydroxyacetone phosphate and Co_2 but about 50% of the glycerol 3-phosphate consumed by rat spermatozoa is converted to an unknown product. Holland and Storey (1981) studied the generation of hydrogen peroxide by rabbit epididymal spermatozoa.

The results of various studies have demonstrated a considerable species variation in the ability of ejaculated spermatozoa to oxidize various substrates, such as pyruvic acid, acetic acid, glycerol, sorbitol, long-chain fatty acids, amino acids, etc. (Mann and Lutwak-Mann 1981). For example, ram and bull spermatozoa oxidize acetic acid in preference to fructose and glucose, but in human, canine, and rabbit spermatozoa the reverse is true. Ram spermatozoa oxizide lower fatty acids with an even number of carbon atoms more efficiently than those with an odd number, whereas dog spermatozoa oxidize acetate and propionate equally well. The spermatozoa of ram and bull oxidize sorbitol, which is not oxidized by those of boar and stallion. The respiration of bull and ram spermatozoa is stimulated markedly by glycerol which has a small effect on spermatozoa of boar, and practically none on spermatozoa of either rabbit or stallion. Pyruvate, lactate, and α-glycerophosphate are rapidly oxidized by boar and guinea-pig spermatozoa, but not by human spermatozoa (Peterson 1982). Variability in the oxidative capacity of the respiratory chain among these species does not appear to be involved because, given an appropriate substrate (e.g. succinate for human spermatozoa), the maximum rate of oxygen uptake by each

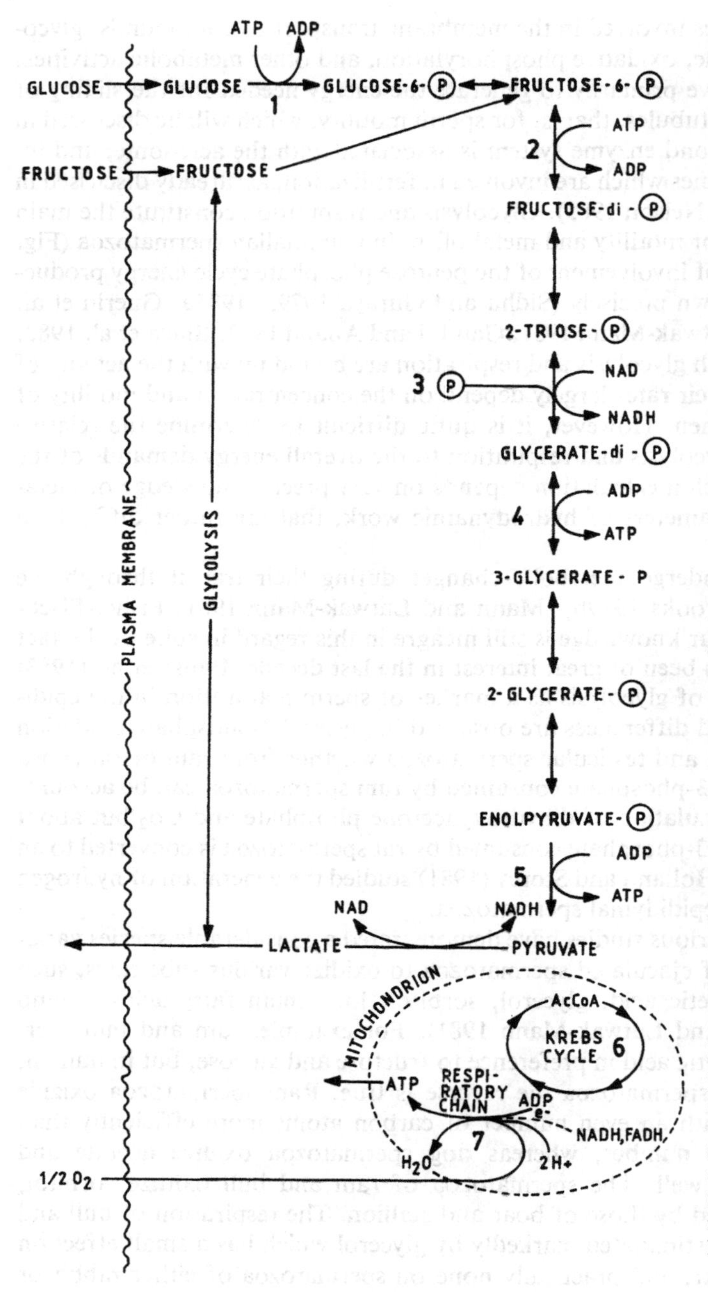

Fig. 74. Pathways of energy metabolism in spermatozoa. As in other cells in spermatozoa, the ATP required for mechanical and chemical work is generated by glycolysis and oxidative phosphorylation. In the abbreviated pathways shown in the figure, the following points are noted. The energy-consuming reactions occur at steps *1* and *2*, catalysed by the enzymes hexokinase and phosphofructokinase; steps *4* and *5* generate ATP. Control of the glycolytic rate in human and other primate sperm appears to involve primarily the enzymes phosphofructokinase and glyceraldehyde phosphate dehydrogenase (steps *2* and *3*); the latter enzyme catalyses the conversion of triose phosphate to diphosphoglyceric acid. Acetyl coenzyme A, formed from the end-product of glycolysis, is oxidized by Krebs cycle enzymes (*6*) in the mitochondria. This series of cyclical reactions generates the reducing equivalents (NADH, FADH) that are oxidized by the components of the respiratory chain (*7*) to produce additional ATP. Marked quantitative differences in the rates of glycolytic metabolism and oxidative metabolism exist among species of spermatozoa (Redrawn from Peterson 1982)

species is comparable. Undoubtedly significant differences in the regulation of Krebs cycle oxidation are observed among different species, and more experimental work is needed to determine the mechanism of regulation. There is a striking difference among species of mammalian spermatozoa in their ability to consume oxygen and in their dependence on oxygen for survival (Mann and Lutwak-Mann 1981, Peterson 1982). The physiological significance of these metabolic differences is not knwon. Metabolic differences between spermatozoa of fertile and infertile men are also reported (Pedron et al. 1979), suggesting some relationship between sperm motility and metabolism.

The effects of various chemicals on sperm metabolism have been studied to determine the mode of their action at the level of metabolic pathways. Hoskins and Casillas (1975b) reviewed the previous observations on the effects of cyclic nucleotides on the metabolism of mammalian spermatozoa. Pulkkinen et al. (1978) demonstrated specific inhibition of spermatozoal energy metabolism by polyamines and oxidized derivatives. Pulkkinen et al. (1977), after studying the mechanism of action of oxidized polyamines on the metabolism of human spermatozoa, suggested that inhibition of the fructose metabolism of intact spermatozoa is due to interaction of the iminoaldehyde with sperm membranes, and not to inhibition of any enzyme of the glycolytic pathway. The metabolism of spermatozoa is greatly altered by α-chlorohydrin (Stevenson and Jones 1982, 1984, Ford and Jones 1983) and gossypol (Wichmann et al. 1983, Stephens et al. 1983). The cellular mechanism by which gossypol exerts its toxic and antifertility effects are yet to be resolved. Its actions appear to be exerted at the level of cell membranes (Reyes et al. 1984). Kelly (1977) observed that 19-hydroxy E prostaglandins depress the respiration of spermatozoa, but do not affect the production of lactate. This effect is not seen with the prostaglandins of the E and F series. The seminal fluid contains a heat-labile component that inhibits the oxidative metabolism of human sperm but does not significantly alter sperm glycolysis (Trifunac and Bernstein 1976). It may be the prostaglandin. Witters and Foley (1976) studied the effect of selected inhibitors and methylene-blue on a possible phosphogluconate pathway in washed boar sperm. The data obtained suggested that a phosphogluconate pathway is activated after treatment of sperm with methylene blue dye. Guerin and Czyba (1976) observed that the respiratory activity is markedly reduced after freezing in liquid N_2. Dop et al. (1978) studied the selective inhibition of pyruvate and lactate metabolism in bovine epididymal spermatozoa by dinitrophenol and α-cyano L_3-hydroxycinnamate. The results of all these studies indicate that sperm metabolism is greatly affected by different drugs. But the mechanism of action need to be defined for most of these drugs. The results of recent studies have provided strong evidence for the localization of various enzymes related to metabolism in the mid-piece of spermatozoa.

a) Enzymes of Glycolysis

Much work has been carried out on the metabolism by spermatozoa of a great number of substrates and particularly on those of the glycolytic pathway. However, there has been relatively little study of the isolation and histochemical localization of the enzymes involved in these metabolic pathways. The enzymes of the glycolytic apparatus shown in Fig. 74 are believed to be present in the mid-piece, although their precise location in this region has not been determined; even their purification and characterization are little worked out. Indeed, the possibility that glycolytic enzymes may occur in regions of the flagellum other than the mid-piece has not been excluded. It is improbable that all glycolytic enzymes exist in free solution in the limited volume of the cytosol of the flagellum. Sidhu and Guraya (1979a) demonstrated that the glycolytic enzymes in buffalo spermatozoa are exclusively present in the mid-piece. Hexokinase, is a rate-controlling enzyme in glycolysis and is strongly bound to the mid-piece of buffalo (Sidhu and Guraya 1979a), bull, ram, and boar (B. E. Morton 1968, R. A. P. Harrison 1971, Storey and Kayne 1975) spermatozoa. Spermatozoa possess two isozymes of hexokinase. Type I hexokinase is the predominant type of caput sperm, whereas the sperm type hexokinase is predominant in caudal sperm. Hexokinase types II and III are absent in both extracts. According to Sidhu and Guraya (1979a) over half of the aldolase activity is strongly bound to the mid-piece of buffalo spermatozoa and this is confirmed by its histochemical localization. Similar observations were also made for rabbit spermatozoa (Storey and Kayne 1975). Chauhan (1973), using biochemical techniques, also estimated aldolase activity in buffalo spermatozoa. Glyceraldehyde-3 phosphate dehydrogenase is loosely bound to the mitochondria (Sidhu and Guraya 1985a). Lactate dehydrogenase (LDH) is strongly bound to the mid-piece of buffalo (Sidhu and Guraya 1979a, Gupta and Srivastava 1981) and other mammalian sperm (Mann 1964, R. A. P. Harrison 1971, Mathur 1971, Storey and Kayne 1975, 1977, 1978, Hutson et al. 1977a, b). Herak et al. (1976) observed that there is an effect of mating season upon the activity of some stallion sperm enzymes, such as beta-glucuronidase, lactate dehydrogenase (LDH), and alpha-hydroxybutyrate dehydrogenase. The activity of beta-glucuronidase and LDH in the stallion's sperms is highest in May.

Some controversy exists about the functional significance and subcellular distribution of LDH-X. The general consensus is that it occurs in the mitochondria and cytosol. Both the intramitochondrial and the extramitochondrial LDH seem to be the same isozyme unique to sperm cells; LDH-X or LDH-C_4 (E. Goldberg 1979, Georgiev 1982, Burkhart et al. 1982, Wheat and Goldberg 1983, Coronel et al. 1983). Bull sperm appear to contain two forms of LDH-C_4 (Georgiev 1982). Van Dop et al. (1977) and Hutson et al. (1977a, b) observed that intramitochondrial LDH functions actively in bovine spermatozoa. Direct oxidation of intramitochondrial pyridine nucleotide by pyruvate in bovine sperm has been shown by Milkowski and Lardy (1977). Coronel et al. (1983) studied the catalytic properties of LDH-X or C_4 from different species. It is a very specialized enzyme, which is believed to be related to metabolic processes supplying energy for sperm motility and survival (C. Burgos et al. 1979, Blanco 1980, Peterson 1982). LDH, particularly the X isozyme (LDH-X), is believed to play an important role in mitochondrial metabolism in spermatozoa. The enzyme, which is present in the cytosol of most other cells, is present in the mitochondria of mammalian spermatozoa. Baccetti et al. (1975), using histochemical techniques also laid emphasis on the mitochondrial localization of LDH-X. It is present in the mitochondrial matrix. Cytoplasmic LDH-X has been localized in the tail sheath of mature intact mouse, human, and rat spermatozoa by histochemical and immunological techniques (Burkhart et al. 1982). The enzyme is also found within the mitochondria. Intact sperm demonstrate a mitochondrial LDH-X. Montamat and Blanco (1976), and Hintz and Goldberg (1977) demonstrated by biochemical and histochemical methods that the cytosol contains a large part of LDH-X of mouse spermatozoa. Montamat and Blanco (1976) quantitated the distribution as about 60% cytosolic and extramitochondrial, and 20% intramitochondrial in these cells. However, Storey and Kayne (1977) estimated that about 90% of the LDH in rabbit spermatozoa is extramitochondrial. The dual localization of LDH enables mammalian spermatozoa to exchange cytosolic and mitochondrial reducing equivalents by means of a lactate/pyruvate shuttle (Blanco et al. 1976, Blanco 1980). The existence of an active intermitochondrial LDH poses a problem to the sperm mitochondrion with regard to pyruvate oxidation. LDH competes effectively for reducing equivalents with respiratory-chain enzymes and permits the direct oxidation of the lactic acid produced by glycolysis. The advantage of this specialization is not clear, although it has been suggested that mitochondrial LDH may participate in shuttling reducing equivalents from the cytosol to the mitochondria, where energy yields are higher (Peterson 1982, Burkhart et al. 1982). Although the effective operation of the 2-oxa 2-hydroxymono carboxylate shuttle remains to be demonstrated, this hypothesis set forth fits the available data and offers an explanation for the possible physiological role of LDH-X in spermatozoa. Gossypol acetic acid affects the activity of LDH-C_4 in human and rabbit spermatozoa (Eliasson and Virji 1982). Storey and Kayne (1978) studied the interactions between lactate, pyruvate, and malate as oxidative substrates for rabbit sperm mitochondria. Burgos et al. (1979) investigated the substrate specificity of LDH-C_4 from human spermatozoa. Goldman-Leikin and Goldberg (1983) characterized monoclonal antibodies to the LDH isozyme in the mouse. Wheat and Goldberg (1983) discussed LDH-C_4 in relation to antigenic structure and immunosuppression of fertility (Wheat et al. 1983, Shelton et al. 1983). A good correlation of LDH-C_4 with the count and motility of spermatozoa has been reported for humans (Gerez de Burgos et al. 1979) and ram (Rao and Pandey 1977). The release of LDH-C_4 from human spermatozoa has been used as an index of membrane permeability (Eliasson and Virji 1982). Skrabei and Mettler (1977) reported a technique which represents an attractive method for the selective isolation of LDH-X.

Studies of total potential activities of the various glycolytic enzymes carried out with homogenized spermatozoa have shown the existence of these enzymes at levels above the requirement of intact spermatozoa for a normal rate of fructolysis (Mann and Lutwak-Mann 1981). Three enzymes, such as hexokinase, phosphofructokinase, and pyruvate kinase, are considered as potential regulators of sperm fructolysis. Phosphofructokinase is a fragile enzyme that is adversely affected by sperm homogenization. Various analyses have also shown that phosphofructokinase and glyceraldehyde-3-phosphate dehydrogenase play key roles in the regulation of glycolysis in bovine, monkey, and human spermatozoa (Peterson 1982). The activities of these enzymes are inhibited by ATP in most cell types including spermatozoa (allosteric effect) and therefore their activities and the rate of glycolysis will decrease when energy reserves in the form of ATP are high. Ford and Harrison (1983) demonstrated changes in the activity of glycerdehyde-3-phosphate dehydrogenase in spermatozoa from different regions of the epididymis in rodents treated with alpha-chlorohydrin or 6-chlorodeoxyglucose. The buffalo spermatozoa contain ketose and aldose reductases which are involved in interconversions of sorbitol and fructose (Gandhi and Anand 1982, Sidhu and Guraya 1985a).

b) Enzymes of Krebs and Pentose Phosphate Cycles

In contrast to the enzymes of glycolysis, those of the Krebs cycle (Fig. 74) occur in the mitochondrial fraction of the cell, mostly in the matrix in proximity to the respiratory enzymes (Harper 1973). Isocitrate, succinate, and malate dehydrogenases in buffalo spermatozoa are strongly bound to the mid-piece (Sidhu and Guraya 1979a). Succinate dehydrogenase activity has also been estimated in buffalo (Ayyagari and Mukherjee 1970, Singh and Sandhu 1972), rat, mouse, and human (Rao et al 1959, Mathur 1971) spermatozoa. It shows a seasonal variability in the buffalo (Sidhu and Guraya 1979b). The levels of malate and succinate dehydrogenases in ram semen are positively correlated to the initial motility, live sperm percentage, and sperm concentration (Rao and Pandey 1977). Glutamate oxaloacetate transaminate is highest in the tail fraction of buffalo spermatozoa (Gupta and Srivastava 1981). Two isozymes of malate dehydrogenase and three of isocitrate dehydrogenase are reported for buffalo (Sidhu and Guraya 1979b).

The contribution of the pentose phosphate pathway to sperm metabolism is controversial and needs to be determined precisely. However, the enzymes of the pentose phosphate cycle, which include glucose-6-phosphate and 6-phosphogluconate dehydrogenases, are strongly bound to the membranes of the sperm mid-piece in boar (Bolton and Linford 1970) and buffalo (Sidhu and Guraya 1979a). Slight activity of glucose-6-phosphate dehydrogenase is also reported in human, mouse, rat, guinea-pig, cat, and dog spermatozoa (Balogh and Cohen 1964). Baccetti (1975) demonstrated cytochrome oxidase activity in the mitochondrial cristae of bull sperm tail. Sidhu and Guraya (1979a) suggested that the pentose phosphate cycle may be involved in the energy production of buffalo spermatozoa. Ito and Amano (1980) observed high levels of two enzymes (glucose-6-phosphate dehydrogenase and 6-phosphogulconate dehydrogenase) associated with the pentose phosphate pathway in human sperm. This is consistent with the conclusion that ribulose-peptide in human sperm is synthesized from D-glucose via the pentose phosphate pathway. Cohn and Jacobs (1980) have demonstrated the involvement of pentose phosphate cycle in ribulose-peptide synthesis and energy production in human spermatozoa. In other tissues, the pentose phosphate cycle serves two important functions: (1) generation of reduced NADP needed for the synthesis of fatty acids or steroid and amino acids by glutamate dehydrogenase and (2) provision of pentoses for nucleotide and nucleic acid synthesis.

c) Phosphatases

Various phosphatases occur in the mammalian spermatozoa, especially in their mid-piece. These phosphatases include Ca^{2+}- and Mg^{2+}-activated ATPase, acid and alkaline phosphatases, thiamine pyrophosphatase, and glucose-6-phosphatase (see Sidhu and Guraya 1985a). The physiological significance of these phosphatases, except ATPase, is obscure. The ATPase (adenosine triphosphatase) activity, that provides energy for sperm motility, has been described in the tail, mostly concentrated in the mid-piece (Mann and Lutwak-Mann 1981, Nelson 1985). However, Mg^{2+}-activated ATPase(s), extractable from mammalian sperm by current methods, is contaminated with mitochondrial ATPase(s) (L. G. Young and Smithwick 1976). The presence of a bivalent cation-dependent ATPase is also shown in human spermatozoa (Abla et al. 1974). Smithwick and Young (1977, 1978) developed methods to produce decapitated, plasmalemmae-free, mitochondria-free, isolated flagella. Dhanottya and Srivastava (1979) assayed ATPase and fructose-1, 6-diphosphatase (FDPase) in the head, mid-piece and tail fraction of buffalo and bull spermatozoa. The highest activity of ATPase is present in the tail fraction, followed by the mid-piece, and a very low activity in the head fraction. The FDPase is localized in the tail fraction; while ATPase is mostly particulate-bound, FDPase appears to be in a highly soluble form. Storey and Kayne (1980) observed that the activities of pyruvate kinase and the flagellar ATPase in a given preparation of hypotonically treated mature rabbit epididymal spermatozoa are comparable. Its spermatozoa have a metabolic strategy which is very similar to muscle cells. It is well established that ATPase brings about the hydrolysis of ATP to ADP with the release of energy. Therefore, the ATPase present in the mid-piece of spermatozoa may be important in supplying the energy for their motility (Mann and Lutwak-Mann 1981, Peterson 1982). ATPase is also involved in permeability and transport processes of the spermatozoa (Guraya and Sidhu 1975a), as evidenced from its presence in the sperm plasma membrane (see Chap. XI). Storey and Kayne (1980) suggested that the major use of the sperm cell's metabolic machinery is in the maintenance of energy for the contractile work of motility, and only minor amounts of metabolic energy appear to be consumed in other reactions including those involed in fertilization.

The spermatozoa of buffalo show acid and alkaline phosphatase activities in the mid-piece, which may be associated with the dephosphorylation and transport of phosphate groups between the seminal plasma and spermatozoa (Guraya and Sidhu 1975a). Acid phosphatase is also known to bring about dephosphorylation of phosphorylcholine present in the semen into choline and orthophosphate. The former is of some physiological significance in the metabolism of phospholipids in the spermatozoa (Mann 1964, 1975, Mann and Lutwak-Mann 1981). The highest number of phosphatase isoenzymes has been reported in boar spermatozoa. Similarly, Sidhu and Guraya (1978a) reported isoenzymes of acid phosphatase in buffalo spermatozoa. Isoenzymes of alkaline phosphatase have also been reported in ram spermatozoa (Petkov and Babadov 1976). Zakhariev et al. (1983) have demonstrated the effect of biologically active substances on the alkaline phosphatase activity, viability, and fertility of bull spermatozoa. The results of various studies have clearly demonstrated the presence of various phosphatases in association with the tail, and especially the mid-piece of spermatozoa. Loewus et al. (1983) have reported the activity of myoinositol-1-phosphate synthase in the epididymal spermatozoa of ram.

d) Esterases

There are relatively few reports on the presence of esterases in the mid-piece of mammalian spermatozoa (Mathur 1971, Meizel and Cotham 1972). Various esterases, such as nonspecific esterases, α-galactosidase, β-glucosidase, carbonic anhydrase, phosphodiesterase, and choline-esterase, have been demonstrated in the mid-piece of buffalo spermatozoa (Sidhu and Guraya 1985a). Nonspecific esterases, α-glactosidase and β-glucosidase may be involved in the hydrolysis of glycosidic linkages and thus facilitate the transport of hydrolysate from seminal fluid to spermatozoa for energy production (Mann 1964, Mann and Lutwak-Mann 1981). Carbonic anhydrase participates in the condensation of Co_2 and H_2O to H_2CO_3, and its dissociation hence helps in maintaining the acid base balance of sperm environment. Phosphodiesterase is involved in the metabolism of cAMP in the spermatozoa, as will be discussed later on. Nonspecific esterases and acetyl-cholinesterase have been studied for their involvement in the regulation of sperm motility, as will be discussed in Chap. XII.

e) Hydroxysteroid Dehydrogenases and Effects of Steroids on Spermatozoa

In recent years, the sperm hydroxysteroid dehydrogenases and the effects of steroids on spermatozoa have formed the subject of many studies. Most of the studies have indicated that the sperm hydroxysteroid dehydrogenases are generally localized in the mid-piece. Wester et al. (1972) found that most of the enzyme for conversion of 17β-oestradiol to oestrone was located in the flagellum. Testosterone at physiological levels enhanced motility and there was significant interaction between testosterone and sperm concentration. Sidhu and Guraya (1980b) made a correlative cytochemical and biochemical study of various hydroxysteroid dehydrogenases (HSDH), viz. 3α-, 3β-, 17β-, Δ^5-3-β-, and 16α-HSDHs in buffalo spermatozoa, which are specifically localized in their mid-pieces. 17β-HSDH (testosterone as substrate) and Δ^5-3β-HSDH (dehydroepiandrosterone as substrate) are readily released by sonication alone. However, most of the activities of 3α-, 3β-, 16α-HSDH are strongly bound, as these are released only after detergent treatment. 17β-HSDH shows more activity with oestradiol than with testosterone as substrate. Similarly, Δ^5-3β-HSDH oxidizes pregnenolone in preference to dehydroepiandrosterone as substrate. Killian and Amann (1974) demonstrated the presence of 3α- and 3β-HSDHs in bull spermatozoa. McGaday et al. (1986) localized 17β-HSDH histochemically in the spermatozoa of the hamster. However, Killian and Amann (1974), using testosterone as substrate could not find 17β-HSDH in bull epididymal or testicular spermatozoa. Interconversions of testosterone and androstenedione by 17β-HSDH were reported in ejaculated spermatozoa from several species of mammals including bulls (Scott et al. 1963, Seamark and White 1964). It was observed that 5β androgens are among the most effective androgens in altering the various metabolic patterns in vitro of bull and buffalo spermatozoa (Voglmayr 1971, Setchell and Waites 1974, Hoskin and Casillas 1975a, Sidhu and Guraya 1978c). Sidhu and Guraya (1980b) suggested that 3α-, 3β-, 17β-, Δ^5-3β-, and 16α-HSDHs demonstrated in the mid-piece of buffalo spermatozoa may be involved in interconversions of androgens which are known to alter various metabolic patterns of spermatozoa (Voglmayr 1975, Sidhu and Guraya 1978c).

Very divergent view exist about (1) the binding of steroids and gonadotrophins by spermatozoa (see references in Amann and Hammerstedt 1976, Cheng et al. 1981a, b, Allag et al. 1983, Isosefi and Lewin 1984, David et al. 1985), (2) the effects of exogenous steroids on the metabolism and fertilizing capacity of spermatozoa (Hammerstedt and Amann 1976a, Hyne et al. 1978, Bhela et al. 1982, Rajalakshmi et al. 1983, Chan et al. 1983) and (3) interconversions of steroids by spermatozoa (Hammerstedt and Amann 1976b, Dow et al. 1985). Voglmayr (1975) suggested that androgens are involved in the maturation process of spermatozoa which in the epididymis may be exposed to an environment of high androgen concentration. Legault et al. (1979) observed that hamster spermatozoa obtained from the cauda epididymidis possess twice the ability to take up sterol sulphates in vitro compared to spermatozoa obtained from the caput. This suggests that a modification of the membrane composition of the spermatozoa evidently occurs during passage through the epididymis. Free steroids are taken up in a similar pattern. Radioautographic studies have shown that for sterol sulphates this uptake occurs selectively in the regions of the head and mid-piece of the spermatozoa, whereas the free sterols are distributed evenly throughout the length of the spermatozoa. The binding of sterol sulphates to spermatozoa seems to involve sites that are unsaturable. Sterol sulphates, previously implicated in membrane stabilization, may play a similar role in spermatozoa. Burck and Zimmerman (1980) recently demonstrated that sterol sulphates inhibit in vitro capacitation of boar spermatozoa by inhibiting their acrosin activity. The effects of androgens on the glycolytic activity of spermatozoa depend on the "maturity" of the spermatozoa and also on the nature of the androgen (Voglmayr 1975). A number of androgens, which stimulate glycolysis of ram and bull spermatozoa, concurrently depress their oxidative metabolism. The effects of steroid hormones on membrane sugar transport in human spermatozoa are indicated (Ballesteros et al. 1983).

The steroid hormones undergo binding as a prelude to metabolic conversion, mostly dependent on enzymatic oxidoreductions. Human spermatozoa bind testosterone more strongly than any other steroids, and they also have in their membranes oestradiol-specific binding sites (Fig. 75) for which testosterone and progesterone compete successfully; the adverse effect of progesterone and certain synthetic contraceptive progestational compounds on the fertilizing ability of human spermatozoa may be due to competition for steroid binding sites on the sperm plasma membrane (Hyne and Boettcher 1978). Striking similarities exist in the binding pattern of antisera to steroid hormones found on the human and monkey spermatozoa (Allag et al. 1983). The concentration of testosterone on the acrosomal and post-acrosomal regions are higher than levels of progesterone and oestrogens, as judged from the intensity of fluorescence. Intact bovine spermatozoa also readily take up a variety of steroids; progesterone binds with the greatest affinity, followed in decreasing .order by oestradiol, dihydrotestosterone, testosterone, androstenediol and etiocholanone (Amann and Hammerstedt 1976). Amann and Hammerstedt (1976) concluded that very few, if any, specific saturable, high-affinity steroid binding sites exist on intact or ejaculated

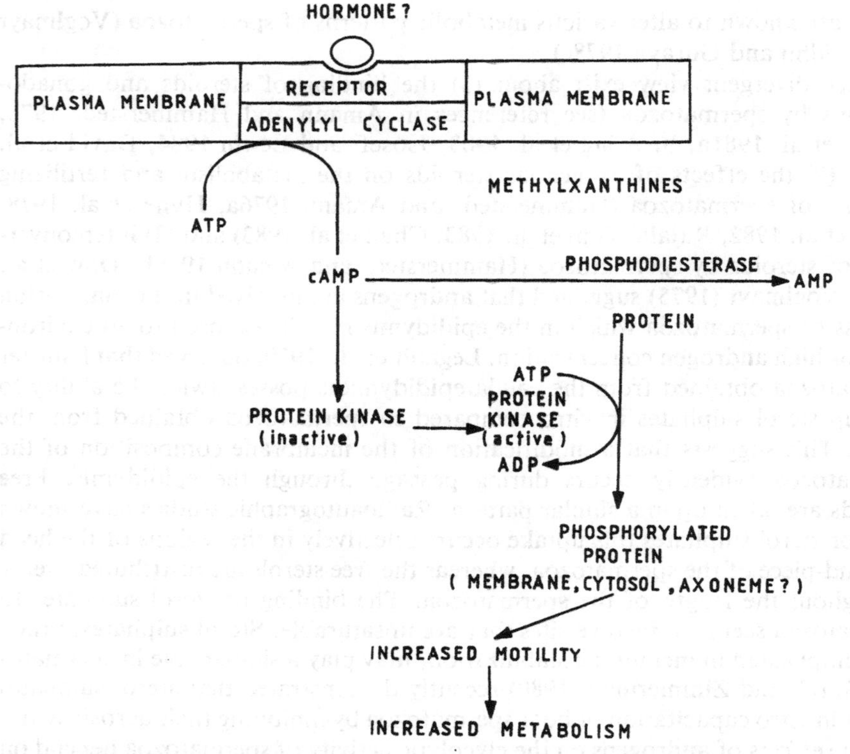

Fig. 75. Role of cyclic AMP in cellular metabolism. The diagram depicts the general scheme for the mechanism of action of cAMP. Membrane-bound adenylyl cyclase, possibly activated by an as yet unidentified hormone, converts ATP to cAMP. The cyclase and the putative hormone receptor are very likely free to move with the fluid liquid bilayer. This reaction is controlled in many cells by calcium and a calcium-binding protein, calmodulin, which recent experiments suggest may also exist in spermatozoa. Cyclic AMP is known to activate protein kinases, which in turn may act to phosphorylate a variety of proteins throughout the cell. These phosphorylated proteins enhance motility by mechanisms that are not known. Caffeine, theophylline, and other phosphodiesterase inhibitors inhibit the breakdown of cAMP, and thereby potentiate its actions (Redrawn from Peterson 1982)

bovine sperm. No single androgen at physiological concentration changes energy metabolism (Hammerstedt and Amann 1976a). Hammerstedt and Amann (1976b) observed few qualitative differences in the steroid interconverting enzyme systems between testicular and ejaculated bovine spermatozoa which contain C_3 and C_{17}-hydroxysteroid oxidoreductases, and probably both sperm-cell types contain a Δ^5-3-ketosteroid 5α-oxidoreductase active on androstenedione. The steroid binding protein acts as a steroid carrier protein across the sperm plasma membrane (David et al. 1985). Hyne et al. (1978) recently studied the metabolism and motility of human ejaculated spermatozoa incubated in vitro with steroids. Progesterone and norethynodrel depresses the respiration, glycolytic metabolism, and the motility of washed spermatozoa. Lynoestrenol does not affect the respiration or glycolysis of the spermatozoa, but does inhibit motility. Oestradiol does not cause any change in the metabolism and motility.

Progesterone and norethynodrel appear to act on the plasma membrane of human spermatozoa to increase its permeability and hence to facilitate the loss of essential co-factors needed for the glycolytic and oxidative processes. Further investigations on the direct effect of steriods on the spermatozoa are needed to reach some definite conclusions. However, Hyne and Boettcher (1978) observed that there is a direct and specific steroid effect on human spermatozoa, as some steroids, such as progesterone, lynestrenol, and norethynodral, markedly inhibit sperm migration and motility, whereas other steroids, such as oestrone, have no detectable effect on sperm migration and motility. There is no conclusive evidence that any of the steriods that are either bound or metabolized by the spermatozoa can be synthesized by the spermatozoa de novo. Steroid hormones differ in this respect from the cyclic nucleotides which the spermatozoa bind, metabolize, and synthesize, as will be discussed here later on.

f) Cyclic Nucleotides and Their Regulatory Enzymes

Garbers et al. (1971b) observed that cyclic nucleotides play a role in the regulation of sperm motility (Fig. 75). Since then it has been well established that mammalian spermatozoa contain cyclic adenosine 3′, 5′-monophosphate (cyclic AMP), and some of the enzymes needed for the synthesis, degradation and expression of action of these cyclic nucleotides (Fig. 75) (R. G. Harrison 1975, Hoskins and Casillas 1975a, b, Purvis et al. 1982, Amann et al. 1982, Tash and Means 1983). The content of cyclic nucleotides varies with the maturational stage of mammalian sperm, e.g. caput vs caudal epididymal, as well with the species. The acknowledged increase in intracellular cAMP levels, which occurs in spermatozoa during epididymal transit, may be a consequence of both increased synthesis (adenylate cyclase) and reduced hydrolysis (phosphodiesterase). Cyclic AMP plays a role in sperm motility, metabolism, maturation, capacitation, acrosome reaction, and fertilization (see also Stengel and Hanoune 1984). A relationship apparently exists between fertilizing ability and cAMP in human spermatozoa (Perreault and Rogers 1982). More recent studies have indicated that cAMP under the proper conditions may be the primary signal for the onset of progressive motility.

The principal enzymes concerned with the synthesis and metabolism of cyclic nucleotides in spermatozoa are adenylate cyclase, guanylate cyclase, AMP-dependent protein kinase, and phosphodiesterase (Fig. 75). Our knowledge of the intracellular localization of these enzymes is not sufficient to enable a precise assignment of their activity to particular structural components of the spermatozoa (Tash and Means 1983) but it is generally believed that a substantial part of this activity is present in the particulate cell fraction and residues in the plasma membrane (Peterson 1982, Tash and Means 1983). However, the plasma membrane may not be the only cellular locus of this enzyme, as a significant portion of adenylate cyclase in bull sperm could be isolated in soluble form, and cytochemical studies at the electron microscope level have shown that a portion of the cellular adenylate cyclase activity in human spermatozoa may be located along the axonemal complex (Peterson 1982). The localization of adenylate cyclase at sites other than the plasma membrane is reported only for a few cell types, and these cytochemical observations must be confirmed by direct enzyme assay using isolated axonemal preparations. Purvis et al. (1982) observed that both cAMP and cGMP phosphodiesterases as well as the adenylate cyclase are all associated primarily with the particulate fraction, and the extent to which these enzymes are associated with the membranes, increases as the spermatozoa pass through the epididymis. Sperm protein carboxylmethylase activity is mainly soluble in all segments of the epididymis. Adenylate cyclase, cAMP phosphodiesterase, and protein carboxymethylase activities are present mainly in sperm tails, while cGMP phosphodiesterase is equally distributed between heads and tails. Cyclic AMP phosphodiesterase has been studied in detail for the buffalo spermatozoa (Bhatnagar and Anand 1981, 1982, Bhatnagar et al. 1979, 1982). Bhatnagar et al. (1979) observed that cyclic AMP-phosphodiesterase of buffalo spermatozoa is distributed in the head, mid-piece, and tail fractions and has multiple forms, 70% of which is in the bound form. Bhatnagar et al. (1982) extended these observations by making a detailed intracellular localization of adenylate cyclase and 3′,5′-cyclic nucleotide phosphodiesterase in different components of buffalo sperm. Adenylate cyclase activity is distributed in heads (8.4%), mid-pieces (16.6%), tail (49.5%), and 5.7% in the soluble supernatant, the total recovery being 81%. In the tail region, adenylate cyclase is found in the plasma membrane and microtubules. Cyclic nucleotide phosphodiesterase in tail is distributed in the plasma membrane (13.7%), microtubules (31.5%), and cytosol (34%), with a total recovery of 80%. The results are discussed in relation to control of cAMP levels in buffalo sperm by adenylate cyclase and cyclic nucleotide phosphodiesterase. In bull spermatozoa 15% of the total adenylate cyclase activity is associated with the plasma membrane and shows highest specific activity compared to other fraction (Herman et al. 1976, Zahler and Herman 1976). However, 30% of its activity is present in the soluble fraction that is believed to be dissociated from the sperm membrane. Cheng and Boettcher (1982) made a study of partial characterization of human spermatozoal phosphodiesterase and adenylate cyclase and also studied the effect of steroids on their activities. None of the steroids was found to have any significant effect. Stephens et al. (1979) studied multiple forms, kinetic properties and changes of the cyclic AMP phosphodiesterase of bovine spermatozoa from the cauda and caput epididymides. Its five separate enzymatic forms have been demonstrated. Shah and Sheth (1978) demonstrated inhibition of phosphodiesterase activity of human spermatozoa by spermine (Casillas et al. 1980).

The enzyme system seems to be a dynamic one, as cyclic AMP is continuously formed and broken down (Fig. 75). The modulation of the activities of either of these enzymes affect the intracellular concentration of cyclic AMP (Hoskins and Casillas 1975b, Tash and Means 1983). Many of the actions of cyclic AMP are believed to be mediated by an interaction of cyclic AMP with cyclic AMP-dependent protein kinases (type I and type II) which catalyse the subsequent phosphorylation of ATP of specific regulatory proteins (Fig. 75), usually enzymes, with a resultant initiation of a physiological event (Tash and Means 1983). However, Tash and Means (1982, 1983) identified a major axonemal protein, namely tubulin, as a phosphoprotein subject to regulation by cAMP. Huacuja et al. (1977) suggested that the cyclic AMP-mediated alterations in the structural and biochemical properties of the human spermatozoa are produced through phosphorylation of specific membrane proteins. The modifications of the tertiary and quaternary structure of the plasma membrane proteins through cyclic AMP-stimulated phosphorylation appear to be requirements for the initial steps of capacitation. Reyes et al. (1979) observed that physiologic levels of capacitation are attained when dibutyryl cyclic adenosine 3′5′-monophosphate is added in the presence of calcium and ionophore A23187 for treatment of spermatozoa. Ca^{2+} is also needed for mouse sperm capacitation and fertilization in vitro (Fraser 1982, Chap. II). Added Ca^{2+} has been suggested to cause an elevation of cyclic AMP concentrations in both hamster (B. E. Morton et al. 1974) and guinea-pig (Hyne and Garbers 1979) spermatozoa. Garbers et al. (1982) have observed a requirement of bicarbonate for Ca^{2+}-induced elevation of cAMP in guinea-pig spermatozoa. Although the biochemical mechanism of the Ca^{2+}-induced elevation of cyclic AMP concentration has not been established in previous studies, the phosphodiesterase inhibitor, 1-methyl-3-isobutylxanthine, acted synergistically with Ca^{2+} to elevate cyclic AMP concentration in guinea-pig spermatozoa (Hyne and Garbers 1979), suggesting that the addition of Ca^{2+} induces an activation of adenylate cyclase. This suggestion is supported by the recent observations of Hyne and Lopata (1982), who have shown that added Ca^{2+} activates adenylate cyclase activity of human ejaculated spermatozoa in broken cell preparations in the presence of Mn^{2+} or Mg^{2+}. Tash and Means (1982, 1983) suggested that Ca^{2+} inhibits motility and the overall incorporation of phosphate into endogenous proteins. The spermatozoa were shown some years ago to contain a Ca^{2+}-binding protein (Brooks and Siegel 1973) and this was subsequently identified as calmodulin, based on its ability to activate cyclic nucleotide phosphodiesterase from tissues other than spermatozoa (H. P. Jones et al. 1978, Tash and Means 1983, Nelson 1985). The sperm calmodulin, however, did not appear to activate cyclic nucleotide phosphodiesterase obtained from spermatozoa (Wells and Garbers 1977). Geremia et al. (1984) have made characterization of a calmodulin-dependent high-affinity cAMP and cGMP phosphodiesterase from male mouse germ-cells. The presence of calmodulin-dependent cyclic nucleotide phosphodiesterase has also been shown in association with demembranated rat caudal epididymal sperm (Wasco and Orr 1984). Tash and Means (1983) and Nelson (1985) have discussed the localization, chemical characterization and possible functional significance of calmodulin, especially in relation to capacitation, sperm motility, and fertilization.

Another calcium-dependent regulator protein similar in action to the intracellular calcium receptor (calmodulin) has been isolated from both human and ram seminal plasma. It is dubbed Calsemin (Bradley and Forrester 1980, 1982). Calsemin modulates the sperm-cell plasma membrane enzyme, a Ca^{2+}-dependent Mg^{2+}-ATPase that is said to be responsible for Ca^{2+} efflux from mammalian ejaculated sperm. Calsemin brings the level of the enzyme activity in cauda epididymal sperm up to that of the ejaculated sperm.

The sperm adenyl cyclase seems to be generally insensitive to stimulation by hormones (for review, see R-G Harrison 1975, Hoskins and Casillas 1975a, b). However, of the various hormones used, including gonadotrophins, steroids, prostaglandins (E_1 and E_2), polypeptide hormones, and catecholamines, only

3,3-5-L-triiodothyronine (T_3) has been observed to stimulate monkey sperm adenyl cyclase (Hoskins and Casillas 1975b, Tash and Means 1983, Nelson 1985). This suggestion that T_3 is physiologically important in vivo needs to be confirmed by the observation that increased adenyl cyclase activity in intact spermatozoa in response to exogenous T_3 is not reflected in increased intracellular cyclic AMP. Berger and Clegg (1983) studied adenylate cyclase activity in porcine sperm in response to female reproductive tract secretions. L. A. Johnson et al. (1983b) made extraction and chromatography of a new sperm protease which activates adenylate cyclases in bull. A low-molecular-weight factor from porcine seminal fluid can activate spermatozoan adenylate cyclase (Okamura and Sugita 1983). Forskolin does not activate sperm adenylate cyclase (Forte et al. 1983). However, the mechanism of its action has been studied by Schneyer et al. (1983). An inhibition of bovine sperm adenylate cyclase by adenosine and adenosine analogues is reported (M. A. Brown and Casillas 1983). Stengel and Hanoune (1984) have described procedure to purify sperm adenylate cyclase.

Very little is known about the enzyme guanyl cyclase which catalyses the synthesis of cyclic GMP (Hoskins and Casillas 1975a, b). Garbers and Radany (1981) have described the characteristics of the soluble and particulate forms of guanylate cyclase. Hanski and Garty (1983) studied the role of guanine nucleotide binding proteins in relation to activation of adenylate cyclase by sperm membranes.

The physiological roles of cyclic AMP as a controlling agent in sperm metabolism and motility (Fig. 75) are still not established and many of the gaps, which remain in our understanding of cyclic AMP metabolism, are fundamental (Tash and Means 1983). Cyclic cAMP-dependent protein kinase, which has been most thoroughly studied and demonstrated in unusually large amounts in sperm extracts, is believed to play a possible direct role in sperm motility (Hoskins and Casillas 1975b, R. G. Harrison 1975, Tash and Means 1983). The enzyme catalyses the phosphorylation of proteins by ATP, and its activity can be stimulated by cyclic AMP (Brandt et al. 1980). Both the plasma membrane and axonemal proteins have been implicated as sites of action of cAMP-activated protein kinases (Peterson 1982, Tash and Means 1982, 1983). Cyclic AMP-dependent protein kinases have been characterized and localized in buffalo ejaculated (Varshney and Anand 1981), bull ejaculated (Pariset et al. 1984), and rat caudal epididymal (Horowitz et al. 1984) spermatozoa. In buffalo, this protein kinase is present in head, mid-piece, and tail fractions. Treatment of head fraction with 0.1% Triton X-100 solubilizes 35% – 40% of protein kinase activity, the remaining 60% – 65% of activity is associated with sperm chromatin. The fact that 50% of protein kinase in buffalo spermatozoa is solubilized with 1M NaCl shows its membrane-bound nature possibly by ionic interaction. Brandt et al. (1980) observed that a significant cAMP-dependent protein phosphorylation in the sperm head plays an important role in capacitation. The opposing reaction to protein phosphorylation is dephosphorylation catalysed by protein phosphatase, which has been shown to be present in sperm (Tash and Means 1983). These observations suggest that protein phosphorylation and dephosphorylation and protein phosphate turnover may be important in the regulation of sperm function. The sperm protein kinase of the flagellar protein complex links ATP breakdown with contractile processes, suggesting the possible existence of a system in which the

ATP breakdown coupled to motility would be directly dependent on catalytic quantities of cyclic AMP. Studies on the differential distribution of cyclic AMP receptors in human spermatozoa have revealed that binding of cyclic AMP is selectively located in the mid-piece tail region. Only 18% of the binding in the whole sperm could be accounted for by the binding of the sperm heads. The distribution is similar to that described for protein kinase of bovine sperm. Blocking membrane sulphydryl groups with p-chloromercury phenyl sulphonate induces a drastic inhibition (65%) of the cyclic AMP binding to the sperm heads, while only moderately reducing (30%) the binding in the sperm tails. The physiological roles of cyclic nucleotides in the various activities or functions of different sperm components still need to be determined more precisely (Tash and Means 1983). But the various observations have shown that spermatozoa have the components known to regulate cAMP and calcium effects on the function of spermatozoa via protein phosphorylation, namely adenyl cyclase, phosphodiesterase, cAMP-dependent protein kinase, calmodulin, myosin light chain kinase, protein phosphatase, and phosphoproteins. These regulatory components and cytoskeletal associated proteins (actin and tubulin) have been demonstrated in specific areas of sperm flagellum and head.

g) Release of Enzymes Under Various Experimental Conditions

The release of enzymes from the spermatozoa under various experimental conditions has been investigated in some previous studies (Mann and Lutwak-Mann 1981). A thorough understanding of the conditions under which the enzymes are lost from the spermatozoa is essential for their proper preservation in various media. The release of acrosomal enzymes has been already discussed in Chap. VII. The release of alcohol, malic, glucose-6-phosphate, isocitric, lactate, sorbitol dehydrogenases, glutamic oxaloacetic transminase (GOT), and glutamic pyruvic transminase (GPT) has been studied in response to cold shock in buffalo spermatozoa (Sidhu and Guraya 1978b). Significant amounts of alcohol, malic and sorbitol dehydrogenases, and GOT were liberated; glucose-6-phosphate, isocitric, lactate dehydrogenases, and GPT were not affected by cold shock. The release of GOT and GPT from bull and ram spermatozoa with cold shock has been reported (Pace and Graham 1970, Graham et al. 1973, Roychaudhury et al. 1974). Chinnaiya et al. (1979) studied the extracellular release of GOT and GPT from buffalo spermatozoa on freezing of semen in various extenders. R. A. P. Harrison and White (1972) studied the leakage of glycolytic enzymes from the spermatozoa of bull, boar, and ram after shock. Hadarag et al. (1976) observed that 20 Xo-glutamate aminoferase and aldolase activities of frozen bull spermatozoa diminish constantly during conservation in liquid nitrogen. Guerin and Czyba (1976) believe that liquid N_2 apparently damages some metabolic structures of cells, affecting their respiratory activity more severely than their motility. Guerin et al. (1979) observed that freezing in liquid nitrogen does not result in appreciable modification of activities of hydrolases or peptidases of human sperm. In contrast, the activities of glucose-6-phosphate dehydrogenase are severely diminished or even abolished after freezing or thawing. This is not the result of the denaturation of enzymes but rather a leak into the surrounding medium.

Kurio and Burlachenko (1976) observed that sucrose-sulphate-yolk-citrate dilutor, with and without glycerol, activates glucosephosphate isomerase of boar spermatozoa and has almost no influence on the activity of aldolase and phosphofructokinase. After deep freezing of the undiluted semen the activity of the enzyme glucosephosphate isomerase is increased in the semen and plasma. The activity of aldolase and phosphofructokinase is increased the least. O'Shea et al. (1979) studied the metabolic changes of boar spermatozoa before, during preparation for, and after storage in liquid nitrogen. Species release differences (Roychaudhury et al. 1974) have been attributed to intrinsic differences in the cells (White and Wales 1960), as there occur species variations in terms of susceptibility to membrane damage caused by rapid cooling to 0 °C (Hammerstedt et al. 1978). The metabolic events in spermatozoa leading to the ageing of the spermcell under various conditions are now under intensive investigations, by both biochemical and ultrastructural methods (Mann and Lutwak-Mann 1975, 1981, R.C. Jones 1975).

Sidhu and Guraya (1979c) estimated the release of various enzymes (GOT, alcohol, malate, lactate, and sorbitol dehydrogenases) during the preservation of buffalo spermatozoa in different dilutors. Enzymes are released progressively with increasing times of storage. The release of various enzymes may be due to changes in various membranous components of spermatozoa, as observed after freezing and thawing (see references in Woolley and Richardson 1978, R. C. Jones and Stewart 1979, Sherman and Liu 1982, Ibrahim and Kovacs 1982, Ibrahim et al. 1982). Sidhu and Guraya (1979c) observed that the supplementation of diluted samples with chloroquine-diphosphate possibly decreases the release of various enzymes, indicating that their release may result from membrane damage due to dilution. Dilution is known to cause damage and increase membrane permeability (Blackshaw and Salisbury 1957, Hood et al. 1970, Darin-Bennett et al. 1973, Tasseron et al. 1977, Talbot 1979). Mueller et al. (1977) stated that the changes of the plasma membrane apparently are primary changes which lead to membrane breakage, acrosomal damage and destruction of mitochondria. These changes are principally caused by two factors: storage of spermatozoa in the undiluted seminal plasma and the osmolarity of the media. The morphological changes of the membranes are apparently due to physico-chemical and biochemical alterations of the plasma membrane (Mann and Lutwak-Mann 1981). Halangk et al. (1980) have suggested that the changes induced in the sperm membrane by cold shock are qualitatively different from those caused by storage and by a slow decrease in temperature.

Abu-Erreish et al. (1978) have made comparison of superoxide dismutase, glutathione peroxidase, and glutathione reductase activities in ram spermatozoa and erythrocytes. Their results have indicated a much lower content of glutathione peroxidase in spermatozoa, suggesting that ineffective H_2O_2 disposal forms the basis of the susceptibility of ram spermatozoa to oxygen toxicity. Mukhtar et al. (1978) demonstrated glutathione S-transferase activities in ram and mouse sperm. These enzymes are believed to function in the metabolism and detoxification of certain electrophilic xenobiotics, if present in sperm.

C. Annulus

1. Morphology

The annulus formerly called the ring centriole or Jensen's ring (Jensen 1887) is present at the junction of the mid-piece and principal piece (Fig. 61) (Phillips 1975a, Harding et al. 1979). The annulus is composed of closely packed filamentous subunits, 3 to 4 nm in diameter (Fawcett 1970). It develops in close association with the plasma membrane and remains firmly adherent to it. In longitudinal sections it is usually represented by two triangular or semicircular profiles on either side of the tail distal to the last mitochondrion (Fawcett 1975b, Phillips 1975a, Holstein and Roosen-Runge 1981). The triangular type is found in bat, dormouse, hamster, and suni antelope (Fawcett 1970); the semicircular pattern in chinchilla, mouse, and ram. In the latter case, the annulus runs in an evagination of the plasma membrane which surrounds the basis of the principal piece. The annulus in the opossum sperm extends as a thin septum caudal to the terminal gyre of the mitochondrial helix and has a concave outer aspect.

2. Origin and Function

The annulus originates in close association to the distal centriole of the early spermatid soon after the formation of the simple primary flagellum and later in development it migrates caudally to its position at the end of the mid-piece (Fawcett 1958, Fawcett and Ito 1965, Ortavant et al. 1969, de Kretser 1969, Holstein and Roosen-Runge 1981). Because of its origin, the classical cytologists considered it to be a centriole of unusual shape. But recent ultrastructural studies have shown that the chromatoid body is, somehow, involved in its formation, as already discussed in Chap. IV.

The origin of the annulus in the spermatozoa of primates needs to be worked out. Its functional significance is still not known. However, Fawcett and Ito (1965) have suggested that the annulus may be involved in the formation of the fibrous sheath of the principal piece. But according to Fawcett (1970), the function of the annulus could be to prevent caudal displacement of the mitochondria.

D. Main-piece

The main-piece of mammalian spermatozoa is surrounded by a fibrous sheath (Fig. 61) which shows a similar basic organization in different species of mammals (Fawcett 1975b, Phillips 1975a, Hughes 1976, Harding et al. 1979, Holstein and Roosen-Runge 1981). However, there are significant variations in shape and size of the columns as well as in the frequency of the anastomosis of the ribs. The fibrous sheath starts immediately behind the annulus which marks the caudal end of the mid-piece (Fig. 61). Therefore, the length of the fibrous sheath defines the principal piece region, which in spermatozoa of most mammalian species is the longest flagellar segment. Initially, the fibrous sheath was believed to be composed of a helicoidal fibre (Schnall 1952, Ånberg 1957, Randall and Friedlander

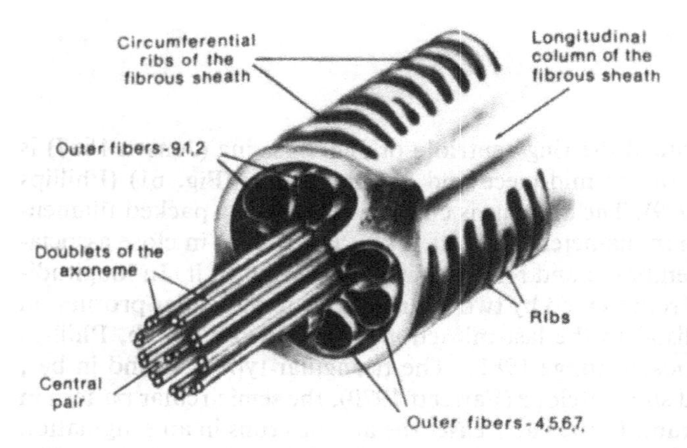

Fig. 76. A segment from the principal piece of a spermatozoon illustrating one of the two longitudinal columns of the fibrous sheath and the associated ribs. Inward prolongations of the longitudinal columns attaching to doublets 3 and 8 divide the tail into two unequal compartments, one containing three outer fibres and the other containing four (From Fawcett 1975b)

1950, Schultz-Larsen 1958). However, recent studies have shown that it is composed of a series of circumferentially oriented ribs (Fig. 76) that extend halfway around the tail to end in two longitudinal columns which run along opposite sides of the sheath for its whole length (Figs. 45B, 57 and 61) (Fawcett 1970, Hughes 1976). The sheath is not attached to the plasma membrane.

In the sperm of some rodents, the ends of ribs of the fibrous sheath are bifid so that a triangular electron-lucent interspace appears between their divergent heads and the longitudinal columns of the sheath (Fawcett 1975b). The ribs of the fibrous sheath in marsupial sperm are hollow near their ends and thus in parasagittal sections these show open rectangular profiles. The columns and ribs of the fibrous sheath in the mammalian spermatozoa show a filamentous structure. However, the subunits of ribs are thinner than those of the longitudinal columns. In its initial part, the fibrous sheath is attached to outer dense fibres 3 and 8. Posterior to the termination of these fibres the adaxial side of the longitudinal column tapers to a thin edge which passes inwards and is apparently attached to a small ridge arising radially from the wall of doublets 3 and 8.

The longitudinal columns extend in the principal piece along the whole length of the fibrous sheath in its dorsal and ventral surfaces (Fig. 76). These are composed of 15 to 20 nm thick longitudinal subunits attached to the dense fibres 3 and 8 (Fawcett and Phillips 1970) and therefore bind the sheath to the axoneme during sperm movement, while the plasma membrane is independent of this complex. Towards the end of the piece, the longitudinal columns progressively reduce in size, meanwhile the ribs become slender (Fig. 61). The abrupt ending of the fibrous sheath marks the junction of the principal and end-piece (Fig. 61). The attenuated remnants of the longitudinal column may extend a short distance into the end-piece. The outer dense fibres and the longitudinal columns of the fibrous sheath are believed to restrict their participation in the sliding movements responsible for the flagellar bending (Fawcett 1975b).

The chemical analysis of the fibrous sheath has received relatively little attention. But it appears to consist of proteins and these are highly resistant to solubilization by acid (Bradfield 1955). The work of Calvin and Bedford (1971), and Bedford and Calvin (1974a) has indicated that the proteins of the fibrous sheath in the fully mature sperm of some mammalian species are stabilized by disulphide bonds. Olson et al. (1976a) and Olson (1979) have isolated the fibrous sheath from spermatozoa of the rat caput and cauda epididymidis and studied its polypeptide composition and subunit substructure. It is composed predominantly of a single polypeptide with a molecular weight of 80,000 having minor components of 90,000, 25,000 and 11,000 MW. Study of its substructure has revealed that the ribs are composed of parallel 5 – 6 nm wide filaments which seem to have a beaded substructure. The fibrous sheath is composed of keratin-like proteins.

One of the striking features of the main-piece is the manner in which the outer longitudinal dense fibres undergo both a reduction in size and systematic reduction in number until these finally disappear altogether (Fig. 61). According to Pedersen (1970a), the first fibre to disappear in the human sperm is number 4, closely followed by numbers 2 and 7. Fibres number 5 and 6 then disappear as numbers 1 and 9 are left to terminate farthest posteriorly. The longitudinal columns are not very conspicuous and do not extend through the whole length of the fibrous sheath; distally are replaced by overlapping of the ribs.

E. End-piece

The end-piece consists of a central pair of axial fibrils and ring of nine doublet fibres, which are surrounded by the plasma membrane (Fig. 61) (Phillips 1975a, Holstein and Roosen-Runge 1981). According to Pedersen (1970a), the 9 + 2 pattern of axial filament complex extends through most part of the tail including the end-piece, but the arrangement of the fibres in the tip of the end-piece is changed. In some sections, the cell membrane encloses hollow tubules with evidence of doublet formation and in decreasing number, suggesting an uncoupling of the doublets and a successive termination of the single subfibres.

Future studies should be directed at biochemical analysis of the different components of spermatozoa, as such studies have not kept pace with the extensive morphological investigations discussed above. Little attention has been paid to the characterization and comparison of total protein constituents within various components of these highly differentiated and organized cells. Very recently, electrophoresis of the SDS-insoluble tail proteins has revealed the presence of at least nine prominent bands with apparent molecular weights between 21,000 and 89,000 (O'Brien and Bellvé 1980a). Electron microscope study has suggested that this fraction contains proteins from the outer dense fibres, fibrous sheath, outer mitochondrial membranes, and the structural elements of the neck region of the sperm tail.

Chapter XI
Plasma Membrane and its Surface Components

Spermatozoa, like other cells, have a continuous limiting membrane or plasma membrane, the nature and properties of which are of special interest as the molecular character of its surface is an important determinant of the spermatozoon's ability to recognize and fuse with the egg (Shapiro 1984, Monroy and Rosati 1983, Peterson et al. 1985, O'Rand et al. 1984) and is the primary determinant of species specificity in fertilization. Therefore, during the past decade, the plasma membrane of spermatozoa as well as their surface components have been extensively studied with modern techniques of cytochemistry, biochemistry, lectins, freeze-fracturing, high resolution electron microscopy, and other physical techniques (Young and Goodman 1980, Holt 1980, 1984, Olson and Gould 1981, Holt and North 1985). These studies have provided basic information about the fine structure and chemical and physical properties of the plasma membrane of spermatozoa and about the regional distribution and density of general classes of compounds or specific activities on cell surfaces. In spite of these advances, our knowledge of membrane organization in spermatozoa is limited. Actually, the spermatozoon plasma membrane is characterized by specific regional specializations of the surface properties (or mosaicism) (Friend 1982, Courtens et al. 1982, Mack and Everingham 1983, Holt 1984).

A. Fine Structure

A continuous triple-layered plasma membrane derived from the original plasma membrane of the spermatid covers the mature spermatozoon. Its outer dense layer is generally 5 nm thick, the intermediate light 3 nm, the inner opaque 2 nm (Baccetti and Afzelius 1976). The head of the mammalian sperm shows during spermateleosis a thicker inner layer, which remains unchanged in the mature sperm in the post-acrosomal region, whereas the acrosomal region restores the normal symmetry by increasing the thickness of the outer leaflet (Bedford et al. 1972).

According to Zamboni and Stefanini (1968), no modifications in plasma membrane structure of sperm are detectable during the transit through the epididymis. However, the results of various other ultrastructural investigations have shown that, in some species, the appearance and topographic configuration of the spermatozoon plasma membrane are altered during epididymal transit (Bedford 1975, Olson and Hamilton 1976, Hammerstedt 1979). Szuki and Nagano (1980) observed that in the initial segment of the epididymis, a floculent material fills the space between the plasma membrane and the outer acrosome membrane of rat spermatozoa. This disappears in the more distal segments. The significance of these changes for membrane functions remains to be established.

The plasma membrane generally remains moulded to the contour of the acrosomal cap (Figs. 58, 59). But it is usually detached and elevated to a different degree from the outer membrane of the acrosomal cap at different places (Jones 1975, Wooding 1975). This loosening appears to be a fixation artifact, as there is not any evidence for this morphology in vivo. With scanning electron microscopy, the acrosomal cap surface is seen to possess droplet-like elevations which, according to Lung and Bahr (1972) may represent vesiculations of the membrane. The elevation and distortion of the plasma membrane are probably due to an artifact of fixation. The degree of detachment of the plasma membrane depends, to some extent, on the osmolarity of the fixative and physiological state of the sperm (M. C. Chang and Hunter 1975); e.g. the plasma membrane separates more easily in uterine spermatozoa than in epididymal and ejaculate sperm, probably because of a change in membrane permeability (Bedford 1964). However, according to Phillips (1972a), the plasma membrane overlying the acrosome of Chinese hamster and mouse sperms is coated with minute vesicles and tusuls which are believed to be of biological importance in attachment during fertilization due to their sticky quality.

Recent electron microscope studies have revealed the presence of some reinforcements in association with the plasma membrane of the spermatozoa in some species of mammals. The brush-tailed phalanger (*Trichosurus*) has a distinct fibre network under the plasma membrane of the sperm mid-piece (Olson 1979, Harding et al. 1979). A nonmitochondrial helical sheath of filaments occurs immediately beneath the plasma membrane of the mid-piece in the marsupial devil (*S. harrisii*) (Hughes 1976). The plasma membrane over most of the mid-piece of opossum spermatozoa has a regularly scalloped contour (Olson et al. 1977). Scanning electron microscopy revealed that the scallops course parallel to the long axis of the flagellum as nearly parallel ridges (Olson et al. 1977). A mat of amorphous or finely filamentous material 25 – 30 nm thick is applied against the cytoplasmic surface of the plasma membrane wherever the membrane is in the scalloped configuration. The plasma membrane of the principal piece shows neither scalloping nor any undercoat on its cytoplasmic surface. The functional role of such additional cytoskeletal elements or reinforcements remains unknown.

B. Macromolecular Organization and Physical Properties

The surface membrane of the spermatozoon is a complex mosaic structure, not only in terms of macromolecular composition that may differ in specific regions of the sperm but also functionally. Its physico-chemical properties are very complex, as revealed by the application of modern techniques. Hammerstedt (1979) provided the schematic molecular representation of a plasma membrane surface (Fig. 77) which is based on results obtained from chemical analysis of isolated membrane components and depicts the generally accepted concept. Three zones

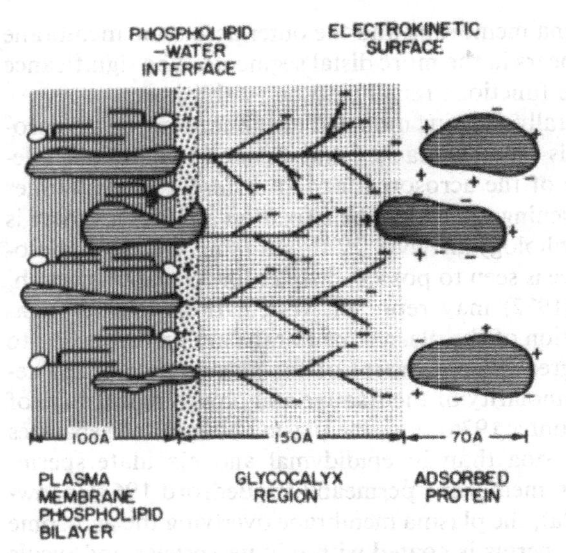

Fig. 77. Schematic representation of the sperm plasma membrane showing zones sample in these experiments. The phospholipid bilayer is composed of polar lipids with their charged groups directed to the surface and proteins (*horizontal lines*) extending partially or completely across the membrane. These proteins can extend out of the membrane into the aqueous solvent around the cell. The phospholipid-water interface has a characteristic electrical charge. The glycocalyx region consists of glycoproteins anchored in the plasma membrane and extending 10 – 15 nm from the phospholipid-water interface. The glycoproteins have a net negative charge, due to sialic acid or sulphate residues attached to the carbohydrate side-chains, which produces an electrokinetic surface to which extracellular proteins may be absorbed. Only fixed electrical charges are represented (From Hammerstedt 1979)

(the phospholipid bilayer, the phospholipid-water interface, and the glycocalyx), representing chemically unique layers of the sperm surface, have been emphasized. The overall functions of the sperm membrane are distributed among the molecules of these zones. The lipid bilayer consists of polar phospholipids oriented with hydrocarbon chains directed to the interior of this zone and the charged polar groups directed towards the polar solvent, which is water. Proteins seen within these lipids are considered either intrinsic (integral and essential for structure) or extrinsic (associated with the membrane but easily removable by mild physical treatment). This zone constitutes a general permeability barrier to charged molecules. Characteristic transport systems are needed for the exchange of intracellular and extracellular molecules. The chemical composition of this zone is at the basis of the ability of the cell to maintain permeability barriers when subjected to physical stress; alterations in compositions lead to changes in bilayer strength. The phospholipid-water interface, which lies between the lipid bilayer and the medium, shows a fixed electrical charge present in the bilayer which represents the mixture of functional groups observed on lipid and protein. Counter-ions present in the solution are attracted by these charges. Alteration either in polar lipids or proteins of the phospholipid bilayer causes a redistribution of the fixed charge at this interface. Biochemical analysis of phospholipids present in spermatozoa has shown that in different species of mammals significant changes in lipid composition occur as their spermatozoa pass through the epididymis, as already discussed in Chap. X (see also Nicolopoulou et al. 1985). This data refer to the bulk lipid composition in spermatozoa, and localization of different classes of phospholipid to different components of the sperm has received very little attention. However, the characterization of a surface-active phospholipase from mouse spermatozoa may be of great significance for the better understanding of these alterations at the molecular level (Thakker and Franson 1983). Finally, some glycoproteins and glycolipids extend out from the lipid bilayer surface and form the glycocalyx zone. Its macromolecules show a negative charge (due to sialic acid or sulphate residues) to be discussed later on, and provide a fixed electrical charge extending possibly 15 nm from the lipid-water interface. Counter-ions to these molecules may be either small ions (hydrogen or metal ions) or macromolecules.

Recent studies on the boar sperm plasma membrane by two-dimensional electrophoresis have shown that more than 250 proteins and glycoproteins could be identified (Russell et al. 1983a). This correlates with the heterogenous capabilities of the spermatozoon membrane system. Wang and Koide (1982), in a brief report, described the chemistry and immunobiology of a human sperm membrane protein. The carbohydrate moieties of glycolipids and glycoproteins present in the plasma membrane of spermatozoa serve as sensors for extraneous substances, such as hormones, antibodies, toxins, and drugs (Mann and Lutwak-Mann 1981). Histones apparently bind specifically to sperm surface receptors sites before agglutination of cells (Majumder 1981a).

Attention is being increasingly given to the role of receptors in the sperm membranes, and to the role of regulatory proteins in capacitation and acrosome reaction (Mann and Lutwak-Mann 1981). The acrosome reaction has recently been shown to be stimulated by epinephrine and norepinephrine (catecholamines). The application of various adrenergic agonists and antagonists has shown that spermatozoa possess β-adrenergic receptors which may be involved in capacitation. A potent fluorescent agent which blocks β-adrenergic receptors has now been used in an attempt to localize these receptors by fluorescent microscopy (Cornett and Meizel 1980). Both α- and β-adrenergic stimulation are needed for fertilization to occur, and these compounds almost certainly act on spermatozoa. Catecholamines may exert similar effects during fertilization in vivo, but there is no information on their concentrations in fluids of the reproductive tract, although these are known to be present in the tissues of the oviduct and uterus. Interactions between catecholamines and receptors in the sperm membrane may lead to an influx of calcium and increased levels of cAMP and so induce capacitation and the acrosome reaction (Nelson 1985).

Friend and Fawcett (1974) showed a concanavalin A and ruthenium red reactivity in the glycocalyx over the guinea-pig acrosome region. The same region in the rabbit sperm contains receptors for prostaglandins (Bartoszewicz et al. 1975). Bartoszewicz et al. (1975), using ferritin-labelled antibody to prostaglandin E_2, followed the distribution of prostaglandin receptors on rabbit spermatozoa. Unwashed, ejaculated spermatozoa show some markers over the plasma membrane overlying the acrosome. The outer regions are not labelled. Washed spermatozoa do not bind antibody at all, indicating that prostaglandin is very loosely bound to their surface. Human spermatozoa do not show reaction in either ejaculated or washed preparations.

Like proteins, sterols are inhomogeneous in distribution in different parts of the guinea-pig spermatozoon plasma membrane (Friend and Elias 1978, Friend et al. 1979). The plasma membrane, where it blankets the acrosomal cap, contains five times more sterol than the post-acrosomal segment. The striated ring, annulus, and zipper are virtually free of such sterols, particularly nonesterified cholesterol. Vijayasarathy and Balaram (1982) revealed more fluid lipid phase in the tail regions of bull spermatozoa. In the head region, fluorescence of lipid phase fluidity is studied for the plasma and acrosomal membranes (Vijayasarathy et al. 1982).

Freeze-etching techniques have revealed spheroidal particles between the outer and the inner leaflets of the plasma membrane of sperm (Olson et al. 1977). Kaczmarski (1978) discussed the organization of integral proteins in the sperm plasma membrane. Sulphur-rich keratin-like proteins are present in the heads of mammalian spermatozoa (Mann and Lutwak-Mann 1981). Mercado et al. (1976), using a new fluorescent probe for SH groups, demonstrated a large concentration of this functional group on the sperm surface, which is believed to be important in sperm binding to the zona pellucida. SH-rich keratinoid quality also forms a feature of coarse fibres and sheaths of the sperm tails. A shift from sulphydryl to disulphide groups within the proteins of the sperm heads and tails forms part of the sperm maturation in the epididymis, and the formation of disulphide linkages also marks the proteins inside the nuclei of mature spermatozoa, as already described in Chap. VII. There are some suggestions about the stabilization of sperm plasma membrane in mammals by these disulphide groups (Calvin 1975, Yoo 1982), probably different in different species. The nucleus, connecting piece, coarse outer fibres, outer membrane of the mitochondria, and fibrous sheath of the principal piece are also stabilized by disulphide cross-link formation. Sattayasai and Panyim (1982) studied the nature of the proteins which form disulphide bonds during the maturation of rat spermatozoa. Interchange reactions between SH and disulphide groups are believed to be involved in motility, metabolism, and survival of spermatozoa in general. The membrane of bovine sperm appears to be strongest as it is more resistant against osmotic shock than the ovine or porcine sperm membrane (Jones 1973). All the regions of the sperm plasma membrane do not possess the same resistance. Triton A-100 affects the plasma membrane in the acrosomal region more than that in the equatorial region of bovine sperm heads (Wooding 1973). This is probably due to inner reinforcements. Wang et al. (1982) studied localization and esterolytic activity of rabbit spermatozoa membrane protein "A".

Chemical characteristics of the sperm plasma membrane are altered under various normal and experimental conditions. Davis and Geregeley (1979) observed that plasma membrane of rat sperm from cauda epididymis undergoes extensive polypeptide hydrolysis during incubation in Krebs Ringer bicarbonate medium. Hydrolysis is markedly increased by addition of 4 mg bovine serum albumin. There occur losses of surface constituents of ejaculated sperm during incubations in vivo and in vitro (Koehler 1978). Davis et al. (1979) obtained evidence for lipid transfer between the plasma membrane of rat sperm and serum albumin during capacitation in vitro. In a further study, Davis et al. (1980) demonstrated that plasma membrane isolated from rat sperm cells after incubation in vitro shows a significantly lower cholesterol/phospholipid mole ratio when the medium contains serum albumin. Transfer of albumin-bound phospholipids to the membrane largely accounts for this effect. The results obtained in this study are broadly consistent with a previously proposed model for albumin-induced destabilization of sperm membrane (capacitation) and its reversal by seminal plasma membrane vesicles (Davis et al. 1979). Albumin also decreases sialic acid and, more specifically, ganglioside level apparently by promoting release of sperm neuraminidase. Cholesterol ester constitutes up to 0.5 mol/ml of cholesterol in these plasma membrane preparations. All these changes in the macromolecular organization can be assumed to affect biological properties of plasma membrane as changes occur in permeability characteristics of the lipoprotein membrane (Amann and Almquist 1962, Glover 1962). Changes in lipid composition accompanying sperm maturation in the epididymis may exert an important influence on the fusion capacity of the sperm.

Hammerstedt et al. (1976, 1978) recommended the use of spin latels and electron spin resonance spectroscopy (ESR) to characterize sperm membranes. The spin labels have been used to evaluate the effects of cold shock and osmolarity on sperm. Hammerstedt et al. (1979) observed that each of the three zones (a lipid bilayer that provides an impermeable barrier to uptake of extracellular polar molecules, a charged lipid-water interface formed by the phospholipids, and membrane-associated proteins and a charged glycoprotein calyx with absorbed proteins) of the plasma membrane of sperm (Fig. 77) changes during epididymal maturation. Further changes probably occur during passage through the female genital tract (Hammerstedt 1979). The physical techniques advanced by Hammerstedt and co-workers will be very useful to work out all phases of the complex events between spermiation and semination by determining a precise quantification of the alterations that occur in the sperm plasma membrane.

There are also other evidences of change in the nature of the membrane or of the character of the sperm-cell surface during epididymal passage (Bedford 1975). Soucek and Vary (1983) observed steady state of fluorescence anisotropy changes of 1,6-diphenyl-1,3,5-hexatriene in plasma membrane of boar epididymal sperm. The ordered pattern of the intramembranous particles of boar spermatozoon plasma membrane is greatly changed as the spermatozoa pass through different zones of the epididymis (Suzuki 1981). However, the organization of the acrosomal membrane particles does not change during the epididymal passage of boar spermatozoa. As rat spermatozoa move into the middle of the caput epididymis, a variable glycocalyx material accumulates upon the outer surface of the plasma membrane covering the acrosome (Suzuki and Nagano 1980). In the distal caput, glycocalyx material begins to separate and becomes completely detached in proximal cauda. Brown et al. (1983a) observed changes in plasma membrane glycoproteins of rat spermatozoa from the caput and cauda epididymis. The protein composition of rat (Jones et al. 1983) and bovine (Yoo et al. 1982) spermatozoa is also changed during sperm maturation. Epididymal maturation of the surface protein structure of mammalian spermatozoa is discussed by Vierula and Rajaniemi (1982). Russell et al. (1983b) studied the binding of basic proteins of boar seminal plasma to sperm plasma membranes and their role in sperm-zona adhesion.

Reyes et al. (1976) reported the participation of membrane sulphydryl groups in the epididymal maturation of human and rabbit spermatozoa. Hipakka and Hammerstedt (1978a) demonstrated changes in 2-deoxyglucose transport during epididymal maturation of ram spermatozoa, suggesting that membrane function is altered during sperm maturation. Hipakka and Hammerstedt (1978b) made a study of 2-deoxyglucose transport and phosphorylation by bovine sperm. Babcock et al. (1979) studied alteration of membrane permeability to calcium ions during maturation of bovine spermatozoa.

C. Intramembranous Particles

Striking regional specialization in the organization of the sperm plasma membrane has also been shown recently by electron microscopy of thin sections, surface replicas, freeze-cleaving, and freeze-etching. This technique has revealed the presence of intramembranous particles which show species differences when corresponding regions of the sperm membranes are compared and contrasted. According to Ocsenyi and Veres (1968), the plasma membrane of the sperm head is not smooth at all but contains pores. Freeze-fracture studies have revealed a variety of ordered particle arrays in the membranes associated with the sperm head of different species (Fig. 78). In deep-etched samples of mouse sperm heads, the external surface of the plasma membrane appears smooth and featureless (Stackpole and Devorkin 1974). Within the cleaved plasma membrane, 9-nm particles are distributed randomly in the head and aggregated in the tail, forming paracrystalline arrays in the mid-piece. The acrosomal membrane shows a periodic arrangement of tightly packed 10-nm particles, as already described in Chap. VII. When the plasma membrane overlying the guinea-pig acrosome is studied by freeze-fracturing, the A-face shows areas of irregular outline having a quilted pattern (Fig. 78B) (Friend and Fawcett 1974, Friend et al. 1979, Friend 1982, Holt 1984). These are separated from one another by narrow strips of a particle-studded membrane of more orthodox appearance. The acrosomal membrane in the rat and guinea-pig also shows a periodic internal structure but with a pattern slightly different from that of the plasma membrane (Friend and Fawcett 1974). Such lattices are absent in the plasma membrane of the mouse (Stackpole and Devorkin 1974) and rabbit (Fléchon 1974b) sperm head. The particles of the plasma membrane covering the acrosome of rat spermatozoa change their arrange-

→

Fig. 78. A Freeze-cleaving preparation of a rat sperm head showing areas of highly ordered structure within the plasma membrane overlying the acrosome. Other areas of the same membrane have the usual random distribution of membrane-intercalated particles; where the plane of fracture has broken across the nucleus a lamellar organization of the condensed chromatin is evident; **B** Freeze-cleaving preparation of the cell membrane overlying the concave surface of a guinea-pig acrosome. Crystalline domains of varying size within the membrane are separated by areas of more orthodox appearance (From Fawcett 1975b)

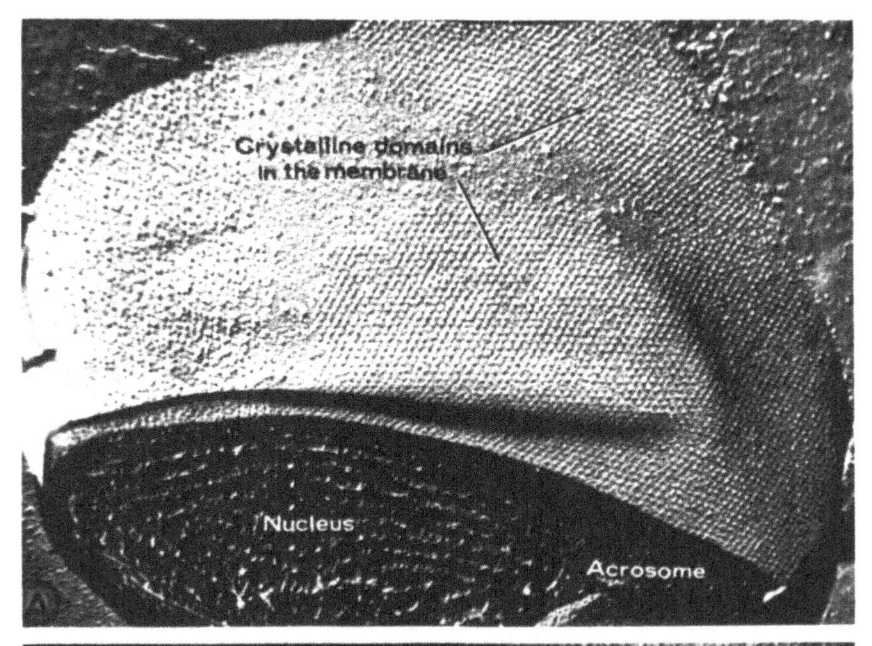
Crystalline domains
in the membrane
Nucleus
Acrosome

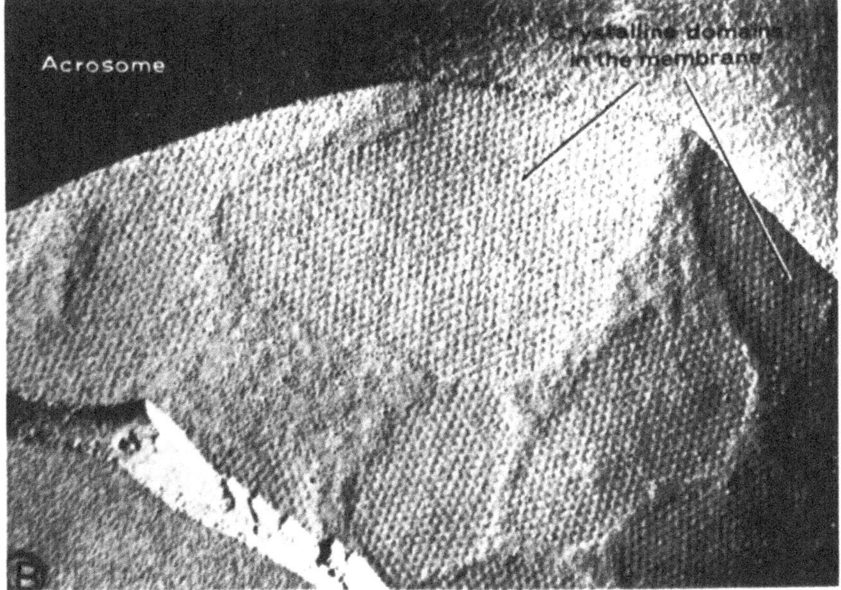
Acrosome
Crystalline domains
in the membrane

Fig. 78

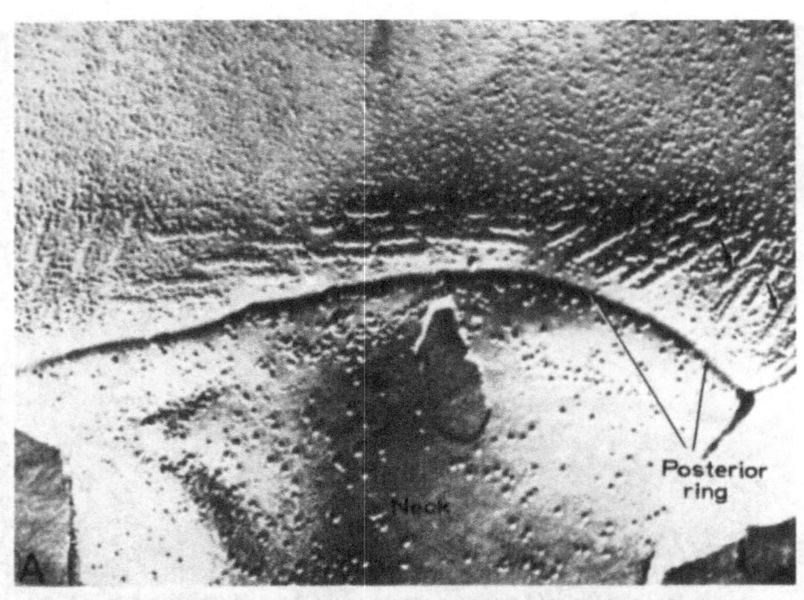
Posterior
ring
Neck
A

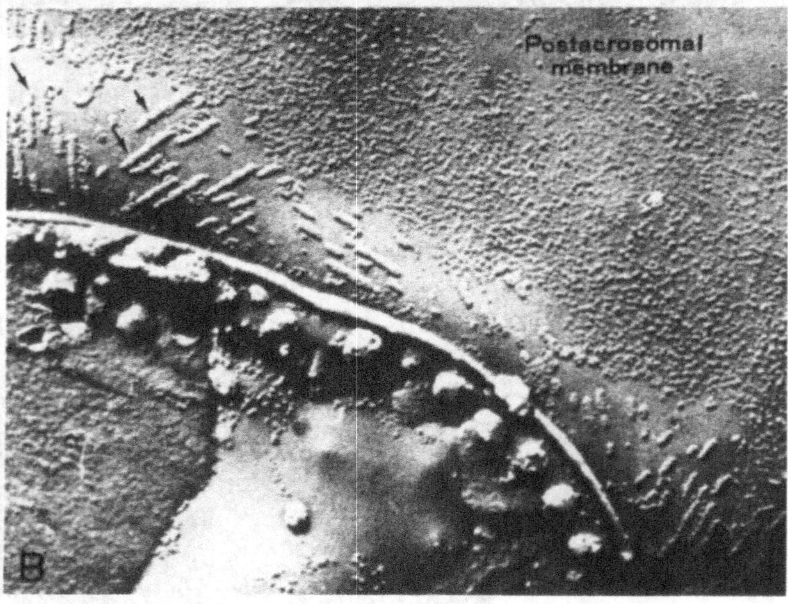
Postacrosomal
membrane
B

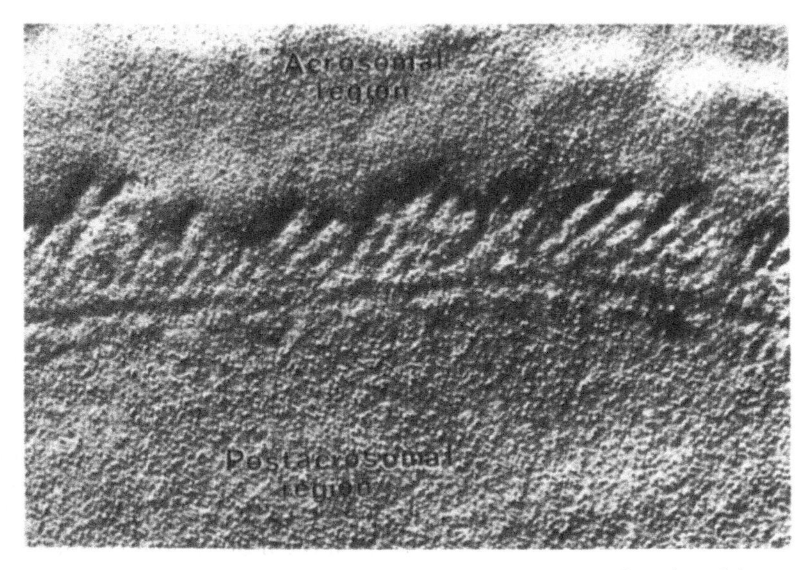

Fig. 80. Freeze-cleave preparation of the A-face of the plasma membrane over the junction of the acrosomal and post-acrosomal regions of a guinea-pig spermatozoon. A palisade of rod-like structures associated with the equatorial segment of the acrosome leave their impression in the overlying membrane. Of particular interest is the higher concentration of intramembranous particles in the membrane of the post-acrosomal region (From Fawcett 1975b)

ment during epididymal passage (Suzuki and Nagano 1980). A hexagonal arrangement of these particles is seen in spermatozoa from the middle of the caput, but the array becomes disorganized as the spermatozoa move through the distal parts of the epididymis. The relations of glycocalyx material on the plasma membrane with the particles are changed during maturation (Olson 1980). Changes also occur in intramembranous particles distribution in the plasma membrane of marsupial (*Didelphis virginiana*) spermatozoa during maturation in the epididymis (Olson 1980).

Crystalline areas have not been observed in the post-acrosomal region of the spermatozoon (Fawcett 1975b, Holt 1984). However, there is a greater number of membrane-intercalated particles per unit area than in the membrane over the acrosome or neck region (Figs. 79 and 80). In the rabbit spermatozoa, the caudal part of the post-acrosomal segment shows conspicuous strands or rods on the

Fig. 79. A Freeze-cleaving preparation showing the intramembranous differentiations of the post-acrosomal region in a rabbit spermatozoon. The great abundance of particles in the post-acrosomal region can be compared with the relatively sparse population in the neck region. The posterior ring appears here as a distinct groove at the boundary between head and neck. Rods or linear aggregations of particles are characteristic of the post-acrosomal membrane near the posterior ring of this species; **B** Another example of the post-acrosomal region of the rabbit sperm. This appears to show the B-face of the membranes, since the posterior ring is seen as a ridge instead of a groove. The coarse projections behind the ring correspond to pores in this region of the nuclear envelope (From Fawcett 1975b)

A-face which consist of closely packed rows of particles (Fléchon 1974b). These particles extend obliquely forward from the posterior ring (Fig. 79A). The same area in the spermatozoa of other species shows very small particles which form well-ordered geometric arrays (Fléchon 1971). The specific role of these unusual specializations in relation to gamete fusion is still not clear. Friend et al. (1977) studied the membrane particle changes accompanying the acrosome reaction in guinea-pig spermatozoa. Their observations support the concepts that membranes become receptive to union at particle-deficient interfaces and that the physiologically created barren areas in freeze-fracture replicas may herald incipient membrane fusion. Fawcett (1975b) believes that the unique features of the post-acrosomal region involved in gamete attachment and fusion may not be present exclusively within the membrane proper, but possibly in the underlying dense postacrosomal sheath or, more likely, in the outer surface of the membrane which is not observed in freeze-cleaving. The various findings have suggested that the capacitation process occurring during the passage of spermatozoa through the female reproductive tract implicates a series of endogenous molecular and physiological alterations of the sperm surface, particularly the rearrangement of intrinsic membrane proteins in the head region, leading to protein-poor areas of high fluidity, thus facilitating multiple fusion events of the plasma membrane within the underlying outer acrosomal membrane (O'Rand 1979).

The bottom of the circumferential groove in the plasmalemma, which is called the striated band or posterior ring, shows a very fine striation with a periodicity of about 10 nm (Pedersen 1972a, Friend and Fawcett 1974, Fléchon 1974b, Stackpole and Devorkin 1974). The plasma membrane of the neck region caudal to the posterior ring shows fewer particles than those of the post-acrosomal region (Fig. 79).

With the exception of mouse and guinea-pig sperm (Koehler 1973b, Friend and Fawcett 1974, Stackpole and Devorkin 1974) the plasma membrane of the mid-piece shows the usually randomly distributed particles (Fig. 81). The external surface of the plasma membrane of the tail in the mouse sperm possesses circumferentially oriented strands in the mid-piece and larger longitudinally oriented "zippers" in the principal piece (Stackpole and Devorkin 1974, Friend et al. 1979). In the guinea-pig mid-piece, the particles associate in linear arrays which wrap circumferentially about the flagellum and precisely overlie the gyres of the mitochondrial sheath (Fig. 81) (Koehler 1973b, Friend and Fawcett 1974). The mid-piece plasma membrane of shrew sperm lacks the concentrically arranged particle "threads" (Koehler 1977), which have been observed in the mid-piece of guinea-pig spermatozoon (Fig. 81) (Koehler 1973b, Friend and Fawcett 1973, 1974, Friend et al. 1979). Linear arrays of particles sometimes arranged in multiple strands are seen on the principal piece of the shrew sperm (Koehler 1977). These appear to represent the membrane-intercalated segment of the surface zipper structures which have been described in guinea-pig (Friend and Fawcett 1973, 1974, Koehler and Gaddum-Rosse 1975, Friend et al. 1979) and mouse (Stackpole and Devorkin 1974) spermatozoa. Friend et al. (1979) advanced a technique for isolating the zipper from the guinea-pig sperm and paved the way for its further characterization by biochemical assays. The lectin-binding results obtained by Enders et al. (1983) suggest the presence of N-linked oligosac-

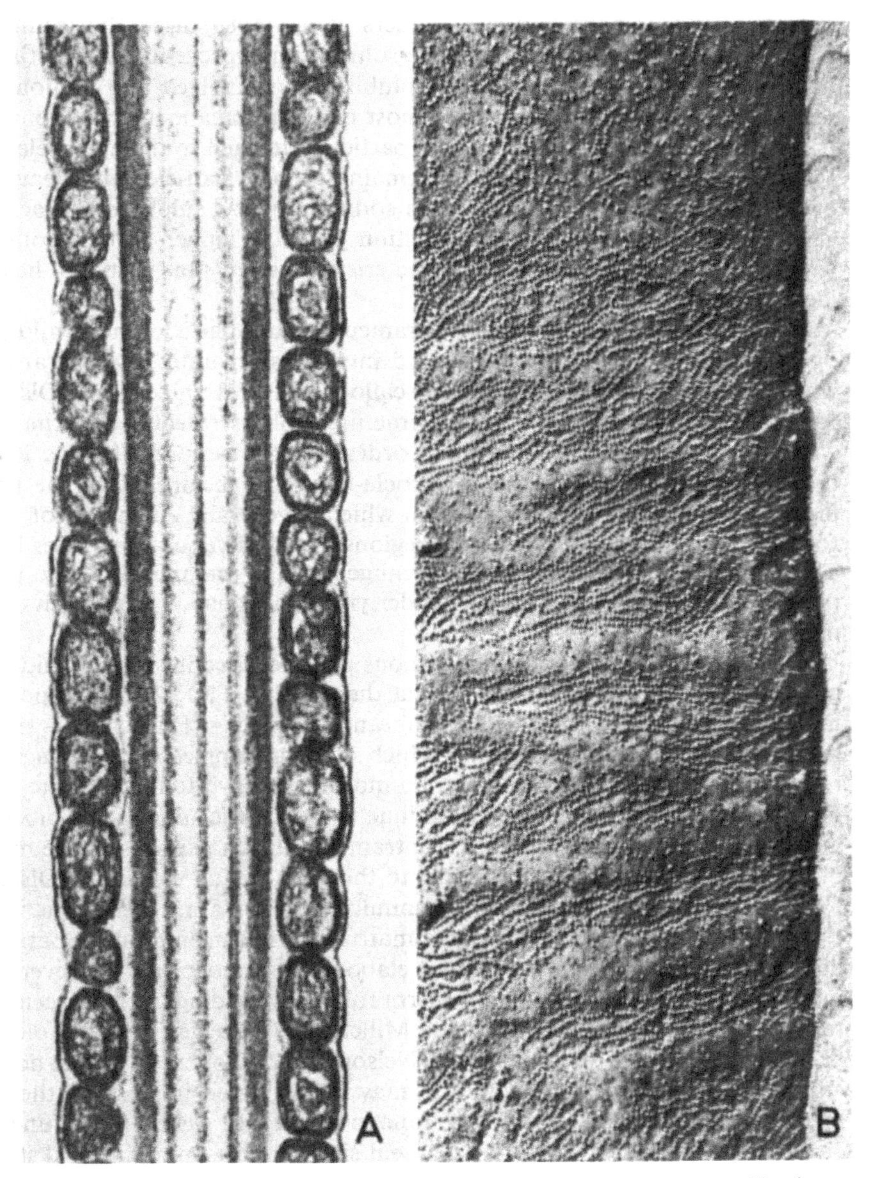

Fig. 81. A Longitudinal thin section of the middle piece of a mammalian spermatozoon. The circumferentially oriented mitochondria are cut transversely. Note how closely the cell membrane is apposed to the underlying mitochondria; **B** Freeze-cleaving preparation of the membrane of the middle piece from a guinea-pig spermatozoon reveals linear arrays of intramembranous particles oriented circumferentially and concentrated over the mitochondria. The membrane over interstices in the mitochondrial sheath is relatively free of particles. These highly ordered arrays of particles are not found in the corresponding region of sperm from other species (From Fawcett 1975b)

charides within zipper particles. Enders et al. (1983) used the lectin-binding results in conjunction with a detergent solubilization procedure to identify potential zipper components. Detergent solubilization involved two nonionic detergents – digitonin, which solubilized most of the plasma membrane, but left approximately two-thirds of the zipper particles attached to the cytoskeleton, and Triton X-100, which solubilized the remaining zipper particles while leaving most other sperm structures intact. Within sodium dodecyl sulphate/polyacrylamide gels of the Triton-X-100-soluble fraction potential zipper particle components with the same lectin-binding characteristics as in situ particles have been identified.

In freeze-fracture replicas, the intramembranous particles in the mid-piece of opossum sperm are closely aggregated into rows separated by several particles which correspond in spacing to the scallops seen in thin sections (Olson et al. 1977). Optical diffraction of freeze-fracture replicas revealed that the particles within the rows are packed in a highly ordered two dimensional lattice. The membrane between rows is relatively particle-free. The membrane of the principal piece shows a longitudinal diffraction which is generally composed of a double row of particles, but which in some regions may be five to six particles in width. The striking and different particle arrangements in the mid-piece and principal piece membranes appear to reflect underlying functional differences in these two membranes.

The linear arrays of intramembranous particles described in the mid-piece of guinea-pig sperm terminate abruptly at the annulus (Fig. 82), which possesses a moderate number of larger intramembranous particles (Friend et al. 1979); the latter are scattered in a background which shows a strippled or roughened texture reminiscent of that associated with desmosomes (Fawcett 1975b). The study of Friend et al. (1979) advanced a technique to isolate the annulus for biochemical analysis. The linear aggregations of intramembranous particles of the mid-piece in the opossum sperm extend distally to the annulus and terminate (Olson et al. 1977). Over specific portions of the annulus, where plasma membrane is closely associated with the underlying dense matrix, the intramembranous particles are densely packed. The function of these elaborations is not clear. However, the immediate immobilization of rodent sperm following binding of fluorescent probes to surface membranes (Edelman and Millette 1971) suggests a vital role for the plasma membrane in sperm motility. Nelson (1972) proposed that the activity of enzymes involved in sperm motility may be regulated in part by the plasma membrane of the tail. The membrane proteins are believed to function as enzymes, ion pumps, receptor attachment sites, or parts of an integral structural framework. The development of techniques for isolating proteins from individual components of the spermatozoon forms the promising research area for determining more about the composition of the membrane, the specialized functions of the specific regions of the membrane, and the membrane organization and regulation in general (Friend et al. 1979, Friend 1982, Holt 1984).

Fig. 82. Freeze-cleave preparation of the A-face of the plasma membrane from the junctional region of middle piece and principal piece of a guinea-pig spermatozoon. The beaded strands of small particles in the middle piece end abruptly at the annulus. The membrane of the principal piece contains a population of randomly distributed larger particles (From Fawcett 1975b)

D. Enzymes

The intramembranous particle arrays, as already described, are probably enzymatic proteins of the spermatozoon's plasma membrane; their flow is, therefore, associated with physiological events, as it has been observed by Koehler and Gaddum-Rosse (1975) after sperm treatment with capacitating media. The intramembranous particles are believed to function in membrane fusion during fertilization, as already stated. Various cytochemical studies demonstrated the presence of phosphatases in the membranes of sperm head in different species of mammals (Teichman and Bernstein 1971, Guraya and Sidhu 1975a, Gordon 1973, 1977, Gordon and Dandekar 1977, Gordon et al. 1978, Mann and Lutwak-Mann 1981, Castellani-Ceresa 1981). Sperm surface galactosyl-transferase is reported for the mouse during in vitro capacitation (Shur and Hall 1982a). The role of this enzyme in sperm binding to the egg zona pellucida is suggested (Shur and Hill 1982b). The regional differentiation of the sperm plasma membrane also finds expression in the very specific localization pattern of several membrane-bound enzymes, such as adenosine triphosphatase (ATPase) and certain other phosphatases (e.g. alkaline and acid phosphatases) (Tash and Means 1983, Soucek and Vary 1984). Casali et al. (1983) made a comparative study of the enzymatic activities of plasma membranes in sperm of bulls of different breeds. Soucek and Vary (1984a, b) studied the properties of alkaline and acid phosphatases in boar sperm plasma membranes. The optimum conditions for their activity were determined, which resulted in manifold stimulation of activity. The observed properties of alkaline and acid phosphatases can be used to distinguish them from other phosphohydrolases in the sperm plasma membrane. This information will also allow the identification of specific enzymes (Casali et al. 1985) that may play important roles in the function of the sperm plasma membrane. Atreja and Anand (1985) have studied the phospholipase and lysophospholipase activities of goat spermatozoa in transit from the caput to cauda epididymidis.

Sperm plasmalemma contains the (Na^+, K^+) ATPase (Tash and Means 1983). Gordon and Barnett (1967) identified Ca^{2+}-ATPase activity on sperm membranes by cytochemical methods. A more direct evidence for the presence of Ca^{2+}-ATPase in the membrane was shown in isolated plasma membranes from spermatozoa (Vijayasarathy et al. 1980, Ashraf et al. 1983). Plasma membrane-bound Ca^{2+}-ATPase in the bull sperm is associated mostly with the tail membranes (Vijayasarathy and Balaram 1982). This Ca^{2+}-ATPase is activated by Ca^{2+} concentration at the millimolar range, and is Mg^{2+}-independent. People have attempted to speculate that this enzyme is coupled with an energy-linked Ca^{2+} transport system in sperm (Tash and Means 1983, Nelson 1985). However, Breitbart et al. (1984a, b) observed that plasma membranes from ram spermatozoa contain Ca^{2+}- or Mg^{2+}-ATPase activities which may be phosphoproteins and are not related to energy-linked Ca^{2+} transport, and Ca^{2+}-stimulated Mg^{2+}-dependent ATPase which relates to the ATP-dependent Ca^{2+} pump. Breitbart et al. (1983) described the detailed properties of the Ca pump or the Ca^{2+}-stimulated Mg^{2+}-dependent ATPase activities (Breitbart and Rubinstein 1982, Tash and Means 1983). A Mg^{2+}-ATPase in hamster sperm head has been characterized during acrosome reaction (Working and Meizel 1982). It could not be determined in which membrane (plasma or outer acrosomal) the enzyme is present. Ashraf et al. (1984) have characterized ($Ca^{2+}-Mg^{2+}$) ATPase activity and Ca^{2+} transport in boar sperm plasma membrane vesicles. Their relationship to phosphorylation of plasma membrane proteins is determined. Meizel (1984) has discussed the significance of Mg^{2+}-ATPase in relation to H^+ pump; enzyme inhibition by Ca^{2+} leads to an influx of H_2O leading to acrosomal swelling, thereby bringing membranes together for fusion. There also exists some evidence against the existence of the Mg^{2+}-ATPase H^+ pump (Meizel 1984) during acrosome reaction. It is involved in pumping H_2O out of the acrosome. Breitbart and Rubinstein (1983) studied the problems of Ca^{2+} transport by spermatozoa plasma membrane. Rufo et al. (1984) have discussed the regulation of Ca^{2+} content in bovine spermatozoa. The effect of calmodulin antagonists on Ca^{2+} uptake is reported for boar spermatozoa (Peterson et al. 1983a).

The (Na^+, K^+) ATPase is an allosteric enzyme, which is activated by Mg^{2+} ions and inhibited by ouabain, undergoes a cycle of conformational changes as it hydrolyses ATP and translocates Na^+ and K^+, thereby balancing the function of the Na/K exchange pump. Meizel (1984) has disussed various cases of evidence in support of and against a role for Na^+, K^+-ATPase and K^+ influx during sperm capacitation and acrosome reaction. Na^+, K^+-ATPase activities, and decreased Mg^{2+}-ATPase (H^+ pump) activity are suggested to be involved in the membrane events of the hamster acrosome reaction or in the events of hamster capacitation, which are needed for the acrosome reaction. K^+ and H^+ channels appear to be formed in the outer acrosomal membrane late during capacitation. The increased K^+ influx may be related to increased glycolysis that may be essential for sperm acrosome reaction. Actually, ATP produced by glycolysis may be required for cAMP synthesis that plays a significant role in capacitation and or the acrosome reaction. In addition to control of ionic exchange reactions. Na^+, K^+-ATPase possibly influences the function of spermatozoa also in other ways. For example, its direct involvement in head-to-head associations between spermatozoa has been envisaged; this type of association is facilitated by Mg, Mn, and Ca ions and inhibited by ouabain (Lindahl 1973). Another possibility is that since the (Na^+-K^+)-ATPase is an enzyme satisfying several criteria for receptor binding, it may be involved facilitating contact between the spermatozoa and extraneous agents, such as certain drugs, hormones, and cyclic AMP (see Mann and Lutwak-Mann 1981). Majumder (1981b, c) has studied enzymic characteristics of ecto-ATPase and ecto-cAMP-dependent protein kinase in rat epididymal spermatozoa. Activities of leucine aminopeptidase and acid phosphatase are also present on the surface of human spermatozoa (Edvinsson et al. 1981). The association of Ca^{2+}-ATPase with the membranes of the sperm head also suggests that these play an important role in making Ca^{2+} available to the acrosomal contents, resulting in the exit of functional acrosomal enzymes (M. H. Johnson 1975, Shapiro and Eddy 1980, Shapiro 1984, Meizel 1984, Nelson 1985). There is as yet no direct or indirect biochemical evidence that inibition or stimulation of Ca^{2+}-ATPase is involved in the acrosome reaction and thus further studies are needed to answer these questions. Nelson and Gardner (1983) reported occupation of Ca^{2+}-binding sites in sperm. Binding of Ca^{2+} antagonists to sperm plasma membrane is reported (Kazazoglan et al. 1984). Evidence has been obtained for a role of calmodulin in the transport of Ca^{2+} by boar spermatozoa (Peterson et al. 1983c). Ashraf et al. (1983) reported the effect of calmodulin antagonists on Ca^{2+} 45 uptake by boar sperm and on binding to sperm plasma membrane. Many other phosphatases (ADPase, AMPase, ATPase, GTPase, p-nitrophenyl phosphatase) have also been demonstrated in the periacrosomal plasma membrane of epididymal mammalian sperm-cells.

Chulavatnatol and Yindepit (1976) observed changes in surface ATPase of rat spermatozoa in transit from the caput to the cauda epididymidis. Rat spermatozoa from the cauda epididymidis show a lower activity of the surface ATPase than the spermatozoa from the caput region. The differences in the enzyme properties of sperm from these two reigons of epididymis suggest that the decline in the activity during maturation of its spermatozoa may reflect changes in the lipids and sulphydryl groups of the sperm membrane. Majumder and Biswas (1979) also demonstrated ecto-ATPase in rat epididymal spermatozoa. The Mg^{2+}-dependent ATPase activity of intact spermatozoa from cauda epididymis hydrolise externally added ATP. This ATPase activity is resistant to the action of proteinases (50 μg/ml-el), i.e. trypsin, chymotrypsin, and pronase, and has a high degree of substrate specificity for ATP. The enzymes cannot be localized after ejaculation because there is an inhibitor in the seminal plasma which is removed in utero (Gordon 1973). Mouse spermatozoa also contain LDH-X on their post-acrosomal surface (R. P. Erickson et al. 1975a, b).

The plasma membrane overlying the acrosomal cap of ejaculated buffalo spermatozoa is associated with 5′-nucleotidase activity which may also be of some biological significance during fertilization besides its role in membrane permeability (Guraya and Sidhu 1975a). This suggestion is also supported by the fact that the release of the acrosome contents is preceded by fusion and vesiculation of the outer acrosomal membrane and the plasma membrane of the apical region (Fig. 83), as already described (see also Yanagimachi and Noda 1970a, Jones 1975, Wooding 1975, Yanagimachi 1978a). Subsequent attachment and fusion of the sperm to the egg surface generally occurs at a medial position along the sperm head at the level of the post-acrosomal sheath or post-acrosomal cap (Yanagimachi and Noda 1970b), indicating that the sperm-cell membrane takes

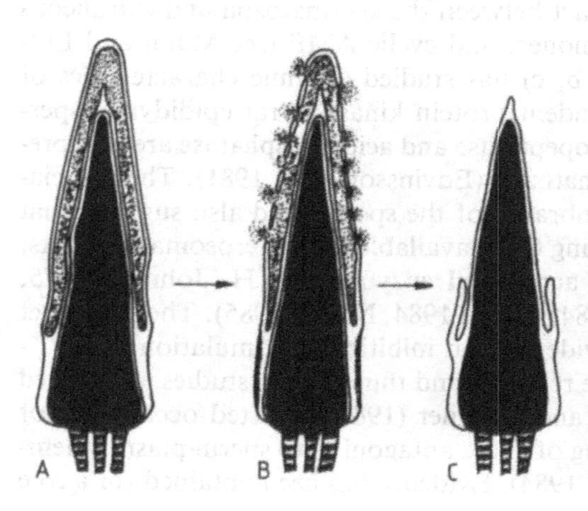

Fig. 83. Schematic representation of the mammalian acrosome reaction. The outer acrosomal membrane fuses with the plasma membrane at multiple sites creating openings through which the enzyme-rich contents of the acrosome escape (*B*). This process leads ultimately to complete loss of the cell membrane over the anterior half of the head (*C*). Thereafter, the inner acrosomal membrane is the limiting membrane of the sperm head in this region. The equatorial segment of the acrosome persists. Its function is still poorly understood (Redrawn from Fawcett 1975b)

active part in fertilization processes. Yanagimachi et al. (1973a), studying the distribution of acid anionic residues on hamster spermatozoa and eggs before and during fertilization, demonstrated that sperm and egg plasma membrane components intermix during gamete fusion (Gabel et al. 1979).

E. Regional Specializations of Surface Properties

The regional specialization of sperm surface properties are expressed by regionally specific intramembranous particles or enzymes, as already discussed, and also by affinities for surface markers that reflect terminal oligosaccharides of intrinsic glycoproteins. These are also shown in different regional charge densities or simply by differential adhesiveness that may cause spermatozoa to agglutinate either head-to-head or tail-to-tail (M. H. Johnson 1975). Here an attempt will be made to describe such regional specialization of the surface in relation to components or structures underlying it at different stages in the life of the sperm. Species differences are also reported.

a) Surface Charge

Changes in the chemical nature of the sperm surface during epididymal maturation are indicated by the study of alterations in the surface charge. Whole-cell electrophoresis studies revealed that intact spermatozoa possess a net negative surface charge (Mudd and Mudd 1929, Nevo et al. 1961) and that there is a marked increase in negative surface charge as sperm mature in the epididymis (Bedford 1975, 1979), Nicolson and Yanagimachi 1979). The immobilized bull and rabbit spermatozoa placed in an electrophoretic field migrate towards the anode with their tails foremost, indicating that these cells have a net negative charge on their surface, with the tail more strongly charged than the head (Nevo et al. 1961). Subsequently, it has been shown that the electrophoretic properties of rabbit spermatozoa change as they became mature, becoming more strongly negative as they pass through the epididymis (Bedford 1975, 1979, Nicolson and Yanagimachi 1979). The polysaccharide nature of the surface coating formed in late spermatid (André 1963) would be in agreement with the observation that the heads of spermatozoa from the caput epididymis of rabbit carry negative surface charge and show little tendency to head-to-head auto-agglutination (Bedford 1975). These observations show that the electric charge carried by the mammalian spermatozoa is a negative one, probably unevenly distributed over the plasma membrane, and subject to environmental changes encountered during sperm passage in the epididymis and as a result of exposure to accessory fluids. Earlier, electrophoretic techniques figured prominently in the determination of the net charge of spermatozoa, in attempts to show that the electric charge on the surface of X and Y spermatozoa differs sufficiently to permit electrophoretic separation of semen into two fractions, which on insemination would produce progeny with a different sex ratio. The results were disappointing, however (Chap. VII).

Recently, there has been a great interest in the surface charges on sperm and especially in morphological methods for their visual demonstration (Koehler 1978, Bedford 1979, Nicolson and Yanagimachi 1979, H. D. M. Moore 1979, Courtens and Fournier-Delpech 1979). Veres (1968), using signal colloids with given positive or negative charges, demonstrated negative charge on the surface of normal mature bull sperm, which is believed to be the cause of collision-free movement of the sperm and to play some role in approaching the ovum. Veres et al. (1976), using AGJ particles as signalizing, investigated the surface charge of bull and boar spermatozoa. The negative surface charge of bull spermatozoa is concentrated at the acrosome complex. On boar spermatozoa, it is equally dispersed on the whole cell. Freezing and thawing of bull spermatozoa do not destroy the negative surface charge. Sialic acid appears to be one of the surface constituents, since the negative charge of bull spermatozoa diminished after treatment with neuraminidase (Fuhrmann et al. 1963, Koehler 1978). The net surface charge has also been determined by isoelectric focussing (H. D. M. Moore 1979).

The surface charge has been investigated by binding pattern of electron-opaque particles of colloidal iron hydroxide (CIH) to the sperm surface followed by study of their localization in electron micrographs of thin sections (Cooper and Bedford 1971, Bedford et al. 1972, Fléchon 1971, 1975, Yanagimachi and Teichman 1972, Yanagimachi et al. 1972, 1973b, Nicolson and Yanagimachi 1979). The CIH-binding technique not only provides information on the distribution of negative charges on the cell surface but also provides insight into the chemical nature of the charged moieties. At low pH (1.8) binding of CIH particles demonstrates the presence of ionized radicals of carboxy sugars, primary phosphate and sulphate, while binding at higher pH (3.0 – 4.6) reveals the presence of amino acid carboxy groups.

Marked differences are observed in the chemistry of the sperm surface between species, as well as between epididymal sperm of differing maturity within a species (Bedford et al. 1972, Yanagimachi et al. 1972, Fléchon 1973, Hammerstedt 1979, Nicolson and Yanagimachi 1979, Bedford 1979). In these studies, changes in the binding of CIH particles are observed between caput epididymal, cauda epididymal, and ejaculated spermatozoa. Bedford et al. (1972) found that the binding of CIH at pH 3.0 – 4.6 to rabbit sperm head is far greater on spermatozoa isolated from cauda epididymidis. Yanagimachi et al. (1972) noticed that CIH binding at pH 1.8 to rabbit sperm tails is increased during epididymal passage of spermatozoa, and Fléchon (1973) demonstrated that CIH particles are bound in a more dense, discontinuous distribution on the ejaculated spermatozoa as compared with spermatozoa obtained from caput or corpus epididymidis.

The distribution of iron particles remains uniform within a particular segment, but abrupt alterations in particle concentration occur sometimes at junctions between segments during the passage of sperm through the epididymis. The patterns of distribution are described to be consistent within species but not between species (Fawcett 1975b, Gwatkin 1976, Koehler 1978, Nicolson and Yanagimachi 1979). For example, in the rabbit the heads remain relatively free of label, the mid-piece lightly labelled, and the remaining part of the tail heavily labelled. In contrast to this, the whole of the surface in the guinea-pig sperm is labelled, but the principal and the end-piece most heavily. The acrosomal region is more heavily labelled than the post-acrosomal segment (Yanagimachi and Teichman 1972). The abrupt alterations seen from one region of the sperm to another are believed to be indicative of some changes in biochemical properties of the membrane at these boundaries.

Direct evidence of a change in the distribution of charge groups at the sperm surface is provided by observation of differences in electrophoretic mobility and head-tail orientation in an electrical field in sperm populations taken from various levels of the epididymis (Bedford 1975, 1979, Nicolson and Yanagimachi 1979). Change in the nature and pattern of distribution of ionized groups on the surface of the sperm head and tail during epididymal passage in monkey and man has also been demonstrated under the electron microscope by using positively charged ferric chloride at different pH (Cooper and Bedford 1971, 1976). This has provided confirmatory evidence of changes in the chemical nature of the sperm surface during epididymal maturation (Nicolson and Yanagimachi 1979). Holt (1980) observed that surface charge of bull spermatozoa is increased during passage between the corpus and cauda epididymidis due to the addition of sialic acid groups to the sperm surface. Differences in surface charge distribution of bull spermatozoa are also revealed with fluorescent probes (Vijayasarathy and Balaram 1982).

Actually, more significant evidence for maturation changes in the surface of epididymal sperm has been obtained from various studies of surface characteristics indicated by adhesiveness, net charge at physiological pH, and charge density over different portions of the sperm surface (Bedford 1975, 1979, Nicolson and Yanagimachi 1979, Courtens and Fournier-Delpech 1979). Courtens and Fournier-Delpech (1979) observed that electronegative charges at pH 1.8, which are uniformly distributed on the heads of ram corpus or cauda epididymal spermatozoa, increase on the flagella of cauda epididymal spermatozoa. The charges are reduced to a great extent after incubation in utero. Electropositive charges at pH 9 are abundant on the post-acrosomal area of both corpus and cauda epididymal sperm. The corpus epididymal sperm membrane becomes heavily electropositive after utero incubation. H. D. M. Moore (1979), using isoelectric focussing, determined changes in net surface charge following sperm maturation, ejaculation, incubation in the female tract, and after enzyme treatment.

b) Antigens

The results of various studies discussed earlier have shown that the chemical characteristics of the surface of one sperm-cell may vary in at least five regions, namely, the acrosome, the post-acrosomal region, post-nuclear cap, the midpiece, and the principal piece of the tail. Somewhat comparable results have been obtained in immunofluorescence studies of human spermatozoa (Hjort and Brogaard 1971), which have brought to light different antigen combinations over the surface of the acrosome, equatorial segment, post-acrosomal region and tail (Koehler 1978, Tung et al. 1982). This work extends the results of earlier agglutination studies (Henle et al. 1938) which pointed to the existence of head- and tail-specific antigens. Surface domains of the guinea-pig sperm have been defined with monoclonal antibodies (Myles et al. 1981, Primakoff and Myles 1983), as also of the human (Hirschel et al. 1984, Hinrichson-Kokane et al. 1985, Villarroga and Scholler 1986) and hamster (Moore et al. 1985) spermatozoa. The observed monoclonal antibody-binding patterns have suggested that the guinea-pig sperma surface is divided into five different antigen domains. These antigens may be restricted to the anterior head, posterior head, whole head, whole tail, or posterior tail cell surface. The surface antigens of bull sperm are also characterized with monoclonal antibodies (Chakraborty et al. 1985). The major polypeptides of boar sperm plasma membrane have also been localized with specific monoclonal antibodies (Hunt et al. 1985).

Actually, during their passage through the epididymidis and vas deferens, the spermatozoa are modified by absorbing materials of epididymal origin, as will be described later. Binding of these substances is associated with changes in the antigenic properties of the sperm surface (Suarez et al. 1981, Vernon et al. 1982, Jones et al. 1984). Some of these absorbed materials are tightly bound and are not readily removed by washing in physiological saline or by treatment with organic solvents. These antigenic components are collectively designated sperm-coating antigens (Dravalnd and Joshi 1981). The functional significance of the spermatozoa-bound substances, of their binding and release during epididymal maturation, and the role these substances play in fertilization, remain obscure and must be determined in future studies (Tung et al. 1980, 1982, Suarez et al. 1981). However, spermatozoa-coating antigens are lost in the female genital tract (Ohara 1981). Jones et al. (1984) have studied the fate of a maturation antigen of rat sperm during fertilization.

Several histocompatibility-related antigens have been found on the surface of spermatozoa: H-Y (E. H. Goldberg et al. 1971, Wachtel 1977), H-2 (Vojtiskova et al. 1969, E. H. Goldberg et al. 1970), HL-A (Fellous and Dausset 1970, Prabhu and Hegde 1982), Ia determinants (Fellous et al. 1976b, Vaiman et al. 1978), F-9 (Gachelin et al. 1976, Jacob 1977), $PCCC_4$ (Gachelin et al. 1977), and probably histocompatibility antigens themselves (R. P. Erickson 1976). Some doubt is expressed about the presence of H-2 antigens (Anderson et al. 1981a, b). The H-Y male-specific antigen is restricted to the cell surface covering the acrosomal cap of the sperm head (Koo et al. 1973). Various serologically defined antigens present on the surface of mammalian spermatozoa are not common to most tissues of the body. These restricted membrane markers are shared only with the nervous system (NS-3, NS-4, Cb1) (Schachner et al. 1975, Weeds 1975). The intact spermatozoon possesses two types of surface antigenic determinants, those integral to the plasma membrane and those secondarily absorbed (M. H. Johnson and Howe 1975, Millette 1979a, b, Vernon et al. 1982).

O'Rand and Metz (1976) observed that fertile spermatozoa have both mobile and nonmobile receptors. O'Rand (1977) stated that the same receptor may be mobile within one surface region (in this case, acrosomal) and nonmobile within another (post-acrosomal and tail). However, using either isoimmune (Koehler 1975a) or heteroimmune (Koehler and Perkins 1974, O'Rand 1977) anti-rabbit whole sperm sera, no significant surface antigen mobility has been observed on mature spermatozoa (Koehler 1978). Consequently, two possible explanations have been put forward (O'Rand 1977): anti-whole sperm IgG cross-links either a sufficient number of mobile surface receptors to immobilize all receptors or a mobile class of surface antigens to a nonmobile class, rendering the mobile class immobile. Mobile receptors might still occur but these would not be detectable due to the overall label with anti-whole sperm antibody. Jager et al. (1976) observed that the percentage of motile spermatozoa found to be coated with antisperm antibodies of the IgG class and the extent of the coating proved to be correlated with the agglutination titre of circulating antisperm antibodies as well as with the initiation of sperm penetration into cervical mucus. The visualization of anti-sperm plasma membrane IgG and fab has been used as a method for localization of boar sperm membrane antigens (Russell et al. 1982). Antisperm antibodies are identified in the sperm of infertile man (Haas et al. 1982). Human sperm antigens are identified to antisperm antibodies (C. N. Lee et al. 1983). Lenzi et al. (1982) developed an enzyme-linked immunosorbent assay for antisperm antibodies detection. The effects of antisperm plasma membrane antibodies on sperm egg binding, penetration, and fertilization in the case of pig are reported (Peterson et al. 1982). Beyler et al. (1983) studied binding of antibodies against synthetic antigens to human and to mouse sperm and their effect on mouse in vitro fertilization. In an excellent review, Bronson et al. (1984) have discussed the role of sperm antibodies in human infertility. Spermatozoal antigens vary greatly and those associated with the plasma membrane appear to be the most important in impairment of fertility. Monoclonal antibodies can be prepared against specific plasma membrane antigens of human spermatozoa and will inhibit fertilization in vitro, and other antibodies against antigens on the surface of spermatozoa will interfere with the binding of spermatozoa to the zona pellucida.

Romrell and O'Rand (1978), using rabbit sperm isoantisera, investigated the relative abundance, mobility, and the ultrastructural localization of sperm surface isoantigens at different developmental stages. Myles and Primakoff (1983) observed a guinea-pig sperm surface antigen that migrates during capacitation. Gaunt (1983) has observed the spreading of a sperm surface antigen within the plasma membrane of the egg after fertilization in the rat. Poulsen (1983a) studied the nature of an isoantigen of the human sperm membrane (Poulsen and Hjort 1981, Hjort and Poulsen 1981). Sperm-specific isoantibodies and auto-antibodies are known to inhibit the binding of human sperm to the human zona pellucida (Bronson et al. 1982b). The other immunological studies have also revealed changes in the antigenic properties of the spermatozoon (Killian and Amann 1973, M. H. Johnson and Howe 1975, Koehler 1978). Usually these immunologically detected changes are attributed either to the loss of specific sperm-coating antigens or to the acquisition of new coating antigens secreted from different regions of the epididymis. Several mechanisms are believed to account for the apparent redistribution of plasma membrane components prior to spermiation (Millette 1979a, b). Receptors for ATBS may be masked nonspecifically by newly adsorbed peripheral surface molecules. Alternatively, the antigenic sites may be selectively removed from specific areas of the developing sperm plasma membrane by proteolysis or by interalization. A third, and more plausible, explanation is that these antigenic determinants are partitioned by lateral translocation in the plane of the membrane.

Rabbit epididymal sperm is more suitable for agglutination than ejaculated sperm. A specific agglutinin recognizing N-acetylglucosamine is involved. Mettler and Skrabei (1979) isolated human spermatozoa surface antigens inhibiting sperm agglutination and sperm immobilization in sera of sterile females (Mettler and Gradl 1977). They concluded that sperm-agglutinating and sperm-immobilizing antibodies react with different sperm antigens. Dacheux et al. (1983) after studying head-to-head agglutination of ram and boar spermatozoa, have produced evidence for an epididymal antagglutinin. Yan et al. (1983) studied monoclonal antibody-inducing human sperm agglutination. Gupta et al. (1985) have observed an incidence of sudden appearance of spermagglutination among normal fertile buffaloes during summer and the rainy season. An inverse relationship is observed between sperm agglutination and percentage dilution of semen as well as percentage motility. Head-to-head agglutination was the predominant type though some spermatozoa showed tail-to-tail agglutination. The reasons for this spermagglutination are obscure. Li et al. (1978) studied the effect of antibody against a purified sperm-coating on human sperm. These results have indicated that the purified sperm-coating antigen (also obtained by Czuppon et al. 1981) is unlikely to be useful as an antifertility antigen for immunologic fertility control. Mettler et al. (1983a, b) have induced high-titre mouse-antihuman spermatozoal antibodies by liposome incorporation of spermatozoal membrane antigens.

Poulsen (1983b) used an improved method for isolation of tritium-labelled auto-antigen-I of the human sperm membrane. Human sperm auto-antigens are also studied in relation to fractionation and analysis of sperm membrane antigenic systems (De Almeida et al. 1983). Guinea-pig sperm auto-antigens have also been localized by direct and indirect immunoferritin techniques in plasma membrane over the entire sperm head, acrosomal contents, fibrous sheath, and outer dense fibres of the tail filament and the inner acrosomal membrane of 50% of acrosome-reacted spermatozoa (Tung et al. 1982). These cellular structures are known to be involved in guinea-pig sperm rouleaux formation, acrosome reaction and interaction of acrosome-reacted sperm with zona pellucida, and with the vitellus of guinea-pig ova. Sperm auto-antigens, which are also reported for human sperm (De Almeida et al. 1982), may be involved in these cellular events. Vernon (1983) observed modification of a mouse sperm surface auto-antigen under in vivo and in vitro capacitation conditions (see also Vernon et al. 1985). A thorough understanding of spermatozoa auto-antigens in animals may lead to a possible way for immunological contraception (Mettler 1982, Voisin 1983, Vernon et al. 1985). The monoclonal antibodies provide probes for immunochemical characterization of sperm antigens and for elucidation of the role of the antigens in sperm functions (Naz et al. 1984, H. D. M. Moore and Hartman 1984, Moore et al. 1985). Monoclonal antibodies are also used for the identification of mobile and fixed antigens on the plasma membrane of rat (Gaunt et al. 1983) and human spermatozoa (Paul et al. 1983). A number of antigens localized to discrete surface regions have been distinguished with monoclonal antibodies on guinea-pig sperm cells (Primakoff and Myles 1983). One of these monoclonal antibodies, PT-1, binds exclusively to the posterior tail region of the sperm-cell surface. Fluorescence redistribution after photobleaching measurements indicate that within its surface domain the PT-1 antigen diffuses rapidly and completely (Myles et al. 1984). Myles et al. (1984) have proposed that the mechanism for localization of the PT-1 antigen may be a barrier to diffusion at the domain boundary. Guinea-pig sperm auto-antigens have also been analyzed immunochemically (Teuscher et al. 1982). Three major protein bands with approximate molecular weights of 69,000, 62,000, and 40,000 are shown.

Mammalian spermatozoa can be immobilized in the presence of complement and antibodies directed against one or more sperm antigens (Tung et al. 1980, 1982, Bronson et al. 1982a). These "immobilizing" antigens seem to be intrinsic sperm plasma membrane antigens (Russo and Metz 1974). Sperm immobilization is believed to result from permeability changes (Drevius 1968, Metz and O'Rand 1975) and, at least in some species, damage to the sperm plasma membrane (Russo et al. 1975, Alexander 1975). O'Rand and Metz (1976) described a method for the isolation of immobilizing antigen from rabbit sperm membranes. This membrane glycoprotein immobilizing antigen has been localized on the head and tail plasma membrane of epididymal, ejaculate, and capacitated sperm by exchange agglutination and immobilization in the presence of guinea-pig complement. Koehler (1976) studied changes in antigenic site distribution on ejaculated and epididymal rabbit spermatozoa after incubation in "capacitating" media. Isotonic and hypertonic media are effective in removing components from the membrane overlying the acrosomal region. The post-nuclear sheath retains a heavy complement of labelled sites. The possible destabilization of head membranes preparatory to the acrosome reaction has been suggested.

c) Coating Substances

Various studies have indicated that the epididymis makes an important contribution to the plasma membrane alterations that occur during transit of spermatozoa through the epididymis. The important questions of whether the membrane alterations acquired in the distal portions of the epididymis are due to absorption and/or insertion of secreted epididymal substances into the sperm plasma membrane, or due to unmasking or chemical alteration of existing plasma membrane constituents by factors in the lower epididymis, are being answered in recent studies discussed in different sections of this chapter. The binding of epididymal secretory products to the sperm surface appears to be common feature of sperm maturation (Dacheux et al. 1984, Muller and Eddy 1984), while the function and mechanism of interaction of these chemical substances with the sperm surface are still not fully understood. Our knowledge about the exact nature, origin, binding specificity, and physiological significance of the various coating substances or surface receptors is still very meagre (Orgebin-Christ 1981, Mann and Lutwak-Mann 1981, Brooks 1983, Bartlett et al. 1984, Brooks and Tiver 1983, 1984, Voglmayr et al. 1985, Moore et al. 1985, Ellis et al. 1985). Components associated with sperm membranes may be removed, added, or altered during sperm maturation in the epididymis (Bostwick et al. 1980, Jones et al. 1981, Mann and Lutwak-Mann 1981, Voglmayr et al. 1980, 1982, 1983, 1985) and sperm capacitation (Fig. 84) (O'Rand 1979, 1982, Esbenshade and Clegg 1980, Vernon et al. 1985). Voglmayr et al. (1982) suggested that internalization and degradation within

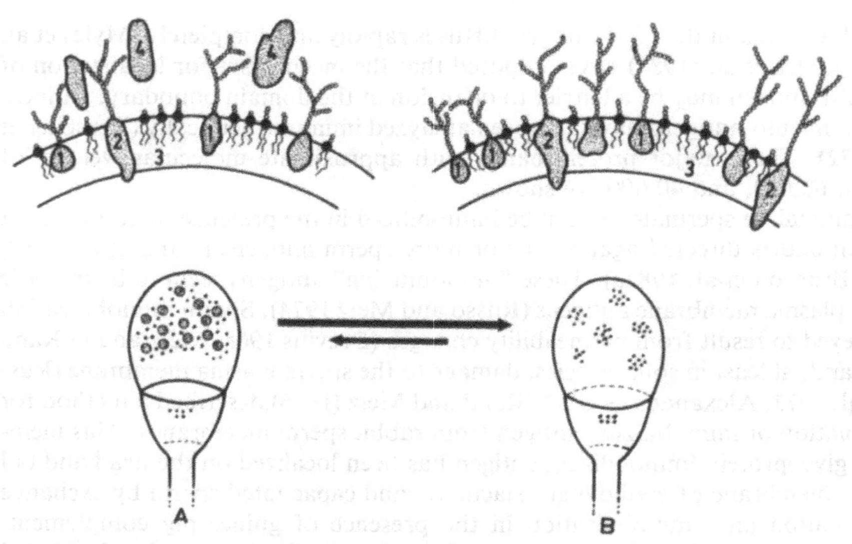

Fig. 84. A model for the surface changes in the peri-acrosomal plasma membrane associated with capacitation. *A* Prior to capacitation; *B* After capacitation. Four classes of molecules are shown: (1) mobile glycoproteins, (2) nonmobile glycoproteins, (3) glycolipids, and (4) peripheral components. The *lower sketches* show the surface pattern expected for intrinsic mobile glycoproteins (class I) in association with peripheral components (class 4) before capacitation (*A*) and without peripheral components after capacitation (*B*). Sperm outlines are not drawn to scale (Redrawn from O'Rand 1979)

the membrane of maturing spermatozoa may be important mechanisms of sperm differentiation in the epididymis. The exposed sugar residues of the plasma membrane may also be modified by many highly active epididymal glycosidases and/or glycosyltransferases (Dacheux and Voglmayr 1983). Epididymal and seminal antigens that are bound tightly to ejaculated sperm are also removed or altered during incubation in the uterus.

The cytochemical techniques have revealed a carbohydrate-rich coat or glycolemma which is developed to different degree in different regions of the rabbit sperm surface (Fléchon 1971, 1973). It varies greatly in its distribution on different portions of the sperm during its transit through the ductular system (Fléchon 1974a). In case of rabbit, there also occurs the modification of the glycoproteins of the acrosome and of the cell surface of the human spermatozoa (Bustos-Obregon 1975). Sialic acid concentration of buffalo spermatozoa varies significantly ($P < 0.01$) in different seasons of the year (Sidhu and Guraya 1979b). The projecting and main areas of acrosome in buffalo spermatozoa show more negative charges, and neutral, acidic, and sulphated mucopolysaccharides, which have a higher polyanion nature than those on the mid-piece (Sidhu and Guraya 1980a). Sulphated mucopolysaccharides are also present on the surface of post-nuclear cap and main-piece of spermatozoa (Fig. 85). Toowicharanont and Chulavatnatol (1983a) made direct assay of bound sialic acids on rat spermatozoa from caput and cauda epididymidis, which are believed to be involved in sperm-zona interaction in the mouse (Lambert and Van Le 1984). The sialoglycoproteins of rat epididymal fluid and spermatozoa have been characterized by periodate-

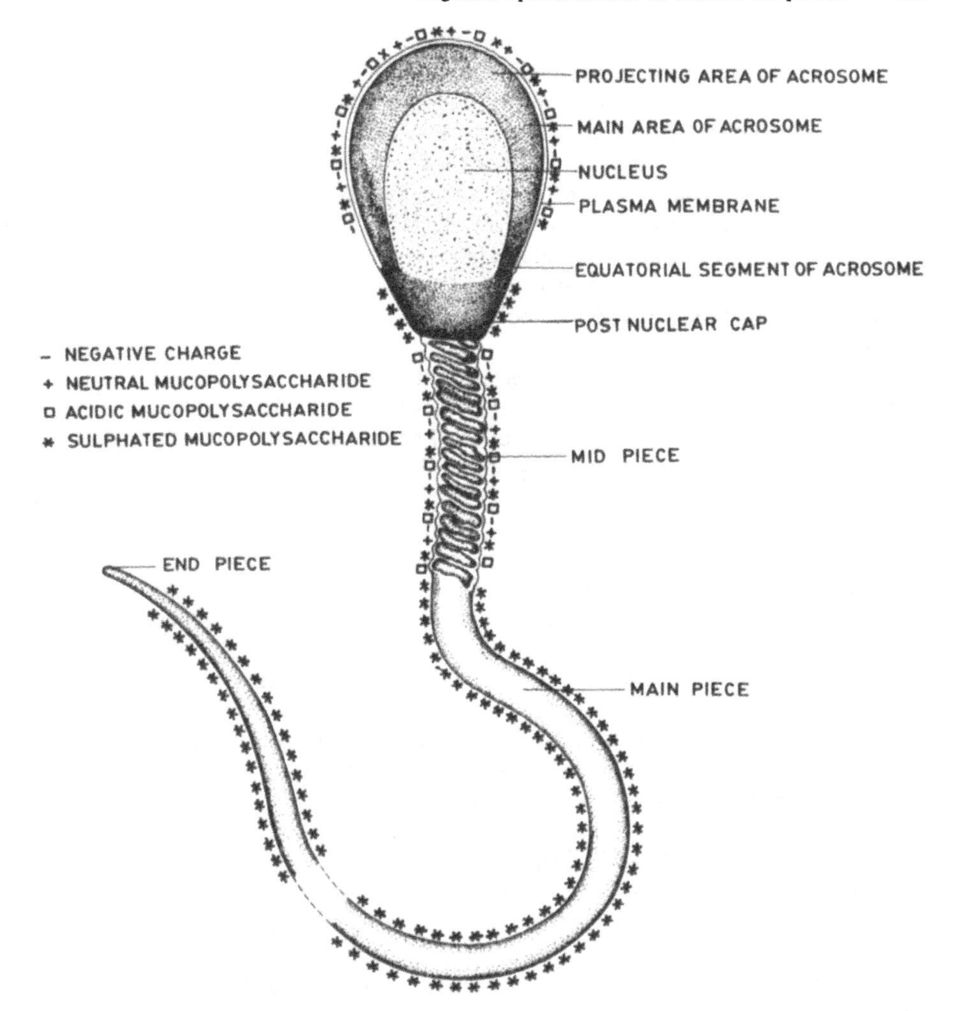

Fig. 85. Buffalo spermatozoon showing distribution of various mucopolysaccharides and negative charges on its surface (From Sidhu and Guraya 1980a)

tritiated borohydride (Toowicharanont and Chulavatnatol 1983b) as also characterized for ram spermatozoa (Malick and Bartoov 1985). Henriet et al. (1976) studied the glycoprotein layer covering the bull spermatozoa from ejaculation to the moment of fertilization. They have suggested that this layer appears to protect spermatozoa against the lysosomic enzymes encountered by them in the female genital tract before the right moment for the acrosomal reaction occurs. Olson and Gould (1981) characterized sperm surface and seminal plasma glycoproteins in the chimpanzee as isolated for the human sperm (Saji et al. 1985). The major glycoprotein on the plasma membrane of rat testicular spermatozoa has 110,000 MW and is replaced by a 32,000 MW protein as the spermatozoa pass through the epididymis (R. C. Jones et al. 1981). The insertion or absorption of this protein of epididymal origin to the plasma membrane of spermatozoa correlates with the acquisition of fertilizing capacity and thus can serve as a good marker for assessing maturation in vitro. Peterson et al. (1983a), using two-dimensional preparative PAGE, have isolated the major acidic glycoprotein of boar sperm plasma membranes. They also isolated specific groups of polypeptides by anion exchange chromatography and lectin affinity chromatography (Peterson et al. 1983b). Kopecny et al. (1984b) made a fine structure autoradiographic study of binding of secreted glycoproteins to spermatozoa in the epididymis. Bartlett et al. (1984) have studied translation of messenger RNA from rat epididymis and identification of poly(A)RNA coding for acidic epididymal glycoproteins.

Gonzalez-Echeverria et al. (1982) identified androgen-dependent glycoproteins in the hamster epididymis and their association with spermatozoa. Dacheux and Voglmayr (1983) studied a sequence of sperm-cell surface differentiation and its relationship to exogenous fluid proteins in the ram epididymis. C. R. Brown et al. (1983b) and Voglmayr et al. (1983) demonstrated surface glycoprotein changes in rat and ram spermatozoa during epididymal maturation as demonstrated by Ellis et al. (1985) and Moore et al. (1985) in hamster sperm with monoclonal antibodies. Zeheb and Orr (1984) characterized a maturation-associated glycoprotein on the plasma membrane of rat caudal epididymal sperm. The most active area for the androgen-regulated glycoprotein secretion is the distal caput, but presumably also other parts have secretory function with typical regional characteristics (Brooks 1983).

Certain spermatozoa-bound substances are known plasma membrane proteins, such as lactoferrin, an iron-bound protein found in the human seminal fluid (Hekman and Rümke 1969). The prominent protein in the plasma membrane of porcine sperm is a 14,000 MW protein exposed on the surface of the sperm-cell (Esbenshade and Clegg 1980). Hunt et al. (1985) have isolated major proteins from boar sperm plasma membrane. Koehler et al. (1980) studied collagen-binding protein on the surface of ejaculated rabbit spermatozoa. Pavelko and Crabo (1976) suggested that inability of spermatozoa to accumulate seminal proteins may be a reason for impaired fertility. The surface proteins of mammalian spermatozoa have been recently distinguished by surface-labelling techniques. Differences in the exposure or display of rat spermatozoa surface proteins during epididymal maturation have been obtained using surface labelling with galactose oxidase or metaperiodate followed by [^{3}H] borohydride reduction (Olson and Hamilton 1978). Olson and Hamilton (1978) investigated the macromolecular composition of a membrane fraction isolated from rat spermatozoa, and used these specific biochemical probes to study the externally oriented plasma membrane glycoproteins of caput and cauda epididymal spermatozoa. Cauda epididymal spermatozoa possess a 37,000 MW glycoprotein on the cell surface, which is not found on caput epididymal spermatozoa (Bostwick et al. 1980). Similarly, radioactive labelling on surface sialo-glycoproteins has revealed that the 37,000 MW glycoprotein could be labelled on cauda epididymal spermatozoa, but is not detected on caput epididymal spermatozoa. White et al. (1983) studied surface proteins of ejaculated human and rhesus monkey spermatozoa, as also studied for rat spermatozoa (Brooks and Tiver 1984). Surface components of ejaculated chimpanzee sperm are also described (Young and Gould 1982). Silvestroni et al. (1983) have developed a new procedure for removing the coating material from viable human spermatozoa. The significance of alterations in proteins and glycoproteins in relation to maturation of spermatozoa in the epididymis has been discussed by Bostwick et al. (1980).

Glycoproteins and proteins, extracted from the corpus epididymis of rams and linked to colloidal gold, bind specifically into the plasma membrane of spermatozoa (Courtens et al. 1982). The sites for the glycoproteins are saturated on spermatozoa after the passage through the corpus epididymis. Sites for proteins are present at the anterior part of testicular spermatozoa, the whole plasma membrane of the head of spermatozoa from the corpus epididymis, and the posterior part of the head, excluding the equatorial segment, of most spermatozoa from the cauda epididymis. Handrow et al. (1984) have studied specific binding of the glycosaminoglycan ^{3}H-heparin to bull, monkey, and rabbit spermatozoa in vitro.

The surface proteins of many cells have now been shown to be labelled effectively with iodination techniques originally reported by Phillips and Morrison (1971). Oliphant and Singhas (1979) outlined methods required for similar labelling of spermatozoa, as also used by other workers (Voglmayr et al. 1982). Differences in the surface components of epididymal and ejaculated sperm so labelled are also reported (Vierula and Rajaniemi 1981). An attempt has been made to reveal the relationship of specific surface proteins to epididymal function and sperm capacitation in the rabbit. Nicolson et al. (1979), using lectoperoxidase-catalysed [^{125}I]-iodination technique and rabbit spermatozoa, reported results similar to those of Olson and Hamilton (1978). Lactoperoxidase-catalysed iodination results in the introduction of [^{125}I] into the exposed tyrosine residues in proteins and glycoproteins of rabbit caput and cauda epididymal spermatozoa, as also reported for the ram (Fournier-Delpech and Courot 1980, Voglmayr et al. 1982) and bull (Vierula and Rajaniemi 1981) spermatozoa. Using these techniques, it is possible to distinguish an approximately 35,000 MW protein component on the surface of rabbit cauda, but not caput epididymal spermatozoa. It is also possible to identify components of approximately 58,000 and 39,000 and some lower MW proteins labelled more heavily on the cauda epididymal spermatozoa. Caput epididymal spermatozoa of bull show a protein peak with molecular weights of 15,000 – 18,000, but this peak is not seen on cauda epididymal spermatozoa (Vierula and Rajaniemi 1981). On caput epididymal spermatozoa the most intensely labelled protein peak is located between 90,000 and 100,000, but on cauda epididymal spermatozoa, the corresponding peak is only weakly labelled and has a MW of 80,000 – 90,000. Surface protein with molecular weights of 42,000 – 47,000 is dominating on cauda epididymal spermatozoa. The exact locations of the iodinated surface proteins on caput and cauda epididymal spermatozoa need to be determined, but the changes described above may occur on spermatozoa head glycoproteins involved in sperm capacitation and egg recognition. Vierula and Rajaniemi (1983) have studied the effect of seminal plasma and calcium on the stability of the surface protein composition of ejaculated bull spermatozoa.

Ji et al. (1981) analysed the surface proteins and glycoproteins of ejaculated bovine spermatozoa. Among some 20 bands initially resolved, 8 are found labelled with [^{125}I]-iodination corresponding to molecular weights of approximately

155,000, 126,000, 93,000, 75,000, 53,000, 29,000, 26,000, and 15,000. Evidence has been given to show that these polypeptide bands are not contaminants from the seminal fluid. Two bands of 26,000 and 15,000 are resistant to digestion by trypsin at the surface; three bands of 29,000 and 15,000 are insensitive to chymotrypsin. The rest of the bands are sensitive to both trypsin and chymotrypsin. A characterization has also been made of sperm surface and seminal plasma glycoproteins in the chimpanzee (Olson and Gould 1981). Spermatozoa labelled by galactose oxidase treatment show a single labelled macromolecular component of 37,000 MW; spermatozoa labelled with sodium metaperiodate sodium boro (^{3}H) hydride treatment show incorporation into macromolecular components of 37,000 and 25,000 MW. The radioactive derivative ^{125}IFC ([^{125}I]-Diiodofluorescein isothiocynate) permits an analysis of the sperm plasma membrane proteins that are responsible for most of the binding (Gabel et al. 1979). R.C. Jones and Brown (1982) have observed the association of epididymal secretory proteins showing alpha-lactalbumin-like activity with the plasma membrane of rat spermatozoa.

Foulkes (1977) have observed that the pattern of labelling of lipids extracted from the washed bovine spermatozoa does not indicate that particular lipids become associated with spermatozoa. However, increases in the specific radioactivity of lipid extracts from washed spermatozoa have lent support to the contention that lipoproteins become firmly bound to the cells (Gordon et al. 1974). Cookson et al. (1984) have investigated immunochemically the interaction of egg-yolk lipoproteins with bovine spermatozoa. An analysis is made of lipid and protein components of ejaculated bull sperm surface and seminal plasma (Young and Goodman 1982). Huacuja et al. (1981) studied exchange of lipids between spermatozoa and seminal plasma in normal and pathological human semen.

Evans and Setchell (1978b) studied the association of exogenous radioactively labelled phospholipids, phosphatidylcholine, phosphatidylethanolamine or phosphatidyllinositol with spermatozoa of boar, bull, and ram. Liposomes of phosphatidyl choline are associated with spermatozoa which form a glycocalyx surrounding the sperm plasma membrane. Epididymal sperm of rat have the ability to associate [^{14}C]-phosphatidylcholine and then convert it to [^{14}C]-glycerylphosphorylcholine (Wang et al. 1981). Davis and Byrne (1980) observed that cholesterol and phosphatidylcholine uptake from dipalmitoyl-phosphatidylcholine liposomes by rabbit spermatozoa show a complex dependence on temperature. Data show the incorporation of these lipids, especially that of cholesterol into sperm plasma membrane. Addition of 10% phosphatidylserine to dipalmitoyl-phosphatidylcholine liposomes containing [^{3}H]-cholesterol or [^{14}C]-phosphatidylcholine, inhibits interactions between liposomes and rabbit spermatozoa (Davis and Byrne 1980). Phosphatidylserine causes an apparent decrease in the inhibitory effect of cholesterol-bearing liposomes on sperm fertilizing ability. Davis (1980), after studying the interaction of lipids with the plasma membrane of sperm-cells, demonstrated the antifertilization action of cholesterol. The results corroborate the view point that alterations in lipid bilayer of sperm plasma membrane significantly influence the fertilizing capacity among mammalian spermatozoa. Langlais et al. (1981), after studying the localization of cholesteryl sulphate in human spermatozoa, have suggested that it acts as a membrane stabilizer and enzyme inhibitor during the maturation of spermatozoa in the epididymis. The cleavage of the sulphate moiety within the female reproductive tract apparently triggers a cascade of events leading to sperm capacitation and fertilization. Uterine fluid proteins bind sperm cholesterol during capacitation in the rabbit (Davis 1982).

For the better understanding of the nature and the physiological significance of coating substances as well as of maturation and capacitation problems, work, utilizing various biological, immuno-histochemical, biochemical, and physical techniques, should be continued. Methods based on electron spin resonance, in conjunction with the techniques of lectin binding, selective partitioning of surface antigens and glycoproteins, and whole-cell isoelectric focussing, are proving increasingly useful in helping to define the regional properties of the plasma membrane in the spermatozoa of various mammalian species and the changes which occur in structural components of the sperm plasmalemma during epididymal maturation, capacitation, and acrosomal reaction (see also Voglmayr et al. 1985). It is also now possible to consider the bulk isolation and purification of most, if not all, of the sperm surface glycoproteins, using newly developed techniques for lectin immobilization and affinity chromatography in buffered detergent solutions (Lotan et al. 1977). High pressure nitrogen gas cavitation provides a simple means for obtaining undamaged, highly purified sperm plasma membranes (Peterson et al. 1980). Specific antibodies made against these components will be very useful in revealing their location, exposure, and mobility in spermatozoa during epididymal transit. Besides, the functional significance of these components in the subsequent steps of capacitation and fertilization can be investigated with the use of antibodies on their fab fragments directed against specific spermatozoa head proteins and glycoproteins (Ellis et al. 1985, Vernon et al. 1985, Moore et al. 1985).

With recent advances in membrane biology, capacitation of spermatozoa, like their maturation, can also be approached as a membrane phenomenon at the molecular level, as recently demonstrated by O'Rand (1979, 1982), who has proposed a molecular model of surface alterations in the peri-acrosomal plasma membrane (Fig. 84). In this model, the plasma membrane prior to capacitation (Fig. 84) is distinguished as having at least four types of molecules involved in the surface change: (1) glycoproteins that are mobile within the plane of the membrane, (2) nonmobile glycoproteins, (3) glycolipids, and (4) peripheral membrane components. After capacitation, the relationship in these four classes of molecule changes (Fig. 84). The endogenous physiological change, therefore, appears to be a membrane alteration related to the rearrangement of intrinsic components. Future investigations for regions of high and low fluidity, as well as the understanding of the specific lipid composition and topography, will lend support to or modify the model proposed. Langlais and Roberts (1985) have discussed the various aspects of molecular membrane model of sperm capacitation and acrosome reaction of mammalian spermatozoa.

d) Lectins as Surface Markers of Spermatozóa

Specialization in the organization of the plasma membrane in different regions of the spermatozoa has also been shown in ultrastructural studies on the binding of plant lectins to the sperm surface. Some of the carbohydrate-binding proteins (lectins), which can be obtained from the seeds of leguminous plants (see Lis and Shron 1973), have been utilized as probes of the spermatozoon surface to determine the saccharide components on the surface of spermatozoa (Susko-Parrish et al. 1985). Two isothiocyanate-conjugated lectins, wheat-germ agglutinin (WGA) and concanavalin A (Con A) have been usually used besides the *Ricinus communis* (Nicolson and Yanagimachi 1979). Lectin-binding sites have been ultrastructurally demonstrated, using lectin peroxidase (Gordon et al. 1975), lectin-ferritin (Nicolson and Yanagimachi 1974, Nicolson et al. 1977), and lectin-homocyanin (Kinsey and Koehler 1976) methods (Nicolson 1978, Nicolson and Yanagimachi 1979). Sinowatz and Friess (1983) have localized lectin receptors on bovine epididymal spermatozoa, using a colloidal gold technique. The results obtained with these lectin labelling and binding techniques so far are rather spotty, and several inconsistencies and peculiarities remain unsolved (Nicolson and Yanagimachi 1979). Much of these variable results appear to be due to species variations in the spermatozoa used, differing degrees of maturation of spermatozoa as well as the purity of the lectins employed, and some statistical uncertainty about the analysis of ultrastructural labelling data, especially when thin sections are used as the sole criterion (Koehler 1978, Nicolson 1978, Nicolson and Yanagimachi 1979). However, some useful information has been obtained about the surface characteristics of spermatozoa with the lectin labelling and binding techniques. The various ultrastructural investigations in this regard have revealed changes in the abundance and the distribution of lectin binding sites on rabbit spermatozoa during epididymal maturation (Gordon et al. 1974, 1975, Nicolson et al. 1977, Courtens and Fournier-Delpech 1979). Lewin et al. (1979) studied difference in Con-A-FITC binding to rat spermatozoa during epididymal maturation and capacitation. Talbot and Chacon (1981) have detected modifications in the tail of capacitated guinea-pig sperm, using lectins. Fournier-Delpech and Courot (1980) studied the relationship of protein having affinity for Con A with the epididymal maturation of ram spermatozoa. Con A receptors from the sperm plasma membrane as quantitated (using [^{3}H]-acetyl-Con A) are decreased in the second part of the epididymis as compared with the epididymal head.

Several of the lectins bind specifically to glucosyl or mannosyl groups on the membrane (Koehler 1978, Nicolson and Yanagimachi 1979). The receptors for these lectins have polar distributions on the surface of the cell. Con A binds predominantly to the acrosomal region of the mouse spermatozoa with much less binding to the mid-piece and tail portions of the cell (Edelman and Millette 1971). Nicolson and Yanagimachi (1972) identified the residual terminal sugars of rabbit and hamster sperm surface by specific agglutinins and found differences in D-mannose, D-galactose, N-acetyl-D-glucosamine, and N-acetyl-D-galactosamine.

Nicolson and co-workers (Nicolson et al. 1977, Nicolson and Yanagimachi 1976) determined that cell surface N-acetylglucosamine-like or N-acetylneuraminic acid-like and galactose-like residues in oligosaccharides, recognized by WGA and *Ricinus communis* agglutinin, are modified after the contact of spermatozoa with seminal plasma, and the spermatozoa become less agglutinable. These workers attributed the apparent decrease in lectin receptors for *Ricinus communis* and WGA during epididymal transit and after ejaculation of rabbit spermatozoa to various cell surface alterations. These include (1) degradation of sperm surface glycoproteins by proteases or changes in oligosaccharides by glycosidases, (2) absorption of coating substances that mask available lectin receptors at the surface, and (3) completion of oligosaccharide chains by glycosyltransferases leading to a change of the lectin-binding sites (for example, addition of sialic acid to terminal galactose residues via a sialotransferase is expected to block binding of *Ricinus communis* agglutinin (see Nicolson 1973).

From the head-to-head sperm agglutination brought about with Sendai or influenza virus, it is believed that N-acetylneuraminic acid is present in abundance on the heads of hamster and rabbit spermatozoa but is less prevalent or absent on the tail (Ericsson et al. 1971, Nicolson and Yanagimachi 1972). The polar distribution of receptors for Con A has been confirmed by quantitative measurements of the number of binding sites on isolated heads and tails from mouse spermatozoa (Gall et al. 1974). Most of the binding sites are found on the isolated head, although the tails show significant binding. Although the density of receptor sites may vary considerably, the total number of binding sites for various lectins remains relatively constant (Nicolson and Yanagimachi 1979). The labelling patterns obtained with various lectins have revealed differences in the post-acrosomal region of the head, the region which first contacts the egg (Zamboni 1971, Yanagimachi 1978a). Kinsey and Koehler (1976) studied the fine structural localization of Con A binding sites on hamster epididymal spermatozoa which show morphological differences over various parts of the head and tail as detected by air-dried replicas and freeze-etching techniques. The results obtained have revealed differences in carbohydrate moieties in different regions of sperm plasma membrane. Courtens and Fournier-Delpech (1979) observed that Con A binding, which is important on ram corpus and cauda epididymal spermatozoa, greatly decreases in most cells from cauda epididymis after in utero incubation.

The chemical analyses have revealed an increase in the abundance of Con A receptors on the surface of rat spermatozoa during epididymal maturation (Fournier-Delpech et al. 1977). Glycoproteins having affinity for Con A have also been analysed chemically for ram sperm plasma membrane in relation to epididymal maturation (Fournier-Delpech and Courot 1980) and for boar spermatozoa surface (Hermann and Keil 1981). The amino acid and carbohydrate composition of Con A receptor of MW 160,000 in the case of boar spermatozoa surface has been determined. This glycoprotein is susceptible to digestion by trypsin or chymotrypsin.

The various observations on lectin-binding sites, as discussed above, further support the concept that spermatozoa are structurally and functionally asymmetric, and their surface heterogeneity apparently indicates the different activities of the plasma membrane in various regions of the cell. The sperm plasma membrane is topographically heterogeneous in terms of lectin binding, antigen distribution, charge characteristics, and intramembranous particle distribution, suggesting the presence of different chemical components in its different regions. However, with the paucity of information about molecular composition of the sperm plasma membrane, very little or nothing is known of the components responsible for its topographic heterogeneity.

Chapter XII
Sperm Motility

The mammalian spermatozoa show motility which is brought about by their tail. The latter consists of various specialized components, as already discussed in detail in Chap. X. Sperm motility forms one of the most important parameters in assessing the fertility potential of a semen specimen, as immotile human sperm cannot penetrate cervical mucus (Amelar et al. 1980). The type of movement also influences fertilizing capacity as the vigorous beating of the sperm tail is necessary for penetration of the sperm head through the corona radiata to fertilize the ovum (Nelson 1985). Under physiological conditions, motility and fertilizing ability of spermatozoa are usually closely related to each other, but there are situations when these two functions of spermatozoa get dissociated (Mann 1975), suggesting that sperm movement alone is not a sufficient criterion for fertilizing capacity. Mature mammalian spermatozoa present within the distal regions of the male reproductive tract show little or no movement in situ, but can become highly motile following their release into suitable salt solutions (Mann 1964, Turner and Howards 1978, Amelar et al. 1980, Mohri and Yanagimachi 1980, Nelson et al. 1980a, b, Mann and Lutwak-Mann 1981). The ultrastructural molecular and functional organization of each component of the tail have been discussed in Chap. X. The axoneme is composed of two principal proteins, the dynein ATPase enzymes and the microtubule protein tubulin, and as many as 100 minor component proteins. All these macromolecules are arranged in the familiar 9+2 cross sectional pattern, as discussed in Chap. X. The various metabolic pathways and the substrates used in the production of energy for sperm motility have also been described in Chap. X. Here an attempt will be made to state briefly the views about the mechanism and energetics of sperm motility in their totality. The effects of various chemical and physical agents on sperm motility will also be reviewed briefly. Goeden and Zenich (1985) have studied the influence of the uterine environment on rat sperm motility and swimming speed. Sidhu et al. (1986b), using a cervical mucus penetration test have made an objective quality assessment of buffalo bull semen.

A. Mechanism of Sperm Motility

The physical, biophysical, and chemical aspects of sperm motility or of flagellar movement have previously been discussed in several excellent reviews and symposia (see Nelson 1967, 1975, Nelson et al. 1970, Warner 1972, Fawcett 1975b, 1977, Brokaw 1975, 1980, Rikmenspoel 1978, Woolley 1979, Satir 1979, Gibbons

1979, Amelar et al. 1980, Nelson et al. 1980a, b, Mann and Lutwak-Mann 1981, Tash and Means 1983). Blum and Lubliner (1973), and Holwill (1977) have provided general reviews of flagellar biophysics. Quantitative assessment has been made of sperm motility (Katz and Overstreet 1979, Atherton 1979, Jouannet 1979, Amann 1979, Wall et al. 1980, Amelar et al. 1980, Phillips 1983). The sperm tail propagates an undulating wave as it moves from the base of the tail to the top and drives the sperm head forward. There are two leading theories for the basis of ciliary and flagellar motility: (1) a localized contraction model in which bending is accomplished by means of contractile units regularly spaced along the axoneme; and (2) a sliding filament model, in which bends are initiated by longitudinal sliding displacements of the doublet microtubules (Warner 1972, Fawcett 1975b, Brokaw 1975, 1980, Woolley 1979, Satir 1979, Gibbons 1979, Linck 1979, Amelar et al. 1980, Mohri and Yano 1982). Neither of these models satisfactorily accounts for the many geometrical complications required for initiation and propagation of bending movements, about which many questions remain unanswered. Woolley (1977, 1979) has made an attempt to obtain evidence for "twisted plane" undulations in golden hamster sperm tails. On the basis of angular displacements between the preferred planes and the finding from the electron microscope, he has presented the following idea as a working hypothesis: if the most proximal plane of bending is topographically determined by peripheral doublet 1, successive distal planes of action are influenced predominantly by doublets 2, 3, etc. in clockwise sequence. Woolley (1977) has discussed the merits and limitations of this hypothesis. He has suggested that it is important now to re-examine the sperm movement pattern in other species, such as the bull and ram, where geometry has been described as elliptically conical.

The specific roles of individual components of axial filament complex are still to be elucidated with regard to the molecular basis of contractile mechanism (Linck 1979, Amelar et al. 1980), as already stated in Chap. X. But the involvement of the dynein arms in the sliding microtubule mechanism of ciliar and flagellar movement is now well established (Gibbons 1979, Satir 1979, 1982, Amelar et al. 1980). According to this hypothesis developed from numerous observations and experiments in many laboratories, the sperm tail moves when the microtubules, powered by dynein arm ATP hydrolysis, slide past one another. For a better understanding of the molecular mechanisms involved in the motility of cilia and flagella, it is important to determine the three-dimensional arrangement of the various structural components associated with the doublet microtubules (e.g. dynein arms, radial spokes, and nexin-fibres), as the interactions between them are of great significance in the development of bending waves (Linck 1979). The analysis of structure of the central-pair microtubule apparatus and its interactions with the radial spokes are of great importance. Biophysical analysis can provide an explanation for sperm propulsion as a result of flagellar bending-wave propagation, and has suggested a plausible explanation for flagellar oscillation and wave propagation (Brokaw 1975, 1980, Rikmenspoel 1978, Katz and Overstreet 1979, Woolley 1979).

Various techniques are being developed to study the patterns of sperm motility. Sidhu and Guraya (1986b) have made a simple turbidimetric analysis of forward motility of buffalo bull sperm. Jouannet et al. (1977) have made a study of light-scattering determination of various characteristic parameters of spermatozoa motility in a series of human sperm (Jouannet 1979). Atherton et al. (1978) have made spectrophotometric quantitation of mammalian spermatozoa motility in humans (Atherton 1979). The stroboscopic illumination has been used for the assessment of hyperactivated motility of mouse spermatozoa (Cooper and Woolley 1982). Sperm motile efficiency has been studied by Ishii et al. (1977). Ericsson (1977) has made a study of isolation and storage of progressively motile human sperm. Methods are now available to work out the details of the relationship between movement of the sperm flagellum and properties of the molecular interaction between filaments which generate flagellar bending (Satir 1979, Gibbons 1979, Linck 1979). Finsy et al. (1979) have done motility evaluation of human spermatozoa by photo correlation spectroscopy. Amann (1979) has computerized measurements of sperm velocity and percentage of motile sperm (Wall et al. 1980). A satisfactory model for this molecular interaction is still not available. Mohri and Yanagimachi (1980), using demembranated models, have made a study of the characteristics of motor apparatus in testicular, epididymal, and ejaculated spermatozoa. Capacitated and activated spermatozoa show very vigorous motility characterized by whiplash-like beating of their flagella. Upon exposure to ATP (and cAMP), demembranated caudal epididymal (hamster and guinea-pig) and ejaculated (hamster and human) spermatozoa begin vigorous whiplash-like movement. Analysis of more realistic structural models of flagella, and more precise analysis of the hydrodynamics of spermatozoan motility, are also needed (Brokaw 1975, Nelson 1975, 1985, Woolley 1979, Amelar et al. 1980).

B. Energetics of Sperm Motility

The mitochondrial sheath is the energy mechanism for mammalian spermatozoon and has been likened to a combustion engine (Katz and Overstreet 1979). In intact spermatozoa, the mitochondria contain the main machinery for trapping, conserving, and supplying energy; the ability of sperm mitochondria to perform these functions depends foremost on the adenine nucleotide pool of ATP, ADP, and AMP. It is still an open question whether in addition to the adenine nucleotides, mammalian spermatozoa contain some other energy-rich phosphorus compounds. But the concentration and the role of ATP, as well as ADP and AMP, has been studied in the spermatozoa of a wide range of mammalian species (see Mann and Lutwak-Mann 1981). To maintain a high ATP/ADP and an adequate adenylate energy charge, the spermatozoa have at their disposal two major mechanisms (see Chap. X). One is the anaerobic glycolysis. In whole semen of many mammals the bulk of carbohydrate available for this purpose is fructose, and the chief source of anaerobically generated energy is fructolysis. The alternative mechanism that allows the spermatozoa to replenish their ATP reserve in the mitochondria is respiration coupled with oxidative phosphorylation, which is linked to oxygen via NAD and cytochrome system. The $NADH_2$-diaphorase activity is located exclusively in the mid-piece of the spermatozoa (Edvinsson et al. 1981). There is a positive correlation between the NADH-diaphorase activity and the spermatozoal motility, density, and morphology. Mann and Lutwak-Mann (1981) have discussed the superiority of respiration and oxidative phosphorylation over glycolysis as a potential source of energy to spermatozoa.

The theory of the role of ATP dephosphorylation as the primary, ultimate, and immediate source of energy for sperm motility has been universally accepted (Voglmayr 1975, Satir 1979, Gibbons 1979, Linck 1979, Amelar et al. 1980, Lindemann 1980, Nelson et al. 1980a, b, Mann and Lutwak-Mann 1981, Nelson 1975, 1985). In other words, the energy to move the sperm tail is supplied by the hydrolysis of ATP to ADP, and phosphoric acid. In the spermatozoon, ATP produced by mitochondria, which are placed near the base of the sperm flagellum, diffuses out towards the tip of the axoneme. An important function of the tail membrane is to maintain, around the axoneme, the proper concentration of ATP and also certain essential ions (magnesium, calcium, manganese, cobalt, nickel, potassium, etc.); the effect of the latter on sperm motility will be discussed in Sect. C. The concentration of ATP determines the frequency of the beat. The relations between motility and ATP are reported for human spermatozoa (Amelar et al. 1980). The suspension of guinea-pig spermatozoa in substrate-free medium results in the cessation of sperm motility and a 94% reduction in ATP level within minutes (Santos-Sacchi and Gordon 1982). ATP depletion is not deleterious to spermatozoa, as motility is fully restored by the addition of pyruvate. According to Lindemann and Rikmenspoel (1972), the bull and human sperms, which are dissected or impalted, retain wave motion in the presence of ADP. Unlike ATP, induced motility required no initiation. The activity observed appeared similar to that of intact cells. Flagellar segments, which were detached from the head and mid-piece, were capable of sustaining coordinated motility.

The results of various studies have indicated the importance in motility of the dynein arm-doublet microtubule interaction and the radial-spoke-central-pair interactions, as discussed in Chap. X. The principal motive force for flagellar bending derives from sliding movements between doublet microtubules generated by the interactions of the dynein ATPase of one doublet microtubule with the adjacent doublet microtubule (see Linck 1979, Lindemann 1980, Amelar et al. 1980, Mann and Lutwak-Mann 1981, Mohri and Yano 1982, Nelson 1985). The flagellar ATPase of bull spermatozoa has been extracted and characterized (Young and Smithwick 1975a, 1983). The second line of research has shown that some form of interaction between the central-pair microtubules and the radial spokes provides a regulatory function in the conversion of doublet-microtubule sliding to bending waves. Summers (1974) has observed ATP-induced sliding of microtubules in bull sperm flagella. The sliding configuration is not seen in the absence of ATP (Satir 1979). But Guerin and Czyba (1977a) have observed that enrichment of the medium with ATP does not lead to any effect on sperm motility. They have discussed this negative result. Voglmayr (1975) has pointed out the differences in the energy (ATP) requirements of relatively immotile testicular spermatozoa and highly motile ejaculated spermatozoa. Katz et al. (1978) have studied the movement characteristics and power obtained from guinea-pig and hamster spermatozoa in relation to activation. The ATP is produced mainly in the mid-piece containing enzymes for metabolic pathways, as already discussed in Chap. X, but is used throughout the tail. It maintains several processes of the sperm-cell, including the motility (Nelson 1975, 1985, Voglmayr 1975, Amelar et al. 1980, Lindemann 1980, Calamera et al. 1982). The relative and absolute concentrations of ATP, Ca^{2+}, Mg^{2+}, and K^+ appear to be quite critical for optimum ATPase activity (Satir 1979, Gibbons 1979). In the intact sperm cell, Mg^{2+} is a less effective activator than Ca^{2+}. Actually, very divergent views have been expressed about the role of divalent cation, Ca^{2+}, or Mg^{2+} in dynein-ATPase activity (Nelson 1975, 1985, Gibbons 1979, Amelar et al. 1980, Nelson et al. 1980b, Tash and Means 1983). Dynein is activated by magnesium, calcium, manganese, cobalt, nickel, and relatively high potassium concentrations. Dynein is inhibited by cadmium, zinc, mercury, and sulphydryl groups (Amelar et al. 1980). The plasma membrane of the sperm, though potentially leaky, maintains, possibly by virtue of outabin-sensitive ion pumps, a differential transmembrane ion distribution, a measurable membrane potential and a resistance (Nelson et al. 1980b, Nelson 1985). Enzymes specifically related to membrane-directed events of spermatozoa include magnesium-dependent Na^+-K^+-activated transport ATPase and acetylcholinesterase (Nelson 1975, 1985, Nelson et al. 1980b); the latter enzyme appears confined to the mitochondrial structure. Its intracellular localization has also been suggested for the spermatozoa of bovines (*Bubalus bubalis* and *Bos taurus*) (Abdou et al. 1977).

The flagellar protein which is mainly involved in converting chemical energy into mechanical movement is dynein (Satir 1979, Gibbons 1979, Amelar et al. 1980). It is a large molecule with a molecular weight of about 500,000, it being a Mg^{2+}- and Ca^{2+}-activated ATPase; dynein ATPase specifically inhibited by vanadate which is also selective inhibitor of Na^+- and K^+-activated ATPase. As in the case of tubulin, it is heterogeneous and can be separated electrophoretically into A_1, A_2, and B fractions (Gibbons and Fronk 1972, Gibbons 1977, 1979). The A fractions, constituting two-thirds of the total, are present in the arms on the doublets, the location of the B fraction being not clear. Pallini et al. (1982) have studied biochemical properties of ciliary, flagellar, and cytoplasmic dyneins. Although the results of various studies have shown that cilia and flagella contain dynein and tubulin (Brokaw and Verdugo 1982), the flagella also contain spermosin and flactin (flagellar myosin and actin) (Burnasheva 1958, Nelson 1966, 1985, Young and Nelson 1968, Tilney 1975, Talbot and Kleve 1978, Clarke and Yanagimachi 1978, Campanella et al. 1979, Clarke et al. 1982). In some of these studies, the presence of the myosin and actin has also been demonstrated in the sperm head (e.g. acrosomal region) (Campanella et al. 1979, Clarke and Yanagimachi 1978). A quantitative physiological relationship has been demonstrated for the in vitro spermosin-flactin complex sperm of flagella (Nelson 1985). Ca^{2+} activates both myosin and spermosin ATPase, whereas Mg^{2+} activates flactospermosin, flactomyosin, and actospermosin ATPase. Bull sperm flactin shows reversible depolymerization in ATP plus Mg^{2+}, the flactin also enhances the ATPase activity of both myosin and spermosin (Young and Nelson 1968). Clarke et al. (1982) agree with the suggestion of Young and Nelson (1968) that the sperm tail actin may be involved in flagellar beat generation. Nelson (1985) feels that the dynein system of sperm tail is more closely related to the intracellular transport of Ca^{2+}, a fact that would strongly bear on flagellar propulsion.

There are at least four substances in the mammalian semen which may be used either directly or indirectly by spermatozoa and hence are potential energy sources for the maintenance of motility (Mann 1975, Mann and Lutwak-Mann 1981). These include fructose, sorbitol, glycerylphosphorylcholine (GPC), and lipids (Brooks 1979b) which are present in the seminal plasma. Lipids are also present in the spermatozoon itself, as already discussed in Chap. X. All these substances can be utilized by mammalian spermatozoa in the presence of oxygen, which would normally be available in most parts of the female tract. However, GPC must first come into contact with an enzyme (phosphodiesterase) in the female tract secretions to liberate choline. The oxygen uptake or respiration of spermatozoon will reflect the overall oxidation of these substances. The results of various studies have suggested that glycolytic metabolism is required to support the development of the whiplash motility pattern associated with fertilizing ability in mouse, hamster, and guinea-pig spermatozoa (see Fraser and Quinn 1981 for references). Fraser and Quinn (1981) have shown that epididymal mouse spermatozoa need exposure to glucose to permit induction of the acrosome reaction and motility changes which are prerequisites for fertilization. In addition to physiological substrates, the spermatozoa can metabolize a large number of related substances which do not occur in semen (see also Suter et al. 1979). Brooks (1979b), and Mann and Lutwak-Mann (1981) have discussed the problems of utilization of various substances, such as reducing sugars, lactic acid, glycerol, inositol, lipids, acetylcarnitine, and amino acids, by the epididymal spermatozoa. Cooper and Brooks (1981) have studied the entry of glycerol into the rat epididymis and its utilization by epididymal spermatozoa. Luminal glycerol is believed to arise from degradation of epididymal lipid. The results of various studies have also indicated that sperm respiration, unlike anaerobic glycolysis, is not restricted to extracellular sugar but can proceed both in the absence of extraneous nutrients (that is, at the expense of endogenous substrates) and in the presence of a wide variety of extracellular oxidizable substrates. But the present knowledge of the oxidative processes which lead to ATP synthesis in sperm mitochondria is inadequate. Gagnon et al. (1980) have reported the deficiency of protein carboxylmethylase in spermatozoa of necrosperm patients. Protein carboxylmethylase is present mainly as an intrinsic component of the sperm, but it also occurs in seminal plasma. This enzyme has been suggested to play some regulatory role in sperm motility.

To sum up, it can be stated that ATP, which plays an important role in muscle contraction, is present in the mammalian spermatozoa and provides a link between the energy yielding reactions and motility. Vilar et al. (1980) have studied the concentration of ATP in spermatozoa of fertile men. Its normal value is $6.5 + 0.34$ SE $\mu g/10^8$ spermatozoa. Pools of spermatozoa show that incubation in seminal plasma at room temperature produces a progressive loss of ATP in the gamete. The hydrolysis of ATP probably furnishes the energy necessary for the sliding movement of adjacent doublet microtubules in much the same way as it does in muscle fibres; the loss of ATP being made good in spermatozoa by the energy-yielding reactions of fructolysis and respiration (Mann 1964, 1975, Mann and Lutwak-Mann 1981). Since fructolysis is significantly correlated with sperm motility, it follows that the plasma membrane of the tail region must offer access to fructose and other metabolites at rates consistent with metabolic requirements. The mechanism of this on-demand regulation, which controls the penetration rate of metabolic substrates across cell membranes, is by no means clear. But carboglutelin, which is a special membrane-linked substrate-binding transport protein, is believed to be involved in binding glucose and other sugars, fructose among them, to the plasma membrane of the sperm tail (Mann and Lutwak-Mann 1981).

C. Effects of Chemical and Physical Agents on Sperm Motility

During recent years, the presence of compounds related to sperm motility and the effects of various chemical and physical agents on sperm motility both in vivo and in vitro systems have been studied extensively to reveal control mechanisms or causative factors of sperm motility more precisely (Bedford 1979, Hoskins et al. 1979, Bavister 1979, see other references in Amelar et al. 1980, Nelson et al. 1980b, Trifunac and Berstein 1982, Tsutomu et al. 1984, Alabi et al. 1986, Goeden and Zenich 1985). Very divergent views have been expressed in this regard. But from the results of various studies, as summarized here, it has become increasingly clear that sperm motility can be beneficially or detrimentally influenced by a wide variety of physical and chemical factors. Several methods are now available to assess sperm motility under various experimental conditions (Amelar et al. 1980, Nelson et al. 1980b). Levin (1980) has described a quantitative method for determining the effects of drugs on spermatozoal motility.

1. Elements and Ions

Elements and ions are known to influence sperm motility directly or indirectly. Generally, cations penetrate the sperm membranes much more slowly than anions, but the normally low permeability to cations of the bovine sperm membrane can be raised significantly by treating spermatozoa with SH reagents, such as mersalyl, showing that membrane-located SH groups are vital to the preservation of the permeability state in the sperm plasma membrane (Mann and Lutwak-Mann 1981). In the whole semen, motile spermatozoa efficiently keep up ionic gradients across the plasma membrane, the concentration of K^+ inside the sperm-cells being higher than outside, while the reverse is true of Na^+. The outward K^+ gradient and the active cation transport are believed to be regulated by an ouabain-sensitive Na/K exchange pump, comparable in action to that of other cells (see also Nelson et al. 1980b). Furthermore, when sperm-cells are stored at low-temperature conditions that slow down resynthesis of depleted ATP, they tend to gain Na^{2+} and lose K^+ into the medium. The presence of various elements has been reported in the spermatozoa of some mammals.

Elemental composition of subcellular structures of human spermatozoa has been studied with energy dispersive analysis of X-ray (Rosado et al. 1977). The mitochondrial spectrum shows the presence of important concentrations of calcium > iron, potassium, phosphorus > magnesium, sulphur > manganese. The membrane is especially rich in calcium, sulphur, and zinc. Potassium is the most concentrated element in both isolated fractions (heads and tails). Sperm heads are richer than tails in sodium, copper, and zinc, while sperm tails have higher concentration of calcium. The zinc concentration of human sperm-cells and their subfractions are considerably higher than the reported zinc concentration in any other human cells or their subfractions. Zinc and magnesium have also been studied in bull and boar spermatozoa (Arver and Eliasson 1980).

Shah et al. (1980a) have studied the effect of spermine on the intracellular levels of calcium, magnesium, sodium, and potassium in human spermatozoa. Spermine decreases the calcium and magnesium levels in a dose-related manner. Alterations in sodium and potassium levels cannot be correlated with spermine concentration. The recent observations of Shah et al. (1980b) have revealed a negative correlation of zinc content of human spermatozoa with prolactin or spermine concentration.

The toxicity to human spermatozoa of seven metals (nickel, palladium, platinum, silver, gold, zinc, and cadmium) and one alloy (brass: 80% copper, 20% zinc) has been assessed in vitro (Holland and White 1980). Only brass and cadmium significantly reduce the percentage of motile unwashed spermatozoa; however, washing the spermatozoa increased the spermicidal effectiveness of both brass and cadmium and also resulted in a significant reduction in motility caused by zinc and silver. Oxygen consumption by once-washed spermatozoa is apparently increased by zinc and brass. But the high rate of oxidation of these metals confounds the interpretation of their effect. Silver caused a decline in the oxygen uptake of spermatozoa. Silver, zinc, brass, and, to a lesser extent, cadmium decrease the quantity of glucose utilized by spermatozoa and also decrease the glucose oxidized. Accumulation of lactate by washed spermatozoa is impaired severely by zinc, and less severely by brass and cadmium. Lindemann et al. (1980) have observed a selective effect of nickel (II) on wave initiation in bull sperm flagella. The bend initiation appears to be inhibited by Ni^{2+} and is a process separate from the sliding microtubule mechanism responsible for wave propagation. Battersby and Chandler (1977) have made a correlation between elemental (sodium, potassium, copper, and zinc) composition and motility of human spermatozoa. They have suggested that subcellular elemental distribution is not a major factor in determining sperm motility in normal human semen. Nelson and Gardner (1986) have composed the binding sites for metals, such as La^{3+}, Ni^{2+}, Zn^{2+} and Mn^{2+}, with those of Ca^{2+} in ejaculated bull sperm.

Membrane properties of mammalian spermatozoa and the ion concentration of the medium have been correlated with certain parameters of motility (McGrady and Nelson 1972, 1973, Nelson 1975, Nelson et al. 1980b, Mann and Lutwak-Mann 1981). In bull spermatozoa, the intracellular-extracellular potassium ratio appears to be the major determinant of membrane potential, with secondary contributions from sodium and chloride. A direct relationship exists between the magnitude of membrane potential and the amplitude of the flagellar wave. Using the fluorescent dye, di S-C_3(5), Rink (1977) has investigated the membrane potential of guinea-pig spermatozoa. The potential does not vary when K^+ replaces Na^+ in the surrounding medium, and in a physiological salt solution the resting potential is found to be about +13 mV. The percentage of progressively moving cells in samples of chimpanzee and bull spermatozoa is influenced by variations in K^+, Na^+, Ca^{2+}, and Cl^- contents of the medium (McGrady et al. 1974), which are well known to affect sperm motility (Nelson 1975, Nelson et al. 1980b). Zimmerman et al. (1979) have made a study of movements of Na^+ and K^+ into epididymal boar spermatozoa (see Wong and Lee 1983). The trace flux analysis has been made of Na^+ and K^+ permeability in differentiating mouse spermatozoa. Pholpramool and Chaturapanich (1979) have studied the effect of Na^+ and K^+ concentrations and pH on the maintenance of motility of rabbit and rat epididymal spermatozoa. The potassium-dependent increases in cytosolic pH stimulate metabolism and motility of mammalian spermatozoa (Babcock et al. 1983). The mechanisms underlying the relationship between pH and motility may be metabolic (Bartoov et al. 1980, Treetisatit and Chulavatnatol 1982). When rat caudal epididymal spermatozoa are suspended in a completely Na^+-free solution, the forward motility suffers a progressive fall and after 3 h is completely suppressed (Wong et al. 1981, Wong and Lee 1983). The role of Na^+ influx in sperm motility maintenance is indicated.

Rikmenspoel et al. (1978) have determined a characterization of the ionic requirements of impaled bull sperm whose motility is maintained by external ATP or ADP. The post-impalement flagellar frequency depends sharply on the external Mg^{2+} concentration. For half-optimal activity with ADP, 0.3 mM Mg^{2+} is required and 0.05 mM Mg^{2+} with ATP as external power source. Mn^{2+} can substitute partially for Mg^{2+} as ionic co-factor. Ca^{2+} cannot substitute for Mg^{2+}, and at concentrations above 0.5 mM it inhibits motility. Zn^{2+} acts only as inhibitor of post-impalement flagellar activity reducing it to zero at concentrations above 1 mM.

Crabo and Zimmerman (1976) studied the factors affecting the movements of Na^+, K^+, Ca^{2+}, and Zn^{2+} into mature and immature boar spermatozoa (see also Crabo et al. 1976). According to Turner et al. (1977), the effect of the absolute values of epididymal Na^+ and K^+ concentration on sperm motility and fertility remain unresolved. It has been suggested that the proteins, which are adsorbed into the sperm surface during epididymal transit and ejaculation, are important for the ion transport mechanisms of the membrane. Semakov and Ryzhkov (1976) have studied the permeability of cytoplasmic membranes in the bull and boar spermatozoa for Na^+, K^+, Ca^{2+}, and P^{3+} under effect of low temperatures. O'Day and Rikmenspoel (1979) investigated the electrical control of flagellar activity in impaled bull spermatozoa. Their results are compatible with an active transport process in bull spermatozoa that controls the flagellar activity in response to current injection by decreasing the internal Mg^{2+} concentrations during the injection of current. Boender (1977) has suggested that under normal physiological conditions the active transport of the cell is able to regulate the K^+/Na^+ ratio but, in the presence of ouabian, active transport mechanism is blocked and the greater permeability results in an accelerated decrease in motility. McGrady (1979b) has studied the effect of ouabain on membrane potential and flagellar wave in ejaculated bull spermatozoa. Membrane potential decreases. Intracellular K^+ decreases, whereas Na^+ increases. Progressive motility decreases. In addition to motility, change produced by ouabain has been identified as a decrease both in beat frequency and in wave amplitude. McGrady (1979d) has further observed that beat frequency is correlated to endogenous

ATP in K^+-supplemented media. Increased Na^+ concentration is related to ATP content but not to beat frequency. With increasing viscosity of the medium, the frequency of flagellum beat is decreased dramatically, while ATP content of the cells remains unchanged. Even though spermatozoa are immobilized immediately upon demembranation by physical or chemical means, these retain their contractile powers and regain propulsive ability, which can be revealed by reactivation in the presence of ATP, K^+ and Mg^{2+} and small amounts of Ca^{2+} (Bishop and Hoffmann-Berling 1959).

Dott et al. (1979) have studied the maintenance of motility and the surface properties of epididymal spermatozoa from bull, rabbit, and ram in homologous seminal and epididymal plasma. They have concluded that the basic parameters of sperm function can be studied more reliably if mature spermatozoa from the cauda epididymidis are used. Dacheux et al. (1979) studied the effects of osmolarity, bicarbonate, and buffer on the metabolism and motility of testicular, epididymal, and ejaculated spermatozoa of boars. They have suggested an increasing motility, glycolysis, and respiration with maturation (Bedford 1975, 1979), but decreased synthetic capacity and increased sensitivity to the fluid environment. Howards et al. (1979) have discussed the fluid environment of maturing spermatozoa. The ionic composition of the medium has a definite effect on the metabolism and motility of the spermatozoa (see Mann 1964, 1975, Nelson 1975, 1985, Evans and Setchell 1978a, b, Nelson et al. 1980a, b, Bavister 1979). Overstreet et al. (1980), after studying motility of rabbit spermatozoa in the secretions of the oviduct, have stated that sperm residing in the oviductal isthmus prior to ovulation apparently can be stimulated to undergo activated motility, while those remaining in the isthmus after ovulation are no longer able to respond.

It has been reported that in several species, exogenous Ca^{2+} is needed for the initiation and control of motility (B. E. Morton et al. 1973, 1978, Yanagimachi and Usui 1974, Hoskins and Casillas 1975b, Yanagimachi 1975, Babcock et al. 1976a, b, Davis 1978, Nelson 1978, 1982, 1985, Nelson et al. 1980b, Heffner et al. 1980, Tash and Means 1983), whereas in other species motility is initiated upon release of sperm from the epididymis, in the absence of Ca^{2+} (Storey 1975, Cascieri et al. 1976, Peterson and Freund 1976, Turner and Howards 1978, Heffner et al. 1980, Tash and Means 1983). Actually, sperm-cells show a pronounced responsiveness to variations in Ca^{2+} content of the environment, which suggests a marked dependence on selective variations in the cationic permeability of the plasma membrane. Boender (1974) has observed that in Ca^{2+}-free media, the motility is most adversely affected; in media supplemented with seminal plasma (containing some free Ca^{2+}) the spermatozoa retain higher motility scores for longer periods. The optimum amount of extracellular calcium essential for motile function in spermatozoa may vary from species to species (Tash and Means 1983). The spermatozoa of the many species move in an environment in which the free calcium concentration is more than that of the interior by four or five orders of magnitude. Actually, deviations from normal values of extracellular calcium concentration lead to the development of different patterns of sperm motility (Tash and Means 1983, Nelson 1985). Morton et al. (1978) have shown that spermatozoa from species with a high Ca^{2+} concentration in the epididymal lumen are motile and show little or no response to added Ca^{2+} in the medium, into which they are released, while spermatozoa from species with low Ca^{2+} content are less motile and respond to less added Ca^{2+}. They have proposed that species with high epididymal levels have little or no dependence on external Ca^{2+} for motility, whereas low epididymal Ca^{2+} results in a requirement for Ca^{2+} to maximize motility (see also Tash and Means 1983). Morton et al. (1979) have further observed a correlation between the amount of free Ca^{2+} surrounding the sperm within the cauda epididymidis of a given species and the level of sperm motility therein. An inverse relationship has also been observed between the free Ca^{2+} concentration in the cauda epididymal fluid of a species and the later inducibility of motility in diluted sperm from that species by Ca^{2+} ions. From the results of various studies, it can be suggested that deprivation of Ca^{2+} depresses sperm motility and excess of extracellular calcium disturbs both membrane potential and normal pattern of sperm-cell movement (Young and Nelson 1974), implicating calcium as an essential factor in the regulation of sperm motility (Tash and Means 1983, Nelson 1985). The pattern of motility undergoes changes with advancing maturation of spermatozoa in the epididymis as a result of changes in their permeability characteristics.

The plasma membrane contributes to the activity of motility as well as to the acrosome reaction by virtue of the control of Ca^{2+} transport across the plasma membrane and possibly by releasing or sequestering Ca^{2+} from plasma membrane storage sites (Tash and Means 1983, Nelson 1985). Hyne and Garbers (1979) have observed calcium-dependent increase in cyclic AMP and the induction of the acrosome reaction in guinea-pig spermatozoa (see also Shapiro and Eddy 1980). The distribution of intracellular Ca^{2+} in mammalian sperm and the mechanisms affecting transmembrane movements of this important regulatory cation form the subject of several recent studies (Babcock et al. 1975, 1976a, b, 1978, 1979, Singh et al. 1978, Stewart and Forrester 1979, Nelson et al. 1980a, b, 1982, Tash and Means 1983, Nelson 1982, 1985). With the development of procedures to isolate and purify highly enzymatically active plasma membrane vesicles from spermatozoa (Gillis et al. 1978), it is now possible to characterize calcium binding to these membranes. The ability of plasma membrane vesicles of boar spermatozoa to bind high concentrations of Ca^{2+} has suggested that the plasma membrane may play an important role in the storage and release of Ca^{2+} in intact spermatozoa (Peterson et al. 1979b). Babcock et al. (1979) have presented evidence that the bovine seminal fluid contains a component that is added to the surface membranes of the sperm at ejaculation which prevents or delays the active uptake of Ca^{2+} by these cells. Nelson et al. (1980b) have suggested that Ca^{2+} transport in sperm may be linked to Na transport and that ouabain sensitivity may be related to small increases in Ca^{2+} content, or else to retention of calcium with the spermatozoon (see also Nelson 1985). One of the initial steps in the chain of events involved in transmembrane transport of Ca^{2+} into spermatozoa is the accumulation of calcium from the bulk, extracellular, phase followed by binding of the calcium to the sperm-cell membrane. The calmodulin localized in both the head and flagellum of sperm (Feinberg et al. 1981) plays an important role in this regard, as its molecule contains Ca^{2+}-binding sites (Nelson 1985, Tash and Means 1983). Calmodulin binding proteins occur in sperm. Calmodulin has also been localized cytochemically in the testis and spermatozoa of guinea-pig (Yamamoto 1985). Trifluoperazine, an inhibitor of calmodulin, inhibits the Ca^{2+}-dependent increase in AC activity of guinea-pig sperm.

Stewart and Forrester (1979) have demonstrated acetylcholine-induced calcium movement in hypotonically washed ram spermatozoa. There exists a model of an acetylcholine-mediated, calcium-regulated mechanism of sperm motility control (Nelson 1985). Recent data obtained by Nelson et al. (1980a, b, 1982), who have investigated neurochemical regulation of Ca^{2+} transport, offer strong support for this concept or model (Nelson 1985). The neurotrophic factors may be involved in regulating transmembrane and intracellular transport of ions in control of sperm cell function. Turner and Howards (1978) have observed that neither reduced osmolarity to iso-osmotic conditions nor dilution of spermatozoa is essential to the initiation of sperm motility. The addition of ionic solutions to cauda epididymal samples is required for the activation of sperm-cells. This ionic requirement is nonspecific.

The results of various studies suggest that there are possibly two categories of Ca^{2+} interactions with mammalian spermatozoa, which are crucial to fertilization (Shapiro and Eddy 1980). One category is those interactions involving Ca^{2+} influx into a cellular compartment. Examples are the Ca^{2+} uptake in guinea-pig sperm which parallels the acrosome reaction (Singh et al. 1978), and Ca^{2+} uptake in bovine sperm which activates motility (Babcock et al. 1976a, 1979). The other category is Ca^{2+} participation in reactions of the sperm plasma membrane surface. An example is Ca^{2+} binding to the plasma membrane at sites from which it can be removed by EGTA (Nelson 1985); this enables the mouse spermatozoa to acquire the ability to bind to mouse zona pellucida. The ability of Ca^{2+} to participate in different reactions (binding to the intact sperm cell to the zona pellucida, followed by the acrosomal reaction at the zona surface, and subsequent sperm penetration of the zona) is of great utility to spermatozoa seeking to fertilize eggs, as Ca^{2+} is ubiquitous in oviducal fluid. But dissection of the fertilization process into its component reactions for study is thereby rendered difficult. In this regard, recent results of Heffner et al. (1980) have demonstrated that mouse sperm have a Ca^{2+} requirement for motility maintenance and that binding to the zona and maintenance of motility are two different Ca^{2+}-dependent processes whose active sites can be distinguished by cation substitution. Nelson (1979) has considered calcium in its modulatory role as a "secondary messenger".

2. *Dilution, Temperature and Osmotic Pressure*

Makler et al. (1979) have studied the influence of sperm dilution on the human sperm velocity and percentage of motility. One of the most pronounced adverse effects on sperm motility arises simply from excessive dilution of spermatozoa, where upon, after an initial very brief burst of vigorous movement, spermatozoa can become completely immotile within a few minutes (Mann 1964). The precise reason for the adverse effect of dilution on sperm motility is obscure (Rikmenspoel 1984), but from observations of spermatozoa it appears that excessive dilution (or washing) of sperm causes loss of cellular components needed for expression of motility (see also Amelar et al. 1980). Bavister (1979) has observed that the loss of sperm motility factor (SMF) and catecholamines from hamster spermatozoa during dilution or washing in solutions lacking these substances accounts for the associated progressive reduction in sperm motility. In other words, both SMF and catecholamines are essential for sustained motility of hamster spermatozoa in vitro (Bavister and Yanagimachi 1977, Bavister 1979).

The effect of temperature (Guerin et al. 1976, Appell and Evans 1978, Makler et al. 1981, Gorus et al. 1982, Rikmenspoel 1984), osmotic pressure (Guerin and Czyba 1977b), urine (Crich and Jequier 1978), and nutrients (Gorus et al. 1982) on sperm motility has also been demonstrated. The effect of temperature decrease on the activity of the cell membrane has been investigated for the ejaculated bull spermatozoa (Halangk et al. 1982). The survival of spermtozoa is variable in different biological media and gaseous compositions (Makler et al. 1984). Preservation of human semen in liquid nitrogen causes a significant impairment of sperm motility (Behrman 1971, Sherman 1973). Unal et al. (1978) have studied the influence of sugars with glycerol on post-thaw motility of bovine spermatozoa in straws (Guay et al. 1981). Hammerstedt and Hay (1980) have used a change in temperature to vary sperm motility in order to determine if changes in glucose consumption and cellular concentration of ATP, ADP, AMP, and cAMP are correlated with the temperature-dependent control of motility. The effect of temperature on the kinetic parameters of adenylate cyclase and cyclic nucleotide phosphodiesterase has also been studied. These studies have indicated that glucose consumption rate is independent of adenine nucleotide concentration or energy charge. The percentage of progressively motile sperm and velocity of motile sperm are independent of mean cAMP concentration. The modulation of progressive motility of sperm may not proceed via alterations in cAMP concentration.

3. *Cyclic AMP, Caffeine, Aminophylline, Theophylline, and Pentoxiphylline*

a) Cyclic AMP

Several workers have implicated cAMP in the initiation of sperm motility (Fig. 75) (Garbers et al. 1971a, 1973b, Hoskins 1973, Lindemann 1978, Tash and Means 1983). The first studies relating sperm motility to cAMP showed that either phosphodiesterase inhibitors (such as caffeine) or dibutrylic cAMP could be employed as in vitro additives to increase the motility of freshly collected bovine epididymal spermatozoa from up to 4 h (Garbers et al. 1971b). Since then the motility of various mammalian spermatozoa has been shown to be increased by the direct addition of cAMP or by activation of endogenous cAMP (Cash and Mann 1973, Tash and Means 1983, Lindemann 1980). The fact that the motility of spermatozoa is only occasionally increased by the addition of cAMP, and not appreciably by other cyclic nucleotides, is probably related to cellular permeability (Hoskins and Casillas 1975b). Recent studies have suggested that one site of cyclic AMP interaction is at the plasma membrane (Peterson and Freund 1976, Huacuja et al. 1977, Peterson et al. 1977b, Tash and Means 1983). This is supported by the fact that specific cAMP receptors have been revealed recently on the surface membranes of intact human spermatozoa (Rosado et al. 1976); the protein conformation of human spermatozoon membranes has been shown to change after treatment with cyclic AMP (Delgado et al. 1976). The nature of the enzyme system(s) upon which cAMP acts needs to be determined more precisely. However, an increase in the percentage motility of the spermatozoa with the addition of cAMP or cGMP is always accompanied by a corresponding increase in cell respiration. Although human spermatozoal metabolism is predominantly a glycolytic process, a significant amount of oxidative metabolism also occurs (Peterson and Freund 1968). This metabolism is greatly affected by cyclic AMP as well as by other substances (Hicks et al. 1972).

More recently, Peterson et al. (1979a) have provided evidence that cAMP plays a role in the transport of ions across the plasma membrane of spermatozoa, which affect sperm motility (Nelson 1975, 1985, Tash and Means 1983). Lindemann (1978) has suggested that cAMP exerts its stimulating action directly on the motile apparatus. Cascieri et al. (1976) have agreed and proposed that the dilution of sperm-cells at ejaculation allows for the activation of the adenylate cyclase system. Chulavatnatol and Haesungcharrern (1977) have made a study of stabilization of adenylate energy charge and its relation to human sperm motility. This study has revealed that spermatozoa can maintain the energy charge above 0.6 under stress. The results of various studies have suggested that the Ca^{2+} and cAMP regulate sperm motility and thus form intracellular messengers (Gorus et al. 1982; see reviews by Shapiro and Eddy 1980, Tash and Means 1983, Nelson 1985).

b) Caffeine

Mammalian sperm motility has been reported as being enhanced and/or prolonged in the presence of caffeine (1,3,7-trimethyl-2,6-dioxypurine) a phosphodiesterase inhibitor (Garbers et al. 1971a, Hicks et al. 1972, Schoenfeld et al. 1975, Tamblyn and First 1977, Ruthkaplan et al. 1978, Shilon et al. 1978, Amelar et al. 1980, Dacheux and Paquignon 1980, Makler et al. 1980, Weeda and Cohen 1982, Serres et al. 1982, Aitken et al. 1983b). Stimulation of the kinetic activity and respiration of spermatozoa after in vitro addition of caffeine were first reported to occur in bovine epididymal and ejaculated spermatozoa (Drevius 1971, Garbers et al. 1971a). Caffeine was later reported to stimulate the motility and forward progression of ejaculated human spermatozoa (Bunge 1973, Haesungcharrern and Chulavatnatol 1973, Schoenfeld et al. 1973, 1975, Aitken et al. 1983a). In some cases, caffeine is able to induce motility in nonmotile spermatozoa, e.g. from the testis (Brackett et al. 1978, Dacheux and Paquignon 1980). Ejaculated human spermatozoa with reduced motility also show an improvement of their motility in the presence of caffeine (Schoenfeld et al. 1973, Johnsen et al. 1974, Homonnai et al. 1976, Makler et al. 1980, Aitken et al. 1983b). Barkay (1979), on the other hand, has observed that the pregnancy rate is significantly enhanced by the addition of caffeine to cryopreserved human spermatozoa prior to insemination. He has also found that children born after such inseminations appear normal and show no chromosomal abnormalities. Aitken et al. (1983b) have studied the influence of caffeine on movement characteristics, fertilizing capacity, and ability to penetrate cervical mucus of human spermatozoa.

Caffeine is an inhibitor of cyclic nucleotide phosphodiesterases (enzymes involved in spermatozoan cyclic nucleotides: see Hardman et al. 1971, Hicks et al.

1972, see also Chap. X) and hence affects the intracellular level of cyclic AMP by inhibiting its enzymatic breakdown (Hoskins and Casillas 1975b); cyclic AMP and/or guanosine 3′-5′-cyclic monophosphate (cyclic GMP) has been implicated in the observed motility increases (Garbers et al. 1971a, b, Hicks et al. 1972, Hoskins and Casillas 1975b, Gorus et al. 1982, Tash and Means 1982), as already discussed.

Tamblyn and First (1977) have made a study of caffeine-stimulated ATP-reactivated motility in a detergent-treated bovine sperm model. Wyker and Howards (1977) have studied the effect of cAMP and caffeine on the motility of the rete testis and epididymal spermatozoa. Caffeine increased motility of epididymal spermatozoa. Dibutryl cAMP and cAMP stimulate caput spermatozoa but have no effect on cauda spermatozoa. Although the exact action of caffeine on spermatozoal motility is still not fully determined, intracellular levels of cyclic AMP are increased within minutes after the addition of caffeine to a semen specimen (Hoskins and Casillas 1975b).

In addition to stimulation of motility, caffeine has also been observed to increase the rates of respiration (Garbers et al. 1971a, b) and glycolysis (Hoskins 1973, Hoskins and Casillas 1975b). More extensive studies may reveal that one of these is the route whereby caffeine accelerates capacitation. Fraser (1979) has provided evidence for accelerated mouse sperm penetration into eggs in vitro, and hence for accelerated capacitation. The participation of cyclic AMP in the maturation and capacitation of spermatozoa has also been suggested (B. E. Morton and Albagli 1973, Toyoda and Chang 1974, Hoskins and Casillas 1975b), as already discussed in Chap. X. This raises the strong possibility that caffeine might have an effect on capacitation, which needs to be determined more precisely (Shilon et al. 1978).

The ability of cAMP-phosphodiesterase inhibitors to stimulate human sperm motility has been suggested to have implications for clinical studies on male infertility. As already stated, the exact mechanism of stimulation of sperm motility by phosphodiesterase inhibitors is still not known. But it has been suggested that cAMP-dependent protein kinase or kinases are activated (Fig. 75) and this is followed by phosphorylation of some protein necessary for the initiation or regulation of motility (Hoskins and Casillas 1975b, Tash and Means 1982). Ca^{2+} is believed to play a critical role in these events as both Ca^{2+} and cAMP + caffeine activate motility in completely quiescent caudal epididymal hamster sperm collected under isotonic sucrose (B. E. Morton et al. 1973). Both cAMP and Ca^{2+} together are involved in a number of biological processes and, in some cases, the physiological response to cAMP needs Ca^{2+} (Rasmussen 1970). Hoskins and Casillas (1975b) have therefore, suggested that cyclic nucleotides may cause activation of a sperm-cyclic AMP-dependent protein kinase, which in turn stimulates a calcium-controlled motility-regulating protein.

c) Aminophylline, Theophylline, and Pentoxiphylline

Three other methyl xanthines, such as aminophylline, theophylline, and pentoxiphylline, have also been used to find out whether they can also increase spermatozoal motility. Both aminophylline and theophylline are shown to make significant improvement in the motility of human spermatozoa in vitro (Haesungcharrern and Chulavatnatol 1973). In an investigation comparing the effects of caffeine to that of theophylline, neither compound shows an increase of spermatozoan motility when spermatozoa are suspended in Tyrode buffer (see Amelar et al. 1980). It is believed that the buffer may have played a significant role in the inability of either compound to stimulate motility in that study. Pentoxiphylline is described to be a phosphodiesterase inhibitor with longer-lasting effects (Stefanovrich 1973) and a higher solubility than either theophylline or caffeine. A later study has shown that although caffeine reverses the immobilizing effects of a Mann-fructose buffer on human spermatozoa, pentoxyiphylline not only neutralizes the effect of this buffer but also results in a significantly higher percentage of forward-moving spermatozoa (De Turner et al. 1979).

4. Kallikrein

Ejaculated human spermatozoa with reduced motility show an improvement of their motility in the presence of pancreatic kallikrein (Schill 1975a, b, Schill et al. 1974, 1976, Hofmann 1981), a kinin-liberating proteinase. Although the manner in which sperm motility is increased in kallikrein is not fully understood at present, the improvement has been shown to be accompanied by a parallel increase in fructose utilization when as little as 1 KU/ml of kallikrein is added to the seminal fluid (Schill 1975c). The increase of motility by kallikrein has been shown to be further improved by the addition of a kininogen source, such as serum (Schill 1975d). Schill et al. (1978) have recently observed that both caffeine and kallikrein stimulate the motility of freshly ejaculated spermatozoa and also improve the motility and fructose utilization (see also Barkay et al. 1977), thus offering a possible means of improving the quality of freeze-preserved human semen. Bratanov et al. (1978, 1983) have also observed that the application of kallikrein exerted considerable stimulation of bull sperm motility in vitro. Very recently, Bratanov et al. (1980) has observed that antikallikrein serum significantly inhibits the motility of bull and boar spermatozoa, but does not exert any marked influence on the motility of ram and human spermatozoa.

5. Carnitine and Acetylcarnitine

Carnitine, a vitamin-like compound and acetylcarnitine are compounds involved in intracellular fat metabolism through the transportation of coenzyme A (Co-A) and activated acyl groups (Amelar et al. 1980). Mammalian semen has been shown to contain L-carnitine and L-O-acetylcarnitine (Frenkel et al. 1974, Lewin et al. 1976, Konengkul et al. 1977, Suter and Holland 1979, Brooks 1979b, Amelar et al. 1980). Both of these compounds have also been observed in the epididymis (Casillas 1972). Calvin and Tubbs (1976) have observed that hypotonically treated spermatozoa (ram and boar) take up L-[^{14}C] carnitine and release it only by a highly temperature-dependent exchange with external L-carnitine, acetyl-L-carnitine, or deoxycarnitine (4-trimethyl-aminobutyrate). The exchange system is probably mitochondrial. Suter and Holland (1979) have determined the concentrations of free L-carnitine and L-O-acetylcarnitine in spermatozoa and seminal fluid of normal, fresh, and frozen human semen. It has also been shown that the distribution of soluble metabolites such as L-carnitine between spermatozoa and seminal fluid is altered by the freezing of semen. After freezing and storage of semen at $-20\,°C$ for 7 days, the intracellular concentrations of free L-carnitine and L-O-acetylcarnitine are decreased to below the limits of assay. Carter et al. (1980) have determined the relationship of carnitine and acetylcarnitines in ejaculated sperm to blood plasma testosterone of dairy bulls. The carnitine content of rabbit epididymal spermatozoa has been determined in organ culture (Casillas and Chaipayungpan 1982).

The role of carnitine and acetylcarnitine in spermatozoan metabolism and motility is controversial (Casillas and Erickson 1975, Milkowski et al. 1976, Tanphaichitr 1977, Brooks 1979b, Amelar et al. 1980, Hinton et al. 1981). Both compounds stimulate sperm motility in undiluted human semen of various sperm densities and various initial motilities at concentrations of 10 mM or 30 mM (Tanphaichitr 1977). The degree of stimulation varies from sample to sample in different donors and even with different samples from the same donor. No correlation is observed between the degree of stimulation of motility with either the initial motility or the sperm density. The acetylcarnitine content appears to reflect the possibility that free carnitine must be transported into the sperm-cells to be acetylated and then released in the seminal fluid. The effect of carnitine and acetylcarnitine on palmitate oxidation has been more contradictory (see Brooks 1979b). Bohmer et al. (1978) have studied carnitine-acetyltransferase in human spermatozoa. The purification and properties of carnitine-acetyltransferases have been studied for the bovine (Huckle and Tamblyn 1983) and ram (Day-Francesconi and Casillas 1982) spermatozoa. The latter authors have also studied its intracellular localization. Casillas and Chaipayungpan (1979) have suggested that the accumulation of carnitine in the rabbit cauda epididymis and its spermatozoa is indicative of its involvement in sperm maturation, as also discussed by Brooks (1979b). Brooks (1979a, b) has not observed any correlation between the content of either total carnitine or the carnitine acyltransferases and the respiratory capacity of spermatozoa. Hinton et al. (1981), after studying the effects of carnitine and some related compounds on the motility of rat spermatozoa from the caput epididymides, have suggested that carnitine may be important in the development by spermatozoa of the potential for motility and also to maintain mature spermatozoa in a quiescent state. It has been suggested that carnitine may act as osmotic balancer (Brooks et al. 1973). On the other hand, it has also been suggested that carnitine and acetylcarnitine may act as buffering agents for acetyl Co-A formed during glycolysis (Casillas and Erickson 1975). A discussion of results obtained in different studies has indicated that the mode of action of these compounds has yet to be revealed fully.

6. Glyceryl Phosphocholine

Turner et al. (1978) have observed that the dilution of epididymal fluid is important along with the requirement for ions. Initiation of sperm motility is significantly inhibited by glyceryl phosphocholine (GPC) and GPC + carnitine ($P < 0.05$). Carnitine alone significantly reduces overall sperm motility ($P < 0.05$), but not the initiation of motility. Addition of 5 mg/ml bovine serum albumin to the test diluents does not protect the spermatozoa from the effects of GPC and carnitine. It has been suggested that the dilution of these compounds that occurs at ejaculation, plus the concomitant addition of electrolytes, allow the initiation of rat sperm motility. Arrata et al. (1978) have observed a significant correlation between motility, progression, and the GPC ratio. Poor motility and progression in the specimens are accompanied by low GPC ratios regardless of the sperm counts. Brooks (1979b) has discussed the formation (or synthesis), accumulation, and metabolism of GPC in the epididymis.

7. *Protein Carboxylmethylase*

Bouchard et al. (1980) have made a study of localization of protein carboxylmethylase (PCM) in sperm tails of rat and rabbit. PCM is found exclusively in the tail fraction, whereas methyl acceptor protein(s) (MAP) is detected in both head and tail fractions. The presence of all the components of the protein carboxylmethylation system in spermatozoa and the localization of PCM and some of its substrates in the sperm tail are consistent with their involvement in sperm motility (Bouchard et al. 1981). Bardin and Gagnon (1982) have reviewed the possible role of protein-carboxyl methylation in sperm motility and capacitation. Sastry and Janson (1983) have observed the depression of human sperm motility by inhibition of enzymatic methylation. Erythro-9-3-(2-hydroxynonyl) adenine arrests sperm motility apparently by inhibiting the axonemal dynein ATPase on which motility depends (Bouchard et al. 1981). The involvement of methyl transfer reactions and S-adenosyl homocysteine in the regulation of bovine sperm motility has been studied by Goh and Hoskins (1986). Gagnon et al. (1984) have provided evidence for a protease involvement in sperm motility.

8. *Epididymal Sperm Motility Factors*

Several recent studies have revealed the importance of epididymal factors in the initiation and maintenance of sperm motility in rodent, bovine, and human sperm (Hoskins and Acott 1976, Bavister et al. 1978, Bedford and Miller 1978, Brandt et al. 1978, Acott and Hoskins 1978, 1981, 1983, Hoskins et al. 1978, B.E. Morton et al. 1978, 1979, Kann and Serres 1980, Acott et al. 1979, Sheth et al. 1981a, b, Baas et al. 1983, Nelson 1985). All evidence now suggests that the factors responsible for forward motility occur in the caudal region of the epididymis (Hoskins et al. 1978, Acott et al. 1979, Kann and Serres 1980, Hinton and Setchell 1980, Egbunike 1982) although the forward motility factor has been observed in spermatozoa liberated from the preceding regions of the epididymis in the boar (Hunter et al. 1976, Egbunike 1980, 1982). The forward motility factor appears to bind to caput sperm as they pass through the epididymis because the continued presence of free forward motility protein is not necessary (Acott and Hoskins 1981). The forward motility protein can be depleted from solution by caput, but not caudal, epididymal sperm. A seminal plasma factor that inhibits the motility of demembranated reactivated spermatozoa has been reported for the rabbit (De Lamirande et al. 1983).

Bavister et al. (1979) have studied the effect of cauda epididymal plasma on motility and the acrosome reaction of hamster and guinea-pig spermatozoa in vitro, and reported the presence of sperm motility factor (SMF), which is observed to be stable to boiling and shows inhibition of the acrosome reaction. The ejaculated buffalo sperm also require the presence of SMF(s) to maintain their viability in vitro (Sidhu et al. 1984, 1985a); SMF(s) has been isolated from the adrenal glands of rats (Meizel and Working 1980). Morton et al. (1978, 1979) have observed a survival factor (SF) in hamster cauda epididymal plasma. The SF is also heat stable and also tolerates pH extremes. However, its activity is lost following trypsin digestion. This protein is believed to protect the spermatozoa from lysis following dilution.

The results of other recent studies have provided evidence for the epididymal origin of bovine sperm forward motility protein (FMP) (Brandt et al. 1978, Hoskins et al. 1978, Acott and Hoskins 1981). The glycoprotein nature of this factor has been shown and it exists in seminal fluid as a member of multiple forms. The molecular weight of the monomeric form of FMP is approximately 37,000. Its specific activity, whether expressed on a volume or protein concentration basis, is far higher in fluid from the cauda epididymis than in any other body fluid. The detailed motility studies have revealed that this compound induces the forward motility in bovine spermatozoa. FMP somehow interacts with sperm as they pass through the epididymis. Hoskins et al. (1979) have concluded that the initiation of motility in the bovine epididymis involves at least two events, an elevation in the intrasperm content of cyclic AMP during epididymal transit and the production by the epididymis of specific FMP. Sheth et al. (1981b) have recently isolated a progressive motility-sustaining factor (PMSF) from human epididymides, which can significantly improve progressive motility of oligozoospermic as well as asthenozoospermic semen. PMSF has been observed to increase cAMP accumulation by the human spermatozoa, indicating its possible mode of action. Fournier-Delpech et al. (1977) have observed that the middle corpus has an essential role in the development of the motility and fertilizing capacity of ram epididymal spermatozoa. Glucosidases secreted by the epididymal epithelium are believed to provide optimal levels of energy for spermatozoal maturation and have been correlated with the percentage of sperm motility in some studies (see Amelar et al. 1980).

Sperm are mechanically immobilized in the cauda epididymis by "immobilin", a high-molecular-weight glycoprotein (Usselman and Cone 1983). Carr and Acott (1984) have observed a sperm motility quiescence factor in the bovine caudal epididymal fluid. Usselman et al. (1984) have observed that rat and bull sperm are immobilized in the epididymis by completely different mechanism, both mechanisms can immobilize the sperm from either species. Bull sperm are quiescent in rat immobilin, and although rat sperm initiate motility in bull cauda epididymal fluid, their motility is greatly suppressed within five minutes after dilution. In contrast to rat and bull, rabbit sperm are vigorously motile in rabbit cauda epididymal fluid. The reasons for this difference needs to be determined. De Lamirande et al. (1984) have studied the characteristics of a seminal plasma inhibitor of sperm motility. The origin and the mode of its action are discussed.

In effect, while some materials are localized in the cauda epididymis and are involved in sperm motility, some extrinsic factors also seem to be involved in the development of sperm motility. According to the results of Norman and Gombe (1975) and Sheth et al. (1980), cyclic AMP may be involved in sperm motility. Also, the anti-malarial drug, chloroquine, stimulates the activity of ejaculated spermatozoa via adenyl cyclase system (Garbers et al. 1971a, b). Chloroquine stimulation of porcine testicular and epididymal spermatozoa also remarkably enhanced sperm motility and the rate of loss of acetylcholinesterase activity in spermatozoa obtained from the cauda portions of the epididymis (Egbunike 1982). Turner and Giles (1982) have proposed that a proteinaceous factor, possibly of testicular origin, is important in preventing the initiation of motility while sperms reside in the distal reproductive tract of the male rat.

9. Albumin and Other Macromolecules

R. A. P. Harrison et al. (1978) have studied the effect of ionic strength, serum albumin, and other macromolecules on the maintenance of motility and surface of mammalian (boar, bull, rabbit, ram and stallion) spermatozoa in a simple medium (Suter et al. 1979). Bovine or human sperm albumin is consistently most effective with regard to its motility and preserving ability. According to Mueller and Kirchner (1978), the motility of rabbit spermatozoa is best maintained in the fractions which contain albumin. Rabbit serum albumin is more effective than bovine or human albumin. A concentration of 4 mg rabbit serum albumin/ml results in markedly decreased motility. Blank et al. (1976) have made a study of adsorption of albumin on rabbit sperm membranes. Chloroquine-diphosphate stimulates the forward motility of buffalo spermatozoa (Sidhu and Guraya 1979c). Motility is depressed in either the presence or absence of glucose and it appears that these effects are produced by an interaction with the sperm plasma membrane. Bavister and Yanagimachi (1977) have studied the effects of sperm extracts and energy sources on the motility and acrosome reaction of hamster spermatozoa in vitro. De Turner et al. (1979) have observed that Mann-fructose fluid results in a decrease in motility and the duration of activity of spermatozoa. Caffeine appears to neutralize the deleterious effect of the buffer, whereas pentoxyphylline and cAMP seem to increase the duration of the activity of spermatozoa. Propranolol results in a dramatic decrease in motility, an effect that cannot be neutralized by the simultaneous addition of cAMP. The inhibition by amiloride of the motility of the rat caudal epididymal spermatozoa is concentration-dependent (Wong et al. 1981). Daugherty et al. (1978) have studied the effect of amylase on sperm motility and viability. The percentage of active spermatozoa and viable cells and the quality of sperm motility are altered in relationship to the amylase levels. Tinneberg et al. (1980) have recently observed the enhancement of sperm motility by treatment with streptokinase-streptodornase (varidase) in boars and men. Peterson et al. (1977a) have suggested that a purine derivative binding site located close to the hexose transport site is involved in the regulation of membrane function and the mechanism of action of cytochalasin B, which inhibits the metabolism and motility of human spermatozoa at low concentrations.

10. Taurine and Hypotaurine

Meizel (1981), after studying the presence of taurine and hypotaurine in sperm and reproductive tract fluids of several mammals, have suggested that they may have in vivo roles in the maintenance of sperm motility and stimulation of capacitation and/or the acrosome reactions (Meizel et al. 1980). Taurine has been observed to maintain and stimulate motility of hamster spermatozoa during capacitation in vitro (Mrsny et al. 1979). Ozasa and Gould (1982) observed the protective effect of taurine from osmotic stress on chimpanzee spermatozoa. Mrsny and Meizel (1985) have produced the first evidence to show inhibition of Na^+, K^+-ATPase activity by taurine and hypotaurine. These observations are discussed in relation to the ability of these compounds to sustain hamster sperm motility and fertility.

11. Steroids

Steroid hormones are also known to affect sperm metabolism and motility, as already discussed in Chap. X. But much more work is needed to reach some definite conclusion about the mode of their action (Warikoo and Das 1983).

12. Cholinergic System

Mammalian spermatozoa have been shown to contain cholinesterase, acetylcholinesterase, acetylcholine, and choline acetyl transferase, which constitute the cholinergic system (Mann 1964, Nelson 1970, 1975, 1985, Sastry 1975, Chakraborty and Nelson 1976, Stewart and Forrester 1978a, Nelson et al. 1980a, b). Bull spermatozoa contain other minor choline esters in smaller quantities than acetylcholine (Bishop et al. 1977). One of the minor choline esters is possibly propionylcholine. Sperm from infertile bulls show low motility and very low levels of acetylcholine. The cholinesterase activity has been studied in seminal plasma and human spermatozoa in normal and infertile subjects (Pedron 1983). The main site of localization of both acetyl- and nonspecific cholinesterases has been found to be the fibrillar component of the sperm tail in humans, chimpanzee, bull, rat, rabbit, and mouse (Chakraborty and Nelson 1976, Nelson 1985). Some reactivity is also seen in the mitochondrial membrane. Cytoplasmic droplets are also reactive (see Chap. IX). Species variations in the amount of product are described there.

McGrady and Nelson (1976) have studied the cholinergic effects on bull and chimpanzee sperm motility. This study has substantiated the role of the acetylcholine-acetylcholinesterase system in control of spermatozoan motility. It is possible that the function of the cholinergic system, consisting of choline acetyltransferase, acetylcholine, acetylcholinesterase, and cholinergic receptor, is coupled with the excitation and relaxation phases of the contractile apparatus (Sastry et al. 1979). Stewart and Forrester (1978b) have observed that spermatozoa contain a nicotinic type acetylcholine receptor. Sastry et al. (1979) have observed that the degree of binding $[^{125}I]$-α bungarotoxin to sperm tails is higher than that to heads or mid-piece. The distribution of the toxin binding is similar to the distribution of choline acetyltransferase and cholinergic receptor in tail>mid-piece>head. This gives further evidence for a nicotinic cholinergic receptor in sperm similar to that present in skeletal muscle. The specific activity for [^{125}I]-α bungarotoxin binding in tail is only twice that in heads. If the nicotinic receptors are to be on the cell surface in bull spermatozoa, the nicotinic receptor proteins of the tails would be dislodged more easily than the intracellular enzymes and thereby cross-contaminate the sperm fractions of heads and mid-pieces. Nelson et al. (1970) have tentatively concluded that a highly stable, true and specific acetylcholinesterase is "advantageously" arrayed in close morphological approximation to the outer coarse longitudinal contractile fibres of the mammalian sperm flagellum. Here it may, by propagated depolarization of the overlying plasma membrane, govern the coordination of the flagellar wave (Nelson et al. 1980a, b, Nelson 1985). McGrady and Nelson (1976), using cinematographic analysis, have investigated the response of ejaculated spermatozoa of bull and chimpanzee to two substrates and three inactivators of the cholinesterase enzymes. Various inhibitors of the two enzymes, acetylcholinesterase and choline acetyltransferase, as well as blockers and depolarizers of the cholinergic receptor, affect either the rate of progressive movement or the pattern of swimming by mammalian spermatozoa (Nelson et al. 1980b). The cholinergic complex in the spermatozoon consists of enzymes for the synthesis and hydrolysis of acetylcholine as well as a receptor which, on binding to acetylcholine, concomitantly undergoes depolarization with changes in ionic permeability. The cholinergic receptor is placed within the confines of the sperm-cell or at least in contact with the cell interior. The cholinergic effect on motility has been identified as a beat frequency variation, and in the case of physostigmine defect of coordination of the wave sperm. The results of this study have further substantiated the role of the acetylcholine-acetylcholinesterase system in control of spermatozoan motility. Nelson et al. (1980b) have stated that spermatozoon is self-propulsive and its swimming behaviour is related to sliding elements which undergo alternate contractile relaxation cycles coupled transiently to labile membrane events; the latter control sequestration and release of Ca^{2+} from intracellular binding sites. Intracellular sites of Ca^{2+} binding have been demonstrated histochemically in thin sections of bull spermatozoa (Nelson et al. 1982). The release of calcium from these sites is possibly triggered by small increments of Ca^{2+} admitted through cholinergically controlled ionophoric channels; the system ultimately undergoes reversion for recycling through action of a possibly ATP-dependent calcium-extrusion pump. Thus, uptake of Ca^{2+} is related to cholinergic regulation of ionophoric channels through the spermatozoon plasma membrane (Nelson et al. 1982, Nelson 1985). More recent studies have also shown that acetylcholinesterase is present in the sperm tail where it is essential for the coordinated propagation involved in sperm motility, and is related to the development of mammalian sperm motility (Egbunike 1980). Spermatozoa lose an enormous amount of acetylcholinesterase activity on the way from the testis to the epididymis (Egbunike and Elemo 1979, Egbunike et al. 1979, Egbunike 1980, 1982). During this transit through the epididymis, the rate of loss in this enzyme is gradual and coincides with the development of sperm motility, which reaches its peak at the caudal portions of the epididymis. Egbunike (1982) has suggested that the physiological maturity of epididymal sperm may be enhanced by chloroquine stimulation, and this suggestion is also supported by the increase in the motility of cauda epididymal sperm by chloroquine.

13. Catecholamines and Tranquillizers

Bavister et al. (1979) have observed that one or more catecholamines are essential co-factors of the "sperm motility factor", and demonstrated that hamster spermatozoa require catecholamines for their motility in vitro. The sperm motility factor(s) isolated from the adrenal glands of rats by Sidhu et al. (1984), and used in the capacitation of buffalo spermatozoa, may be the catecholamines. Some agents, which have been observed to act on the central nervous system (the so-called psychoactive drugs such as the phenothiazines), decrease spontaneous motor activity, cause strong adrenergic-blocking effects, and inhibit frog sperm motility (see Nelson 1975). Cates and Jozefowicz (1970) have observed that 60 μm chloropromazine causes moderate inhibition of frog sperm motility, and exposure to six times that amount results in total inhibition. Other tranquillizers also cause slight-to-moderate inhibition, but only the chloropromazine effect can be reversed.

14. Arginine

Some contradictory reports have been published about the effect of arginine, an amino acid, on sperm motility (Amelar et al. 1980). Its oral administration is observed to be beneficial in the treatment of oligospermia and asthenospermia (Amelar et al. 1980 for references). But these results have not been supported by later studies. However, Keller and Polakoski (1975) have observed that addition of L-arginine to ejaculated human spermatozoa in vitro stimulates their motility in specimens having an initially low motility. The stimulation of forward motility from initial levels appears to be dose-related; the optimal stimulation occurs at a concentration of 0.004 M L-arginine at either 23° or 37 °C, and the percentage of stimulation is inversely proportional to the initial motility. D-arginine, L-homoarginine, L-nitroarginine, and L-ornithine do not produce any stimulatory activity. Both the terminal and guanidino group and the length of the carbon chain appear to be significant for the effect of L-arginine in vitro. The mode of action of arginine needs to be determined more precisely. However, L-arginine has been shown to accumulate in ejaculated rabbit spermatozoa by some highly specific transport mechanism; respiration rates are stimulated significantly above those observed for endogenous levels of L-arginine.

In a nutshell, it may be stated that a variety of chemical substances (or compounds), alone or in combination, affect sperm motility. But the mechanisms of their action at the molecular level for several of them are still to be determined more precisely. These may act directly or indirectly on the sperm-cell by inducing changes in the permeability of the sperm membrane to certain substances and/or enzymes (Nelson 1985). Fraioli et al. (1984) have discussed the possible role of beta-endorphin, met-encephalin and calcitonin of human semen in sperm motility.

References

Aaseth EP, Saack RG (1985) J Reprod Fertil 74:473
Abdel-Aziz MT, El-Hagger S, Tawadrous GA, Hamada T, Shawky MA, Amin KS (1983) Andrologia 15:259
Abdou MSS, El-Guindi MM, Bayoumi MT (1977) Zentralbl Veterinaermed R A 24:636
Abla A, Mrouen A, Durr IF (1974) J Reprod Fertil 37:121
Abraham KA, Bhargava PM (1963) J Biochem 86:308
Abraham SK, Franz J (1983) Mutat Res 108:373
Abu-Erreish G, Magnes L, Li TK (1978) Biol Reprod 18:554
Acott TS, Hoskins DD (1978) J Biol Chem 253:6744
Acott TS, Hoskins DD (1981) Biol Reprod 24:234
Acott TS, Hoskins DD (1983) J Submicrosc Cytol 15:77
Acott TS, Johnson DJ, Brandt H, Hoskins DD (1979) Biol Reprod 20:247
Adams CS, Johnson AD (1977) Comp Biochem Physiol B 58:409
Adimoelja A, Hariadi R, Amitaba IGB, Adisetya P, Soeharno (1977) Andrologia 9:289
Adiyodi KG, Adiyodi RG, (eds) (1983) Reproductive biology of invertebrates, vol II: Spermatogenesis and sperm function. Wiley, New York Chichester
Afzelius BA, Eliasson R, Johnsen O, Lindholmer C (1975) J Cell Biol 66:225
Ahluwalia B, Holman RT (1969) J Reprod Fertil 18:431
Ahmad N, Haltmeyer GC, Eik-Nes K (1975) J Reprod Fertil 44:103
Aitken RJ, Wang YF, Liu J, Best F, Richardson DW (1983a) Int J Androl 6:180
Aitken RJ, Best F, Richardson DW, Schats R, Simm G (1983b) J Reprod Fertil 67:19
Aizawa S, Nishimune Y (1979) J Reprod Fertil 56:99
Alabi NS, Whanger PD, Wu ASH (1986) Biol Reprod 33:911
Alexander NJ (1975) Fed Proc Fed Am Soc Exp 34:1692
Alfert M (1956) J Biophys Biochem Cytol 2:109
Allag IS, Das RP, Roy S (1983) J Androl 4:415
Allison AC, Hartree EF (1970) J Reprod Fertil 21:501
Almeida de M, Lefroit-Joliy M, Righenzi S (1982) Am J Reprod Immunol 2:270
Almeida de M, Lefroit-Joliy M, Righenzi S (1983) In: Shulman S, Dondero F (eds) Immunological factors in human contraception. Acta Medica, Rome, pp 39–46
Alonso-Lancho MT, Garcia-Diez LC, Gonzalez-de-Buitrago JM, Meza-Mendoza S, Miralles-Garcia JM (1981) Rev Clin Esp 162:193
Alvarez JG, Storey BT (1982) Biol Reprod 27:1102
Alvarez JG, Storey BT (1983a) J Androl 4:40
Alvarez JG, Storey BT (1983b) Biol Reprod 28:1129
Alvarez JG, Storey BT (1983c) Biol Reprod 28 Suppl 1:156
Alvarez JG, Storey BT (1983d) Biol Reprod 29:548
Alvarez JG, Storey BT (1984) Biol Reprod 30:323
Alvarez JG, Storey BT (1985) Biol Reprod 32:342
Amann RP (1962) Am J Anat 110:69
Amann RP (1979) In: Fawcett DW, Bedford JM (eds) The spermatozoon. Urban & Schwarzenberg, Baltimore München. pp 431–436
Amann RP (1984) Congr Proc 10th Int Congr Anim Reprod Artif Insem, vol IV. Univ Ill, Urbana Champaign, USA, p II, 28
Aman RP, Almquist JO (1962) J Reprod Fertil 3:260

Amann RP, Hammerstedt RH (1976) Biol Reprod 15:670
Amann RP, Hay SR, Hammerstedt RH (1982) Biol Reprod 27:723
Ambadkar PN, George JC (1964) J Histochem Cytochem 12:587
Amelar RD (1966) Infertility in men. Diagnosis and treatment. Davis Phil, Penn
Amelar RD, Dubin L, Schoenfeld C (1980) Fertil Steril 34:197
Amir D (1984) Congr Proc 10th Int Congr Anim Reprod Artif Insem, vol II. Univ Ill, UrbanaChampaign, USA, p 33
Amir D, Esnault C, Nicolle JC, Courot M (1977) J Reprod Fertil 51:453
Amos LA, Klug A (1974) J Cell Sci 4:523
Amos LA, Linck RW, Klug A (1976) In: Goldman RD, Pollard TD, Rosenbaum JL(eds) Cell motility, Book C. Cold Spring Harbor, New York, pp 847–867
Ånberg A (1957) Acta Obstet Gynecol Scand 36, Suppl 2:I
Anderson K (1978) Anat Histol Embryol 7:164
Anderson RA Jr, Beyler SA, Mack SR, Zaneveld LJ (1981a) Biochem J 199:307
Anderson RA Jr, Oswald C, Zaneveld LJ (1981b) J Med Chem 24:1288
Anderson WA, Personne P (1969) J Microsc (Paris) 8:87
André J (1962) J Ultrastruct Res 6 Suppl 3:1
André J (1963) In: Mazia D, Tyler A (eds) General physiology of cell specialization. McGraw-Hill, New York, pp 91–115
Antonyuk VS (1976) VIIIth Congr Int Reprod Anim Insem Artif, Krako, vol IV, pp 866–868
Aoki A (1980) Eur J Cell Biol 22:467
Aoki A, Fawcett DW (1978) Biol Reprod 19:144
Appell RA, Evans PR (1978) Fertil Steril 30:436
Arrata WSM, Burt T, Corder S (1978) Fertil Steril 39:329
Arthur SH Wu (1968) In: Thibault C (ed) VI Congrès de reproduction et insemination artificielle, vol I. Inst Nat Rech Agron (Paris), pp 217–219
Arver S, Eliasson R (1980) J Reprod Fertil 60:481
Ashraf M, Peterson RN, Russell LD (1983) Fed Proc Fed Am Soc Exp 42:615
Ashraf M, Peterson RN, Russell LD (1984) Biol Reprod 31:1061
Atherton RW (1979) In: Fawcett DW, Bedford JM (eds) The spermatozoon. Urban & Schwarzenberg, Baltimore Munich, pp 421–425
Atherton RW, Radany EW, Polakoski KI (1978) Biol Reprod 18:624
Atreja SK, Anand SR (1985) J Reprod Fertil 74:687
Aughey E, Orr PS (1978) J Reprod Fertil 53:341
Aumüller G, Schenk B, Neumann F (1975) Andrologia 7:317
Austin CR (1951) Aust J Biol Sci B4:581
Ax RL, Dickson K, Lenz RW (1985) J Dairy Sci 68:387
Ayyagari VB, Mukherjee DP (1970) J Reprod Fertil 22:375
Baas JW, Molan PC, Shannon P (1983) J Reprod Fertil 68:275
Babcock DF, First NL, Lardy HA (1975) J Biol Chem 250:6488
Babcock DF, First NL, Lardy HA (1976a) J Biol Chem 251:3881
Babcock DF, First NL, Hutchinson T (1976b) Soc Stud Reprod, Abstr, 9th Annu Meet, p 46
Babcock DF, Stamerjohn DM, Hutchinson T (1978) J Exp Zool 204:391
Babcock DF, Singh JP, Lardy HA (1979) Dev Biol 69:85
Babcock DF, Rufo GA Jr, Lardy HA (1983) Proc Natl Acad Sci USA 80:1327
Baccetti B (1975) In: Duckett JG, Racey PA (eds) The biology of the male gamete. Academic Press, London New York, pp 317–319
Baccetti B (1979) In: Fawcett DW, Bedford JM (eds) The spermatozoon. Urban & Schwarzenberg, Baltimore Munich, pp 305–329
Baccetti B (1982) Symp Soc Exp Biol 35:521
Baccetti B, Afzelius, BA (1976) The biology of the sperm cell. Karger, Basel
Baccetti B, Pallini V, Burrini AG (1973) J Submicrosc Cytol 5:237
Baccetti B, Burrini AG, Pallini V (1974) Proc 8th Int Congr Electr Microsc, Canberra, vol II, pp 92–93
Baccetti B, Pallini V, Burrini AG (1975) Exp Cell Res 90:183
Baccetti B, Pallini V, Burrini AG (1976a) J Ultrastruct Res 54:261
Baccetti B, Pallini V, Burrini AG (1976b) J Ultrastruct Res 57:289

Baccetti B, Renieri T, Rosati F, Selmi MG, Casanova S (1977) Andrologia 9:255
Baccetti B, Bigliardi E, Burrini AG (1978) Dev Biol 63:187
Baccetti B, Bigliardi E, Burrini AG (1980) J Ultrastruct Res 71:272
Baccetti B, Burrini AG, Pallini V, Renieri T (1981) J Cell Biol 88:102
Bae DS (1984) Korean J Anim Sci 26:509
Bae DS, Kim JW (1981) Korean J Anim Sci 23:235
Bairati A, Perotti ME, Gioria MR (1980) J Submicrosc Cytol 12:599
Baker TG (1971) In: Diczfalusy E, Standley CC (eds) Use of non-human primates for research on problems of human reproduction. Acta Endocrinol (Kbh) Suppl 166, pp 18–41
Balázs A, Blazsek I (1979) Control of cell proliferation by endogenous inhibitors. Elsevier/North-Holland Biomedical Press, Amsterdam New York
Balbontin J, Bustos-Obregon E (1982) Arch Androl 9:159
Baldwin J, Temple-Smith P, Tidemann C (1974) Biol Reprod 11:377
Balhorn R, Gledhill BL, Wyrobek AJ (1977) Biochemistry 16:4074
Ballesteros LM, Delgado NM, Rosado A, Hernandez O (1983) Arch Androl 11:95
Balogh K, Cohen RB (1964) Fertil Steril 15:35
Bane A, Nicander L (1963) Int J Fertil 8:865
Baradi AF, Rao NS (1983) Arch Androl 10:219
Barcellona WJ, Brackeen RB, Brinkley BR (1974) J Reprod Fertil 39:41
Bardin CW, Gagnon C (1982) Prog Clin Biol Res 87:217
Bardin CW, Sherins RJ (eds) (1982) Cell biology of the testis. Ann N Y Acad Sci vol 383
Barham SS, Berlin JD (1974) Cell Tissue Res 148:159
Barham SS, Berlin JD, Brackeen RB (1976) Cell Tissue Res 166:497
Barkay J (1979) 1st Pan-American Conf Andrology, Toronto, Can, March 13–16, 1979
Barkay J, Zuckerman H, Sklan D, Gordon S (1977) Fertil Steril 28:175
Barlow P, Vosa CG (1970) Nature (London) 226:961
Barnstable CJ, Bodmer WF, Brown G, Galfre G, Milstein C, Williams AF, Ziegler A (1978) Cell 14:9
Barr AB, Moore DJ, Paulsen CA (1971) J Reprod Fertil 25:75
Bartlett RJ, Lea OA, Murphy EC, French FS (1984) In: Catt KJ, Dufau ML (eds) Hormone action and testicular function. Ann NY Acad Sci vol 438:pp 29–38
Bartmanska J, Clermont Y (1983) Cell Tiss Kinet 16:135
Bartoov B, Messer GY (1976) J Ultrastruct Res 57:68
Bartoov B, Bar-Sagie D, Mayevsky A (1980) Int J Androl 3:602
Bartoov B, Reis I, Fisher J (1981) J Reprod Fertil 61:295
Bartoszewicz W, Dandekar P, Glass RH, Gordon M (1975) J Exp Zool 191:151
Battellino LJ, Ramos Jaime F, Blanco A (1968) J Biol Chem 243:5185
Battersby S, Chandler JA (1977) Fertil Steril 28:557
Bavdek S, Glover TD (1970) J Reprod Fertil 22:371
Bavister BD (1979) In: Fawcett DW, Bedford JM (eds) The spermatozoon. Urban & Schwarzenberg, Baltimore Munich, pp 169–172
Bavister BD, Yanagimachi R (1977) Biol Reprod 16:228
Bavister BD, Yanagimachi R, Teichman RJ (1976) Biol Reprod 14:219
Bavister BD, Rogers BJ, Yanagimachi R (1978) Biol Reprod 19:358
Bavister BD, Chen AF, Fu PC (1979) J Reprod Fertil 56:507
Bawa SR (1975) In: Duckett JG, Racey PA (eds) The biology of the male gamete, vol VII. Academic Press, London New York, pp 275–278
Bayers SW, Glover TD (1984) J Reprod Fertil 71:23
Bearer EL, Friend DS (1980) Proc Natl Acad Sci USA 77:6601
Bearer EL, Friend DS (1981) J Cell Biol 92:604
Beatty RA (1970) Biol Rev 45:73
Beatty RA (1972) In: Beatty RA, Gluecksoh S (eds) The genetics of the spermatozoon. Proc Int Symp Edinburgh, Aug 16–20, 1971
Beatty RA (1975) In: Duckett JG, Racey PA (eds) The biology of the male gamete. Academic Press, London New York, pp 291–299
Beaumont HM, Mandl AM (1963) J Embryol Exp Morphol 11:715
Bebe Han LP, Tung KSK (1979) Biol Reprod 21:99
Beckman JK, Gray ME, Coniglio JG (1978) Biochim Biophys Acta 530:367

Bedford JM (1964) Proc 5th Int Congr Anim Reprod, Toronto, vol VII, pp 286–288
Bedford JM (1967) Am J Anat 121:329
Bedford JM (1970) Biol Reprod Suppl 2:129
Bedford JM (1972) Am J Anat 133:213
Bedford JM (1974) Contrib Primatol 3:97
Bedford J (1975) In: Hamilton DW, Greep RO (eds) Handbook of physiology, sect VII: Endocrinology, vol V. Physiol Soc, Bethesda, pp 303–317
Bedford JM (1979) In: Fawcett DW, Bedford JM (eds) The spermatozoon. Urban & Schwarzenberg, Baltimore Munich, pp 7–21
Bedford JM (1983) Biol Reprod 28:108
Bedford JM, Calvin HI (1974a) J Exp Zool 188:137
Bedford JM, Calvin HI (1974b) J Exp Zool 187:181
Bedford JM, Miller RP (1978) Biol Reprod 19:396
Bedford JM, Nicander L (1971) J Anat 108:527
Bedford JM, Copper GW, Calvin HI (1972) In: Beatty RA, Gluecksohn S (eds) The genetics of spermatozoon. Proc Int Symp, Edinburgh, Aug. 16–20, 1971, pp 69–89
Bedford JM, Bent MJ, Calvin HI (1973) J Reprod Fertil 33:19
Bedford JM, Moore HDM, Franklin LE (1979) Exp Cell Res 119:119
Bedford MM, Dudkiewicz AB, Penny GS, Dyckes DF, Burleigh BD, Woolley RE, McRorie RA (1981) Am J Vet Res 42:1082
Behnke D, Forer A (1967) J Cell Biol 2:169
Behrman SJ (1971) In: Ingelman-Sundberg A, Lunell NO (eds) Current problems in fertility. Plenum Press, New York, pp 10–16
Bellvé AR (1979) In: Finn CA (ed) Oxford reviews of reproductive biology. Clarendon Press, Oxford, pp 159–261
Bellvé AR (1982) In: Amann RP, Seidel GE Jr (eds) Prospects for sexing mammalian sperm. Colorado Assoc Univ Press, Boulder, pp 69–102
Bellvé AR, Feig LA (1984) Recent Prog Horm Res 40:531
Bellvé AR, Moss SB (1983) Biol Reprod 28:1
Bellvé AR, O'Brien DA (1979) In: Fawcett DW, Bedford JM (eds) The spermatozoon. Urban & Schwarzenberg, Baltimore Munich, pp 379–383
Bellvé AR, O'Brien DA (1983) In: Hartman JF (ed) Mechanism and control of fertilization. Academic Press, London New York, pp 56–138
Bellvé AR, Anderson E, Hanley-Bowdoin L (1975) Dev Biol 47:349
Bellvé AR, Cavicchia JC, Millete CF, O'Brien DA, Bhatnagar YM, Dym M. (1977a) J Cell Biol 74:68
Bellvé AR, Millette CF, Bhatnagar YM, O'Brien DA (1977b) J Histochem Cytochem 25:480
Bellvé R, Feig L, Klagsbrun M (1983) Acta Endocrinol 103, Suppl 256:55 (Abstr)
Benahmed M, Reventos J, Tabone E, Saez JM (1984) In: Catt KJ, Dufau ML (eds) Hormone action and testicular function. Ann NY Acad Sci, vol 438, pp 684–687
Bendich A, Borenfreund E, Witkin SS, Beju D, Higgins PJ (1976) Prog Nucleic Acid Res Mol Biol 17:43
Bennett MD (1977) Philos Trans R Soc London Ser B 277:201
Bennett WI, Gall AM, Southard JL, Sidman RL (1971) Biol Reprod 5:30
Berg Vande JL, Cooper DW, Close PJ (1976) J Exp Zool 198:231
Berg Vande JL, Lee CY, Goldberg E (1981) J Exp Zool 217:935
Bergada C, Mancini NE (1973) J Clin Endocrinol Metab 37:935
Berger T, Clegg ED (1983) Gamete Res 7:169
Bergh A (1983a) Inter J Androl 6:73
Bergh A (1983b) Inter J Androl 6:57
Bergh A (1985) Inter J Androl 8:80
Bergh A, Damber JE (1983) Acta Endocrinol 103 Suppl 26:247 (Abstr)
Bergh A, Damber JE (1984) Inter J Androl 7:409
Berman MI, Sairam MR (1982) In: Bardin CW, Sherins RJ (eds) The cell biology of the testis. Ann NY Acad Sci, vol 383, pp 426–427
Berndston WE, Desjardins C (1974) Am J Anat 140:167
Bernstein MH, Teichman RJ (1968) J Cell Biol 39:14A
Bernstein MH, Teichman RJ (1972) Am J Anat 133:165

Bernstein MH, Teichman RJ (1973) J Reprod Fertil 33:239
Berrios M, Fléchon JE, Barros C (1978) Am J Anat 151:39
Berruti G (1980) Arch Androl 5:267
Berruti G (1981) Comp Biochem Physiol B Comp Biochem 69:323
Berruti G, Martegani E (1982) J Exp Zool 222:149
Berruti G, Martegani E (1984) J Histochem Cytochem 32:526
Berry RE, Mayer DT (1960) Exp Cell Res 20:116
Berthelsen JG, Holm PB, Rasmussen SW (1980) Carlsberg Res Commun 45:25
Betlach CJ, Erickson RP (1973) Nature (London) 242:114
Betlach CJ, Erickson RP (1976) J Exp Zool 198:49
Beyler SA, Zaneveld LJ (1982) J Reprod Fertil 66:425
Beyler SA, Wheat T, Goldberg E (1983) Biol Reprod 28 Suppl 1:160
Bhalla VK, Tillman WL, Williams WL (1973) J Reprod Fertil 34:137
Bhatnagar SK, Anand SR (1981) Biochim Biophys Acta 674:212
Bhatnagar SK, Anand SR (1982) Biochim Biophys Acta 716:133
Bhatnagar SK, Chaudhry PS, Anand SR (1979) J Reprod Fertil 56:133
Bhatnagar SK, Chaudhry PS, Anand SR (1982) Biochim Biophys Acta 716:424
Bhatnagar YM (1985) Biol Reprod 32:957
Bhatbagar YM, Romrell LJ, Bellvé AR (1985) Biol Reprod 32:599
Bhattacharya BC (1976) 8th Congr Int Reprod Anim Insem Artif, Krako, vol IV, pp 876–879
Bhattacharyya A, Bhattacharyya AK (1985) Experientia 41:62
Bhattacharyya AK, Zaneveld LJD (1976) Andrologia 8:1
Bhattacharyya AK, Zaneveld LJD (1978) Fertil Steril 30:70
Bhattacharyya AK, Goodpasture JC, Zaneveld LJD (1979) Am J Physiol 237:E40
Bhattacharyya AK, Sarkar SR, Basu J (1980) J Androl 1:73
Bhela SL, Sengupta BP, Rajdan MN (1982) Indian J Anim Sci 52:147
Bielanski W, Dudek E, Bittmar A, Kosiniak K (1982) J Reprod Fertil Suppl 32:21
Bilaspuri GS, Guraya SS (1980a) Reprod Nutr Dev 20:975
Bilaspuri GS, Guraya SS (1980b) Indian J Exp Biol 18:205
Bilaspuri GS, Guraya SS (1981a) Proc Indian Acad Sci (Anim Sci) 90:357
Bilaspuri GS, Guraya SS (1981b) Indian J Anim Sci 51:1038
Bilaspuri GS, Guraya SS (1982) Reprod Nutr Dev 22:505
Bilaspuri GS, Guraya SS (1983a) Indian J Anim Sci 53:605
Bilaspuri GS, Guraya SS (1983b) Indian J Anim Sci 53:824
Bilaspuri GS, Guraya SS (1983c) Indian J Anim Sci 53:818
Bilaspuri GS, Guraya SS (1983d) Gegenbaurs Morphol Jahrb Leipzig 129:495
Bilaspuri GS, Guraya SS (1983e) J Agric Sci Camb 101:457
Bilaspuri GS, Guraya SS (1983f) Folia Biol (Praha) 29:429
Bilaspuri GS, Guraya SS (1984a) J Endocrinol 101:359
Bilaspuri GS, Guraya SS (1984b) J Agric Sci Camb 103:359
Bilaspuri GS, Guraya SS (1984c) J Agric Sci Camb 102:269
Bilaspuri GS, Guraya SS (1985) Folia Biol (Praha) 31:241
Bilaspuri GS, Guraya SS (1986a) Theriogenology 25:485
Bilaspuri GS, Guraya SS (1986b) Indian J Anim Sci 56: (in press)
Bilaspuri GS, Guraya SS (1986c) Acta Vet Hung 34:121
Bilaspuri GS, Guraya SS (1986d) Arch Androl (in press)
Bilaspuri GS, Guraya SS (1986e) (submitted)
Bishop DW (1968) In: Diamond D (ed) Perspectives in reproduction and sexual behavior. Indian Univ Press, Bloomington, pp 261–268
Bishop DW, Hoffmann-Berling H (1959) J Cell Comp Physiol 53:445
Bishop MWH, Walton A (1960) In: Parkes AS (ed) Marshall's physiology of reproduction, vol I (2). Longmans Green, New York, pp 1–29
Bize I, Santander G (1985) J Exp Zool 235:261
Blackshaw AW, Elkington JSH (1970) Biol Reprod 2:268
Blackshaw AW, Salisbury GW (1957) J Dairy Sci 40:1099
Blanco A (1980) Johns Hopkins Med J 146:231
Blanco A, Gutierrez M, Henquin de CG, Burgos de HMG (1969) Science 164:835

Blanco A, Burgos C, Gerez de Burgos MM, Montamat EE (1976) Biochem J 153:165
Blank JL, Desjardins C (1984) Biol Reprod 30:410
Blank M, Soo L, Britten JS (1976) J Membr Biol 29:401
Bleil JD, Wassarman PM (1983) Dev Biol 95:317
Blom E, Birch-Andersen A (1965) Nord Veterinaermed 17:193
Bloom G, Nicander A (1961) Z Zellforsch Mikrosk Anat 55:833
Blum JJ, Lubliner J (1973) Annu Rev Biophys Bioeng 4:181
Boccabella AV (1963) Endocrinology 72:787
Bode J, Willmitzer L, Opatz K (1977) Eur J Biochem 72:3
Boel EJ (1985) J Exp Zool 234:107
Boender J (1974) Proc K Ned Akad Wet Ser C Biol Med Sci 77:125
Boender J (1977) Proc K Ned Akad Wet Ser C Biol Med Sci 80:35
Bohmer T, Johansen L, Kjekshus E (1978) Int J Androl 1:262
Boitani C, Geremia R, Rossi R, Monesi V (1979) Cell Differ 9:41
Boitani C, Palombi F, Stefanini M (1983) Cell Biol Int Rep 7:383
Bolton AE, Linford E (1970) J Reprod Fertil 21:353
Bomstein RA, Steberl EA (1957) Exp Cell Res 12:254
Bongso TA (1983) Arch Androl 11:13
Borders CL, Raftery MA (1968) J Biol Chem 243:3756
Borhan GS, Schleuning WD, Tschesche H, Fritz H (1976) Hoppe-Seylers Z Physiol Chem 357:671
Borlin G, Chessa G, Cavaggion G, Marchiori F, Mueller-Esterol W (1981) Hoppe-Seylers Z Physiol Chem 362:1435
Borst P (1969) In: Lima-De-Faria A (ed) Handbook of molecular cytology, chap 35. North-Holland Publ Comp, Amsterdam London, pp 914–942
Bostwick EF, Bentley MD, Hunter AG, Hammer R (1980) Biol Reprod 23:161
Bottenstein I, Hayashi I, Hutchings S, Masui H, Mather J, McClure DB, O'Hara S, Rozzino A, Sato G, Serrero G, Wolf R, Wu R (1979) Methods Enzymol 58:94
Bouchard P, Claude P, Phillips DM, Bardin CW (1980) J Cell Biol 86:417
Bouchard P, Penningrowth SM, Cheung A, Gagnon C, Bardin CW (1981) Proc Natl Acad Sci USA 78:1033
Bouvier D (1977) Cytobiologie 15:420
Boyse EA, Old LJ (1969) Annu Rev Genet 3:269
Brachet J, Hulin N (1969) Nature (London) 222:481
Brackett BG, Hall JL, Oh Y (1978) Fertil Steril 29:571
Braden AWH (1957) Anat Rec 127:270
Bradfield JRG (1955) Symp Soc Exp Biol 9:306
Bradford MM, Dudkiewicz AB, Penny GS, Dyckes DF, Bradley FM, Meth BM, Bellvé AR (1981) Biol Reprod 24:691
Bradley MP, Forrester IT (1980) In: Siegel FL, Carafoli E, Kretsinger RH, Maclennan DH, Wasserman RH (eds) Calcium-binding proteins: structure and function. Elsevier/North Holland Biomedical Press, Amsterdam New York, pp 205–206
Bradley MP, Suzuki N, Garbers DL (1984) In: Catt KJ, Dufau ML (eds) Hormone action and testicular function. Ann NY Acad Sci, vol 438, pp 142–155
Bradley MP, Forrester IT (1982) Arch Androl 9:78
Bragg PW, Handel MA (1979) Biol Reprod 20:333
Brandt H, Acott TS, Johnson DJ, Hoskins DD (1978) Biol Reprod 19:830
Brandt H, Brehmer L, Hoskins DD (1980) In: Steinberger A, Steinberger E (eds) Testicular development, structure and function. Raven, New York, pp 463–472
Branham JM (1970) J Reprod Fertil 22:469
Branson RE, Grimes SR, Yonuschot G, Irvin JL (1975) Arch Biochem Biophys 168:403
Bratanov K, Somlev B, Doycheva M, Tornyov A, Efremova V (1978) Int J Fertil 23:73
Bratanov K, Tornyov A, Efremova V, Somlev B (1980) Int J Fertil 25:67
Bratanov K, Mollova M, Somlev B, Efremova V (1983) Adv Exp Med Biol 156:1187
Breed WG (1983) Cell Tissue Res 229:611–625
Breed WG (1984) Gamete Res 10:31
Breitbart H, Rubinstein S (1982) Arch Androl 9:147
Breitbart H, Rubinstein S (1983) Biochim Biophys Acta 732:464

Breitbart H, Stern B, Rubinstein S (1983) Biochim Biophys Acta 728:349
Breitbart H, Stern B, Rubinstein S (1984a) Congr Proc 10th Int Congr Anim Reprod Artif Insem, vol II. Univ Ill, Urbana-Champaign, USA, p 37
Breitbart H, Rubinstein S, Nass-Arden L (1984b) Int J Androl 7:439
Bremner WJ, Matsumoto AM, Sussman AM, Paulsen CA (1981) J Clin Invest 68:1044
Breucker H, Schäfer E, Holstein AF (1985) Cell Tissue Res 240:303
Brezani P, Kalina I (1982) Folia Biol (Praha) 28:16
Bril-Petersen E, Westenbrink HGK (1963) Biochim Biophys Acta 76:152
Brock WA (1977) J Exp Zool 202:69
Broer K-H, Winkhaus I, Sombroek H, Kaiser R (1976) Int J Fertil 21:181
Broer K-H, Weber D, Kaiser R (1977) Fertil Steril 28:1077
Brokaw CJ (1975) In: Duckett JG, Racey PA (eds) The biology of the male gamete. Academic Press, London New York, pp 423–439
Brokaw CJ (1980) In: Steinberger A, Steinberger E (eds) Testicular development, structure and function. Raven, New York, pp 447–453
Brokaw CJ, Verdugo P (eds) (1982) Mechanisms and control of ciliary movement. Liss, New York
Brökelmann J (1963) Z Zellforsch Mikrosk Anat 59:820
Bronson RA, Cooper GW, Rosenfeld DL (1982a) Am J Reprod Immunol 2:222
Bronson RA, Cooper GW, Rosenfeld DL (1982b) Fertil Steril 38:724
Bronson RA, Cooper GW, Rosenfeld (1984) Fertil Steril 42:171
Brooks DE (1979a) J Reprod Fertil 56:667
Brooks DE (1979b) In: Fawcett DW, Bedford JM (eds) The spermatozoon. Urban & Schwarzenberg, Balitmore Munich, pp 23–34
Brooks DE (1983) Gamete Res 7:367
Brooks DE (1984) In: Lamming GE, Parkes A (eds) Marshall's physiology of reproduction, 4th edn, vol II, chap 6. Churchill Livingstone, London
Brooks DE, Tiver K (1983) J Reprod Fertil 69:851
Brooks DE, Tiver K (1984) J Reprod Fertil 71:249
Brooks DE, Hamilton DW, Mallek AH (1973) Biochem Biophys Res Commun 52:1354
Brooks JC, Siegel FC (1973) Biochem Biophys Res Commun 55:710
Brown CR (1975) J Reprod Fertil 45:537
Brown CR (1982) J Reprod Fertil 64:457
Brown CR (1983) J Reprod Fertil 69:289
Brown CR, Harrison RAP (1978) Biochim Biophys Acta 52:202
Brown CR, Hartree EF (1976a) J Reprod Fertil 46:249
Brown CR, Hartree EF (1976b) Hoppe-Seylers Z Physiol Chem 357:57
Brown CR, Andani Z, Hartree EF (1975) Biochem J 149:133
Brown CR, Glos von KI, Chulavatnatol M (1983a) J Reprod Fertil 67:275
Brown CR, Glos von KI, Jones R (1983b) J Cell Biol 96:256
Brown MA, Casillas ER (1983) Fed Proc Fed Am Soc Exp 42:1852
Brown SW, Chandra RS (1973) Proc Natl Acad Sci USA 70:195
Brown SW, Hartree EF (1974) J Reprod Fertil 36:195
Brunish R, Hogberg B (1968) C R Trav Lab Carlsberg 32:35
Bryan J (1974) Fed Proc Fed Am Soc Exp Biol 33:152
Bryan JHD, Umithan RR (1973) Histochemie 33:169
Bucci LR, Brock WA, Meistrich ML (1982) Expl Cell Res 140:111
Bullough WS (1975) Biol Rev 50:99
Bunge RG (1973) Urology 1:371
Burck PJ, Zimmerman RE (1980) J Reprod Fertil 58:121
Burgos C, Gerez de Burgos NM, Coronel CE, Blanco A (1979) J Reprod Fertil 55:101
Burgos C, Coronel CE, Burgos de NM, Rovai LE, Blanco A (1982) Biochem J 208:413
Burgos MH, Fawcett DW (1955) J Biophys Biochem Cytol 1:287
Burgos MH, Vitale-Calpe R, Aoki A (1970a) In: Johnson AD, Gomes WR, Vandermark NL (eds) The testis vol I, Academic Press, London New York, pp 451–649
Burgos MH, Sacerdote FL, Vitale-Calpe R, Beri D (1970b) In: Rosenberg E, Paulsen CA (eds) The human testis. Plenum, New York London, pp 369–379
Burkhart JG, Ansari AA, Malling HV (1982) Arch Androl 9:115

Burnasheva SA (1958) Biokhimiya 23:558
Busch H, Smetana K (1970) The nucleolus. Academic Press, London New York
Bustos-Obregon R (1975) Anat Rec 181:322 (Abstr)
Bustos-Obregon E, Courot M, Fléchon JE, Hochereau-de-Reviers MT, Holstein AF (1975) Andrologia 7:141
Byrd W (1981) J Exp Zool 215:35
Byskov AG (1978) Ann Biol Anim Biochem Biophys 18:327
Byskov AG (1983) Acta Endocrinol 103 Suppl 256:54 (Abstr)
Calamera JC, Brugo S, Vilar O (1982) Andrologia 14:239
Calvin HI (1975) In: Duckett JG, Pacey PA (eds) The biology of the male gamete. Academic Press, London New York, pp 257–273
Calvin HI (1976) Biochim Biophys Acta 434:377
Calvin HI (1979a) In: Fawcett DW, Bedford JM (eds) The spermatozoon. Urban & Schwarzenberg, Baltimore Munich, pp 387–389
Calvin HI (1979b) Biol Reprod 21:873
Calvin HI (1981) J Reprod Fertil 61:65
Calvin HI, Bedford JM (1970) In: Holstein AF, Horstmann E (eds) Morphological aspects of andrology. Grosse, Berlin, pp 77–80
Calvin HI, Bedford JM (1971) J Reprod Fertil (Suppl) 13:65
Calvin HI, Cooper GW (1979) In: Fawcett DW, Bedford JM (eds) The spermatozoon. Urban & Schwarzenberg, Baltimore Munich, pp 135–140
Calvin HI, Kosto B, Williams-Ashman HG (1967) Arch Biochem Biophys 118:670
Calvin HI, Yu CC, Bedford JM (1973) Exp Cell Res 81:333
Calvin HI, Hwang FHF, Wohlrab H (1975) Biol Reprod 13:228
Calvin HI, Cooper GW, Wallace E (1981) Gamete Res 4:139
Calvin J, Tubbs PK (1976) J Reprod Fertil 48:417
Calvin J, Tubbs PK (1978) Eur J Biochem 89:315
Cameron DF, Murray TF, Drylie DD (1985) Anat Rec 213:53
Campanella C, Gabbiani G, Baccetti B, Burrini AG, Pallini V (1979) J Submicrosc Cytol 11:53
Carr DW, Acott TS (1984) Biol Reprod 30:913
Carr I, Clegg EJ, Meek GA (1968) J Anat 102:501
Carranco A, Reyes R, Magdaleno VM, Huacuja L, Hernandez, O'Rosado A, Merchant H, Delgado NM (1983) Arch Androl 10:213
Carter AL, Hutson SM, Stratman FW, Hanning RV (1980) Biol Reprod 23:820
Casali E, Lindner L, Farruggia G, Spisni A, Masotti L (1983) IRCS Med Sci 11:528
Casali E, Farruggia G, Spisni A, Pasquali-Ronchetti I. Masotti L (1985) J Exp Zool 235:397
Cascieri M, Amann RP, Hammerstedt RH (1976) J Biol Chem 251:787
Cash JS, Mann T (1973) Proc R Soc London Ser B 184:109
Casillas ER (1972) Biochim Biophys Acta 280:545
Casillas ER, Chaipayungpan S (1979) J Reprod Fertil 56:439
Casillas ER, Chaipayungpan S (1982) J Reprod Fertil 65:247
Casillas ER, Elder CM, Hoskins DD (1980) J Reprod Fertil 59:297
Casillas ER, Erickson BJ (1975) Biol Reprod 12:275
Castellani-Ceresa L (1981) Arch Androl 7:361
Castellani-Ceresa L, Berruti G, Colombo R (1982) Biol Cell 45:156
Castellani-Ceresa L, Berruti G, Colombo R (1983) J Exp Zool 227:297
Castro AE, Seiguer AC, Mancini RE (1970) Proc Soc Exp Biol Med 133:582
Castro AE, Alonso A, Mancini RE (1972) J Endocrinol 52:129
Cates NR, Jozefowicz RH (1970) Res Commun Chem Pathol Pharmacol 1:223
Cavicchia JC, Dym M (1978) Biol Reprod 18:219
Cawood AH, Breckon G (1983) Mutat Res 122:149
Cechova D, Fritz H (1977) Andrologia 9:28A
Cechova D, Jonakova V, Sedlakova E, Mach O (1979a) Hoppe-Seylers Z Physiol Chem 360:1753
Cechova D, Jonakova V, Havranova M, Sedlakova E, Mach O (1979b) Hoppe-Seylers Z Physiol Chem 360:1759
Chakraborty J (1979) Gamete Res 2:25
Chakraborty J, Nelson L (1974) Biol Reprod 10:85

Chakraborty J, Nelson L (1976) Biol Reprod 15:579
Chakraborty J, Constaninou A, Mccorquodale M (1985) Anim Reprod Sci 9:101
Chan SY, Tang LC, Tang GW, Chan PH (1983) Contraception 28:481
Chandley AC (1978) Ann Biol Anim Biochem Biophys 18:359
Chang CH, Angellis D, Fishman WH (1980) Gamete Res 80:1506
Chang JP, Yokoyama M, Brinkley BR, Mayahara H (1974) Biol Reprod 11:601
Chang MC (1951) Nature (London) 168:697
Chang MC (1984) J Androl 5:45
Chang MC, Hunter RHF (1975) In: Hamilton DW, Greep RO (eds) Handbook of physiology, Sect 7: Endocrinology, vol V. Am Physiol Soc, Washington, DC, pp 339–352
Chang TSK, Zirkin BR (1978) J Exp Zool 204:283
Channing CP, Segal SJ (eds) (1982) Intraovarian control mechanisms. Plenum, New York
Chauhan RAS (1973) JNKVV Res J 6:40
Chauviere M (1977) Exp Cell Res 108:127
Chemes HE, Podesta E, Rivarola MA (1976a) Biol Reprod 14:332
Chemes HE, Rivarola MA, Bergada C (1976b) J Reprod Fertil 46:283
Chemes HE, Fawcett DW, Dym, M (1978) Anat Rec 192:493
Chemes HE, Dym M, Roy HGM (1979) Biol Reprod 21:241
Chen C-L, Mather JP, Morris PL, Bardin CW (1983) Abstr 8th Testis Worksh Hormone action and testicular function. Bethesda, MD, Abstr 59
Chen SYW (1985) Cell Biol 9:127
Cheng CY, Boettcher B (1982) Int J Androl 5:253
Cheng CY, Boettcher B, Rose RJ, Kay DJ, Tinneberg HR (1981a) Int J Androl 4:1
Cheng CY, Rose RJ, Boettcher B (1981b) Int J Androl 4:304
Cherr GN, Lambert H, Meizel S, Katz DF (1986) Dev Biol 114:1190
Chevalier M, Dufaure JP (1982)In: Bardin CW, Sherins RJ (eds) The cell biology of the testis. Ann N Y Acad Sci, vol 383, pp 433–434
Chinnaiya GP, Ganguli NC (1980) Zentralbl Veterinaermed 27:339
Chinnaiya GP, Kakar SS, Ganguli NC (1979) Zentralbl Veterinaermed R A 26:402
Chiu ML, Irvin JL (1978) Biol Reprod 19:984
Chiu ML, Irvin JL (1983) Biochim Biophys Acta 740:342
Chowdhury AK (1979) J Endocrinol 82:331
Chowdhury AK, Marshall G (1980) In: Steinberger A, Steinberger E (eds) Testicular development, structure and function. Raven, New York, pp 129–139
Chowdhury AK, Steinberger E (1975) Biol Reprod 12:609
Chowdhury AK, Steinberger E (1976) Annat Rec 185:155
Chowdhury AK, Tcholakian RK (1979) Steroids 34:151
Chowdhury AK, Steinberger A, Steinberger E (1975) Andrologia 7:297
Chubb C, Lopez MJ (1983) Biol Reprod 28 Suppl 1:102
Chulavatnatol M, Haesungcharrern A (1977) J Biol Chem 252:8088
Chulavatnatol M, Yindepit S (1976) J Reprod Fertil 48:91
Chulavatnatol M, Panyim S, Wititsuwannakul D (1982) Biol Reprod 26:197
Churg A, Zaneveld LJD, Schumacher GEB (1974) Biol Reprod 10:429
Clarke JR, Esnault C, Nicolle JC (1980) Reprod Nutr Dev 20:183
Clarke GN, Yanagimachi R (1978) J Exp Zool 205:125
Clarke GN, Clarke FM, Wilson S (1982) Biol Reprod 26:319
Clausen OPF, Purvis K, Hansson V (1977) Biol Reprod 17:555
Clegg ED (1983) In: Hartmann JF (ed) Mechanism and control of animal fertilization. Cell Biol Ser. Academic Press, London New York
Clegg ED, Morre DJ, Lunstra DD (1975) In: Duckett JG, Racey PA (eds) The biology of the male gamete. Academic Press, London New York, pp 321–335
Cleland KW, Rothschild L (1959) Proc R Soc London Ser B 150:24
Clermont Y (1956) J Biophys Biochem Cytol Suppl 2:119
Clermont Y (1962) Am J Anat 111:111
Clermont Y (1963) Am J Anat 112:35
Clermont Y (1967) Arch Anat Microsc Morphol Exp Suppl to 3 and 4:7
Clermont Y (1969) Am J Anat 126:57

Clermont Y (1972) Physiol Rev 52:198
Clermont Y, Antar M (1973) Am J Anat 136:153
Clermont Y, Bustos-Obregon E (1968) Am J Anat 122:237
Clermont Y, Harvey SC (1965) Endocrinology 76:80
Clermont Y, Harvey SC (1967) In: Wolstenholme GE, Connor MO (eds) Endocrinology of the testis. Churchill, London, pp 173–189
Clermont Y, Hermo L (1975) Am J Anat 142:159
Clermont Y, Leblond CP (1955) Am J Anat 96:229
Clermont Y, Mauger A (1974) Cell Tissue Kinet 7:165
Clermont Y, Mauger A (1976) Cell Tissue Kinet 9:99
Clermont Y, Rambourg A (1978) Am J Anat 151:191
Clermont Y, Clegg RE, Leblond CP (1955) Exp Cell Res 8:453
Clermont Y, Lalli M, Rambourg A (1981) Anat Rec 201:613
Coelingh JP, Rozijn RH (1975) In: Duckett JG, Racey PA (eds) The biology of the male gamete. Academic Press, London New York, pp 245–256
Coelingh JP, Rozijn TH, Montfoort CH (1969) Biochim Biophys Acta 188:353
Coelingh JP, Montfoort CH, Rozijn TH, Gevers Leuven JA, Schiphof R, Steynparve EP, Braunitzer G, Schrank B, Ruhfus S (1972) Biochim Biophys Acta 285:1
Cohn HY, Jacobs RD (1980) Biol Reprod 22 (Suppl 1):55A
Cole LJ, Waletzky E, Shakelford M (1940) J Heredt 31:501
Comhaire FH, Vermeulen A (1978) Ann Biol Anim Biochim Biophys 18:457
Comings DE, Okada T (1972) J Ultrastruct Res 39:15
Connell CJ (1977) Am J Anat 148:149
Connell CJ (1980) In: Steinberger A, Steinberger E (eds) Testicular development, structure and function. Raven, New York, pp 71–78
Connell CJ (1984) In: Catt KJ, Dufau ML (eds) Hormone action and testicular function. Ann NY Acad Sci, vol 438, pp 472–475
Connors EC, Greenslade FC, Davanzo JP (1974) Biol Reprod 9:57
Conti M, Geremia R, Monesi V (1979) Mol Cell Endocrinol 13:137
Cookson AD, Thomas AN, Foulkes JA (1984) J Reprod Fertil 70:599
Cooper GW, Bedford JM (1971) Anat Rec 169:300 (Abstr)
Cooper GW, Bedford JM (1976) J Cell Biol 69:415
Cooper TG, Brooks DE (1981) J Reprod Fertil 61:163
Cooper TG, Woolley DM (1982) J Exp Zool 223:291
Cornelish M, Schiphof R, Rozijn T, Steyn-Parve EP (1973) Biochim Biophys Acta 173:3221
Cornett LE, Meizel S (1978) Proc Natl Acad Sci USA 75:4954
Cornett LE, Meizel S (1979) Biol Reprod 20:925
Cornett LE, Meizel S (1980) J Histochem Cytochem 28:462
Coronel CE, Burgos C, Gerez de Burgos NM, Rovai LE, Blanco A (1983) J Exp Zool 225:379
Corson SL, Batzer FR, Alexander NJ, Schlaff S, Otis C (1984) Fertil Steril 42:756
Corteel JM, Paquignon M (1984) Congr Proc 10th Int Congr Anim Reprod Artif Insem, vol IV. Univ Ill Urbana-Champaign, USA, p II–20
Couchie J, Mann T (1957) Nature (London) 179:1190
Courot M (1970) In: Rosenberg E, Paulsen CA (eds) The human testis. Plenum, New York, pp 353–364
Courot M (1971) Thèse Doct Sci, Univ Paris
Courot M (1976) Andrologia 8:187
Courot M (1980) 9th Int Congr Anim Reprod Artif Insem, Madrid, vol II, pp 155–162
Courot M, Courtens JL (1982) In: Bardin CW, Sherins RJ (eds) The cell biology of the testis. Ann NY Acad Sci, vol 383, pp 338–339
Courot M, Fléchon J (1966) Ann Biol Anim Biochem Biophys 6:479
Courot M, Loir M (1968) VI Congr Int Reprod Anim Insem Artif, Paris, vol I, pp 125–127
Courot M, Ortavant R (1981) J Reprod Fertil Suppl 30:47
Courot M, Hocherau-de-Reviers MT, Ortavant R (1970) In: Johnson AD, Gomes WR, Vandemark NL (eds) The testis, vol I. Academic Press, London New York, pp 339–432
Courot M, Ortavant R, Reviers de M (1971) Exp Anim 4:201

Courot M, Hochereau-de-Reviers MT, Monet-Kuntz C, Locatelli A, Pisselet C, Blanc MR, Dacheaux JL (1979) J Reprod Fertil Suppl 26:165
Courot M, Hochereau-de-Reviers MT, Pisselet C, Kilgour RJ, Dubois MP, Sairam MR (1984) In: Courot M (ed) The male in farm animal reproduction. The Hague, pp 75–79
Courtens JL (1978a) J Ultrastruct Res 65:173
Courtens JL (1978b) J Ultrastruct Res 65:182
Courtens JL (1978c) Ann Biol Anim Biochim Biophys 18:1455
Courtens JL (1979a) Ann Biol Anim Biochim Biophys 19:989
Courtens JL (1979b) Ann Biol Anim Biochim Biophys 19:597
Courtens JL (1982a) Gamete Res 5:137
Courtens JL (1982b) Reprod Nutr Dev 22:825
Courtens JL (1984) Recent progress in cellular endocrinology of the testis, vol 123. INSERM, pp 57–84
Courtens JL, Courot M (1980) Anat Rec 197:143
Courtens JL, Fournier-Delpech S (1979) J Ultrastruct Res 68:136
Courtens JL, Loir M (1975a) J Microsc Biol Cell 24:249
Courtens JL, Loir M (1975b) J Microsc Biol Cell 24:259
Courtens JL, Loir M (1981a) J Ultrastruct Res 74:322
Courtens JL, Loir M (1981b) J Ultrastruct Res 74:327
Courtens JL, Loir M (1981c) Reprod Nutr Dev 21:467
Courtens JL, Courot M, Fléchon JK (1976) J Ultrastruct Res 57:54
Courtens JL, Amir D, Durand J (1980) J Ultrastruct Res 71:103
Courtens JL, Rozinek J, Fournier-Delpech S (1982) Andrologia 14:509
Courtens JL, Delaleu B, Dubois MP, Lanneau M, Loir M, Rozinek J (1983) Gamete Res 8:21
Courtens JL, Nunes JF, Corteel JM (1984) Gamete Res 9:287
Crabo BG, Zimmerman KJ (1976) VIII Congr Int Reprod Anim Insem Artif, Krako, vol IV, pp 698–700
Crabo BG, Zimmerman KJ, Moore R, Thornhurgh FH (1976) Biochem Biophys Acta 444:875
Crich J, Jequier AM (1978) Fertil Steril 30:572
Cruvellier P, Morales CR, Clermont Y (1985) Biol Reprod 32 (Suppl I):125
Cummins JM, Teichman RJ (1974) Biol Reprod 10:555
Cummins JM, Wooddall PF (1985) J Reprod Fertil 75:183
Cummins JM, Bernstein MH, Teichman RJ (1974) J Reprod Fertil 41:75
Cunningham GR, Huckins C (1979) Endocrinology 105:177
Cunningham GR, Tindall DJ, Huckins C, Means AR (1978) Endocrinology 102:16
Cunningham GR, Schill WB, Hafez ESE (eds) (1980) Regulation of male fertility. Nijhoff, The Hague
Currie RW, White FP (1981) Science 214:72
Cusan L, Gordeladze JO, Parvinen M, Clausen OPF, Hansson V (1981) Biol Reprod 25:925
Czaker R (1984) Experientia 40:960
Czaker R (1985a) Andrologia 17:42
Czaker R (1985b) Andrologia 17:547
Czaker R (1985c) J Ultrastruct Res 90:26
Czuppon AB, Mettler L, Schauer R, Pawassarat V (1981) Hoppe-Seylers Z Physiol Chem 362:963
Dacheux JL, Paquignon M (1980) In: Steinberger A, Steinberger E (eds) Testicular development, structure, function. Raven, New York, pp 513–521
Dacheux JL, Voglmayr JK (1983) Biol Reprod 29:1033
Dacheux JL, O'Shea TO, Paquignon M (1979) J Reprod Fertil 55:297
Dacheux JL, Paquignon M, Combarnous Y (1983) J Reprod Fertil 67:181
Dacheux JL, Paquignon M, Lanneau M (1984) In: Catt KJ, Dufau ML (eds) Hormone action and testicular function. Ann NY Acad Sci, vol 438, pp 526–529
Daentl DL, Erickson RP, Betlach CJ (1977) Differentiation 8:159
D'Agostino A, Geremia R, Monesi V (1976) Arch Ital Anat Embriol 81 Suppl:35
D'Agostino A, Geremia R, Monesi V (1978) Cell Differ 7:175
D'Agostino A, Monaco L, Stefanini M, Geremia R (1984) Exp Cell Res 150:430
Dalcq AM (1965) C R Assoc Anat 138:371
Dalcq AM (1967) C R H Acad Sci Ser D 264:2386

Dalcq AM (1973) Cytomorphologie normale du testicule et spermatogenèse chez les mammiféris. Mem Acad R Med Belg, vol 46, Bruxelles
Damme van MP, Robertson DM, Marana R, Ritzen EM. Diczfalsuy E (1980) In: Steinberger A, Steinberger E (eds) Testicular development, structure and function. Raven, New York, p 169
Danilova AV (1976) Izv Akad Nauk SSSR Ser Biol 2:281
Danzo BJ, Eller BD (1978) J Steroid Biochem 9:209
Daoust R, Clermont Y (1955) Am J Anat 96:255
Darin-Bennet A, Pulos A, White IG (1973) J Dairy Sci 40:1099
Darzynkiewicz Z, Gledhill BL, Ringertz NR (1969) Exp Cell Res 58:435
Das D, Sidhu NS (1975) J Reprod Fertil 44:333
Daugherty KA, Cockett ATK, Urry RL (1978) J Urol 120:425
David G, Boyce JA, Schwartz D (1977) J Reprod Fertil 50:377
David GFX, Koehler JK, Brown JA, Petra PH, Farr AG (1985) Biol Reprod 37:503
David JC, Vinson D, Loir M (1982) Exp Cell Res 141:357
Davies AG (1981) Arch Androl 7:97
Davies AG, Courot M, Gresham P (1974) J Endocrinol 60:37
Davis BK (1978) Proc Soc Exp Biol Med 157:54
Davis BK (1980) Arch Androl 5:249
Davis BK (1981) Proc Natl Acad Sci USA 78:7560
Davis BK (1982) Experientia 38:1063
Davis BK, Byrne R (1980) Arch Androl 5:263
Davis BK, Geregeley AF (1979) Biochem Biophys Res Commun 88:613
Davis BK, Byrne R, Hungund B (1979) Biochim Biophys Acta 558:2757
Davis BK, Byrne R, Bedigian KP (1980) Proc Natl Acad Sci USA 77:1546
Davis JR, Firlit CF (1965) Annu J Physiol 209:425
Davis JR, Firlit CF (1970) In: Rosemberg E, Paulsen CA (eds) The human testis. Plenum, New York, pp 315-332
Dawra RK, Sharma OP, Makkar HP (1983) Int J Fertil 28:231
Day-Francesconi M, Casillas ER (1982) Arch Biochem Biophys 215:206
Dean PN, Pinkel D, Mendelsohn ML (1978) Biophys J 23:7
Dedman JR, Means AR (1980) In: Steinberger E, Steinberger A (eds) Testicular development structure and function. Raven, New York, pp 315-321
Deenen van LLM, Gier de J, Demel RA (1972) In: Ganguly J, Smellie RMS (eds) Proc Int Symp Lipids, Bangalore. Academic Press, London New York, pp 377-382
Delgado NM, Huacuja RL, Pancarido RM, Merchant H, Rosado A (1976) Fertil Steril 27:413
Desclin J, Ortavant R (1963) Ann Biol Anim Biochem Biophys 3:329
Dhanottya RS, Srivastava RK (1979) Zentralbl Veterinaermed R 26:810
Diasio R, Glass R (1971) Fertil Steril 22:303
Dickmann Z (1964) J Exp Biol 41:177
Dickmann Z, Dziuk PJ (1964) J Exp Biol 41:603
Dietert SE (1966) J Morphol 120:317
Dietl T, Kruck J, Schill WB, Fritz H (1976) Hoppe-Seylers Z Physiol Chem 357:1333
Dietrich AJJ, Deboer P (1983) Genetica 61:119
Dietrich AJJ, Mulder RJP, Tates AD (1983) Genetica 61:113
Distel RJ, Kleene KC, Hecht NB (1984) Science 224:68
DiZerega GS, Sherins RJ (1981) In: Burger H, Kretser de D (eds) The testis. Raven, New York, pp 127-140
Dodgson KS, Spencer B, Wynn CH (1956) Biochem J 62:500
Dooher GB, Bennett D (1973) Am J Anat 136:339
Dooher GB, Bennett D (1974) J Embryol Exp Morphol 32:749
Dooher GB, Bennett D (1977) Biol Reprod 17:269
Dop CV, Hutson SM, Lardy HA (1978) Arch Biochem Biophys 187:235
Dop van C, Hutson SM, Lardy HA (1977) J Biol Chem 252:1303
Dorrington JH, Fritz IB (1975) Endocrinology 96:879
Dorrington JH, Fritz IB, Armstrong DT (1976) Mol Cell Endocrinol 6:117
Dorrington JH, Fritz IB, Armstrong DT (1978a) Int J Androl (Suppl) 2:53
Dorrington JH, Fritz IB, Armstrong DT (1978b) Biol Reprod 18:55
Dott HM (1973) Adv Reprod Physiol 6:231

Dott HM, Dingle JT (1968) Exp Cell Res 52:523
Dott HM, Harrison RAP, Foster GCA (1979) J Reprod Fertil 55:113
Dow PR, Davis JC, Gow JG, Wade AP, Rose ME (1985) J Steroid Biochem 22:371
Dravland JE, Joshi MS (1981) Biol Reprod 25:649
Dravland JE, Ilanos MN, Munn RJ, Meizel S (1984) J Exp Zool 232:117
Drevius LO (1968) Exp Cell Res 51:362
Drevius LO (1971) J Reprod Fertil 24:427
Dudenhausen E, Talbot P (1982) Gamete Res 6:257
Dudkiewicz AB (1982) Immunol Commun 11:59
Dudkiewicz AB (1983a) Biol Reprod 28 Suppl 1:156
Dudkiewicz AB (1983b) Gamete Res 8:183
Dudkiewicz AB, Powledge AG (1982) Andrologia 14:306
Dufau ML, Charreau EH, Catt KJ (1973) J Biol Chem 248:6973
Duijn van C (1961) Fertil Steril 12:509
Dumontier AJ, Sheridan WF (1977) Chromosoma (Berlin) 60:81
Dupressoir T, Sautière P, Lanneau M, Loir M (1985) Exp Cell Res 161:63
Dustin P (1984) Microtubules. Springer, Berlin Heidelberg New York
Duve de C (1963) In: Reuck de AVS, Cameron MP (eds) Ciba foundation symposium on lysosomes. Little Brown, Boston, pp 1–35
Dym M (1973) Anat Rec 175:639
Dym M (1977) In: Yates RD, Gordon M (eds) Male reproductive system. Masson, St. Paul, MN, pp 155–169
Dym M, Clermont Y (1970) Am J Anat 128:265
Dym M, Fawcett DW (1970) Biol Reprod 3:308
Dym M, Fawcett DW (1971) Biol Reprod 4:195
Dym M, Cavicchia JC (1977) Biol Reprod 17:390
Dym M, Madhwa Raj HG (1977) Biol Reprod 17:676
Dym M, Pladellorens M (1977) Anat Rec 187:571 (Abstr)
Dym M, Madhwa Raj HG, Chemes HE (1977) In: Troen Ph, Nankin HR (eds) The testis in normal and infertile men. Raven, New York, pp 97–124
Dziuk PJ, Dickmann Z (1965) J Exp Zool 158:237
Ebert L,. Freundl G, Hofmann N (1982) Andrologia 14:229
Eddy EM (1970) Biol Reprod 2:114
Eddy EM (1974) Anat Rec 178:731
Edelman GM (1976) Science 192:218
Edelman GM, Millette CF (1971) Proc Natl Acad Sci USA 68:2436
Edvinsson A, Heyden G, Steen Y, Nilsson S (1981) Int J Androl 4:297
Egbunike GN (1980) Int J Androl 3:459
Egbunike GN (1982) Andrologia 14:503
Egbunike GN, Elemo AO (1979) IRCS Med Sci Key Rep Cell Mol Biol 7:360
Egbunike GN, Elemo AO, Dede TI (1979) IRCS Med Sci Key Rep Cell Mol Biol 7:278
Eikvar L, Levy FO, Attramadal H, Jutte NHPM, Ritzen EM, Horn R, Hansson V (1984) In: Catt KJ, and Dufau ML (eds) Hormone action and testicular function. Ann N Y Acad Sci, vol 438, pp 563–565
Eisenberg F, Bolden AH (1964) Nature (London) 202:599
Elce JS, McIntyre EJ (1982) Can J Biochem 60:8
Elder FFB, Hsu TC (1981) Cytogenet Cell Genet 30:157
Elfvin LG (1968) J Ultrastruct Res 24:259
Elias PM, Friend DS, Goerke J (1979) J Histochem Cytochem 27:1247
Eliasson R, Virji N (1982) Acta Physiol Scand 118:24A
Elkington JSH, Fritz IB (1980) Endocrinology 107:970
Ellis LC (1970) In: Johnson AD, Gomes WR, Vandermark NL (eds) The testis, vol III. Academic Press, London New York, pp 333–376
Ellis LC, Cosentino MJ (1982) Archs Androl 9:81
Ellis DH, Hartman TD, and Moore HDM (1985) J Reprod Immunol 7:299
Enders GC, Friends DS (1985) Am J Anat 173:241
Enders GC, Zena W, Friend DS (1983) J Cell Sci 60:303

Erickson BH, Hall GG (1983) Mut Res 108:317
Erickson RP (1976) In: Edidin M, Johnson MH (eds) Immunobiology of gametes. Cambridge Univ Press, London New York, pp 85–114
Erickson RP (1978) Fed Proc 37:2517
Erickson RP, Friend DS, Tennenbaum D (1975a) Exp Cell Res 91:1
Erickson RP, Hoppe PC, Tennenbaum D, Spielmann H, Epstein CJ (1975b) Science 188:261
Erickson RP, Erickson JM, Betlack CJ, Meistrich ML (1980a) J Exp Zool 214:13
Erickson RP, Kramer JM, Rittenhouse J, Salkeld A (1980b) Proc Natl Acad Sci USA 77:6086
Erickson RP, Susan EL, Martin B (1980c) J Reprod Immunol 3:195
Ericsson RJ (1977) Andrologia 9:11
Ericsson RJ, Buthala DA, Norland JF (1971) Science 173:54
Ericsson RJ, Langevin CN, Nishino M (1973) Nature (London) 246:421
Esbenshade KL, Clegg ED (1980) Biol Reprod 23:530
Escalier D (1983) Biol Cell 48:65
Esnault C, Nicolle JC (1976) Ann Histochim 21:187
Esnault C, Courot M, Ortavant R (1974) In: Hafez ESE, Thibault C (eds) Transport, survie et pouvoir fecondant des spermatozoides chez les vertébres. INSERM Coll, Nouzilly, Nov 4–7, 1973
Espevik T, Elgsaeter A (1978) J Reprod Fertil 54:203
Esponda P, Avila J (1983) Eur J Cell Biol 30:313
Evans HJ (1972) In: Beatty RA, Gluecksohn-Waelsch S (eds) The genetics of the spermatozoon. Proc Int Symp. Edinburgh New York, pp 144–159
Evans HJ, Buckland RA, Pardue ML (1974) Chromosoma 48:405
Evans RW, Setchell BP (1978a) J Reprod Fertil 52:15
Evans RW, Setchell BP (1978b) J Reprod Fertil 53:357
Evans RW, Setchell BP (1979a) J Reprod Fertil 57:189
Evans RW, Setchell BP (1979b) J Reprod Fertil 57:197
Evans RW, Weaver DE, Clegg ED (1980) J Lipid Res 21:223
Evenson DP, Witkin SS, Harven de E, Bendich A (1978) J Ultrastruct Res 63:178
Ewing LL, Robaire B (1978) Annu Rev Pharmacol Toxicol 18:167
Ezzatollah K, Storey BT (1973) Biochim Biophys Acta 305:577
Faiman C, Winter JSD, Reyes FI (1981) In: Burger H. Kretser de D (eds) The testis. Raven, New York, pp 81–106
Fain-Maurel MA, Dadoune JP, Reger JF (1984) Anat Rec 208:375
Faltas S, Smith M, Stambaugh R (1975) Fertil Steril 26:1070
Farooqui AA (1983) Int J Biochem 15:463
Farooqui AA, Srivastava PN (1979a) Biochem J 181:331
Farooqui AA, Srivastava PN (1979b) Int J Biochem 10:745
Farriaux JP, Frontaine G (1967) Ann Biol Clin 25:815
Favard P (1969) In: Lima-De-Faria A (ed) Handbook of molecular cytology, chap 41. North-Holland Publ Comp, Amsterdam, pp 1130–1135
Fawcett DW (1958) Int Rev Cytol 7:195
Fawcett DW (1959) In: Rudnick D (ed) Developmental cytology. Ronald, New York, pp 161–189
Fawcett DW (1965) Z Zellforsch Mikrosk Anat 67:279
Fawcett DW (1970) Biol Reprod 2 (Suppl 2):90
Fawcett DW (1972) In: Beatty RA, Gluecksohn-Waelschs (eds) The genetics of the spermatozoon. Proc Int Symp, Edinburgh, Aug 16–20, 1971, pp 37–68
Fawcett DW (1974) In: Mancini RE, Martini L (eds) Male fertility and sterility. Academic Press, London New York, p 13
Fawcett DW (1975a) In: Greep RO (ed) Handbook of physiology, Sect 7, Endocrinology, vol V, Male reproductive system. Williams & Wilkins, Baltimore, pp 21–55
Fawcett DW (1975b) Dev Biol 44:394
Fawcett DW (1975c) In: Markert L, Papanconstantinou J (eds) The developmental biology of reproduction. Academic Press, New York, pp 25–53
Fawcett DW (1977) In: Birnkley BR, Porter KR (eds) International cell biology. Rockefeller Univ Press, New York, pp 581–601
Fawcett DW (1979) In: Alexander NJ (ed) Animal models for research on contraception and fertility. Harper & Row, New York, pp 84–104

Fawcett DW, Burgos MH (1956) Ciba Found Coll Ageing 2:66
Fawcett DW, Chemes HE (1979) Tissue Cell 11:147
Fawcett DW, Hollenberg RD (1963) Z Zellforsch Mikrosk Anat 60:276
Fawcett DW, Ito S (1958) J Biophys Biochim Cytol 4:135
Fawcett DW, Ito SA (1965) Am J Anat 116:567
Fawcett DW, Phillips DM (1967) J Cell Biol 35:152 A
Fawcett DW, Phillips DM (1969a) Anat Rec 165:153
Fawcett DW, Phillips DM (1969b) J Reprod Fertil Suppl 6:405
Fawcett DW, Phillips DM (1970) In: Baccetti B (ed) Comparative spermatology. Accad Naz Lincei, Rome, pp 13–29
Fawcett DW, Ito S, Slautterback D (1959) J Biophys Biochem Cytol 5:453
Fawcett DW, Eddy E, Phillips DM (1970) Biol Reprod 2:129
Fawcett DW, Anderson WA, Phillips DM (1971) Dev Biol 26:220
Feig LA, Bellvé AR, Horbach-Erickson N, Klagsbrun M (1980) Proc Natl Acad Sci USA 77:4774
Feig LA, Klagsbrun M, Bellvé A (1983) J Cell Biol 97:1435
Feinberg J, Weinman J, Weinman S, Walsh MP, Harricane MC, Gabrion J, Demaille JG (1981) Biochim Biophys Acta 673:303
Felici de M, McLaren A (1983) Exp Cell Res 144:417
Fellous M, Dausset J (1970) Nature (London) 225:191
Fellous M, Erickson RP, Gachelin G, Dubois P, Jacob F (1976a) Folia Biol (Praha) 22:381
Fellous M, Erickson RP, Gachelin G, Jacob F (1976b) Transplantation 22:441
Ficsor G, Ginsberg LC, Oldford GM, Snoke RE, Becker RW (1983) Fertil Steril 39:548
Fink E, Wendt V (1978) J Reprod Fertil 53:75
Fink E,. Schiessler H, Arnhold M, Fritz H (1972) Hoppe-Seylers Z Physiol Chem 353:1633
Finsy R, Peetermans J, Lekkerker H (1979) Biophys J 27:187
Fisher J, Bartoov B (1980) Arch Androl 4:157
Fisher-Fischbein J, Gagnon C, Bardin CW (1986) Int J Androl 8:403
Fléchon JE (1971) J Microsc 11:53
Fléchon JE (1973) Thèse 3. Cycle, Univ Paris
Fléchon JE (1974a) In: Hafez ESE, Thibault C (eds) Transport, survive, et pourvoir fecondant des spermatozoides chez les vertébrates. INSERM Coll, Nouzilly, Nov 4–7, 1973
Fléchon JE (1974b) J Microsc 19:59
Fléchon JE (1975) In: Hafez ESE, Thibault CC (eds) The biology of spermatozoa. Karger, Basel, pp 36–45
Fléchon JE (1976) VIII Congr Int Reprod Anim Insern Artif, Krako, vol IV, pp 892–895
Fléchon JE, Dubois MP (1975) C R Helid Seances Acad Sci 280:877
Fléchon JE, Huneau D, Brown CR, Harrison RAP (1977) Ann Biol Anim Biochem Biophys 17:749
Fleming AD, Armstrong DT (1985) J Exp Zool 233:93
Fleming AD, Kuehl TJ (1985) J Exp Zool 233:405
Fleming AD, Yanagimachi R (1984) J Exp Zool 229:485
Flickinger C, Fawcett DW (1967) Anat Rec 158:207
Florke S, Phi-Van L, Muller-Esterl W, Scheber HP, Engel W (1983) Differentiation 24:250
Florke-Gerloff S, Topfer-Petersen E, Muller-Esterl W, Schill WB, Engel W (1983) Human Genet 65:61
Fock-Nuzel R, Lottspeich F, Henschen A, Muller-Esterl W, Fritz H (1980) Hoppe-Seylers Z Physiol Chem 361:1823
Foote RN, Swierstra EE, Hunt WL (1972) Anat Rec 173:341
Ford WC (1981) Comp Biochem Physiol B Comp Biochem 68:289
Ford WC, Harrison A (1983) J Reprod Fertil 69:147
Ford WC, Jones AR (1983) Contraception 28:565
Forster MS, Smith WD, Lee WI, Berger RE, Karp LE, Stenchever MA (1983) Fertil Steril 40:655
Forte LR, Rylund DB, Zahler WL (1983) Mol Pharmacol 24:42
Foulkes JA (1977) J Reprod Fertil 49:277
Fouquet JP (1968) C R Acad Sci Ser D 267:545
Fournier-Delpech S, Courot M (1980) Biochem Biophys Res Commun 96:756
Fournier-Delpech S, Colas G, Courot M (1977) Ann Biol Anim Biochem Biophys 17:987
Fraioli F, Fabbri A, Gnessi L, Silvestroni L, Moretti C, Redi F, Isidori A (1984) In: Catt KJ, Dufau ML (eds) Hormone action and testicular function. Ann NY Acad Sci, vol 438, pp 365–370

Franchimont P, Croze F, Demoulin A, Bologne R, Hustin J (1981) Acta Endocrinol 98:312
Franke WW, Grund C, Fink A, Weber K, Jorkusch BM, Zentgraf H, Osborn M (1978a) Biol Cell 31:7
Franke WW, Schmid E, Osborn M, Weber K (1978b) Proc Natl Acad Sci USA 75:5034
Franke WW, Grund C, Schmidt E (1979) Eur J Cell Biol 19:269
Franklin LE (1968) Anat Rec 161:149
Franklin LE (1974) In: Afzelius BA (ed) The functional anatomy of the spermatozoon. Pergamon, Oxford, pp 89–95
Franklin LE, Barros C, Fussel EN (1970) Biol Reprod 3:180
Fraser LR (1979) J Reprod Fertil 57:377
Fraser LR (1982) J Androl 3:412
Fraser LR (1983a) J Reprod Fertil 69:419
Fraser LR (1983b) J Reprod Fertil 69:539
Fraser LR (1984) J Reprod Fertil 72:372
Fraser LR, Quinn PJ (1981) J Reprod Fertil 61:25
Fredricsson B, Waxegard G, Brege S, Lundberg I (1977) Fertil Steril 28:168
Free MJ, Schultze GA, Jaffe RA (1976) Biol Reprod 14:481
Frenkel G, Peterson RN, Davis JE, Freund M (1974) Fertil Steril 25:84
Freund M (1968) In: Behrman SJ, Kistner RW (eds) Progress in infertility. Little Brown, Boston, pp 593–627
Friend DS (1980) In: Gilula NB (ed) Membrane-membrane interactions. Raven, New York, pp 153–165
Friend DS (1982) J Cell Biol 93:243
Friend DS, Elias PM (1978) J Cell Biol 79:216a
Friend DS, Heuser JE (1981) Anat Rec 199:159
Friend DS, Fawcett DW (1973) J Cell Biol 59:105A
Friend DS, Fawcett DW (1974) J Cell Biol 63:641
Friend DS, Rudolf I (1974) J Cell Biol 63:466
Friend DS, Orci L, Perrelet A, Yanagimachi R (1977) J Cell Biol 74:561
Friend DS, Elias PM, Rudolf I (1979) In: Fawcett DW, Bedford JM (eds) The spermatozoon. Urban & Schwarzenberg, Baltimore Munich, pp 157–188
Friend GF (1936) Q J Microsc Sci 78:419
Fritz H, Förg-Brey B, Fink E, Meier M, Schiessler H, Schirren C (1972) Hoppe-Seylers Z Physiol Chem 353:1943
Fritz IB (1973) In: Horeker BL, Stadtman ER (eds) Current topics in cellular regulation, vol VII. Academic Press, London New York, pp 129:174
Fritz IB (1978) In: Litwack G (ed) Biochemical actions of hormones, vol V. Academic Press, London New York, p 249
Fritz IB, Rommerts FFG, Louis BG, Dorrington JR (1976) J Reprod Fertil 46:17
Fritz IB, Parvinen M, Karmally K, Lacroix M (1982) In: Bardin CW, Sherins RJ (ed) The cell biology of the testis. Ann NY Acad Sci, vol 383, pp 447:448
Froman DP, Amann RP, Riek PM, Olar TT (1984) J Reprod, Fertil 70:301
Fuhrmann GF, Granzer E, Bey E, Ruthenstroth-Bauer G (1963) Z Naturforsch 180:236
Fujimoto H, Erickson RP (1982) Biochem Biophys Res Commun 108:1369
Fulton C, Kane RE, Stephens EF (1971) J Cell Biol 50:762
Fuster CD, Farrell D, Stern FA, Hecht NB (1977) J Cell Biol 74:698
Gabel CA, Eddy EM, Shapiro BM (1979) J Cell Biol 82:742
Gachelin G, Fellous M, Guenet J-L, Jacob F (1976) Dev Biol 50:310
Gachelin G, Kemler R, Kelly F, Jacob F (1977) Dev Biol 57:199
Gaddum-Rosse P, Blandau RJ (1977) Am J Anat 149:423
Gagnon C, Sherins RJ, Mann T, Bardin CW, Amelar RD, Dubin L (1980) In: Steinberger A, Steinberger E (eds) Testis development, structure and function. Raven, New York, pp 491:495
Gagnon C, De Lamirande E, Belles-Isles M (1984) In: Catt KJ, Dufau ML (eds) Hormone action and testicular function. Ann NY Acad Sci, vol 438, pp 535–536
Galdieri M, Monaco L (1983) Cell Differ 13:49
Galdieri M, Monesi V (1974) Exp Cell Res 85:287
Galdieri M, Zani B, Stefanini M (1981) In: Frajese G, Hafez ESE, Fabbrini A, Conti C (eds) Oligospermia. Raven Press, New York, p 496

Galdieri M, Zani B, Monaco L (1982) In: Bardin CW, Sherins RJ (eds) The cell biology of the testis. Ann NY Acad Sci, vol 383, pp 449–450
Galdieri M, Zani BM, Monaco L, Ziparo E, Stefanini M (1983) Exp Cell Res 145:191
Gall WE, Millette CF, Edelman GM (1974) In: Diczfalusy E (ed) Immunological approaches to fertility control. Trans 7th Symp, Geneva, July 29–31, 1974, pp 154–172
Gandhi KK, Anand SR (1982) J Reprod Fertil 64:145
Garbers DL, Radany EW (1981) Adv Cyclic Nucleotide Res 14:241
Garbers DL, Wakabayashi T, Reed PW (1970) Biol Reprod 3:327
Garbers DL, First NL, Sullivan JJ, Lardy HA (1971a) Biol Reprod 5:336
Garbers DL, Lust WD, First NL, Lardy HA (1971b) Biochemistry 10:1825
Garbers DL, First NL, Lardy HA (1973a) J Biol Chem 248:875
Garbers DL, First NL, Gorman SK, Lardy HA (1973b) Biol Reprod 8:599
Garbers DL, Tubb DJ, Hyne RV (1982) J Biol Chem 257:6980
Gardner PJ (1966) Anat Rec 155:235
Garner DL (1984) Congr Proc 10th Int Congr Anim Reprod Artif Insem, vol IV. Univ Ill, Urbana-Champaign, USA, p X–9
Garner DL, Easton MP (1977) J Exp Zool 200:157
Garner DL, Easton MP, Munson ME, Doane MA (1975) J Exp Zool 191:127
Garner DL, Reamer SA, Johnson LA, Lessley BA (1977) J Exp Zool 201:209
Garner DL, Gledhill BL, Pinkel D, Lake S, Stephenson D, Dilla van MA, Johnson LA (1983) Biol Reprod 28:312
Gatenby JB, Beams HW (1935) QJ Microsc Sci 78:1
Gaunt SJ (1982) Dev Biol 89:92
Gaunt SJ (1983) J Embryol Exp Morphol 75:259
Gaunt SJ, Brown CR, Jones R (1983) Exp Cell Res 144:275
Gecse A, Ottlecz A, Torok L, Telegdy G, Morvay J, Sas M (1979) Arch Androl 2:311
Georgiev GH (1982) Cell Mol Biol 28:271
Geremia R, Galdieri M, D'Agostino A, Boitani C, Proietti F, Ferracin A, Monesi V (1976) Boll Zool 43:139
Geremia R, Boitani C, Conti M, Monesi V (1977a) Cell Differ 5:343
Geremia R, Boitani A, D'Agostino A, Ferracinconti M, Monesi V (1977b) Boll Zool 44:113
Geremia R, D'Agostino A, Monesi V (1978) Exp Cell Res 111:23
Geremia R, Boitani C, Conti M, Mocini D, Monesi V (1981) Cell Biol Int Rep 5:1071
Geremia R, Rossi P, Mocini D, Pezzotti R, Conti M (1984) Biochem J 217:693
Gerez de Burgos NM, Burgos C, Montamat EE, Moreno J, Blanco A (1978) Biochem Biophys Res Commun 81:644
Gerez de Burgos NM, Burgos CE, Coronel A, Bertarelli de Camusso A, Pigini T, Blanco A (1979) J Reprod Fertil 55:107
Gerton GL, Millette CF (1983) Fed Proc Fed Am Soc Exp 41:2051
Gerton GL, Millette CF (1984) J Cell Biol 98:619
Ghosal SK, Mukherjee BB (1971) Can J Genet Cytol 13:672
Gibbons BH (1977) In: Brinkley BR, Porter KR (eds) International cell biology. Rockefeller Univ Press, New York, pp 348–357
Gibbons BH (1979) In: Fawcett DW, Bedford JM (eds) The spermatozoon. Urban & Schwarzenberg, Baltimore Munich, pp 91–97
Gibbons IR, Fronk E (1972) J Cell Biol 54:365
Gibbons IR, Fronk E, Gibbons BH, Ogawa K (1976) In: Goldman RD, Pollard TD, Rosenbaum JL (eds) Cell motility, Book C. Cold Spring Harbor Lab, Cold spring Harbor, New York, pp 915–932
Gill SK, Guraya SS (1980) Indian J Exp Biol 18:1351
Gillis G, Peterson RN, Hook L, Russell L, Freund M (1978) Prep Biochem 8:363
Glätzer KH (1975) Chromosoma 53:371
Gledhill BL (1971) J Reprod Fertil Suppl 13:77
Gledhill BL (1975) In: Duckett JG, Racey PA (eds) The biology of the male gamete. Academic Press, London New York, pp 215–226
Gledhill BL (1983a) J Dairy Sci 66:2623
Gledhill BL (1983b) Fertil Steril 40:572

Gledhill BL (1984) In: Catt KJ, Dufau ML (eds) Hormone action and testicular function. Ann N Y Acad Sci, vol 438, pp 189–205
Gledhill BL (1986) Gamete Res 12:423
Gledhill BL, Cambell GL (1972) In: Thaer AA, Sernetz M (eds) Fluorescence techniques in cell biology. Springer, Berlin Heidelberg New York, p 151
Gledhill BL, Geldhill MP, Rigler R, Ringertz NR (1966) Exp Cell Res 41:652
Glover TD (1962) Int J Fertil 7:1
Gluecksohn-Waelsch S (1972) In: Beatty RA, Gluecksohn-Waelsch S (eds) Proc Int Symp Genet Spermatozoon, Edinburgh, pp 306–309
Go KJ, Wolf DP (1983) Adv Lipid Res 20:317
Go KJ, Wolf DP (1985) Biol Reprod 32:145
Go KJ, Mastroianni L Jr, Stambaugh R (1982) Gamete Res 4:15
Goeden HM, Zenich H (1985) J Exp Zool 233:247
Goh P, Hoskins DD (1986) Gamete Res 12:399
Gold B, Stern L, Bradley EM, Hecht NB (1983a) J Exp Zool 225:123
Gold B, Fujimoto H, Kramer JM, Erickson RP, Hecht NB (1983b) Dev Biol 98:392
Goldberg E, Wheat TE, Gonzales-Prevatt V (1983) In: Gill TJ III, Wegmann TG (eds) Reproductive immunology. Oxford Univ Press, New York, pp 492–504
Goldberg EH (1977) In: Rattazzi MC, Scandalios JC, Litt GS (eds) Isozymes: current topics in biological and medical research, vol I. Liss, New York, pp 79–124
Goldberg EH (1979) In: Talwar GP (ed) Recent advances in reproduction and regulation of fertility. Elsevier/North-Holland, Biomedical Press, Amsterdam New York, pp 281–290
Goldberg EH (1983) In: Zatuchni GI (ed) Research frontiers in fertility regulation, vol II, 6. North-Western Univ Press, Chicago, pp 1–11
Goldberg EH, Aoki T, Boyse EA, Bennett D (1970) Nature (London) 228:570
Goldberg EH, Boyse EA, Bennett D, Scheid M, Carswell EA (1971) Nature (London) 232:478
Goldberg EH, Sberna D, Wheat TE, Urbanski GJ, Morgoliash E (1977) Science 196:1010
Goldberg RB, Hawtrey C (1967) J Exp Zool 164:309
Goldberg RB, Wheat TE (1976) J Exp Zool 164:309
Goldberg RB, Wheat TE (1976) In: Spilman CH, Lobl TJ, Kirton KT (eds) Regulatory mechanism of male reproductive physiology. Elsevier, Amsterdam New York, p 133
Goldberg RB, Geremia R, Bruce WR (1977) Differentiation 7:167
Goldman-Leikin RE, Goldberg E (1983) Proc Natl Acad Sci USA 80:3774
Gomes WR (1970) In: Johnson AD, Gomes WR, Vandermark NL (eds) The testis, vol III. Academic Press, London New York, pp 483–554
Gondos B, Zemjanis R (1970) J Morphol 131:431
Gonzales LW, Meizel S (1973) Biophys Biochim Acta 320:166
Gonzalez-Echeverria FM, Cuasnicu PS, Blaquier JA (1982) J Reprod Fertil 64:1
Goodall J, Roberts AM (1976) J Reprod Fertil 48:433
Goodenough DA, Revel JP (1970) J Cell Biol 45:272
Goodpasture JC, Reddy JM, Zaneveld LJD (1981a) Biol Reprod 25:44
Goodpasture JC, Zavos PM, Cohen MR, Zaneveld LJD (1981b) J Reprod Fertil 63:397
Gordeladze JD, Cusan L, Abyholm T, Hansson V (1982a) Arch Androl 8:265
Gordeladze JD, Parvinen M, Clausen OPF, Hansson V (1982b) Arch Androl 8:43
Gordon M (1966) J Exp Zool 165:111
Gordon M (1969) J Reprod Fertil 19:367
Gordon M (1972a) Z Zellforsch Mikrosk Anat 131:15
Gordon M (1972b) J Ultrastruct Res 39:364
Gordon M (1973) J Exp Zool 185:111
Gordon M (1977) In: Yates RD, Gordon M (eds) Male reproductive system. Masson, New York
Gordon M, Barnett RJ (1967) Exp Cell Res 48:395
Gordon M, Benesch KG (1968) J Ultrastruct Res 24:33
Gordon M, Dandekar PV (1977) J Reprod Fertil 49:155
Gordon M, Dandekar PV, Bartoszewicz W (1974) J Reprod Fertil 36:211
Gordon M, Dandekar PV, Bartoszewicz W (1975) J Ultrastruct Res 50:199
Gordon M, Dandekar PV, Eager PR (1978) Anat Rec 191:123
Gorus FK, Finsy R, Pipeleers DG (1982) Am J Physiol 242:C304

Gospodarowicz D, Moran JS, Mescher AL (1978) In: Papaconstantinou J, Rutter WJ (eds) Molecular control of proliferation and differentiation. Academic Press, London New York, pp 33–63
Gould KG (1980) Int Rev Cytol 63:323
Gould SF, Bernstein MH (1973) J Cell Biol 59:119A
Gouranton J, Folliot R, Thomas D (1979) J Submicrosc Cytol 11:379
Graham EF, Pace MM (1967) Cryobiology 4:75
Graham EF, Crabo BG, Schmehl MKL (1973) Proc 8th Int Zootech Symp, Milan, p 95
Gravis CJ (1978) Anat Rec 190:406 (Abstr)
Gravis CJ (1979) Z Mikrosk Anat Forsch 93:321
Gray P, Fraken DR, Slabber CF, Potgieter GM (1981) Andrologia 13:127
Green DPL (1978a) J Cell Sci 32:137
Green DPL (1978b) J Cell Sci 32:153
Green DPL (1978c) J Cell Sci 32:165
Green DPL (1982) J Cell Sci 54:161
Green DPL, Hockaday AR (1978) J Cell Sci 32:177
Green JA, Dryden GL (1976) Biol Reprod 14:327
Greep RO, Fevold HL, Hisaw FL (1936) Anat Rec 65:261
Greesh M, Madan ML, Razdan MN (1977) Trop Agric 54:21
Grell RF, Oakberg EF, Generoso RE (1980) Proc Natl Acad Sci USA 77:6720
Grimes SR Jr, Henderson N (1983) Arch Biochem Biophys 221:108
Grimes SR Jr, Chae CB, Irvin JL (1975a) Biochem Biophys Res Commun 64:911
Grimes SR Jr, Platz RD, Meistrich M, Hnilica LS (1975b) Exp Cell Res 110:31
Grimes SR Jr, Chae C-B, Irvin JL (1975c) Arch Biochem Biophys 169:425
Grimes SR Jr, Meistrich ML, Platz RD, Hnilica LS (1977) Exp Cell Res 110:31
Grinsted J, Byskov AG, Andreasen MP (1979) J Reprod Fertil 56:653
Grippo P, Orlando P, Locorondo G, Geremia R (1982) Prog Clin Biol Res 85:389
Griswold MD, Mably ER, Fritz IB (1976) Mol Cell Endocrinol 4:139
Griswold MD, Solari A, Tung PS, Fritz IB (1977) Mol Cell Endocrinol 7:151
Grocock CA, Clarke JR (1975) J Reprod Fertil 43:461
Grogan WM, Lam JW (1982) Lipids 17:604
Grogan WM, Huth EG (1983) Lipids 18:275
Grogan WM, Farnham WF, Szoplak BA (1981) Lipids 16:401
Grootegoed JA, Grolle-Hey AH, Rommerts FFG, Molen van der HJ (1977a) Biochem J 168:23
Grootegoed JA, Peters MJ, Mulder E, Rommerts FFG. Molen van der HJ (1977b) Mol Cell Endocrinol 9:159
Grootegoed JA, Jutte NH, Rommerts FF, Molen van der HJ, Ohno S (1982a) Exp Cell Res 139:472
Grootegoed JA, Jutte NH, Rommerts FF, Molen van der HJ (1982b) In: Bardin CW, Sherins RJ (eds) The cell biology of the testis. Ann NY Acad Sci, vol 383, pp 454–455
Grootegoed JR, Jansen R, Molen van der HJ (1984) In: Catt KJ, Dufau ML (eds) Hormone action and testicular function. Ann NY Acad Sci, vol 438, pp 557–560
Guay P, Rondeau M, Boucher S (1981) Equine Vet JO:1977
Guerin JF, Czyba JC (1976) C R Seances Soc Biol 170:395
Guerin JF, Czyba JC (1977a) C R Seances Soc Biol 171:370
Guerin JF, Czyba JC (1977b) C R Seances Soc Biol 171:822
Guerin JF, Czyba JC, Souchier C (1976) C R Seances Soc Biol 170:1218
Guerin JF, Menezo Y, Czyba JC (1979) Cryobiology 16:443
Guerin JF, Czyba JC, Menezo Y (1980) Bull Assoc Anat 64:547
Gunn SA, Gould TC (1970) In: Johnson AD, Gomes WR, Vandemark NL (eds) The testis, vol III. Academic Press, London New York, pp 377–481
Gupta RC, Srivastava RK (1981) Cell Mol Biol 27:539
Gupta BC, Sidhu KS, Sundhey R, Guraya SS (1985) Buffalo Bull 4:32
Guraya SS (1962) Experientia 18:167
Guraya SS (1963) Experientia 19:94
Guraya SS (1965) Cellule 65:367
Guraya SS (1968) Acta Morphol, Neerl Scand 7:15
Guraya SS (1971a) Z Zellforsch Mikrosk Anat 114:321
Guraya SS (1971b) Acta Anat 79:120

Guraya SS (1973) Acta Anat 84:552
Guraya SS (1974) Int Rev Cytol 37:121
Guraya SS (1980) Int Rev Cytol 62:187
Guraya SS (1985) Biology of ovarian follicles. Springer, Berlin Heidelberg New York
Guraya SS, Bilaspuri GS (1976a) Gegenbaurs Morphol. Jahrb (Leipzig) 122:147
Guraya SS, Bilaspuri GS (1976b) Ann Biol Anim Biochem Biophys 16:137
Guraya SS, Bilaspuri GS (1976c) Indian J Anim Sci 8:388
Guraya SS, Gill SK (1977) Indian J Exp Biol 15:1036
Guraya SS, Gill SK (1978) Andrologia 10:278
Guraya SS, Sidhu KS (1975a) Acta Histochem 54:307
Guraya SS, Sidhu KS (1975b) J Anim Sci 45:923
Guraya SS, Sidhu KS (1975c) Indian J Anim Sci 45:923
Guraya SS, Sidhu KS (1976) Proc VIII Int Congr Anim Reprod Artif Insem, Krako, vol I, pp 900–903
Guraya SS, Sidhu KS (1986) Indo-USA Worksh Development of pre-implantation embryos, New Delhi, March 3–14, 1986. ICMR, pp 3–4
Gutierrez M, Burgos de NMG, Burgos C, Blanco A (1972) Comp Biochem Physiol (A) 43:47
Gwatkin RBL (ed) (1976) Fertilization mechanisms in man and mammals. Plenum, New York
Gwatkin, RBL (1979) In: Talwar GP (ed) Recent advances in reproduction and regulation of fertility. Elsevier/North-Holland Biomedical Press, Amsterdam New York, pp 115–122
Haas GG, Weisswik R, Wolf DP (1982) Fertil Steril 38:54
Hadarag E, Buruiana LM, Coltau G, Barbulescu I (1976) VIII Congr Int Reprod Anim Insem Artif, Krako, vol IV, pp 802:804
Hadek R (1963a) J Ultrastruct Res 9:10
Hadek R (1963b) J Ultrastruct Res 8:161
Hadek R (1965) Int Rev Cytol 18:29
Hadley MA, Dym M (1983) Anat Rec 205:381
Hadley MA, Byers SW, Suarez-Quian CA, Kleinman HK, Dym M (1985) J Cell Biol 10:1511
Haesungcharrern A, Chulavatnatol M (1973) Fertil Steril 24:662
Hagopian A, Jackson J, Carlo DJ, Limjuco GR, Eylar EH (1975) J Immunol 155:1731
Hagopian A, Limjuco GA, Jackson JJ, Carlo DJ, Eylar EH (1976) Biochim Biophys Acta 434:354
Halangk W, Bohnensack R, Kunz W (1980) Acta Biol Med Ger 39:791
Halangk W, Frank K, Bohnensack R (1982) Acta Biol Med Ger 41:715
Hall PF, Mita M (1984) Biol Reprod 31:863
Hall PF, Nakamura M (1981) J Reprod Fertil 63:373
Hammerberg C, Klein J (1975) Nature (London) 253:137
Hammerstedt RH (1979) In: Fawcett DW, Bedford JM (eds) The spermatozoon. Urban & Schwarzenberg, Baltimore Munich, pp 205–210
Hammerstedt RH, Amann RP (1976a) Biol Reprod 15:678
Hammerstedt RH, Amann RP (1976b) Biol Reprod 15:686
Hammerstedt RH, Hay SR (1980) Arch Biochem Biophys 199:427
Hammerstedt RH, Amann RP, Rucinsky T, Morse II PD, Lepock J, Snipes LW, Keith AD (1976) Biol Reprod 14:381
Hammerstedt RH, Keith AD, Snipes W, Amann RP, Arruda D, Griel Jr LC (1978) Biol Reprod 18:686
Hammerstedt RH, Keith AD, Hay S, Deluca N, Amann RP (1979) Arch Biochem Biophys 196:7
Hancock JL (1952) J Exp Biol 29:445
Hancock JL (1957) J R Microsc Soc 76:94
Hancock JL (1966) Adv Reprod Physiol 1:125
Hancock JL, Tevan DJ (1957) J R Microsc Soc 76:77
Handel MA (1979) J Embryol Exp Morphol 51:73
Handrow RR, Boehm SK, Lenz RW, Robinson JA, Ax RL (1984) J Androl 5:51
Haneji T, Nishimune Y (1982) Endocrinology 94:43
Haneji T, Maekawa M, Yoshitake N (1982) Biochem Biophys Res Commun 108:132
Haneji T, Maekawa M, Nishimune Y (1983) J Nutr 113:1119
Haneji T, Maekawa M, Nishimune Y (1984) Endocrinology 114:801
Hanski E, Garty NB (1983) FEBS Lett 162:447
Hansson V, Ritzen EM, French FS, Nayfeh SN (1975) In: Greep RO, Astwood EB (eds) Handbook of physiology, Sect 7: Endocrinology, vol V. Am Physiol Soc, Washington, DC, pp 173–202

Harding HR, Carrick FN, Shorey CD (1979) In: Fawcett DW, Bedford JM (eds) The spermatozoon. Urban & Schwarzenberg, Baltimore Munich, pp 289–303

Hardman JG, Robinson GA, Sutherland FW (1971) Annu Rev Physiol 33:311

Harper HA (1973) Review of physiological chemistry. Lange, Los Altos, California

Harris ME, Bartke A, Weisz J, Watson D (1977) Fertil Steril 28:113

Harris WF (1976) S Afr J Sci 72:82

Harrison RAP (1971) Biochem J 124:741

Harrison RAP (1975) In: Duckett JG, Racey PA (eds) The biology of the male gamete. Academic Press, London New York, pp 301–316

Harrison RAP (1982) J Reprod Immunol 4:231

Harrison RAP (1983) In: Andre J (ed) The sperm cell: fertilizing power, surface properties, motility, nucleus and acrosome evolutionary aspects. Nijhoff, The Hague

Harrison RAP, Brown CR (1979) Gamete Res 2:75

Harrison RAP, Fléchon JE (1980) Reprod Nutr Dev 20:1801

Harrison RAP, White IG (1969) Biochem J 111:36P

Harrison RAP, White IG (1972) J Reprod Fertil 30:105

Harrison RAP, Dott HM, Foster GC (1978) J Reprod Fertil 52:65

Harrison RG (1975) In: Hamilton DW, Greep RO (eds) Handbook of physiology, Sect 7: Endocrinology, vol V. Am Physiol Soc, Washington DC, pp 219–224

Hartree EF (1971) In: Monoel J, Borek E (eds) Of microbes and life. Columbia Univ Press, New York, pp 271–303

Hartree EF (1977) Biochem Soc Trans 5:375

Hartree EF, Mann T (1961) J Endocrinol 34:247

Hartree EF, Srivastava PN (1965) J Reprod Fertil 9:47

Hathaway RR, Hartree EF (1963) J Reprod Fertil 5:225

Hawtrey C, Goldberg E (1968) Ann NY Acad Sci 151:611

Heath E, Jeyendran RS, Graham EF (1983) Anat Histol Embryol 12:245

Hecht NB, Williams JL (1978) Biol Reprod 19:573

Hecht NB, Farrell D, Williams JL (1979) Biochim Biophys Acta 561:358

Hedger MP, Robertson DM, Browne CA, Kretser de DM (1984) In: Catt KJ, Dufau ML (eds) Hormone action and testicular function. Ann NY Acad Sci, vol 438, pp 371–381

Hedrick JL, Wardrip NJ, Urich UA (1985) Dev Growth Differ 27:175

Heffner LJ, Saling PM, Storey BT (1980) J Exp Zool 212:53

Hegde UC, Shastry PR, Rao SS (1978) J Reprod Fertil 53:403

Heindel JJ, Berkowitz AS, Bartke A (1984) In: Catt KJ, Dufau ML (eds) Hormone action and testicular function. Ann NY Acad Sci, vol 438, pp 569–571

Hekman S, Rumke P (1969) Fertil Steril 20:312

Heller CG, Clermont Y (1963) Science 140:184

Heller CG, Clermont Y (1964) Recent Progr Hormone Res 20:545

Henderson AS (1963) Nature (London) 200:1235

Henderson AS (1964) Chromosoma 15:345

Henderson AS, Warburton D, Atwood KC (1972) Proc Natl Acad Sci USA 69:3394

Henderson AS, Eicher EM, Yu MT, Atwood KC (1974) Chromosoma 49:155

Hendrikse J, Holst v. d. W, Best AP (1976) VIII Congr Int Reprod Anim Insem Artif, Krako, p 106 (Abstr)

Henle W, Henle G, Chambers LA (1938) J Exp Med 68:335

Henning W (1967) Chromosoma 22:294

Henning W, Meyer GF, Henning I, Leoncini O (1974) Cold Spring Harber Symp Quant Biol 38:673

Henriet L, Horuat F, Lehoucq JC (1976) VIII Congr Int Reprod Anim Insem Artif, Krako, p 107 (Abstr)

Herak M, Herak M, Sukalić M, Premzl B, Emanović D (1976) VIII Congr Int Reprod Anim Insem Artif, Krako, vol IV, pp 904–907

Herman CA, Azhler WL, Doak GA, Cambell BJ (1976) Arch Biochem Biophys 177:622

Hermann J, Keil B (1981) Biochim Biophys Acta 643:30

Hermo L, Lalli M (1978) Anat Rec 190:420 (Abstr)

Hermo L, Clermont Y, Rambourg A (1979) Anat Rec 193:243

Hermo L, Rambourg A, Clermont Y (1980) Am J Anat 157:357

Herz Z, Northey D, Lawyer M, First NL (1985) Biol Reprod 32:1163
Hicks JJ, Martinez-Mannautou J, Pedron N, Rosado A (1972) Fertil Steril 23:172
Hillman N, Nadijcka M (1978a) J Embryol Exp Morphol 44:243
Hillman N, Nadijcka M (1978b) J Embryol Exp Morphol 44:263
Hilscher B, Hilscher W, Maurer W (1969) Z Zellforsch Mikrosk Anat 94:593
Hilscher B, Maurer W, Wichmann HE, Hilscher W (1985) Acta Histochemica, Suppl 31:139
Hilscher W (1964) Beitr Pathol Anat Allg Pathol 130:69
Hilscher W (1967) Arch Anat Microsc 56 (Suppl 3–4):75
Hilscher W (1983) Bibliotheca Anat 24:1
Hilscher W, Hilscher B (1969) Z Zellforsch Mikrosk Anat 96:625
Hilscher W, Grover R, Adolf B, Hilscher B (1985) Bas Appl Histochem 29:245
Hilscher W, Makoski HB (1968) Z Zellforsch Mikrosk Anat 86:327
Hinrichsen MJ, Blaquier JA (1980) J Reprod Fertil 60:291
Hinrichsen-Kohane AC, Hinrichsen MJ, Schill W-B (1985) Fertil Steril 43:279
Hinton BT, Setchell BP (1980) J Androl 1:831 (Abstr)
Hinton BT, Brooks DE, Dott HM, Setchell BP (1981) J Reprod Fertil 61:59
Hintz M, Goldberg E (1977) Dev Biol 57:375
Hipakka RA, Hammerstedt RH (1978a) Biol Reprod 19:1030
Hipakka RA, Hammerstedt RH (1978b) Biol Reprod 19:368
Hirschel MD, Isahakia MA, Alexander NJ (1984) In: Catt KJ, Dufau ML (eds) Hormone action and testicular function. Ann NY Acad Sci, vol 438, pp 508–511
Hjort T, Brogaard K (1971) Clin Exp Immunol 8:9
Hjort T, Poulsen F (1981) J Clin Lab Immunol 6:61
Hochereau MT (1967) Arch Anat Microsc Morphol Exp 56 (Suppl 3–4):85
Hochereau-de-Reviers MT (1970) D Sci Thesis, Faculté Sci Univ Paris
Hochereau-de-Reviers MT (1976) Andrologia 8:137
Hochereau-de-Reviers MT (1981) In: Mckerns KW (ed) Reproductive processes and contraception. Plenum, New York, pp 307–331
Hochereau-de-Reviers MT, Courot M (1978) Ann Biol Anim Biochim Biophys 18:573
Hochereau-de-Reviers MT, Loir M, Pelletier J (1976) J Reprod Fertil 46:203
Hochereau-de-Reviers MT, Blanc MR, Courot M, Garnier DH, Pelletier J, Poirier JC (1980) In: Steinberger E, Steinberger A (eds) Testicular development, structure and function. Raven, New York, pp 237–247
Hochereau-de-Reviers MT, Bindon BM, Courot M, Lafortune E, Land RB, Lincoln GM, Ricordeau G (1984a) In: Courot M (ed) The male in farm animal reproduction. Nijhoff, The Hague, pp 69–74
Hochereau-de-Reviers MT, Land RB, Perreau C, Thompson R (1984b) J Reprod Fertil 70:157
Hochereau-de-Reviers MT, Perreau C, Lincoln GA (1985) J Reprod Fertil 74:329
Hofgartner FJ, Krone W, Jain K (1979a) Hum Genet 47:329
Hofgartner FJ, Schmid M, Krone W, Zenzes MT, Engel W (1979b) Chromosoma 71:197
Hofmann N (1981) Andrologia 13:265
Holland MK, Storey BT (1981) Biochem J 198:273
Holland MK, White IG (1980) Fertil Steril 34:483
Holland MK, Alvarez JG, Storey BT (1982) Biol Reprod 27:1109
Holm PB, Rasmussen SW (1977a) Carlsberg Res Commun 42:282
Holm PB, Rasmussen SW (1977b) In: Chapelle de la A, Sorsa M (eds) Chromosomes today. Elsevier/North Holland Biomedical Press, Amsterdam New York, pp 83–93
Holm PB, Rasmussen SW, Wettstein von D (1982) Mutat Res 95:45
Holmes RS, Cooper DW, Vandeberg JL (1973) Endocrinology 113:1916
Holmes SD, Bucci LR, Lipschultz LI, Smith RG (1983) Endocrinology 113:1916
Holstein AF (1976) Andrologia 8:157
Holstein AF, Roosen-Runge EC (1981) Atlas of human spermatogenesis. Grosse, Berlin
Holstein AF, Schäfer E (1978) Cell Tissue Res 192:359
Holstein AF, Schirren C (1979) In: Fawcett DW, Bedford JM (eds) The spermatozoon. Urban & Schwarzenberg, Baltimore Munich, pp 341–353
Holstein AF, Bustos-Obregon E, Hartmann M (1984) Cell Tissue Res 236:35
Holt WV (1979) J Ultrastruct Res 68:58
Holt WV (1980) Biol Reprod 23:847

Holt WV (1984) Int Rev Cytol 87:159
Holt WV, Moore HD (1984) J Anat 138:175
Holt WV, North RD (1985) J Reprod Fertil 73:285
Holwill ME (1977) Adv Microb Physiol 16:1
Holzmann A, Strahl H, Jabn J, Bamberg E (1978) Zuchthygiene 13:21
Homonnai ZT, Pas G, Sofer A, Kraicer PF, Harell A (1976) Int J Fertil 21:163
Hood RD, Foley CW, Mastin TG (1970) J Anim Sci 30:91
Hoppe PC (1976) Biol Reprod 15:39
Horowitz JA, Toeg H, Orr GA (1984) J Biol Chem 259:832
Horstmann E (1961) Z Zellforsch Mikrosk Anat 54:68
Horstmann E (1970) In: Holstein AF, Horstmann E (eds) Morphological aspects of andrology, vol I. Grosse, Berlin, pp 24–28
Hoskins DD (1973) J Biol Chem 248:1135
Hoskins DD, Acott TS (1976) V Int Congr Endocrinol, Hamburg, July 18–24, 1976, p 235
Hoskins DD, Casillas ER (1975a) In: Thomas JA, Singhal RL (eds) Molecular mechanisms of gonadal hormone action. Advances in sex hormone research, vol I. Univ Park Press, Baltimore, pp 283–329
Hoskins DD, Casillas ER (1975b) In: Hamilton DW, Greep RO (eds) Handbook of physiology, Sect 7: Endocrinology, vol V. Am Physiol Soc, Washington DC, pp 453–460
Hoskins DD, Casillas ER, Stephens DT (1972) Biochem Biophys Res Commun 48:1331
Hoskins DD, Brandt H, Acott TS (1978) Fed Proc Fed Am Soc Exp 37:2534
Hoskins DD, Johnson D, Brandt H, Acott TS (1979) In: Fawcett DW, Bedford JM (eds) The spermatozoon. Urban & Schwarzenberg, Baltimore Munich, pp 43–53
Hotta Y, Ito M, Stern H (1966) Proc Natl Acad Sci 56:1184
Hotta Y, Chandley AC, Stern H (1977) Chromosoma 62:255
Howards S, Lechene C, Vigersky R (1979) In: Fawcett DW, Bedford JM (eds) The spermatozoon. Urban & Schwarzenberg, Baltimore Munich, pp 35–41
Hruban Z, Martan J, Aschenbrenner I (1971) J Morphol 135:87
Hrudka F (1968a) Folia Morphol (Praha) 16:48
Hrudka F (1968b) Folia Morphol (Praha) 16:60
Hrudka F (1978) J Ultrastruct Res 63:1
Huacuja L, Delgado NM, Merchant H, Pancarde RM, Rosado A (1977) Biol Reprod 17:89
Huacuja L, Delgado NM, Caizada L, Wens A, Reyes R, Pedron N, Rosado A (1981) Arch Androl 7:343
Huang HFS, Hembree WC (1979) Biol Reprod 21:891
Huang HFS, Nieschlag E (1984) J Reprod Fertil 70:31
Huang TTF Jr, Hardy D, Yanagimachi H, Teuscher C, Tung K, Wild G, Yanagimachi R (1985) Biol Reprod 32:45
Huckins C (1971a) Cell Tissue Kinet 4:139
Huckins C (1971b) Anat Rec 169:533
Huckins C (1971c) Cell Tissue Kinet 4:313
Huckins C (1971d) Cell Tissue Kinet 4:335
Huckins C (1978a) Am J Anat 153:97
Huckins C (1978b) Biol Reprod 19:747
Huckins C (1978c) Anat Rec 190:905
Huckins C, Oakberg EF (1978a) Anat Rec 192:519
Huckins C, Oakberg EF (1978b) Anat Rec 192:529
Huckins C, Mills N, Besch P, Means AR (1973) Endocrinology (Suppl 92):94
Huckle WR, Tamblyn TM (1983) Arch Biochem Biophys 226:94
Huggenvik J, Sylvester SR, Griswold MD (1984) In: Catt KJ, Dufau ML (eds) Hormone action and testicular function. Ann NY Acad Sci, vol 438, pp 1–7
Hughes RL (1976) VIII Congr Int Reprod Anim Insem Artif, Krako. vol IV, pp 908–910
Hundeau D, Harrison RAP, Fléchon JE (1984) Gamete Res 9:425
Hunt DM, Johnson DR (1971) J Embryol Exp Morphol 26:111
Hunt WP, Peterson RN, Saxena NK, Saxena N, Arthur R, Russell LD (1985) Prep Biochem 15:9
Hunter RHF, Holtz W, Henfrey PJ (1976) J Reprod Fertil 46:463
Hutson SM, Dop van C, Lardy HA (1977a) J Biol Chem 252:1309
Hutson SM, Dop van C, Lardy HA (1977b) Arch Biochem Biophys 181:345

Hyne RV (1984) Biol Reprod 31:312
Hyne RV, Boettcher B (1978) Fertil Steril 30:322
Hyne RV, Garbers DL (1979) Proc Natl Acad Sci USA 76:5699
Hyne RV, Lopata A (1982) Gamete Res 6:81
Hyne RV, Murdoch RN, Boettcher B (1978) J Reprod Fertil 53:315
Hyne RV, Higginson RE, Kohlman D, Lopata A (1984) J Reprod Fertil 70:83
Ibrahim MA, Kovacs L (1982) Acta Vet Acad Sci Hung 30:243
Ibrahim MA, Kovacs L, Toth BL (1982) Acta Vet Acad Sci Hung 30:235
Ichikawa S, Takehara Y, Shibata T, Tamada H, Oda K, Murro S (1984) J Reprod Fertil 72:515
Ierardi LA, Moss SB, Bellvé AR (1983) J Cell Biol 96:1717
Illison L (1968) Ph D Thesis, Univ Sydney, Aust
Irons MJ (1980) Ph D Thesis, McGill Univ, Montreal
Irons MJ (1983) J Ultrastruct Res 82:27
Irons MJ, Clermont Y (1979) Cell Tissue Kinet 12:425
Irons MJ, Clermont Y (1982a) Anat Rec 202:463
Irons MJ, Clermont Y (1982b) Am J Anat 165:121
Ishihara M, Ishihara K, Matsumoto Y, Sasaki T (1983) Acta Endocrinol 103 Suppl 256:99 (Abstr)
Ishii N, Mitsukawa S, Shirai M (1977) Andrologia 9:55
Isosefi S, Lewin LM (1984) Andrologia 16:509
Itagaki G, Takahashi M (1977) Zool Mag (Tokyo) 86:85
Ito M, Amano H (1980) Jpn J Fertil Steril 25:242
Iversen OH (1981) In: Baserger R (ed) Handbook of experimental pharmacology, vol 57. Born GVR, Farah A, Herken H, Welch AD (eds) Tissue growth factors. Springer, Berlin Heidelberg New York, pp 491–550
Iype PT, Abraham KA, Bhargava PM (1963) J Reprod Fertil 5:151
Jackson JJ, Hogpian A, Carlo DJ, Limjuco GA, Eylar EH (1975) J Biol Chem 250:6141
Jackson JJ, Hagopian A, Carlo DJ, Limjuco GA, Eylar EH (1976) Biochim Biophys Acta 427:251
Jacob F (1977) Immunol Rev 33:3
Jager S, Kremer J, Slochteren-Draaisma TV (1976) Int J Fertil 23:12
Jagiello G, Tantravahi U, Fang JS, Erlanger BF (1982) Exp Cell Res 141:253
Jagiello G, Sung WK, Vant-Hof J (1983) Exp Cell Res 146:281
Jain YC, Anand SR (1975) J Reprod Fertil 42:129
Jain YC, Anand SR (1976) J Reprod Fertil 47:261
James MD, Clegg ED, Lunstra B, Mollenhauer HH (1974) Proc Soc Exp Biol Med 154:1
Jamil K (1982) Arch Androl 9:195
Jamil K, White IG (1981) Arch Androl 7:283
Jean P, Hartung M, Mirre C, Stahl A (1983) Anat Rec 205:375
Jegou B (1976) Thèse Univ P. M. Curie, Paris IV
Jensen OS (1887) Arch Mikrosc Anat 330:379
Ji L, Yoo BY, Ji TH (1981) Biol Reprod 24:617
Joel CA (1966) Fertil Steril 17:374
Johnsen O, Eliasson R (1978) Int J Androl 1:485
Johnsen O, Eliasson R, Abdel-Kader MM (1974) Andrologia 6:53
Johnsen O, Mas Diaz J, Eliasson R (1982a) Int J Androl 5:636
Johnsen O, Eliasson R, Lofman CO (1982b) Acta Physiol Scand 114:475
Johnsen SG (1964) Acta Endocrinol Suppl 90
Johnsen SG (1970) Acta Endocrinol 64:193
Johnson HA, Hammond HD (1983) Exp Cell Res 31:608
Johnson LA (1984) In: Catt KJ, Dufau ML (eds) Hormone action and testicular function. Ann NY Acad Sci, vol 438, pp 217–223
Johnson LA (1985) Biol Reprod 32:1181
Johnson LA, Gerrits RJ, Young EP (1969) Biol Reprod 1:330
Johnson LA, Pursel VG, Gerrits RJ (1972) J Anim Sci 35:398
Johnson LA, Amann RP, Pickett BW (1979) Fertil Steril 29:209
Johnson LA, Petty CS, Neaves WB (1981) Biol Reprod 25:217
Johnson LA, Garner DL, Truitt-Gilbert AJ, Lessley BA (1983a) J Androl 4:222
Johnson LA, Minuth M, Schultz G, Jakobs KH (1983b) Fed Proc Fed Am Soc Exp 42:1651

Johnson LA, Petty CS, Neaves WB (1983c) Biol Reprod 29:207
Johnson LA, Zane RS, Petty CS, Neaves WB (1984a) Biol Reprod 31:785
Johnson LA, Lebovitz RM, Samson WK (1984b) Anat Rec 209:501
Johnson LA, Petty CS, Porter JC, Neaves WB (1984c) Biol Reprod 31:779
Johnson MH (1970) J Pathol 102:131
Johnson MH (1975) J Reprod Fertil 44:167
Johnson MH, Howe CWS (1975) In: Duckett JG, Racey PA (eds) The biology of the male gamete. Academic Press, London New York, pp 205–214
Johnson WL, Hunter AG (1972) Biol Reprod 5:13
Johnsonbaugh RE, Ritzen EM, Hall K, Parvinen M, Wright WW (1982) Programme 2nd Eur Worksh Mol Cell Endocrinol Testis, Rotterdam, The Netherlands, p 12
Jones HP, Bradford MM, McRorie RA, Cormier MJ (1978) Biochem Biophys Res Commun 82:1264
Jones HP, Lenz RW, Palevitz BA, Cormier MJ (1980) Proc Natl Acad Sci USA 77:2772
Jones RC (1970) 7th Congr Int Microsc Electr 3:641
Jones RC (1971) J Reprod Fertil (Suppl) 13:51
Jones RC (1973) J Reprod Fertil 33:179
Jones RC (1975) In: Duckett JG, Racey PA (eds) The biology of the male gamete. Academic Press, London New York, pp 343–365
Jones RC, Brown CR (1982) Biochem J 206:161
Jones RC, Mann T (1977a) J Reprod Fertil 50:255
Jones RC, Mann T (1977b) J Reprod Fertil 50:261
Jones RC, Stewart DL (1979) J Reprod Fertil 56:233
Jones RC, Mann T, Sherins A (1979) Fertil Steril 31:531
Jones RC,Pholpramool C, Setchell BP, Brown CR (1981) Biochem J 202:457
Jones RC, Glos von KI, Brown CR (1983) J Reprod Fertil 67:299
Jones RC, Brown CR, Glos von KI, Gaunt SJ (1984) Exp Cell Res 156:31
Jorge Sans P, Soledad Berris EM, Fontecilla E (1977) Andrologia 9:271
Jouannet P (1979) In: Fawcett DW, Bedford JM (eds) The spermatozoon. Urban & Schwarzenberg, Baltimore Munich, pp 427–430
Jouannet P, Volochine B, Deguent P, Seres C, David G (1977) Andrologia 9:36
Jutte NHPM, Hansson V (1984) In: Catt KJ, Dufau ML (eds) Hormone action and testicular function. Ann NY Acad Sci, vol 438, pp 566–568
Jutte NHPM, Koolen LM, Jansen R Grootegoed JA, Rommerts FFG, Molen van der HJ (1981a) Int J Androl Suppl 3:59 (Abstr)
Jutte NHPM, Grootegoed JA, Rommerts FFG, Molen van der HJ (1981b) J Reprod Fertil 62:399
Jutte NHPM, Jansen R, Grootegoed JA, Rommerts FFG, Clausen OPF, Molen van der HJ (1982) J Reprod Fertil 65:431
Jutte NHPM, Jahnsen T, Attramadal H, Erichsen A, Froysa A, Tvermyr M, Eikvar L, Levy FO, Horn R, Hansoon V (1983) Acta Endocrinol 103 Suppl 256:56 (Abstr)
Jutte NHPM, Johnsen R, Grootegoed JA, Rommerts FFG, Molen van der HJ (1985) J Exp Zool 233:285
Kaczmarski F (1978) Postepy Biol Komorki 5:301
Kaiser R, Broer KH, Gitober P (1974) 8th World Congr Fertil Steril, Buenos Aires, Argentina (Abstract 579)
Kaiser R, Hermmann WP, Broer KH, Fischer W (1977) Andrologia 9:23
Kaneko S, Moriwaki C (1981) J Pharmacobiodyn 4:20
Kaneko S, Kobanawa K, Oshio S, Kobayashi T, Iizuka R (1983) J Reprod Immunol Suppl 26, 27 (Abstr)
Kaneko S, Oshio S, Kobayashi T, Iizuka R, Mohri H (1984) BBRC 124:950
Kann ML, Serres C (1980) Reprod Nutr Dev 20:1739
Kano Y, Maeda S, Sugiyama T (1976) Chromosoma 55:37
Kato S, Iritani A, Yoshida S, Nishikawa Y (1979) Jpn J Zootech Sci 50:150
Katz DF, Overstreet JW (1979) In: Fawcett DW, Bedford JM (eds) The spermatozoon. Urban & Schwarzenberg, Baltimore Munich, pp 413–420
Katz DF, Pedrotti L (1977) J Theor Biol 67:723
Katz DI, Yanagimachi R, Dresdner RD (1978) J Reprod Fertil 52:167
Kaur AJ, Guraya SS (1982) Arch Androl 9:58

Kaur S, Guraya SS (1981a) Int J Androl 4:196
Kaur S, Guraya SS (1981b) Int J Fertil 26:8
Kaur S, Guraya SS (1982a) Andrologia 14:543
Kaur S, Guraya SS (1982b) Arch Androl 9:85
Kaur S, Guraya SS (1983) Int J Fertil 28:43
Kaya M, Harrison RG (1976) J Anat 121:279
Kazazoglon T, Schackmann RW, Shapiro BM (1984) J Cell Biol 99:258a
Kazuyoshi H, Kubota N (1980) Jpn J Anim Reprod 26:50
Keenan TV, Nyquist SE, Mollenhauer HH (1972) Biochim Biophys Acta 270:433
Keller DW, Polakoski KL (1975) Biol Reprod 13:154
Kelly RW (1977) Biol Reprod 50:217
Kennedy WP, Polakoski KL (1981) Biochemistry 20:2240
Kennedy WP, Parrish RF, Polakoski KL (1981) Biol Reprod 25:197
Kennedy WP, Swift AM, Parrish RF, Polakoski KL (1982) J Biol Chem 257:3095
Kennedy WP, Van der HH, Straus JW, Bhattacharyya AK, Waller DP, Zaneveld LJD, Polakoski KL (1983) Biol Reprod 29:999
Kerr JB, Kretser de DM (1975) J Reprod Fertil 43:1
Kerr JB, Kretser de DM (1981) In: Burger H, Kretser de DM (eds) The testis. Raven, New York, pp 141–196
Keulen van CJG, Rooij de DG (1975) Cell Tissue Kinet 8:543
Keyhani E, Storey BT (1973) Biochim Biophys Acta 305:557
Khortin AY, Vikha IV, Milishnikov AN (1973) FEBS Lett 31:107
Kido A, Oya M, Tsutsume H, Fujisawa K (1983) Nagoya J Med Sci 45:107
Kierszenbaum AL, Tres LL (1974a) J Cell Biol 60:39
Kierszenbaum AL, Tres LL (1974b) J Cell Biol 63:923
Kierszenbaum AL, Tres LL (1975) J Cell Biol 65:258
Kierszenbaum AL, Tres LL (1978) Fed Proc Fed Am Soc Exp 37:2512
Kille JW, Wheat TE, Mitchell G, Goldberg E (1978) J Exp Zool 204:259
Killian GJ, Amann RP (1973) Biol Reprod 9:489
Killian GJ, Amann RP (1974) J Reprod Fertil 38:59
Kim JW, Kitts WD, Ahmed MS, Krishnamurti CR (1979) Can J Zool 57:924
Kim SH, Ko SG, Kim K (1979) Korean J Dairy Sci 1:13
Kinsey WH, Koehler JK (1976) J Supramol Struct 5:185
Kissinger C, Skinner MK, Griswold MD (1982) Biol Reprod 27:233
Kistler WS, Geroch ME (1975) Biochem Biophys Res Commun 63:378
Kistler WS, Williams-Ashman HG (1975) In: French FS, Hansson V, Ritzen EM, Nayfeh SN (eds) Hormonal regulation of spermatogenesis. Plenum, New York, pp 423–432
Kistler WS, Geroch ME, Williams-Ashman HG (1973) J Biol Chem 248:4532
Kistler WS, Geroch ME, Williams-Ashman HG (1975a) Invest Urol 12:346
Kistler WS, Geroch ME, Williams-Ashman HG (1975b) Biochem Biophys Res Commun 63:378
Kistler WS, Noyes C, Hsu R, Heinrikson RL (1975c) J Biol Chem 250:1847
Kistler WS, Keim PS, Heinrikson RL (1976) Biochim Biophys Acta 427:752
Kleene KC, Distel RJ, Hecht NB (1983) Dev Biol 98:455
Kleiss C, Liebich HG (1983) Anat Histol Embryol 12:230
Kluin PM, Rooij de DG (1981) Int J Androl 4:475
Knudsen O (1954) Ph D Thesis, Lund
Knudsen O (1958) Int J Fertil 3:389
Koehler JK (1966) J Ultrastruct Res 16:359
Koehler JK (1970a) J Ultrastruct Res 33:598
Koehler JK (1970b) In: Baccetti B (ed) Comparative spermatology. Accad Naz Lincei, Rome, pp 515–522
Koehler JK (1972) J Ultrastruct Res 39:520
Koehler JK (1973a) J Ultrastruct Res 44:355
Koehler JK (1973b) J Microsc 18:263
Koehler JK (1974) In: Afzelius NA (ed) The functional anatomy of the spermatozoon. Pergamon, Oxford, pp 105–114
Koehler JK (1975a) J Ultrastruct Res 44:355

Koehler JK (1975b) In: Duckett JG, Racey PA (eds) The biology of the male gamete. Academic Press, London New York, pp 337–342
Koehler JK (1976) Biol Reprod 15:444
Koehler JK (1977) Am J Anat 149:135
Koehler JK (1978) Int Rev Cytol 54:73
Koehler JK, Gaddum-Rosse P (1975) J Ultrastruct Res 51:106
Koehler JK, Perkins WD (1974) J Cell Biol 60:789
Koehler JK, Nudelman ED, Hakomori S (1980) J Cell Biol 86:529
Kofman-Alfaro S, Chandley AC (1970) Chromosoma 31:404
Kofman-Alfaro S, Chandley AC (1971) Exp Cell Res 69:33
Koichi K. Hoshi K, Saito A, Tsuiki A, Momono K, Hoshiai H, Suzuki M (1984) Jpn J Fertil Steril 29:33
Kojima Y, Tshikawa T (1963) Jpn J Vet Res 11:152
Kolk A, Samuel T (1975) Biochem Biophys Acta 393:307
Konengkul S, Tanphaichitr V, Muangmun V, Tanphaichitr N (1977) Fertil Steril 28:1333
Koo GC, Stackpole CW, Boyse EA, Hammerling Y, Lardis MP (1973) Proc Natl Acad Sci USA 70:1502
Koo GC, Boyse EA, Wachtel SS (1977a) In: Edidin M, Johnson MH (eds) Clinical and experimental immunoreproduction. II. Immunology of the gamete. Cambridge Univ Press, pp 73–84
Koo GC, Mittl LR, Boyse EA (1977b) Immunogenetics 4:213
Kopecny V, Fléchon JE (1981) Biol Reprod 24:201
Kopecny V, Sedlakov E, Cechovo D, Pivko J, Stanek R (1980) Histochemistry 68:67
Kopecny V, Cechova D, Zelezna B. Fléchon JE, Motlik J, Pech V (1984a) Histochem J 16:419
Kopecny V. Fléchon JE, Pivko J (1984b) Anat Rec 208:197
Koren E, Milkovic S (1973) J Reprod Fertil 32:349
Kornblatt MJ (1979) Can J Biochem 57:255
Kornblatt MJ, Knapp A, Levine M, Schachter H, Murray RK (1974) Can J Biochem 52:689
Kornblatt MJ, Klugerman A, Nagy F (1983) Biol Reprod 29:157
Kotite NJ, Nayfeh SN, French FS (1978) Biol Reprod 18:65
Kotonski B, Hutny J, Dubiel A (1980) Med Wet 36:590
Koulischer L, Hustin J, Demoulin A, Franchimont P, Debry JM (1982) Cytogenet Cell Genet 84:78
Kramer JM, Erickson RP (1982) J Reprod Fertil 64:139
Kramer MF (1981) Dev Biol 87:30
Kramer MF, Rooij de DG (1970) Virchows Arch Abt. B. Zellpathol 4:276
Kretser de DM (1969) Z Zellforsch Mikrosk Anat 98:477
Kretser de DM, Catt KJ, Burger HG, Smith GC (1969) J Endocrinol 43:105
Kretser de DM, Catt KJ, Paulsen CA (1971) Endocrinology 80:332
Kretser de DM, Burger HG, Hudson B (1974) J Clin Endocrinol Metab 38:787
Kretser de DM, Kerr JB, Rich KA, Risbridger G, Dobos M (1980) In: Steinberger A, Steinberger E (eds) Testicular development, structure and function. Raven, New York, pp 107–116
Krimer DB, Esponda P (1978) Mikroskopie 34:55
Krimer DB, Esponda P (1979) Eur J Cell Biol 20:156
Krueger PM, Hodgen GD, Sherius RJ (1974) Endocrinology 95:955
Kuehl FA, Pantanelli DJ, Tarnoff J, Humes JL (1970) Biol Reprod 2:154
Kula K (1983) Acta Endocrinol 103 Suppl 256:248 (Abstr)
Kulka P, Nissen HP, Kreysel HW (1984) Andrologia 16:48
Kullander S, Rausing A (1975) Int J Fertil 20:33
Kumar RA (1985) IRCS J Med Sci 13:677
Kumaroo KK, Irvin JL (1980) Biochem Biophys Res Commun 94:49
Kumaroo KK, Jahnke G, Irvin JL (1975) Arch Biochem Biophys 168:413
Kurio JG, Burlachenko LW (1976) VIII Congr Int Reprod Anim Insem Artif, Krako, vol IV, pp 1013–1015
Kvist U, Afzelius A, Nilsson L (1980) Growth Differ 22:543
Kyono K, Hoshi K, Saito A, Tsuiki A, Hoshiai H, Suzuki M (1984) J Exp Med 144:257
Kyono K, Hoshi K, Saito A, Momono K, Tsuiki A (1985) Acta Obstet Gynecol Jpn 37:910
Lacroix M, Fritz IB (1980) J Cell Biol 87:152a
Lacroix M, Smith FE, Fritz IB (1977) Mol Cell Endocrinol 9:227

Lacroix M, Parvinen M, Fritz IB (1981) Biol Reprod 25:143
Lacy D (1962) Br Med Bull 18:205
Lago A, Rolandi dal MT, Bortolussi M, Galli S (1975) J Reprod Fertil 43:123
Lalli M (1973) Diss Abstr 33:2900
Lalli M, Clermont Y (1981) Am J Anat 160:419
Lalli M, Tang XM, Clermont Y (1984) Biol Reprod 30:493
Lam DMK, Bruce WR (1971) J Cell Physiol 78:13
Lambert H, Le van A (1984) Gamete Res 10:153
Lambert H, Overstreet JW, Morales P, Hanson FW, Yanagimachi R (1985) Fertil Steril 43:325
Lambovot-Manzur M. Tishler PV, Atkins L (1972) Clin Genet 3:103
Lamirande de E, Bardin CW, Gagnon C (1983) Biol Reprod 28:788
Lamirande de E, Belles-Isles M, Gagnon C (1984) In: Catt KJ, Dufau ML (eds) Hormone action and testicular function. Ann NY Acad Sci, vol 438, pp 125–131
Landon GV, Pryor JP (1981) Virchows Arch Abt Pathol Histol 392:355
Langlais J, Roberts KD (1985) Gamete Res 12:183
Langlais J, Plante ZL, Chapdelaine A, Bleau G, Roberts KD (1981) Proc Natl Acad Sci USA 78:7266
Lanneau M, Loir M (1982) J Reprod Fertil 65:163
Lanoiselee-Perrin-Houdon A, Hochereau-de-Reviers MT (1980) Bull Assoc Anat 64:509
Leathem JH (1975) In: Hamilton DW, Greep RO (eds) Handbook of physiology, Sect 7: Endocrinology, vol V. Am Physiol Soc, Washington DC, pp 225–232
Leblond CP, Clermont Y (1952a) Ann NY Acad Sci 55:548
Leblond CP, Clermont Y (1952b) Am J Anat 90:167
Lee CN, Lenz RW, Ax RL (1983) J Anim Sci 57 Suppl 1:352
Lee CN, Handrow RR, Lenz RW, Ax RL (1986) Gamete Res 12:345
Lee CYG, Lum V, Wong E, Menge AC, Huang YS (1983) Am J Reprod Immunol 3:183
Lee IP (1983) Am J Ind Med 4:135
Lee ML, Storey BT (1985) Biol Reprod 33:235
Legault Y, Bleau G, Chapdelaine A, Koberts KD (1979) Steroids 34:89
Leidl W (1968) Z Tierzuecht Zuechtungsbiol 84:273
Lenz RW, Ax RL (1982) J Anim Sci (Abstr) 55 Suppl 1:366
Lenz RW, Cormier MJ (1982) Ann NY Acad Sci 383:85
Lenz RW, Ball GD, Lohse JK, First NL, Ax RL (1983) Biol Reprod 28:683
Lenzi A, Gandini L, Claroni F, Lombardo F, Misurca E, Cerasaro M, Dondero F (1982) Acta Eur Fertil 13:221 (Abstr)
Letts PJ, Hunt RC, Shirely MA, Pinterie L, Schachter H (1978) Biochim Biophys Acta 541:59
Leuchtenberger C (1960) J Dairy Sci 43 Suppl 31:31
Levin RM (1980) Fertil Steril 33:631
Levinger LF, Carter CW Jr, Kumaroo KK, Irvin JL (1978) J Biol Chem 253:5232
Lewin LM, Beer R, Lunenfeld B (1976) Fertil Steril 27:9
Lewin LM, Weissenberg R, Sobel JS, Marcus Z, Nebel L (1979) Arch Androl 2:279
Lewin LM, Nevo Z, Gabsu A, Weissenberg R (1982) Int J Androl 5:37
Li SSL, Feldmann RJ, Okabe M, Pan YCE (1983) J Biol Chem 258:7017
Li TS, Pelosi MA, Gowda VV, Caterini H, Kaminetzky HA (1978) Int J Fertil 23:38
Libbus BL, Schuetz AW (1978) Cell Tissue Kinet 11:377
Libbus BL, Schuetz AW (1979) Biol Reprod 21:353
Libbus BL, Schuetz AW (1980) J Reprod Fertil 60:1
Lifschytz E, Lindskey DL (1972) Proc Natl Acad Sci USA 69:182
Lifschytz E, Lindskey DL (1974) Genetics 78:323
Lima-de-Faria A, German J, Ghatnekar M, McGovern J, Anderson L (1968) Hireditas 60:249
Linck RW (1973) J Cell Sci 14:551
Linck RW (1976) J Cell Sci 20:405
Linck RW (1979) In: Fawcett DW, Bedford JM (eds) The spermatozoon. Urban & Schwarzenberg, Baltimore Munich, pp 99–115
Linck RW, Amos LA (1974) J Cell Biol 63:387
Lindahl PE (1958) Nature (London) 181:784
Lindahl PE (1973) Exp Cell Res 81:413
Lindemann CB (1978) Cell 13:9

Lindemann CB (1980) In: Steinberger A, Steinberger E (eds) Testicular development, structure and function. Raven, New York, pp 473–479
Lindemann CB, Rikmenspoel R (1972) Exp Cell Res 73:255
Lindemann CB, Fentie I, Rikmenspoel R (1980) J Cell Biol 87, Pt I:420
Lindholmer C (1974) Andrologia 6:7
Lingwood C, Schachter H (1981) J Cell Biol 89:621
Lipshultz LI, Murthy L, Tindall DJ (1982) J Clin Endocrinol Metab 55:228
Lis H, Shron N (1973) Am Rev Biochem 42:575
Llanos MN, Meizel S (1983) Biol Reprod 28:1043
Lobl TJ, Mathews J (1978) J Reprod Fertil 52:275
Loewenstein WR (1981) Physiol Rev 61:829
Loewus MW, Wright RW Jr, Bondioli KR, Bedgar DL, Karl A (1983) J Reprod Fertil 69:215
Loir M (1970) C R hebd Seances Acad Sci Paris 271:1634
Loir M (1972a) Ann Biol Anim Biochem Biophys 12:203
Loir M (1972b) Ann Biol Anim Biochem Biophys 12:411
Loir M, Courtens J-L (1979) J Ultrastruct Res 67:309
Loir M, Hochereau-de-Reviers MT (1972) J Reprod Fertil 31:127
Loir M, Lanneau M (1975) Exp Cell Res 92:509
Loir M, Lanneau M (1978a) Exp Cell Res 115:231
Loir M, Lanneau M (1978b) Biochem Biophys Res Commun 80:975
Loir M, Lanneau M (1984) J Ultrastruct Res 86:262
Loir M, Bouvier D, Fornells M, Lanneau M, Subirana JA (1985) Chromosoma 92:304
Lok D, Rooij de DG (1983a) Cell Tissue Kinet 16:7
Lok D, Rooij de DG (1983b) Cell Tissue Kinet 16:31
Lok D, Jansen MT, Rooij de DG (1983) Cell Tissue Kinet 16:19
Lok D, Jansen MT, Rooij de DG (1984) Cell Tissue Kinet 17:135
Long WF, Williamson FB (1983) Biochem Biophys Res Commun 117:319
Lostroh AJ (1969) Endocrinology 85:438
Lotan R, Beattie G, Hubbell W, Nicolson GL (1977) Biochemistry 16:1787
Lu C, Steinberger A (1977) Biol Reprod 17:84
Ludvigson MA, Waites GM, Hamilton DW (1982) Biol Reprod 26:311
Lung B (1972) J Cell Biol 52:179
Lung B, Bahr GF (1972) J Reprod Fertil 31:317
Lyon MF (1971) Nature (London) 232:229
MacGregor HC, Walker MH (1973) Chromosoma 40:243
Machado de Domenech E, Domenech C, Aoki A, Blanco A (1972) Biol Reprod 6:136
Mack S, Everingham J (1983) Biol Reprod 28 Suppl 1:157
Mack S, Bhattacharyya AK, Juyce C, Ven vander H, Zaneveld LJ (1983) Biol Reprod 28:1032
Mackinnon EA, Abraham JP (1972) Z Zellforsch Mikrosk Anat 124:1
Mackinnon EA, Abraham JP, Svatek A (1973) Z Zellforsch Mikrosk Anat 136:447
MacLaughlin JC, Turner C (1973) Biochem J 133:635
MacLeod J (1965) Clin Obstet Gynecol 8:115
Mahowald AP (1971) Z Zellforsch Mikrosk Anat 118:162
Maillet PL, Gouranton J (1965) C R Seances Biol 261:1417
Majumder GC (1981a) J Reprod Fertil 62:459
Majumder GC (1981b) Biochem J 105:111
Majumder GC (1981c) Biochem J 195:103
Majumder GC, Biswas R (1979) Biochem J 183:737
Majumder GC, Tarkington RW (1974) Biochemistry 13:2857
Makler A, Blumenfeld Z, Brandes JM, Paldi E (1979) Fertil Steril 32:443
Makler A, Makler E, Itzkovitz J, Brandes JM (1980) Fertil Steril 33:624
Maker A, Deutch M, Vilensky A, Palti Y (1981) Int J Androl 4:559
Makler A, Fisher M, Murillo O, Laufer N, Cherney de A, Naftolin F (1984) Fertil Steril 41:428
Males JL, Turkington RW (1970) J Biol Biochem 245:5329
Malick Z, Bartoov B (1985) Biol Cell 54:93
Mancini RE (1968) In: Mancini RE (ed) Testiculo humano. Panamericana, Buenos Aires, pp 11–45
Mancini RE, Panhos JC, Izquierdo IA, Heinrich JJ (1960) Proc Soc Exp Biol Med 104:699

Mancini RE, Alonso A, Barguet J, Alvares B, Nemirovsky N (1964) J Reprod Fertil 8:325
Mann T (1964) The biochemistry of semen and of the male reproductive tract. Methuen, London
Mann T (1968) VI Congr Int Reprod Anim Insem Artif, Paris, vol I, pp 3–30
Mann T (1975) In: Hamilton DW, Greep RO (eds) Handbook of physiology, Sect 7: Endocrinology, vol V. Am Physiol Soc, Washington DC, pp 461–472
Mann T (1978) Int J Fertil 23:133
Mann T, Lutwak-Mann C (1975) In: Blandau RJ (ed) Aging gametes. Karger, Basel, pp 122–150
Mann T, Lutwak-Mann C (1981) Male reproductive function and semen. Springer, Berlin Heidelberg New York
Mann T, Rootenberg DA (1966) J Endocrinol 34:257
Mann T, Jones R, Sherins R (1980) In: Steinberger A, Steinberger E (eds) Testicular development, structure and function. Raven, New York, pp 497–501
Mansouri A, Phi-van L, Geithe HP, Engel W (1983) Differentiation 24:149
Mao-Chi, Damjanov I (1985) Anat Rec 212:282
Margulis L (1973) Int Rev Cytol 34:333
Markewitz M, Graffs S, Veenema RJ (1967) Nature (London) 214:402
Marshall GR, Nieschlag E (1984) In: Catt KJ, Dufau ML (eds) Hormone action and testicular function. Ann NY Acad Sci, vol 438, pp 549–550
Marshall GR, Wickings EJ, Nieschlag E (1983) Acta Endocrinol 103 Suppl 256:250
Martino de C, Floridi A, Marcante ML, Malorni W, Scorza Barcellon P, Bellocci M. Silvestrini B (1979) Cell Tissue Res 196:1
Marushige Y, Marushige K (1974) Biochim Biophys Acta 340:498
Marushige Y, Marushige K (1975a) Biochim Biophys Acta 403:180
Marushige Y, Marushige K (1975b) J Biol Chem 250:39
Marushige Y, Marushige K (1978) Biochim Biophys Acta 518:440
Marushige Y, Marushige K (1983) Biochim Biophys Acta 748:461
Marzowski, Sylvester SR, Gilmont RR, Griswold MD (1985) Biol Reprod 32:1237
Marzucki ZM, Coniglio JG (1982) Biol Reprod 27:312
Mather JP (1980) Biol Reprod 23:243
Mather JP, Salomon YS, Liotta AS, Margioris A, Bardin CW, Krieger DT (1982a) Progr 2nd Eur Worksh Mol Cell Endocrin Testis, Rotterdam, The Netherlands
Mather JP, Zhuang L-Z, Perez-Infante V, Phillips DM (1982b) In: Bardin CW, Sherins RJ (eds) The cell biology of the testis. Ann NY Acad Sci, vol 383, pp 44–68
Mather JP, Gunsalus GL, Musto NA, Cheng CY, Parvinen M, Wright W, Perez-Infante V, Margioris A, Liotta A, Becker R, Krieger DT, Bardin CW (1983a) J Steroid Biochem 19:41
Mather JP, Wolpe SD, Gunsalus GL, Bardin CW, Phillips DM (1983b) Acta Endocrinol 103 Suppl 256:247 (Abstr)
Mather JP, Wolpe SD, Gunsalus GL, Bardin CW, Phillips DM (1984) In: Catt KJ, Dufau ML (eds) Hormone action and testicular function. Ann NY Acad Sci, vol 438, pp 572–575
Mathur RS (1971) J Reprod Fertil 27:5
Matsumoto AM, Karpas AE, Paulsen CA, Bremner WJ (1983) J Clin Invest 72:1005
Matsumoto AM, Bremner WJ (1985) J Androl 6:137
Matsumoto AM, Paulsen CA, Bremner WJ (1984) J Clin Endocrinol Metab 59:882
Matsumoto AM, Karpas AE, Bremner WJ (1986) J Clin Endocrinol Metab 62, in press
Matsuyama S, Yokoki Y, Ogasa A (1971) Natl Inst Anim Health Quart 11:46
Maunoury R, Hill A-M (1980) C R Hebd Seances Acad Sci Ser D 291:425
Mayer JF Jr, Zirkin BR (1979) J Cell Biol 81:403
Mayer JF Jr, Chang TSK, Zirkin BR (1981) Biol Reprod 25:1041
McCann SM, Dhindsa DS (eds) (1983) Role of peptides and proteins in control of reproduction. Elsevier/North Holland Biomedical Press, Amsterdam New York
McComb RB, Bowers GN Jr, Posen S (1979) Alkaline phosphatase. Plenum, New York
McGaday J, Baillie AH, Ferguson MM (1966) Histochemie 7:211
McGinley DM, Posalaky Z, Porvaznik M, Russell L (1979) Tissue Cell 11:741
McGinley DM, Posalaky Z, Porvaznik M (1981) Tissue Cell 13 :337
McGrady A (1979a) J Cell Physiol 99:223
McGrady A (1979b) J Reprod Fertil 56:549
McGrady A (1979c) J Cell Physiol 99:223

McGrady A (1979d) Arch Androl 2:381
McGrady A, Nelson L (1972) Exp Cell Res 73:192
McGrady A, Nelson L (1973) Exp Cell Res 76:349
McGrady A, Nelson L (1976) Biol Reprod 15:248
McGrady A, Nelson L, Ireland M (1974) J Reprod Fertil 40:71
McIntosh JR, Sisken JE, Chu LK (1979) J Ultrastruct Res 66:40
McRorie RA, Williams WL (1974) Ann Rev Biochem 43:777
McRorie RA, Turner RB, Bradford MM, Williams WL (1976) Biochem Biophys Res Commun 71:492
Means AR (1974) In: Moudgal NR (ed) Gonadotropins and gonadal function. Academic Press, London New York, pp 485–499
Means AR (1977) In: Johnson AD, Gomes WR (eds) The testis, vol IV. Academic Press, London New York, pp 163–188
Means AR, MacDougall E, Soderling TR, Corbin JD (1974) J Biol Chem 249:1231
Means AR, Tash JS, Chafouleas JG, Lagace L, Guerriero V (1982) In: Bardin CW, Sherins RJ (eds) The cell biology of the testis. Ann NY Acad Sci, vol 383, pp 69–84
Meistrich E, Marushige Y, Wong TK (1976) Biochemistry 15:2047
Meistrich ML (1972) J Cell Physiol 80:299
Meistrich ML, Bruce WR, Clermont Y (1973) Exp Cell Res 79:213
Meistrich ML, Reid BO, Barcellona WJ (1975) J Cell Biol 64:211
Meistrich ML, Trostle PK, Frapart M, Erickson RP (1977) Dev Biol 60:428
Meistrich ML, Brock WA, Grimes SR, Platz RD, Hnilica LS (1978) In: Saunders GF (ed) Cell differentiation and neoplasia. Raven, New York, pp 403–412
Meistrich ML, Longtin J, Brock WA, Grimes SR Jr, Mace ML (1981) Biol Reprod 25:1065
Meizel S (1981) J Exp Zool 217:443
Meizel S (1984) Biol Rev 59:125
Meizel S, Cotham J (1972) J Reprod Fertil 28:303
Meizel S, Lui CW (1976) J Exp Zool 195:137
Meizel S, Mukerji SK (1975) Biol Reprod 13:83
Meizel S, Mukerji SK (1976) Biol Reprod 14:444
Meizel S, Turner KO (1983a) J Exp Zool 226:171
Meizel S, Turner KO (1983b) Fed Proc Fed Am Soc Exp 42:1823
Meizel S, Turner KO (1984) J Exp Zool 231:283
Meizel S, Turner KO (1986) J Exp Zool 237:137
Meizel S, Working PK (1980) Biol Reprod 22:211
Meizel S, Lui CW, Working PK, Mrsny RJ (1980) Dev Growth Differ 22:483
Meizel S, Turner KO, Thomas P (1985) In: Ben-Jonathan N, Bahr J (eds) Catecholamines as hormone regulators. Serono, vol 18. Raven, New York, pp 329–341
Melner MH, Puett D (1984a) In: Puett D (ed) Human fertility, health and food: impact of molecular biology and biotechnology. United Nations Fund for Population Activities, New York, pp 129–140
Melner MH, Puett D (1984b) Arch Biochem Biophys 232:197
Menezo Y, Testart J, Khartchadourian C, Frydman R (1984) Int J Fertil 29:61
Menger H, Menger S (1981a) Arch Exp Veterinaermed 35:359
Menger H, Menger S (1981b) Equine Vet J 13:177
Mercado E, Carvajal G, Reyes A, Rosado A (1976) Biol Reprod 14:632
Mettler L (1982) Prog Clin Biol Res 112:Pt B:3
Mettler L, Gradl T (1977) Eur J Obstet Gynecol Reprod Biol 7:5
Mettler L, Skrabei H (1979) Int J Fertil 24:44
Mettler L, Czuppon AB, Buchheim W, Baukloh V, Ghyczy M, Etschenberg J, Holstein AF (1983a) Am J Reprod Immunol 4:127
Mettler L, Baukloh V, Paul S (1983b) Geburtshilfe Frauenheilkd 43:288
Metz CB (1973) Fed Proc Fed Am Soc Exp 32:2057
Metz CB (1978) Curr Top Dev Biol 12:107
Metz CB, O'Rand MG (1975) J Exp Zool 191:301
Meyer GE, Henning W (1974) In: Afzelius BA (ed) The functional anatomy of the spermatozoon. Pergamon, Oxford, pp 69–75
Meyhöfer W (1963) Arch Klin Exp Dermatol 216:556
Meyhöfer W (1964) Arch Klin Exp Dermatol 219:925

Milkowski AL, Lardy HA (1977) Arch Biochem Biophys 181:270
Milkowski AL, Babcock DF, Lardy HA (1976) Arch Biochem Biophys 176:250
Miller MJ, Vincent NR, Mawson CA (1961) J Histochem Cytochem 9:111
Miller OJ, Miller DA, Warburton D (1973) In: Steinberg AG, Bearn AG (eds) Progress in medical genetics, vol IX. Grune & Stratton, New York
Millette CF (1976) In: Edidin M, Johnson MH (eds) Immunobiology of gametes. Cambridge Univ Press, London New York, pp 51–71
Millette CF (1979a) Curr Top Dev Biol 13:1
Millette CF (1979b) In: Fawcett DW, Bedford JM (eds) The spermatozoon. Urban & Schwarzenberg, Baltimore Munich, pp 177–186
Millette CF, Bellvé AR (1977) J Cell Biol 74:86
Millette CF, Bellvé AR (1980) Dev Biol 79:309
Millette CF, Moulding CT (1981a) J Cell Sci 48:367
Millette CF, Moulding CT (1981b) Gamete Res 4:317
Millette CF, Scott BK (1982) J Cell Biol 9:175a
Millette CF, Scott BK (1984) J Cell Sci 65:233
Millette CF, Spear PG, Gall WE, Edelman GM (1973) J Cell Biol 58:662
Millette CF, O'Brien DA, Moulding CT (1980) J Cell Sci 43:279
Mills NC, Means AR (1972) Endocrinology 91:147
Mills NC, Van NT, Means AR (1977) Biol Reprod 17:760
Mita M, Hall PF (1982) Biol Reprod 26:445
Mita M, Bortland K, Price JM, Hall PF (1985) Endocrinology 116:987
Moens PB, Go VLW (1971) Z Zellforsch Mikrosk Anat 127:201
Moens PB, Go VLW (1972) Z Zellforsch Mikrosk Anat 73:89
Moens PB, Hugenholtz AD (1975) J Cell Sci 19:487
Moens PB, Hugenholtz AD (1976) In: Cairnie AB, Lala PK, Osmond DG (eds) Stem cells of renewing cell populations. Academic Press, London New York, pp 303–310
Mohri H (1968) Nature (London) 217:1053
Mohri H, Mohri T (1965) Exp Cell Res 38:217
Mohri H, Yanagimachi R (1980) Exp Cell Res 127:191
Mohri H, Yano Y (1982) Prog Clin Biol Res 80:243
Mohri H, Mohri T, Ernster L (1965) Exp Cell Res 38:217
Mohsenian M, Syner FN, Moghissi KS (1982) Fertil Steril 37:223
Molen van der HJ, Rommerts FFG (1981) In: Burger M, Kretser de D (eds) The testis. Raven, New York, pp 213–238
Mollenhauer HH, Morre DJ (1977) Am J Anat 150:381
Mollenhauer HH, Morre DJ (1981) Eur J Cell Biol 25:340
Mollenhauer HH, Morre DJ (1983) Protoplasma 116:187
Mollenhauer HH, Zebrun W (1960) J Biophys Biochem Cytol 8:761
Mollenhauer HH, Hass BS, Morre DJ (1976) J Microsc Biol Cell 27:33
Mollenhauer HH, Morre DJ, Hass BS, (1977) J Ultrastruct Res 61:166
Monakhova MA, Abbas AL (1979) Biol Nauk (Moscow) 0 (1):85
Monesi V (1962) J Cell Biol 14:1
Monesi V (1964) J Cell Biol 22:521
Monesi V (1965a) Exp Cell Res 39:197
Monesi V (1965b) Chromosoma 17:11
Monesi V (1967) Arch Anat Microsc Morphol Exp (Suppl 3–4):56,61
Monesi V (1971) J Reprod Fertil (Suppl) 13:1
Monesi V (1974) In: Mancini RE, Martini L (eds) Male fertility and sterility. Academic Press, London New York, p 59
Monesi V, Geremia R, D'Agostino A, Boitani C (1978) Curr Top Dev Biol 12:11
Monet-Kuntz C, Terqui M, Locatelli A, Hochereau-de-Reviers MT, Courot M (1977) C R Acad Sci Ser D 283:1763
Monfoort GH, Schiphof R, Rozijn TH, Stay-Parves EP (1973) Biochim Biophys Acta 322:173
Moniem KA, Glover TD (1972a) J Reprod Fertil 29:65
Moniem KA, Glover TD (1972b) J Anat 111:437

Monn E, Desautal M, Christiansen RO (1972) Endocrinology 91:716
Monroy A, Rosati F (1983) Gamete Res 71:85
Montamat EE, Blanco A (1976) Exp Cell Res 103:241
Montamat EE, Moreno J, Blanco A (1978) J Reprod Fertil 53:117
Moore DJ, Clegg ED, Lunstra DD, Mollenhauer HH (1974) Proc Soc Exp Biol Med 145:1
Moore GPM (1971) Exp Cell Res 68:462
Moore GPM (1972) In: Beatty RA, Gluecksohn-Waelsch S (eds) The genetics of the spermatozoon. Proc Int Symp, Edinburgh, Aug 16–20, 1971, pp 90–96
Moore GPM (1975) In: Mulcaphy DK (ed) Gamete competition in plants and animals. Elsevier/North-Holland Biomedical Press, Amsterdam New York, pp 69–73
Moore HDM (1979) Int J Androl 2:449
Moore HDM (1981) Int J Androl 4:321
Moore HDM, Hartman TD (1984) J Reprod Fertil 70:175
Moore HDM, Hartman TD, Brown AC, Smith CA, Ellis DH (1985) Expl Clin Immunogenet 2:84
Morales C, Clermont Y (1982) Anat Rec 203:233
Morales C, Clermont Y (1985) Biol Reprod 32 (Suppl I) :124
Morales C, Clermont Y, Maddler NJ (1984) Anat Rec 208:120A
Morisawa M, Mohri H (1972) Exp Cell Res 70:311
Morisawa M, Mohri H (1974) Exp Cell Res 83:87
Moroz LG, Shapiev SH, Korban NV, Sharobaiko VI, Mashanskii VF (1977) S KH Biol 11:850
Morre DJ, Mollenhauer HH, Eppler CM (1980) Cell Tissue Res 211:65
Morton BE (1968) J Reprod Fertil 15:113
Morton BE, Albagli L (1973) Biochem Biophys Res Commun 50:697
Morton BE, Harrigan-Lum J, Albagli L, Jooss T (1974) Biochem Biophys Res Commun 56:372
Morton BE, Harrigan-Lum J, Jooss T (1973) Biol Reprod 9:71 (Abstr)
Morton BE, Sagadraca R, Fraser C (1978) Fertil Steril 29:695
Morton BE, Fraser CF, Albagli CB (1979) Fertil Steril 32:99
Morton DB (1973) Trans Biochem Soc 1:385
Morton DB (1975a) Front Biol 45:203
Morton DB (1975b) J Reprod Fertil 45:375
Morton DB (1976) Lysosomes Biol Med 5:203
Morton DB, McAnulty PA (1979) J Reprod Immunol 1:61
Moruzzi JF (1979) J Reprod Fertil 57:319
Moses MJ (1968) Annu Rev Genet 2:363
Mrsny RJ, Meizel S (1981) J Cell Biol 91:77
Mrsny RJ, Meizel S (1985) Life Sci 36:27
Mrsny RJ, Waxman L, Meizel S (1979) J Exp Zool 210:123
Mudd S, Mudd EBH (1929) J Immunol 17:39
Mueller B, Kirchner C (1978) J Reprod Fertil 54:167
Mueller H, Sterba G, Brueckner G, Hofmann I, Lehmann J (1977) Biol Zentralbl 96:571
Muffly KL, Turner TT, Brown M, Hall PF (1985) Biol Reprod 33:1245
Mujica A, Alonso R, Hernandez-Montes H (1978) Int J Fertil 23:112
Mukerji SK, Meizel S (1979) Biol Chem 254:11721
Mukherjee AB, Cohen MM (1968) Nature (London) 219:489
Mukherjee DP, Singh SP (1965) Indian J Vet Sci 35:213
Mukhtar H, Lee IP, Bend JR (1978) Biochem Biophys Res Commun 83:1093
Mulder E, Peters MJ, Beurden WMO, Molen van der HJ (1974) FEBS Lett 47:209
Muller CH, Eddy EM (1984) In: Catt KJ, Dufau ML (eds) Hormone action and testicular function. Ann NY Acad Sci, vol 438, pp 533–534
Muller-Esterl W, Schill W-B (1982) Andrologia 14:309
Muller-Esterl W, Zippel B, Fritz H (1980) Hoppe-Seylers Z Physiol Chem 361:1381
Muller-Esterl W, Wendt V, Leidl W, Dann O, Shom E, Wagner G, Fritz H (1983) J Reprod Fertil 67:13
Multamaki S, Niemi M (1972) Int J Fertil 17:43
Muramatsu M, Utakoji T, Sugano H (1968) Exp Cell Res 53:278
Murphy KM, Goodman RR, Synder SH (1983) Endocrinology 113:1299
Murthy HMS, Moudgal NR (1983) J BioSci 5:115

Mushkambarov NN, Volkova NP (1983) Ontogenez 14:89
Mushkambarov NN, Volkova NP, Nikolaev SYA (1982) Tistologiya 24:719
Musick WDL, Rossmann MG (1979a) J Biol Chem 254:7611
Musick WDL, Rossmann MG (1979b) J Biol Chem 254:7621
Musto NA, Gunsalus GL, Bardin CW (1978) In: Hansson V, Ritzen EM, Purvis K, French FS (eds) Endocrine approach to male contraception. Scriptor, Copenhagen, pp 424–433
Musto NA, Larrea F, Cheng S-L, Kotite N, Gunsalus G, Bardin CW (1982) In: Bardin CW, Sherins RJ (eds) Cell biology of the testis. Ann NY Acad Sci, vol 383, pp 343–359
Myles DG, Primakoff P (1983) Biol Reprod 28 Suppl 1:70
Myles DG, Primakoff P, Bellvé AR (1981) Cell 23:433
Myles DG, Primakoff P, Koppel DE (1984) J Cell Biol 98:1905
Nakamura M, Hall PF (1978) Biochem Biophys Res Commun 85:756
Nakamura M, Hall PF (1980) J Biol Chem 255:2907
Nakamura M, Kato J (1981) Dev Growth Differ 23:255
Nakamura M, Hino A, Yasumasu I, Kato J (1981) J Biochem 89:1309
Nakamura M, Fujiwara A, Yasumasu I, Okinaga S, Aria K (1982) J Biol Chem 257:13945
Nakamura M, Fujiwara A, Yasumasu I, Okinaga S, Arai K (1983) Dev Growth Differ 25:281
Nakhla AM, Jänne OA, Mather JP, Bardin CW (1984) In: Catt KJ, Dufau ML (eds) Hormone action and testicular function. Ann NY Acad Sci, vol 438, pp 588–590
Narayan P, Scott BK, Millette CF, Wolf de WG (1983) Gamete Res 7:227
Nasr H, Soliman FA, Farahat AA, Aboul Fadle WA, Nemetalla BR (1979) Biol Zentralbl 98:107
Navarro J, Vidal F, Guitart M, Egozcue J (1981) Hum Genet 9:419
Naz RK, Rosenblum BB, Menge AC (1984) Proc Natl Acad Sci USA 81:857
Neaves WB (1975) Contraception 11:571
Neaves WB (1977) In: Johnson AD, Gomes WR (eds) The testis, vol IV. Academic Press, London New York, pp 126–162
Neaves WB, Johnson L (1985) Biol Reprod 32 (Suppl)1:86
Nebel BR, Coulson EM (1962) Chromosoma 13:272
Neill AR, Masters CJ (1973) J Reprod Fertil 34:279
Nelson L (1966) J Cell Physiol 68:113
Nelson L (1967) In: Metz CB, Monroy A (eds) Fertilization, vol I. Academic Press, London New York, pp 27–88
Nelson L (1970) In: Baccetti B (ed) Comparative spermatology. Academic Press, London New York, pp 465–474
Nelson L (1972) Biol Reprod 6:319
Nelson L (1975) In: Hamilton DW, Greep RO (eds) Handbook of physiology, Sect 7: Endocrinology, vol V. Am Physiol Soc, Washington DC, pp 421–435
Nelson L (1978) Fed Proc Fed Am Soc Exp 37:2543
Nelson L (1979) In: Pepe F (ed) Motility in cell function. Academic Press, London New York, pp 453–459
Nelson L (1982) Biol Bull 163:492
Nelson L (1985) In: Metz CB, Monroy A (eds) Biology of fertilization, vol II. Academic Press, London New York, pp 215–234
Nelson L, Gardner ME (1983) Fed Proc Fed Am Soc Exp 42:294
Nelson L, Gardner ME (1986) Gamete Res (in press)
Nelson L, McGrady AV, Fangboner ME (1970) In: Baccetti B (ed) Comparative spermatology. Accad Naz, Dei Lincei, Rome, pp 465–474
Nelson L, Chakraborty J, Young M, Goodwin A, Kock E, Gardner ME (1980a) In: Steinberger A, Steinberger E (eds) Testicular development, structure and function. Raven, New York, pp 503–511
Nelson L, Young MJ, Gardner ME (1980b) Life Sci 26:1739
Nelson L, Gardner ME, Young MJ (1982) Cell Motil 2:225
Nevo AC, Michaeli I, Schindler H (1961) Exp Cell Res 23:69
Nicander L (1957) Acta Morphol Neerl Scand 1:99
Nicander L (1968) VI Congr Int Reprod Anim Insem Artif, Paris, vol I, pp 89–108
Nicander L (1970) In: Baccetti B (ed) Comparative spermatology. Accad Naz Dei Lincei, Rome, pp 47–56
Nicander L, Bane A (1962) Z Zellforsch Mikrosk Anat 57:390

Nicander L, Bane A (1966) Z Zellforsch Mikrosk Anat 72:496
Nicander L, Plöen L (1969) Z Zellforsch Mikrosk Anat 99:221
Nicolle JC, Esnault C, Courot M (1980) Ann Biol Anim Biochim Biophys 19:1817
Nicolopoulou M, Soucek DA, Vary JC (1985) Biochim Biophys Acta 815:486
Nicolson GL (1973) J Natl Cancer Inst 50:1443
Nicolson GL (1976) Biochim Biophys Acta 458:1
Nicolson GL (1978) In: Koehler JK (ed) Advanced techniques in biological electron microscopy. Springer, Berlin Heidelberg New York, pp 1–18
Nicolson GL, Yanagimachi R (1972) Science 177:276
Nicolson GL, Yanagimachi R (1974) Science 184:1294
Nicolson GL, Yanagimachi R (1979) In: Fawcett DW, Bedford JM (eds) The spermatozoon. Urban & Schwarzenberg, Baltimore Munich, pp 187–194
Nicolson GL, Usui N, Yanagimachi R, Yanagimachi H, Smith JR (1977) J Cell Biol 74:950
Nicolson GL, Brodginski AB, Beattie G, Yanagimachi R (1979) Gamete Res 2:153
Niemi M, Kormano M (1965) Anat Rec 151:159
Nikkanen V, Söderström K-O, Parvinen M (1978) J Reprod Fertil 53:255
Nissen HP, Kreysel HW, Schirren C (1981) Andrologia 13:444
Nistal M, Paniagua R, Esponda P (1980) Acta Anat 108:238
Norman C, Gombe S (1975) J Reprod Fertil Suppl 18:65
Novi AM (1968) Virchows Arch Abt B Zellpathol 1:346
Novikoff AB, Mori M, Quintana N, Yam A (1977) J Cell Biol 75:148
Nuzzo NA, Joyce C, Zaneveld LJD (1983) Biol Reprod 28 Suppl 1:38
Nykanen M (1979) Cell Tissue Res 198:441
Nyquist SE. Acuff K, Mollenhauer HH (1973) Biol Reprod 8:119
Oakberg EF (1956) Am J Anat 99:391
Oakberg EF (1959) Radiat Res 11:700
Oakberg EF (1964) Jpn J Genet 40:119
Oakberg EF (1971a) Anat Rec 169:515
Oakberg EF (1971b) Mutat Res 11:1
Oakberg EF (1974) Jpn J Genet Suppl 40:119
Oakberg EF (1975) In: Hamilton DW, Greep RO (eds) Handbook of physiology, Sect 7: Endocrinology, vol V. Am Physiol Soc, Washington DC, pp 233–243
Oakberg EF (1978) Mut Res 50:927
Oakberg EF, Crosthwait CD (1983) Mutat Res 108:337
Oakberg EF, Huckins C (1976) In: Cairnie AB, Lala PK, Osmond DG (eds) Stem cells of renewing cell populations. Academic Press, London New York, pp 287–302
Oakberg EF, Crosthwait CD, Raymer GD (1982) Mutat Res 94:165
O'Brien DA, Bellvé AR (1980a) Dev Biol 75:386
O'Brien DA, Bellvé AR (1980b) Dev Biol 75:405
O'Brien DA, Millette CF (1984) Dev Biol 101:307
Očsenyi A, Veres I (1968) Mikroskopie 23:238
Odartchenko N, Pavillard M (1970) Science 167:1133
O'Day PM, Rikmenspoel R (1979) J Cell Sci 35:123
Ohara M (1981) Jpn J Zootech Sci 52:507
Ohju E, Yanagimachi R (1982) J Exp Zool 224:259
Ohwada S, Tamate H (1975) Tohoku J Agric 26:131
Ohwada S, Tamate H (1980) J Reprod Fertil 58:51
Okamura N, Sugita Y (1983) J Biol Chem 258:13056
Oko RJ, Costerton JW, Coulter GH (1976) Can J Zool 54:1326
Olds PJ (1971) J Anat 109:31
Olds-Clarke P (1984) In: Catt KJ, Dufau ML (eds) Hormone action and testicular function. Ann N Y Acad Sci, vol 438, pp 206–216
Oliphant G,Singhas CA (1979) Biol Reprod 21:937
Olson GE (1979) In: Fawcett DW, Bedford JM (eds) Spermatozoon. Urban & Schwarzenberg, Baltimore Munich, pp 395–400
Olson GE (1980) Anat Rec 197:471
Olson GE, Gould KG (1981) J Reprod Fertil 62:185

Olson GE, Hamilton DW (1976) Anat Rec 186:387
Olson GE, Hamilton DW (1978) Biol Reprod 19:26
Olson GE, Linck RW (1977) J Ultrastruct Res 61:21
Olson GE, Sammons DW (1980) Biol Reprod 22:319
Olson GE, Winfrey VP (1985) J Ultrastruct Res 90:9
Olson GE, Hamilton DW, Fawcett DW (1976a) Biol Reprod 14:517
Olson GE, Hamilton DW, Fawcett DW (1976b) J Reprod Fertil 47:293
Olson GE, Lifsics M, Fawcett DW, Hamilton DW (1977) J Ultrastruct Res 61:21
Onuma H, Nishikawa Y (1963) Bull Natl Inst Anim Ind (Chiba, Jpn) 1:125
O'Rand MG (1977) Dev Biol 55:260
O'Rand MG (1979) In: Fawcett DW, Bedford JM (eds) The spermatozoon. Urban & Schwarzenberg, Baltimore Munich, pp 195–204
O'Rand MG (1982) In: Bardin CW, Sherins RJ (eds) The cell biology of the testis. Ann NY Acad Sci, vol 383, pp 393–404
O'Rand MG, Metz CB (1976) Biol Reprod 14:586
O'Rand MG, Romrell LJ (1977) Dev Biol 55:347
O'Rand MG, Romrell LJ (1980) Dev Biol 75:431
O'Rand MG, Romrell LJ (1981) Dev Biol 84:322
O'Rand MG, Emery JJ, Wolch JE, Fisher S-J (1984) J Cell Biol 99:395a
O'Rand MG. Mathews JE, Welch JE, Fisher SJ (1985) J Exp Zool 235:423
Orgebin-Crist MC (1981) Progr Reprod Biol 8:80
Ortavant R (1956) Arch Anat Microsc Morphol Exp 45:1
Ortavant R (1958) Thesis, Univ Paris
Ortavant R (1959) In: Cole HH, Cupps PT (eds) Reproduction in domestic animals, vol II, Chap 1. Academic Press, London New York, pp 1–46
Ortavant R, Courot M, Hochereau MT (1969) In: Cole HH, Cupps PT (eds) Reproduction in domestic animals, 2nd edn. Academic Press, London New York, pp 251–276
Ortavant R, Courot M, Hochereau-de-Reviers MT (1972) J. Reprod Fertil 31:451
Orth J, Christensen AK (1978) Endocrinology 103:1944
O'Shea T, Dacheux J-L, Paquignon M (1979) J Reprod Fertil 55:277
Otto FJ, Zante U, Hacker J, Zante J, Schumann J, Goehde W. Meistrich ML (1979) Histochemistry 61:249
Oud JL, Rooij de DG (1977) Anat Rec 187:113
Oura C (1971) Monit Zool Ital 5:253
Overstreet JW, Katz DF, Johnson LL (1980) Biol Reprod 22:1083
Ozasa H, Gould KG (1982) Arch Androl 9:121
Pace MM, Graham EF (1970) Biol Reprod 3:140
Paddock SW, Woolley DM (1980) Exp Cell Res 126:199
Pallini V, Bacci E (1979) J Submicrosc Cytol 11:165
Pallini V, Baccetti B, Burrini AG (1979) In: Fawcett DW, Bedford JM (eds) The spermatozoon. Urban & Schwarzenberg, Baltimore Munich, pp 141–151
Pallini V, Bugnoli M, Mencarelli C, Scapigliati G (1982) Symp Soc Exp Biol 35:339
Pan YCE, Sharief FS, Okabe M, Huang S, Li SS (1983) J Biol Chem 258:7005
Panse GT, Jayaraman S, Sheth AR (1983) Arch Androl 11:137
Pariset CC, Weinman JS, Escaig FT, Guyot YM, Ifto de FC, Weinman SJ, Demaille JG (1984) Gamete Res 10:443
Parks JE, Meacham TN, Saacke RG (1981) Biol Reprod 24:399
Parmentier M, Inagami T, Pochet R, Desclin JC (1983) Endocrinology 112:1318
Parrish JJ, Susko-Parish JL, First NL (1985) Biol Reprod 32:211
Parrish RF, Polakoski KL (1977) Biol Reprod 17:417
Parrish RF, Polakoski KL (1979) Int J Biochem 10:391
Parrish RF, Straus JW, Polakoski KL, Dombrose FA (1978) Proc Natl Acad Sci USA 75:149
Parshad VR, Guraya SS (1984) Proc Indian Natn Sci Acad 50:559
Parvinen L-M, Parvinen M (1978) Ann Biol Anim Biochem Biophys 18:585
Parvinen L-M, Jokelainen T, Parvinen M (1978a) Hereditas 88:75
Parvinen L-M, Söderström KO, Parvinen M (1978b) Exp Pathol 15:85
Parvinen M (1973) Virchows Arch Abt B Zellpathol 13:38
Parvinen M (1982) Endocrinol Rev 3:404

Parvinen M (1983) Acta Endocrinol 103 Suppl 256:53 (Abstr)
Parvinen M, Jokelainen PT (1974) Biol Reprod 11:85
Parvinen M, Parvinen L-M, (1979) J Cell Biol 80:621
Parvinen M, Ruokonen A (1982) J Androl 3:211
Parvinen M, Söderström K-O (1976) Nature (London) 239:235
Parvinen M, Vanha-Perttula T (1972) Anat Rec 174:435
Parvinen M, Byskov AG, Andersen CY, Grinsted J (1982) in: Bardin CW, Sherins RJ (eds) The cell biology of the testis. Ann NY Acad Sci, vol 383, pp 483–484
Parvinen M, Wright WW, Phillips DM, Mather JM, Musto AN, Bardin CW (1983) Endocrinology 112:115
Parvinen M, Nikula H, Huhtaniemi I (1984) In: Catt KJ, Dufau ML (eds) Hormone action and testicular function. Ann NY Acad Sci, vol 438, pp 681–683
Patanelli DJ (1975) In: Hamilton DW, Greep RO (eds) Handbook of physiology, Sect 7: Endocrinology, vol V. Am Physiol Soc Washington DC, pp 245–258
Patil DS, Hukeri VB, Deshpande BR (1981) Indian J Anim Reprod 1:90
Paul HE, Paul MF, Kopko F, Bender RC, Everett G (1953) Endocrinology 53:585
Paul S, Baukloh V, Czuppon AB, Mettler L, (1983) In: Shulman S, Dondero F (eds) Immunobiological factors in human contraception. Acta Medica, Rome, pp 103–112
Pavelko MK, Crabo BG (1976) VIII Congr Int Reprod Anim Insem Artif, Krako, vol IV, pp 1045–1047
Pawlowitzki IH, Bosse HG (1971) Hum Genet 13:338
Payne AH, O'Shaughnessy PJ, Chase DJ, Dixon GEK, Christensen AK (1982) in: Bardin CW, Sherins RJ (eds) The cell biology of the testis. Ann NY Acad Sci, vol 383, pp 174–203
Pearson PL, Bobrow M, Vosa C (1970) Nature (London) 226:78
Pearson PL, Geraedts JPM, Ploeg van der M, Pawlowitzki IH (1975) In: Duckett JG, Racey PA (eds) The biology of the male gamete. Academic Press, London New York, pp 279–289
Pedersen H (1969) Z Zellforsch Mikrosk Anat 94:542
Pedersen H (1970a) In: Baccetti B (ed) Comparative spermatology. Accad Naz Dei Lincei, Rome, pp 133–142
Pedersen H (1970b) J Ultrastruct Res 33:457
Pedersen H (1972a) J Ultrastruct Res 40:366
Pedersen H (1972b) J Reprod Fertil 31:99
Pedersen H (1972c) Z Zellforsch Mikrosk Anat 123:305
Pedersen H (1974a) Dani Med Bull 21 Suppl 1
Pedersen H (1974b) The human spermatozoon. Costers Bogtrykheri, Copenhagen
Pedersen H, Rebbe H (1974) J Reprod Fertil 37:51
Pedersen H, Rebbe H (1975) Biol Reprod 12:541
Pedron H (1983) Arch Androl 10:249
Pedron N, Giner J, Hicks JJ (1979) Int J Fertil 23:65
Pelletier RM (1983) Biol Reprod 28 Suppl 1:69
Pelletier RM, Friend DS (1983a) Am J Anat 167:119
Pelletier RM, Friend DS (1983b) Am T Anat 168:213
Pellicciari C, Hosokawa Y, Fukuda M, Manfredi-Romanini MG (1983) J Reprod Fertil 68:371
Perreault SD, Rogers BJ (1982) J Androl 3:396
Perreault SD, Zirkin BR (1982) J Exp Zool 224:253
Perreault SD, Zirkin BR, Rogers BJ (1982) Biol Reprod 26:343
Persona L, Bustos-Obregon E (1983) Arch Androl 10:113
Peter Z, Zao R, Meizel S, Talbot P (1985) J Exp Zool 234:63
Peterson RN (1982) In: Zaneveld LJD, Chatterton RT (eds) Biochemistry of mammalian reproduction. I: Gametes and genital tract fluids. II: Reproductive endocrinology. Wiley Interscience, New York, pp 153–174
Peterson RN, Freund M (1968) J Reprod Fertil 17:357
Peterson RN, Freund M (1976) Fertil Steril 27:1301
Peterson RN, Freund M (1977) Fertil Steril 28:257
Peterson RN, Bundman D, Freund M (1977a) Biol Reprod 17:198
Peterson RN, Bundman D, Freund M (1977b) J Cell Biol 75:207a
Peterson RN, Seyler D, Bundman D, Freund M (1979a) J Reprod Fertil 55:395

Peterson RN, Russell L, Bundman D, Freund M (1979b) Biol Reprod 21:583
Peterson RN, Russell L, Bundman D, Freund M (1980) Biol Reprod 23:637
Peterson RN, Robi JM, Dziuk PJ, Russell LD (1982) J Exp Zool 223:79
Peterson RN, Hunt WP, Russell LD, Russell TA, Arthur RD (1983a) J Androl 4:28
Peterson RN, Russell LD, Hunt W, Bundman D, Freund M (1983b) J Androl 4:71
Peterson RN, Ashraf M, Russell LD (1983c) Fed Proc Fed Am Soc Exp 42:2244
Peterson RN, Ashraf M, Russell LD (1983d) Biochem Biophys Res Commun 114:28–33
Peterson RN, Russell LD, Hunt WP (1985) J Exp Zool 231:137
Petkov ZZ, Bobadov ND (1976) VIII Congr Int Reprod Anim Insem Artif, Krako, p 198 (Abstr)
Pihlaja DJ, Roth LE (1973) J Ultrastruct Res 44:293
Philip de RM, Kierszenbaum AL (1982) Proc Natl Acad Sci USA 79:6551
Philip de MF, Austin Spruill W, French FS, Kienzenbaum AL (1982) In: Bardin CW, Sherins RJ (eds) Cell biology of the testis. Ann NY Acad Sci, vol 383, pp 360–371
Philippe M, Chevaillier P (1976) Biochem Biophys Acta 447:188
Phillips DM (1970a) J Ultrastruct Res 33:366
Phillips DM (1970b) U Ultrastruct Res 39:381
Phillips DM (1972a) J Ultrastruct Res 38:591
Phillips DM (1972b) J Cell Biol 53:561
Phillips DM (1974) Spermiogenesis. Academic Press, London New York
Phillips DM (1975a) In: Hamilton DW, Greep RO (eds) Handbook of physiology, Sect 7: Endocrinology, vol V. Am Physiol Soc, Washington DC, pp 405–420
Phillips DM (1975b) J Exp Zool 191:1
Phillips DM (1977a) J Ultrastruct Res 58:144
Phillips DM (1977b) Biol Reprod 16:128
Phillips DM (1980) J Ultrastruct Res 72:103
Phillips DM (1983) J Submicrosc Cytol 15:29
Phillips DM, Bedford JM (1985) J Exp Zool 235:119
Phillips D, Morrison M (1971) Biochem J 10:1766
Phillips DM, Olson G (1975) In: Afzelius BA (ed) The functional anatomy of the spermatozoon. Pergamon Press, Oxford, pp 117–126
Phi-Van L, Muller-Estrel W, Florke S, Schmid M, Engel W (1983) Biol Reprod 29:479
Pholpramool C, Chaturapanich G (1979) J Reprod Fertil 57:245
Picheral B, Bassez Jh (1971a) J Microsc 12:107
Picheral B, Bassez Jh (1971b) J Microsc 12:341
Pikó L (1969) In: Metz CB, Monory A (eds) Fertilization. Academic Press, London New York, pp 325–404
Pikó L, Tyler A (1964) Proc 5th Int Congr Reprod, Toronto, vol II, p 372
Pinatel MC, Czyba JC, Guerin JF (1980) Acta Eur Fertil 11:281
Pinkel D, Gledhill BL, Lake S, Stephenson D, Dilla van MA (1982) Science 218:904
Pinkel D, Garner DL, Gledhill BL, Lake S, Stephenson D, Johnson LA (1983a) J Anim Sci 57 Suppl 1:366 (Abstr)
Pinkel D, Geldhill BL, Dilla van MA, Lake S, Wyrobek AJ (1983b) Radiat Res 95:550
Plachot M, Mandelbaum J, Junca AM (1984) Fertil Steril 42:418
Plattner H (1971) J Submicrosc Cytol 3:19
Platz RD, Grimes SR, Meistrich ML, Hnilica LS (1975) J Biol Chem 250:5791
Plöen L (1973a) Virchows Arch Abt B Pathol Anat Physiol 14:159
Plöen L (1973b) Virchows Arch Abt B Pathol Anat Physiol 14:185
Plöen L, Ekwall H, Afzelius BA (1979) J Ultrastruct Res 68:149
Plowmann KM, Nelson L (1962) Biol Bull 123:478
Pogany GC, Corzett M, Weston S, Balhorn R (1981) Exp Cell Res 136:127
Poirier GR (1975) J Reprod Fertil 43:495
Polakoski KL (1974) Fed Proc Fed Am Sec Exp 33:1308 (Abstr)
Polakoski KL, Parrish RF (1977) J Biol Chem 252:1888
Polakoski KL, Zaneveld LJD (1976) Meth Enzyme 45:325
Polakoski KL, Zaneveld LJD, Williams WL (1972) Biol Reprod 6:23
Polakoski KL, McRorie RA, Williams WL (1973a) J Biol Chem 248:8178
Polakoski KL, McRorie RA, Williams WL (1973b) J Biol Chem 248:8183

Polakoski KL, Zahler WL, Paulson JD (1977) Fertil Steril 28:668
Polakoski KL, Clegg ED, Parrish RF (1979) Int J Biochem 10:483
Pond FR, Tripp MJ, Wu AS, Whanger PD, Schmitz JA (1983) J Reprod Fertil 69:411
Pongswasdi P, Svasti J (1976) J Biochim Biophys Acta 434:462
Posalaki Z, Szabo D, Bacsi E, Okros I (1968) J Histochem Cytochem 16:249
Poulos A, Voglmayr JK, White IG (1973a) J Reprod Fertil 32:309
Poulos A, Darin-Bannett AC, White IG (1973b) Comp Biochem Physiol 46B:541
Poulos A, Brown PD, Cox R, White IG (1974) J Reprod Fertil 36:442
Poulsen F (1983a) J Reprod Immunol 5:49
Poulsen F (1983b) J Clin Lab Immunol 10:59
Poulsen F, Hjort T (1981) J Clin Lab Immunol 6:69
Prasad RL, Singh LN (1980) Indian J Anim Sci 50:379
Premkumar E, Bhargava PM (1972) Nature New Biol 240:139
Premkumar E, Bhargava PM (1973) Indian J Biochem Biophys 10:239
Price JM (1973) J Cell Biol 59:272a (Abstr)
Price JM (1974) Thesis, Division of medical sciences. Harvard Univ, Cambridge, MA
Primakoff P, Myles DG (1983) Dev Biol 98:417
Primakoff P, Myles DG, Bellvé AR (1980) Dev Biol 80:324
Prubhu G, Hegde UC (1982) J Reprod Immunol 2:243
Pruslin FH, Romani M, Rodman RC (1980) Exp Cell Res 128:207
Pudney J, Fawcett DW (1985) J Androl Suppl 6:P-75
Pulkkinen P (1979) Contraception 17:423
Pulkkinen P, Sinervirta R, Jänne J (1977) J Reprod Fertil 51:399
Pulkkinen P, Puk K, Koso P, Jänne J (1978) Acta Endocrinol 87:845
Purvis K, Cusan L, Attramandal H, Ege A, Hansson V (1982) J Reprod Fertil 65:381
Puwaravutipanich T, Panyim S (1975) Exp Cell Res 90:153
Quinlivan WL, Sullivan H (1974) Fertil Steril 25:315
Quinn PJ, White IG (1967) Aust J Biol Sci 20:1205
Quinn PJ, White IG, Cleland KW (1969) J Reprod Fertil 18:209
Quinn PJ, Chow PYW, White IG (1980) J Reprod Fertil 60:403
Radu I, Voisin GA (1975) Differentiation 3:107
Rahi H, Srivastava PN (1984) Gamete Res 10:57
Rahi H, Sheikhnejade G, Srivastava PN (1983) Gamete Res 7:215
Rajalakshmi M (1985) J Biosci 7:191
Rajalakshmi M, Sehgal A, Pruthi JS, Anand Kumar TC (1983) Steroids 41:587
Ramaswami LS (1983) In: Goel SC, Bellairs R (eds) Developmental biology: an afro-asian perspective. Indian Soc Dev Biol, Poona, pp 215–233
Randall JT, Friedlander MHG (1950) Exp Cell Res 1:1
Rao BR, Pandey JN (1977) J Anim Sci 47:469
Rao SS, Sheth AR, Phadke AM (1959) Indian J Med Sci 13:302
Rasmussen H (1970) Science 170:404
Rattner JB (1972) J Ultrastruct Res 40:498
Rattner JB, Brinkley BR (1970) J Ultrastruct Res 32:316
Rattner JB, Brinkley BR (1971) J Ultrastruct Res 36:1
Raviola E, Gibila NB (1973) Proc Natl Acad Sci USA 70:1677
Reddy JM, Audhya TK, Goodpasture JC, Zaneveld LJ (1982) Biol Reprod 27:1076
Reddy PRK, Villee CA (1975) Biochem Biophys Res Commun 63:1063
Redi CA, Hilscher B, Winking H (1983) Andrologia 15:322
Reger JF, Fain-Maurel MA, Cassier P (1977) J Ultrastruct Res 60:84
Reichert LE Jr, Dias JA, Fletcher PW, O'Neill WC (1982) In: Bardin CW, Sherins RJ (eds) The cell biology of the testis. Ann NY Acad Sci, vol 383, pp 135–150
Reid BL (1977) J Reprod Fertil 49:169
Reissenweber NJ (1970) Histochemie 21:73
Rensburg van SWJ, Rensburg van SJ, Vos de WH (1966) Onderstepoort J Vet Res 33:169
Restall BJ, Wales RG (1968) Aust J Biol Sci 21:491
Retzius G (1909) Biol Untersuch NFH 14:163
Reviers de MT, Courot M (1976) VIII Congr Int Reprod Anim Insem Artif, Krako, vol III, pp 84–87

Reviers de M, Hochereau-de-Reviers MT, Blanc MR (1980) Reprod Nutr Dev 20:241
Reyes A, Oliphant G, Brackett BG (1975) Fertil Steril 26:148
Reyes A, Mercado E, Goicoechea B, Rosado A (1976) Fertil Steril 27:1452
Reyes A, Goicoechea B, Rosado A (1979) Fertil Steril 29:451
Reyes J, Allen J, Tanphaichitr N, Bellvé AR, Benos DJ (1984) J Biol Chem 259:9607
Richards JM, Witkin SS (1978) J Reprod Fertil 54:43
Rikmenspoel R (1978) Biophys J 23:177
Rikmenspoel R (1984) J Exp Biol 108:205
Rikmenspoel R, Orris SE, O'Day PM (1978) Exp Cell Res 111:253
Ringertz NR, Bolund L, Darzynkiewicz Z (1970) Exp Cell Res 63:233
Rink TJ (1977) J Reprod Fertil 51:155
Risbridger GP, Hodgson YM, Kretser de DM (1981) In: Burger H, Kretser de D (eds) The testis. Raven, New York, pp 195–212
Ritzen EM (1983) J Steroid Biochem 19:499
Ritzen EM, French FS (1974) J Steroid Biochem 5:151
Ritzen EM, Dobbins MC, Tindall DJ, French FS, Nayfeh SN (1973) Steroids 21:593
Ritzen EM, Hagenäs L, Ploen L, French FS, Hansson V (1977) Mol Cell Endocrinol 8:335
Ritzen EM, Hansson V, French FS (1981) In: Burger H, Kretser de D (eds) The testis. Raven, New York, pp 171–194
Ritzen EM, Boitani C, Parvinen M, French F, Feldman M (1982) Mol Cell Endocrinol 25:25
Rivarola MA, Podesta FJ, Chemes HE, Calandra RS (1977) In: James VHT (ed) Endocrinology, vol I. Excerpta Medica, Amsterdam, pp 307–313
Rob O, Rezinek J (1976) VIII Congr Int Reprod Anim Insem Artif, Krako, vol IV, pp 746–747
Roberts AM (1972) Nature (London) 238:223
Roberts ML, Scouten WH, Nyquist SE (1976) Biol Reprod 14:421
Robinson R, Fritz IB (1979) Can J Biochem 57:962
Robinson R, Fritz IB (1981) Biol Reprod 24:1032
Rodman TC, Litwin SD, Romani M, Vidali G (1979) J Cell Biol 80:605
Rodman TC, Pruslin FH, Allfrey VG (1982) J Cell Sci 53:227
Rodriguez-Rigau LJ, Zuckman Z, Weiss DB, Smith KD, Steinberger E (1980) In: Steinberger A, Steinberger E (eds) Testicular development, structure and function. Raven, New York, pp 139–146
Rogers BJ (1979) Gamete Res 1:165
Rogers BJ, Morton BE (1973) J Reprod Fertil 35:477
Rogers BJ, Yanagimachi R (1975) J Reprod Fertil 44:135
Rogers BJ, Yanagimachi R (1976) Biol Reprod 15:614
Rogers BJ, Chang L, Yanagimachi R (1979) J Exp Zool 207:107
Rohde W, Porstmann T, Dorner G (1973) J Reprod Fertil 33:167
Rommerts FFG, Krüger-Sewnarain BC, Woerkom-Blik A, Grootegoed van JA, Molen van der HJ (1978) Mol Cell Endocrinol 10:39
Romrell LJ, O'Rand MG (1978) Dev Biol 63:76
Romrell LJ, Ross MH (1979) Anat Rec 193:23
Romrell LJ, Bellvé AR, Fawcett DW (1976) Dev Biol 49:119
Rooij de DG (1970) In: Holstein AF, Horstmann E (eds) Morphological aspects of andrology, vol I. Grosse, Berlin, pp 13–16
Rooij de DG (1973) Cell Tissue Kinet 6:281
Rooij de DG (1980) Virchows Arch Abt B Cell Pathol 33:67
Rooij de DG, Kramer MF (1968) Z Zellforsch Mikrosk Anat 85:206
Rooij de DG, Kramer MF (1970) Virchows Arch Abt B Zellpathol 4:267
Roomans GM (1975) Exp Cell Res 96:23
Roomans GM, Afzelius BA (1975) J Submicrosc Cytol 7:61
Roosen-Runge EC (1955) Z Zellforsch Mikrosk Anat 41:221
Roosen-Runge EC (1962) Biol Rev 37:343
Roosen-Runge EC (1973) J Reprod Fertil 35:339
Roosen-Runge EC (1977) The process of spermatogenesis in animals. Cambridge Univ Press, Cambridge
Rosado A, Huacuja L, Delgado N, Hicks J, Pancardo R (1976) Life Sci 17:1707
Rosado A, Huacuja L, Delgado NM, Merchant H, Pancardo RM (1977) Life Sci 20:647

Ross A, Chrishe Sh, Kerr KG (1971) J Reprod Fertil 24:99
Ross A, Christie Sh, Edmond P (1973) J Reprod Fertil 32:243
Ross MH (1976) Anat Rec 186:79
Rothschild L (1960) Nature (London) 187:253
Rowley MJ, Berlin JD, Heller CG (1971) Z Zellforsch Mikrosk Anat 112:139
Rowson LEA (1975) In: Blandau RJ (ed) Aging gametes. Karger, Basel, pp 249–264
Roychaudhury PN, Pareek PK, Gowda HC (1974) Andrologia 4:315
Rudland PS, Durbin H, Clingan D, Asua de LJ (1977) Biochem Biophys Res Commun 75:556
Rudolph NS, Shovers JB, Berg van de JL (1982) J Reprod Fertil 65:39
Rufo GA Jr, Schoff PK, Lardy HA (1984) J Biol Chem 259:2547
Ruiter-Bootsma de AL, Karmer MF, Rooij de DG, Davis JAG (1976) Radiation Res 67:56
Russell LD (1977a) Am J Anat 148:313
Russell LD (1977b) Biol Reprod 17:184
Russell LD (1977c) Tissue Cell 9:475
Russell LD (1979a) Anat Rec 194:213
Russell LD (1979b) Anat Rec 194:233
Russell LD (1980a) Anat Rec 197:21
Russell LD (1980b) Gamete Res 3:179
Russell LD (1984) In: Blerkom van J, Motta PM (eds) Ultrastructure of Reproduction. Martinus Nijhoff, Boston, pp 46–66
Russell LD, Clermont Y (1977) Anat Rec 187:347
Russell LD, Frank B (1978) Anat Rec 190:79
Russell LD, Peterson RN (1985) Int Rev Cytol 94:177
Russell LD, Myers P, Ostenberg J, Malone J (1980a) In: Steinberger AS, Steinberger E (eds) Testicular development, structure and function. Raven, New York, pp 55–64
Russell LD, Peterson RN, Freund M (1980b) Anat Rec 198:449
Russell LD, Malone JP, MacCurdy DS (1981) Tissue Cell 13:349
Russell LD, Peterson RN, Russell TA (1982) J Histochem Cytochem 30:1217
Russell LD, Peterson RN, Russell TA, Hunt WP (1983a) Biol Reprod 28:393
Russell LD, Peterson RN, Hunt WP (1983b) J Androl 4:22
Russell LD, Tallon-Doran M, Weber JE, Wong V, Peterson RN (1983c) Am J Anat 167:181
Russell LD, Lee IP, Ettlin R, Malone JP (1983d) Tissue Cell 15:291
Russo E, Giancotti V, Crane-Robinson C, Geraci G (1983) Int J Biochem 15:487
Russo J, Metz CB (1974) Biol Reprod 10:293
Russo J, Metz CB, Dunbar BS (1975) Biol Reprod 13:136
Ruthkaplan PG, Yedwab G, Homonnai ZT, Kraicer PF (1978) Int J Androl 1:145
Saji F, Minagawa Y, Negor T, Nakamuro K, Tanizawa O (1985) Acta Obstet Gynec Jpn 37:801
Sakai Y, Yamamoto N, Yasuda K (1979) Acta Histochem Cytochem 12:151
Saling PM, Sowinski J, Storey BT (1979) J Exp Zool 209:229
Sanborn BM, Wagle JR, Steinberger A (1984) In: Catt KJ, Dufau ML (eds) Hormone action and testicular function. Ann NY Acad Sci, vol 438, pp 586–587
Sandoz D (1970) J Microsc 9:535
Sandoz D (1972) J Microsc 15:403
Sandoz D, Roland JC (1976) J Microsc Biol Cell 27:207
Santos-Sacchi J, Gordon M (1982) J Androl 3:108
Sapsford CS (1962a) Aust J Zool 10:178
Sapsford CS (1962b) Aust J Agric Res 13:487
Sapsford CS, Rae CA, Cleland KW (1967) Aust J Zool 15:881
Sapsford CS, Rae CA, Cleland KW (1969) Aust J Zool 17:195
Sarafis V, Lambert RW, Breed WG (1981) J Reprod Fertil 61:399
Sarmah BC, Kaker ML, Razdan MN (1984) Theriogenology 22:621
Sarvamangala BS, Krishnan KA, Jayaraman S, Sheth AR (1983) Arch Androl 10:233
Sastry BV, Michael R, Bishop R, Kau ST (1979) Biochem Pharmacol 28:1271
Sastry R (1975) Paper Conf Andrology, Wayne, State Univ, April 1975
Satir P (1974) Sci Am 231:44
Satir P (1979) In: Fawcett DW, Fawcett JM (eds) The spermatozoon. Urban & Schwarzenberg, Baltimore Munich, pp 81–90

Satir P (1982) Progress in protozoology. Proc VI Int Congr Protozool. Spec Congr Vol Acta Protozool Pt 1, pp 131–140
Sato N, Oura C (1985) Okajimas Folica Anat Jpn 61:267
Satoru K, Oshio S, Kobayashi T, Ilzuka R, Mohri H (1984) Biochem Biophys Res Commun 124:950
Sattayasai N, Panyim S (1982) Int J Androl 5:337
Saxena VB, Tripathi SS (1980) Indian J Anim Health 19:89
Schachner M, Wortham KA, Carter LD, Chaffee JK (1975) Dev Biol 44:313
Schiessler H, Fritz H, Arnhold M, Fink E, Tschiesche H (1972) Hoppe-Seylers Z Physiol Chem 353:1638
Schill WB (1974) Fertil Steril 25:703
Schill WB (1975a) Andrologia 7:229
Schill WB (1975b) Int J Fertil 20:61
Schill WB (1975c) Andrologia 7:105
Schill WB (1975d) Andrologia 7:135
Schill WB, Fritz H (1976) Arch Dermatol Res 256:1
Schill WB, Braun-Falco O, Haberland GL (1974) Int J Fertil 19:163
Schill WB, Wallner O, Palm S, Fritz H (1976) In: Hafez ESE (ed) Human semen and fertility regulation in men. Mosby, St. Louis, pp 442–451
Schill WB, Leidl W, Prinzen R, Wallner O (1978) In: Proc 4th Eur Steril Congr, Madrid, Spain, pp 440–444
Schill WB, Feifel M, Fritz H, Hammerstein J (1981) Int J Androl 4:25
Schilling E (1971) In: Kiddy CA, Hafs HD (eds) Sex ratio at birth-prospects for control. Am Soc Anim Sci, Albany, pp 76–84
Schilling E, Thormäehlen D (1975) Zuchthyg Fortpflanz Besam Haustiere 10:1880
Schilling E, Thormäehlen D (1976) VII Congr Int Reprod Anim Artif, Krako, vol IV, pp 931–934
Schilling E, Thormäehlen D (1977) Andrologia 9:106
Schirren C, Kaukel H, Ketels-Harken H (1977) Andrologia 9:313
Schleiermacher E, Schmidt W (1973) Humangenetik 19:75
Schleuning W, Hell R, Fritz H (1976) Hoppe-Seylers Z Physiol Chem 357:671
Schleuning W, Hell R, Fritz H (1977) Andrologia 9:28A
Schmid M, Löser C, Schmidtke J, Engel W (1982) Chromosoma 86:149
Schmid M, Müller HJ, Stasch S, Engel W (1983) Hum Genet 64:363
Schmidt FC (1964) Z Zellforsch Mikrosk Anat 63:707
Schmidt M, Krone W (1976) Chromosoma 56:327
Schmidt M, Hofgärtner FJ, Zenzes MT, Engel W (1977) Hum Genet 38:279
Schmitt DE, Polakoski KI (1980) Biol Reprod 22 Suppl 1 55 A Abstr: 10
Schnall M (1952) Fertil Steril 3:62
Schneyer CR, Pineyro MA, Gregerman RI (1983) Life Sci 33:275
Schoenfeld C, Amelar RD, Dubin L (1973) Fertil Steril 24:772
Schoenfeld C, Amelar RD, Dublin L (1975) Fertil Steril 26:158
Schroeder V (1941) Z Tierzucht Zucht Biol 50:1
Schuler HM, Gier HT (1976) J Exp Zool 197:1
Schultz MC, Hermo L, Leblond CP (1984) Am J Anat 171:41
Schultz-Larsen J (1958) Acta Pathol Microbiol Scand Suppl 128:1
Schulze C (1979) Cell Tissue Res 198:191
Schulze C (1981) Fortschr Androl 7:58
Schulze W (1978) Andrologia 10:307
Schulze W (1982) Andrologia 14:200
Schulze W, Rehder U (1984) Cell Tissue Res 237:395
Schulze W, Schulze C (1981) Cell Tissue Res 217:259
Schumacher GFB, Zaneveld LJD (1972) VII Int Congr Anim Reprod Artif Insem München, p 430 (Abstr)
Scott JE, Persaud TVN (1978) Int J Fertil 23:282
Scott TW (1973) J Reprod Fertil Suppl 18:65
Scott TW, Dawson RMC (1968) Biochem J 108:457
Scott TW, Dawson RMC, Rowlands IW (1963) Biochem J 87:507
Scott TW, Voglmayr JK, Setchell BP (1967) Biochem J 102:456

Seamark RF, White IG (1964) J Endocrinol 30:307
Segal SJ (ed) (1985) Gossypol. Plenum, New York
Seiguer AC, Castro AE (1972) Biol Reprod 7:31
Selivonchick DP, Schmidt PC, Natarajan V, Schmidt HO (1980) Biochem Biophys Acta 618:242
Semakov VG, Ryzhkov TF (1976) S Kh Biol 11:200
Semczuk M (1977) Z Mikrosk Anat Forsch 9:31
Serres C, Feneux D, David G (1982) Andrologia 14:45
Setchell BP (1974) J Reprod Fertil 37:165
Setchell BP, Waites GMH (1974) In: Hafez ESE, Thibault CG (eds) Transport, Survie et pouvoir fécondant des spermatozoides chez les vertébres. INSERM, Paris, pp 11-34
Setchell BP, Voglmayr JK, Waites GMH (1969) J Physiol 200:73
Setchell BP, Davies RV, Main SJ (1977) In: Johnson AD, Gomes SR (eds) The testis, vol IV. Academic Press, London New York, pp 189-238
Setchell BP, Laurie MS, Fritz IB (1980) In: Steinberger A, Steinberger E (eds) Testicular development, structure and function. Raven, New York, pp 65-70
Seyedin SM, Kistler WS (1979) Biochemistry 18:1371
Shah GV, Sheth AR (1978) Experientia 34:980
Shah GV, Gunjikar AN, Sheth AR, Raut SJ (1980a) Arch Androl 4:145
Shah GV, Gunjikar AN, Sheth AR, Raut SJ (1980b) Andrologia 12:207
Shapiro BM (1984) Ciba Found Symp 103:86
Shapiro BM, Eddy EM (1980) Int Rev Cytol 66:257
Sharma DP, Venkitasubramanian TA (1977) Indian J Anim Sci 45:812
Sharpe RM (1983) Acta Endocrinol 103 Suppl 256:57 (Abstr)
Sharpe RM (1984) Biol Reprod 30:29
Sharpe RM, Fraser HM, Cooper I, Rommerts FFG (1981) Nature (London) 290:785
Shastry PR, Hegde UC, Rao SS (1977) Nature (London) 269:58
Shelton JA, Abe M, Goldberg E (1983) J Reprod Immunol Suppl 86 (Abstr)
Shepherd RW, Millette CF, Wolf de WC (1981) Gamete Res 4:487
Sherman JK (1973) Fertil Steril 24:397
Sherman JK, Liu KC (1982) Cryobiology 19:503
Sheth AR, Shah GV, Gunjikar AN (1980) J Androl 1:84 (Abstr)
Sheth AR, Sucheta DP, Seethalakshmi N (1981a) Andrologia 13:232
Sheth AR, Gunjikar AN, Shah GV (1981b) Andrologia 13:142
Shi QH, Friends DS (1985) J Androl 6:45
Shilon M, Paz G, Homonnai ZT, Schoenbaum M (1978) Int J Androl 1:416
Shinohara H. Yanagimachi R, Srivastava PN (1985) Gamete Res 11:19
Shires A, Carpenter MP, Chalkley R (1975) Proc Natl Acad Sci USA 72:2714
Shires A, Carpenter M, Chalkley R (1976) J Biol Chem 251:4155
Shirley MA, Schachter H (1980) Can J Biochem 58:1230
Shur BD, Hall NG (1982a) J Cell Biol 95:567
Shur BD, Hall NG (1982b) J Cell Biol 95:574
Sidhu KS (1980) Biol Reprod 22 (Suppl 1) 133 A
Sidhu KS, Guraya SS (1977) Acta Morphol Acad Sci Hung 25:161
Sidhu KS, Guraya SS (1978a) Indian J Exp Biol 16:1185
Sidhu KS, Guraya SS (1978b) Ann Biol Anim Bioch Biophys 18:283
Sidhu KS, Guraya SS (1978c) J Steroid Biochem 9:824
Sidhu KS, Guraya SS (1979a) J Reprod Fertil 57:205
Sidhu KS, Guraya SS (1979b) Indian J Anim Sci 49:884
Sidhu KS, Guraya SS (1979c) Ann Biol Anim Bioch Biophys 19:45
Sidhu KS, Guraya SS (1980a) Indian J Exp Biol 18:48
Sidhu KS, Guraya SS (1980b) J Steroid Biochem 13:997
Sidhu KS, Guraya SS (1982) Buffalo Sem Reprod Meat Prod, Jan 15-17, 1982, Tanuku, A P, India, pp 27-28
Sidhu KS, Guraya SS (1985a) Buffalo Sem Morphol Physiol Methodol, USG, Ludhiana, India
Sidhu KS, Guraya SS (1985b) Buffalo J 1:37
Sidhu KS, Guraya SS (1986) Int J Androl (in press)
Sidhu KS, Sundhey R, Guraya SS (1984) Int J Androl 7:324

Sidhu KS, Bassi A, Gupta BC, Guraya SS (1986a) Int J Androl 9:(in press)
Sidhu KS, Sundhey R, Guraya SS (1986b) Aust J Biol Sci (in press)
Sidhu KS, Sundhey R, Guraya SS (1986c) Life Sci Adv (in press)
Sidhu KS, Sandhu JK, Guraya SS (1986d) Proc IV Int Symp Vet Lab Diagn 1:4
Sidhu KS, Bassi A, Singh N, Guraya SS (1986e) Theriogenology (in press)
Siegel MS, Polakoski KL (1985) Biol Reprod 32:713
Silver LM, White M (1982) Dev Biol 91:423
Silvestroni L, Sartori C, Modesti A, Frajese G (1983) In: Shulman S, Dondero F (eds) Immunological factors in human contraception. Acta Medica, Rome, pp 67–71
Simpson ME, Evans HM (1946) Endocrinology 39:281
Singer R, Sagiv M, Barnet M, Segenreich E, Allalouf D, Landau B, Servadio C (1980a) Andrologia 12:92
Singer R, Barnet M, Allalouf D, Servadiao C (1980b) Arch Androl 5:195
Singer R, Sagiv M, Allalouf D, Levinski H, Barnet M, Landau B, Segenreich E, Servadio C (1980c) Int J Androl 5:195
Singer R, Sagiv M, Barnet M, Allalouf D, Landau B, Segenreich E, Servadio C (1981) Andrologia 13:256
Singer R, Sagiv M, Allalouf D, Levinski H, Barnet M, Landau B, Segenreich E, Servadio C (1982) Int J Fertil 27:176
Singer R, Sagiv M, Allalouf D, Levinsky H, Barnet M, Landau B, Servadio C (1983) Int J Fertil 28:119
Singer R, Joshua H, Hazaz B, Sagir M, Laner H, Levinsky H, Servadio C (1986) Arch Androl 15:11
Singh B, Sandhu DP (1972) J Dairy Sci 25:91
Singh JP, Babcock DF, Lardy HA (1978) Biochem J 172:549
Singleton CL, Killian GJ (1983) J Androl 4:150
Sinowatz F, Friess AE (1983) Histochemistry 79:335
Sinowatz F, Wrobel KH (1981) Cell Tissue Res 219:511
Sipski ML, Wagner TE (1977) Biol Reprod 16:428
Skinner MK, Griswold MD (1980) J Biol Chem 255:9523
Skinner MK, Griswold MD (1983) Biol Reprod 28:1225
Skinner MK, Tung PS, Fritz IB (1985) J Cell Biol 100:1941
Skrabei H, Mettler L (1977) Res Exp Med 171:297
Skrzypczak J, Pisarski T, Biczysko W (1981) Folia Morphol 40:435
Smith AJ, Clausen OPF, Kirkhus B, Jahnsen T, Mollen OM, Hansson V (1984) J Reprod Fertil 72:453
Smith BVK, Lacy D (1959) Nature (London) 184:249
Smith FE, Berlin JD (1977) Cell Tissue Res 176:235
Smith KD, Stultz DR, Steinberger E (1975) In: Blandau RJ (ed) Aging gametes. Karger, Basel, pp 265–277
Smithwick EB, Young LG (1977) Biol Reprod 17:443
Smithwick EB, Young LG (1978) Biol Reprod 19:280
Sobhon P, Tanphaichitr N, Chuttatape C, Vongperybal P, Panuwatsuk W (1982) J Exp Zool 223:277
Söderström K-O (1976a) Chromosome activities during spermatogenesis in the rat. Diss, Univ Turku, Kirjapaino Pika Oy Turku
Söderström K-O (1976b) Exp Cell Res 102:237
Söderström K-O (1977) Cell Tissue Res 184:411
Söderström K-O (1979) Z Mikrosk Anat Forsch 92:417
Söderström K-O (1981) Cell Tissue Res 215:425
Söderström K-O, Anderson LC (1981) Exp Mol Pathol 35:332
Söderström K-O, Parvinen M (1976a) Hereditas 82:25
Söderström K-O, Parvinen M (1976b) J Cell Biol 70:239
Söderström K-O, Parvinen M (1976c) Mol Cell Endocrinol 5:181
Söderström K-O, Parvinen M (1976d) Cell Tissue Res 168:335
Söderström K-O, Parvinen M (1978) Acta Anat 100:557
Söderström K-O, Sege K, Anderson LC (1982) J Immunol 128:1671
Sohval AR, Suzuki Y, Gabrilove JL,. Churg J (1971) J Ultrastruct Res 34:83
Solari AJ (1964) Exp Cell Res 36:160
Solari AJ (1969) J Ultrastruct Res 27:289
Solari AJ (1970) Chromosoma 29:217

Solari AJ (1971) Chromosoma 34:99
Solari AJ (1974) Int Rev Cytol 38:273
Solari AJ, Bianchi NO (1975) Chromosoma 52:11
Solari AJ, Tres L (1967) Exp Cell Res 47:86
Sosa A, Altamirano E, Hernandez P, Rosado A (1972) Life Sci 11:493
Soucek DA, Vary JC (1983) Biol Reprod 28 Suppl 1:157
Soucek DA, Vary JC (1984a) Congr Proc 10th Int Congr Anim Reprod Artif Insem, vol II. Univ Ill Urbana-Champaign, USA, p 46
Soucek DA, Vary JC (1984b) Biol Reprod 31:687
Speed RM, deBoer P (1983) Cytogenet Cell Genet 35:257
Srivastava PN (1973) Biol Reprod 9:84
Srivastava PN, Hussain AI (1977) Biochem J 161:193
Srivastava PN, Ninjoor V (1982) Biochem Biophys Res Commun 109:63
Srivastava PN, Adams CE, Hartree EF (1965) Nature (London) 205:498
Srivastava PN, Zaneveld LJD, Williams WL (1970) Biochem Biophys Res Commun 39:575
Srivastava PN, Munnell JF, Yang CH, Foley CW (1974) J Reprod Fertil 36:363
Srivastava PN, Brewer JM, White RA Jr (1982) Biochem Biophys Res Commun 108:1120
Srivastava PN, Tandon HO, Kumar V (1984) Gamete Res 9:239
Stackpole CW, Devorkin D (1974) J Ultrastruct Res 49:167
Stambaugh R (1978) Gamete Res 1:65
Stambaugh R, Buckley J (1971) In: Kiddy CA, Hafs HD (eds) Sex ratio at birth-prospects for control. Am Soc Anim Sci, Albany, pp 59–68
Stambaugh R, Buckley J (1972) Biochem Biophys Acta 284:473
Stambaugh R, Mastroianni L Jr (1980) J Reprod Fertil 59:479
Stambaugh R, Smith M (1978) J Exp Zool 203:135
Stambaugh R, Smith M, Faltas S (1975) J Exp Zool 193:199
Stanislavov R, Petrova V, Protich M, Nalbanski B (1978) Akush Ginekol (Sofiia) 17:339
Steeno O, Adimoelja A, Steeno J (1975) Andrologia 7:95
Stefanini M, Oura C, Zamboni L (1969) J Submicrosc Cytol 1:1
Stefanini M, Martino de C, Agostino A-D, Agrestini A, Monesi V (1974) Exp Cell Res 86:166
Stefanov S, Penkov V (1979) God Sofii Univ Biol Fak 69:35
Stefanovrich V (1973) Commun Chem Pathol Pharmacol 5:665
Steinberger A, Steinberger E (1966) Exp Cell Res 44:429
Steinberger A, Steinberger E (1967) J Reprod Fertil Suppl 2:117
Steinberger A, Steinberger E (1977) In: Johnson AD, Gomes WR (eds) The testis, vol IV. Academic Press, London New York, pp 371–399
Steinberger A, Steinberger E (eds) (1980) Testicular development, structure and function. Raven, New York
Steinberger A, Walther J, Heindel JJ, Sanborn BM, Tsai YH, Steinberger E (1979) In vitro 15:23
Steinberger AE, Sud BN (1970) J Reprod Fertil 56:201
Steinberger E (1971) Physiol Rev 51:1
Steinberger E, Duckett GE (1965) Endocrinology 76:1184
Steinberger E, Steinberger A (1972) In: Balin H, Glasser S (eds) Reproductive biology. Excerpta Medica, Amsterdam, pp 144–267
Steinberger E, Steinberger A (1975) In: Hamilton DW, Greep RO (eds) Handbook of physiology, Sect 7: Endocrinology, vol V. Am Physiol Soc, Washington DC, pp 1–20
Steinberger E, Steinberger A, Perloff WH (1964) Endocrinology 74:788
Steinberger E, Steinberger A, Ficher M (1970) Recent Prog Horm Res 26:547
Steinberger E, Root A, Ficher M, Smith KD (1973) J Clin Endocrinol Metab 37:746
Stengel D, Hanoune J (1984) In: Catt KJ, Dufau ML (eds) Hormone action and testicular function. Ann NY Acad Sci, vol 438, pp 18–28
Stephens DT, Wang J-J, Hoskins DD (1979) Biol Reprod 20:483
Stephens DT, Critchlow LM, Hoskins DD (1983) J Reprod Fertil 69:447
Stephens RE (1974) In: Sleigh MA (ed) Cilia and flagella. Academic Press, London New York, pp 39–76
Stern L, Gold B, Hecht NB (1983a) Biol Reprod 28:483
Stern L, Keene KC, Gold B, Hecht NB (1983b) Exp Cell Res 143:247

Steven FS, Griffin MM, Chantler EN (1982) Int J Androl 5:401
Stevenson D, Jones AR (1982) Aust J Biol Sci 35:595
Stevenson D, Jones AR (1984) Int J Androl 7:79
Stewart TA, Forrester T (1978a) Biol Reprod 19:271
Stewart TA, Forrester T (1978b) Biol Reprod 19:965
Stewart TA, Forrester T (1979) Biol Reprod 21:109
Stewart TA, Bellvé AR, Leder P (1984) Science 226:7070
Stockert JC, Colman OD, Gimenez-Martin G, Esponda P (1975) Mikroskopie 31:36
Storey BT (1975) Biol Reprod 13:1
Storey BT, Kayne FJ (1975) Fertil Steril 26:1257
Storey BT, Kayne FJ (1977) Biol Reprod 16:540
Storey BT, Kayne FJ (1978) Biol Reprod 18:527
Storey BT, Kayne FJ (1980) J Exp Zool 211:361
Straus JM, Polakoski KI (1982a) Biochem J 201:657
Straus JM, Polakoski KI (1982b) J Biol Chem 257:7962
Struck E (1974) 8th World Congr Fertil Steril, Buenos Aires Abstr 497
Sturzebecher J (1981) Acta Biol Med Ger 40:1519
Sturzebecher J, Markwardt F (1980) Hoppe-Seylers Z Physiol Chem 361:25
Suarez-Quian CA, Dym M (1984) In: Catt KJ, Dufau ML (eds) Hormone action and testicular function. Ann NY Acad Sci, vol 438, pp 476–480
Suarez-Quian CA, Hadley MA, Dym M (1984) In: Catt KJ, Dufau ML (eds) Hormone action and testicular function. Ann NY Acad Sci, vol 438, pp 417–434
Suarez SS, Hinton BT, Oliphant HG (1981) Biol Reprod 25:1091
Suarez SS, Katz DF, Meizel S (1984) Gamete Res 10:253
Subirana JA (1975) In: Duckett JG, Racey PA (eds) The biology of the male gamete. Academic Press, London New York, pp 239–244
Sud BN (1961) Q Microsc Sci 102:495
Sullivan JL, Smith FA, Garman RH (1979) J Reprod Fertil 56:201
Sumi H, Toki N (1980) Experientia 36:1103
Sumner AT, Robinson JA (1976) J Reprod Fertil 48:9
Sumner AT, Robinson JA, Evans HJ (1971) Nature New Biol 229:231
Summers KE (1974) J Cell Biol 60:321
Summers KE, Gibbons IR (1971) Proc Natl Acad Sci USA 68:3092
Summers KE, Gibbons IR (1973) J Cell Biol 58:618
Sundhey R, Sidhu KS, Guraya SS (1984) Indian J Anim Reprod 5:47
Susi FR (1970) Anat Rec 166:386 (Abstr)
Susi FR, Clermont Y (1970) Am J Anat 129:177
Susi FR, Leblond CP, Clermont Y (1971) Am J Anat 130:251
Susko-Parrish JS, Hammerstedt RH, Senger PL (1985) Biol Reprod 32:129
Suter D, Holland MK (1979) Fertil Steril 31:54
Suter D, Chow PYW, Martin ICA (1979) Biol Reprod 20:505
Suzuki F (1981) Anat Rec 199:361
Swerdloff RS, Heber D (1981) In: Burger H, Kretser de D (eds) The testis. Raven, New York, pp 107–126
Swierstra EE, Foote RH (1963) J Reprod Fertil 5:309
Swierstra EE, Foote RH (1965) Am J Anat 116:401
Swierstra EE, Gebauer MR, Pickett BW (1974) J Reprod Fertil 40:113
Syed V, Khan SA, Ritzen EM (1983) Acta Endocrinol 103 Suppl 256:249 (Abstr)
Sylvester SR, Griswold MD (1984) In: Catt KJ, Dufau ML (eds) Hormone action and testicular function. Ann NY Acad Sci, vol 438, pp 561–562
Sylvester SR, Griswold MD (1984) Biol Reprod 31:195
Syner FN, Kuras RA, Moghissi KA (1979) Biol Reprod 21:857
Szöllösi D, Hunter RHF (1978) J Anat 127:33
Szuki N, Nagano T (1980) Biol Reprod 22:1219
Szuki N, Withers HR (1978) Science 202:1214
Tait AJ, Johnson E (1982) J Reprod 65:53
Takanari H, Pathak S, Hsu TC (1982) Chromosoma 86:359

Takei GH, Fleming AD, Yanagimachi R (1984) Andrologia 16:38
Taketo T, Koide S (1981) Dev Biol 84:61
Taketo T, Thau RB, Adeyemo O, Koide SS (1984) Biol Reprod 30:189
Talbot P (1979) J Reprod Fertil 55:9
Talbot P (1984) J Exp Zool 229:309
Talbot P, Chacon R (1981) J Exp Zool 21:435
Talbot P, Franklin LE (1974a) J Exp Zool 189:311
Talbot P, Franklin LE (1974b) J Exp Zool 189:321
Talbot P, Franklin LE (1976) J Exp Zool 198:163
Talbot P, Kleve MG (1977) J Cell Biol 75:G726, 170a
Talbot P, Kleve MG (1978) J Exp Zool 204:131
Talbot P, Summers RG, Hylander BL, Keough EM, Franklin LE (1976) J Exp Zool 198:383
Tamblyn TM, First NL (1977) Arch Biochem Biophys 181:208
Tang XM, Lalli MF, Clermont Y (1982) Am J Anat 163:283
Tanphaichitr N (1977) Int J Fertil 22:85
Tanphaichitr N, Sobhon P, Taluppeth N, Chalermisarachai P (1978) Exp Cell Res 117:347
Tanphaichitr N, Chen LB, Bellvé AR (1984) Biol Reprod 31:1049
Tanphaitr N, Bellvé AR (1985) In: Segal SJ (ed) Gossypol. Plenum, New York, pp 119–141
Tash JS, Means AR (1982) Biol Reprod 26:745
Tash JS, Means AR (1983) Biol Reprod 28:75
Tasseron F, Amir D, Schindler H (1977) J Reprod Fertil 51:461
Teichman RJ, Bernstein MH (1971) J Reprod Fertil 27:243
Teichman RJ, Cummins JM, Takei GH (1974) Biol Reprod 10:565
Temple-Smith PD, Bedford JM (1976) Am J Anat 147:471
Tesarik J (1985) J Reprod Fertil 74:383
Teuscher C, Wild GC, Tung KSK (1982) Biol Reprod 26:218
Thakkar JK, Franson RC (1983) Fed Proc Fed Am Soc Exp 42:1905
Thakkar JK, East J, Seyler D, Franson RC (1983) Biochim Biophys Acta 754:44
Thibault C (1969) In: Grasse PP (ed) Traité de zoologie, anatomie, systématique, biologie, vol XVI. Masson, Paris, pp 716–798
Thormas TF, Huang JR, Hardy D, Yanagimachi H, Teuscher C, Tung K, Wild G, Yanagimachi R (1985) Biol Reprod 32:451
Thumann A, Bustos-Obregon E (1982) Andrologia 14:35
Thumann A, Carboni R, Bustos-Obregon E (1981) Andrologia 13:583
Tice LW, Barnett RJ (1963) Anat Rec 147:43
Tierney WJ, Daly IW, Abbatiello ER (1979) Int J Fertil 24:206
Tijole DY, Steinberger E (1966) Naturwissenschaften 53:486
Tillman WL (1972) Thesis, Univ Georgia, Athens
Tilney LG (1975) In: Inoue S, Stephens RE (eds) Molecules and cell movement. Raven, New York, pp 339–386
Tilney LG, Bryan J, Bush DJ, Fujievara K, Mooseker MS, Morphy B, Synder DH (1973) J Cell Biol 59:67
Tindall DJ, Rowley DR, Lipshultz LI (1983) In: Lipshultz LI, Howards SS (eds) Infertility in the male. Churchill, Livingstone London, pp 71–98
Tindall DJ, Rowley DR, Murthy L, Lipshultz LI, Chang CH (1985) Int Rev Cytol 94:127
Tinneberg H-R, Cheng C-Y, Boettcher B, Kay DJ, Buxton J (1980) Fertil Steril 33:94
Tobias PS, Schumacher GFB (1977) Biochem Biophys Res Commun 74:434
Tobita T, Nomoto M, Tsutsumi H, Nakano M, Ando T (1979) Proc 11th Int Congr Biochem, Toronto, p 46 (Abstr)
Tobita T. Nomoto M, Nakano M, Ando T (1982) Biochem Biophys Acta 707:252
Tobita T, Suzuki H, Soma K, Nakano M (1983) Biochim Biophys Acta 748:461
Tokuyasu KT (1974) J Ultrastruct Res 48:284
Toowicharanont P, Chulavatnatol M (1983a) J Reprod Fertil 67:275
Toowicharanont P, Chulavatnatol M (1983b) J Reprod Fertil 67:133
Töpfer-Petersen E, Schill W-B (1981) Andrologia 13:174
Töpfer-Petersen E, Schill W-B (1983) Int J Androl 6:375
Töpfer-Petersen E, Schmoeckel C, Schill W-B (1983a) Andrologia 15:197

Töpfer-Petersen E, Schmoeckel C, Schill W-B (1983b) Andrologia 15:62
Töpfer-Petersen E, Weiss A, Krassnigg F, Schill W-B (1984) Congr Proc 10th Int Congr Anim Reprod Artif Insem, vol II. Univ Ill, Urbana-Champaign, USA, p 45
Töpfer-Petersen E, Heissler E, Schill W-B (1985) Andrologia 17:224
Toppari J, Brown WRA, Parvinen M (1984) In: Catt KJ, Dufau ML (eds) Hormone action and testicular function. Ann NY Acad Sci, vol 438, pp 515–518
Toullet F, Voisin GA (1974) J Reprod Fertil 37:299
Toullet F, Voisin GA, Nemirovsky M (1973) Immunology 24:635
Toyama Y (1976) Anat Rec 186:477
Toyoda Y, Chang MC (1974) J Reprod Fertil 36:175
Treetisatit N, Chulavatnatol M (1982) Exp Cell Res 142:495
Tres LL, Kierszenbaum AL (1975) In: French FS, Hansson V, Ritzen EM, Nayfeh SN (eds) Hormonal regulation of spermatogenesis. Plenum, New York, pp 455–478
Tres LL, Kierszenbaum AL (1983) Proc Natl Acad Sci USA 80:3377
Tres LL, Solari AJ (1968) Z Zellforsch Mikrosk Anat 91:75
Triana LR, Babcicz DF, Lorton SP, First NL, Lardy HA (1980) Biol Reprod 23:47
Trifunac NP, Bernstein GS (1976) Fertil Steril 27:1295
Trifunac NP, Bernstein GS (1982) Contraception 25:69
Troen P (1978) In: Hansson V, Ritzen EM, Purvis K, French FS (eds) Endocrine approach to male contraception. Scriptor, Copenhagen, pp 432–433
Tschesche H, Wittig B, Decker G, Muller-Esterl W, Fritz H (1982) Eur J Biochem 128:99
Tso WW (1985) In: Segal S (ed) Gossypol, a potential contraceptive for man reproductive biology. Plenum, New York, pp 245–256
Tso WW, Lee CS (1982) Arch Androl 8:148
Tsong S-D, Phillips DM, Halmi N, Krieger D, Bardin CW (1982a) Biol Reprod 27:755
Tsong S-D, Phillips DM, Halmi N, Liotta AS, Margioris A, Bardin CW, Krieger DT (1982b) Endocrinology 110:2204
Tsutomu H, Otsuka T, Kanematsu S (1984) Jpn J Anim Sci 30:159
Tung KSK, Fritz IB (1978) Dev Biol 64:297
Tung KSK, Han LPB, Evan AP (1979) Dev Biol 68:224
Tung KSK, Okada A, Yanagimachi R (1980) Biol Reprod 23:877
Tung KSK, Yanagimachi H, Yanagimachi R (1982) Anat Rec 202:241
Tung PS, Skinner MK, Fritz IB (1984) In: Catt KJ, Dufau ML (eds) Hormone action and testicular function. Ann NY Acad Sci, vol 438, pp 435–446
Turkington RW, Majumder GC (1975) J Cell Physiol 85:495
Turner TT, Giles RD (1982) Am J Physiol 242:R199
Turner TT, Howards SS (1978) Biol Reprod 18:571
Turner TT, Hartmann PK, Howards SS (1977) Fertil Steril 28:191
Turner TT, D'Addario D, Howards SS (1978) Biol Reprod 19:1095
Turner de EA, Aparicio N, Turner D, Schwarzstein L (1979) Fertil Steril 29:328
Uhlenbruck G, Sprenger I, Schumacher GFB, Zaneveld LJD (1972) Naturwissenschaften 59:124
Ulvik NM (1983a) Int J Androl 6:367
Ulvik NM (1983b) Int J Androl 6:469
Ulvik NM, Dahl E (1981) Cell Tissue Res 221:311
Unal MB, Berndtson WE, Pickett BW (1978) J Dairy Sci 61:83
Underwood EF (1977) Trace elements in human and animal nutrition. 4th edn, Chap 8. Academic Press, London New York, pp 196–241
Unni E, Rao MRS, Ganguly J (1983) Indian J Exp Biol 21:180
Unterberger F (1930) Dtsch Med Wochenschr 56:304
Upadhyay S, Zamboni L (1982) Proc Natl Acad Sci USA 79:6584
Urch UA, Wardrip NJ, Hedrick JL (1985a) J Exp Zool 236:239
Urch UA, Wardrip NJ, Hedrick JL (1985b) J Exp Zool 233:479
Urena F, Malavasi J (1978) Rev Biol Trop 26:371
Uschewa A, Avramova Z, Isanev R (1982) FEBS Lett 138:50
Usselman MC, Cone RA (1983) Biol Reprod 29:1241
Usselman MC, Carr DW, Acott TS (1984) In: Catt KJ, Dufau ML (eds) Hormone action and testicular function. Ann NY Acad Sci, vol 438, pp 530–532

Utakoji T (1966) Exp Cell Res 42:585
Uzu G, Courtens JL, Courot M (1976) VIII Congr Int Reprod Anim Insem Artif, Krako, vol IV, pp 748–750
Vaiman M, Fellous M, Wiels J, Renard C, Lecointre J, Messildu Busson du F, Dausset J (1978) J Immunogenet 5:135
Vale Filho VR, Megale F, Garcia OS (1976) VIII Congr Int Reprod Anim Insem Artif, Krako, vol IV, pp 752–755
Vandemark NL, Free MJ (1970) In: Johnson AD, Gomes WR, Vandemark NL (eds) The testis, vol III. Academic Press, London New York, pp 233–312
Vanha-Perttula T (1978) Ann Biol Anim Biochim Biophys 18:633
Vanha-Perttula T, Mather JP, Bardin CW, Moss SB, Bellvé AR (1985) Biol Reprod 33:870
Varshney GC, Anand SR (1981) Indian J Biochem Biophys 18:144 (Abstr)
Vaughn JC (1966) J Cell Biol 31:257
Vegni-Talluri M, Menchini-Fabris F, Renieri T (1977) Andrologia 9:315
Ven van der HH, Kaminski J, Baur L, Zaneveld LJD (1985) Fertil Steril 43:609
Vera JC, Brito M, Zuvic T, Burzio LO (1984) J Biol Chem 259:597
Veres I (1968) Mikroskopie 23:166
Veres I, Očsenyi A (1968) Mikroskopie 23:48
Veres I, Muller E, Očsenyi A, Turanyi J, Szasz F, Bernroider G, Kronroif AJ (1976) VIII Congr Int Reprod Anim Insem Artif, Krako, vol IV, pp 857–858
Vernon RB (1983) Anat Rec 205:206A
Vernon RB, Muller CH, Herr JC, Fenchter FA, Eddy EM (1982) Biol Reprod 26:523
Vernon RB, Hamilton MS, Eddy EM (1985) Biol Reprod 32:669
Vernon RG, Go VLW, Fritz IB (1971) Can J Biochem 49:761
Vernon RG, Go VLW, Fritz IB (1975) J Reprod Fertil 42:77
Veselsky L, Cechova D (1980) Hoppe-Seylers Z Physiol Chem 361:715
Vierula M, Rajaniemi H (1981) Int J Androl 4:314
Vierula M, Rajaniemi H (1982) Med Biol 60:323
Vierula M, Rajaniemi H (1983) Andrologia 15:436
Viguier-Martinez MC, Hochereau-de-Reviers MT (1977) Ann Biol Anim Biochim Biophys 17:1069
Viguier-Martinez MC, Hochereau-de-Reviers MT, Perreau C (1985) Acta Endocrinol 109:550
Vijayasarathy S, Balaram P (1982) Biochem Biophys Res Commun 108:760
Vijayasarathy S, Shivaji S, Balaram P (1980) FEBS Lett 114:45
Vijayasarathy S, Shivaji S, Balaram P (1982) Biochem Biophys Res Commun 108:760
Vilar O (1973) In: Hafez ESE, Evans TN (eds) Human reproduction, conception and contraception. Harper & Row, New York, pp 12–37
Vilar O, Paulsen CA (1967) Anat Rec 157:336 (Abstr)
Vilar O, Paulsen A, Moore DJ (1970) In: Rosemberg ER, Paulsen CA (eds) The human testis, vol X. Plenum, New York, pp 63–74
Vilar O, Glovenco P, Calamera JC (1980) Andrologia 12:225
Villarroya S, Scholler R (1986) J Reprod Fertil 76:435
Vogl AW, Soucy LJ (1985) J Cell Biol 100:814
Vogl AW, Dym LM, Fawcett AW (1983a) Am J Anat 168:83
Vogl AW, Linck RW, Dym M (1983b) Am J Anat 168:99
Vogl AW, Soucy LJ, Lew GJ (1985a) Anat Rec 213:63
Vogl AW, Soucy LJ, Foo V (1985b) Am J Anat 172:75
Voglmayr JK (1971) Acta Endocrinol 68:793
Voglmayr JK (1973) Prostaglandins 4:673
Voglmayr JK (1975) In: Hamilton DW, Greep RO (eds) Handbook of physiology, Sect 7: Endocrinology, vol V. Am Physiol Soc, Washington DC, pp 437–452
Voglmayr JK, Fairbank G, Jackowitz MA, Colella JR (1980) Biol Reprod 22:655
Voglmayr JK, Fairbanks G, Vespa DB, Colella JR (1982) Biol Reprod 26:483
Voglmayr JK, Fairbanks G, Lewis RG (1983) Biol Reprod 29:767
Voglmayr JK, Sawyer RF Jr, Dacheux JL (1985) Biol Reprod 33:165
Voisin GA (1983) In: Sulman S, Dondero F (eds) Immunological factors in human contraception. Acta Medica, Rome, pp 23–38

Voisin GA, Toullet F, Almeida MD (1974) In: Diczfalusy E (ed) Immunological approaches to fertility control. Trans 7th Symp, Geneva, July 29–31, 1974, pp 173–301
Vojtiskova M, Polackova M, Pokorna Z (1969) Folia Biol 15:322
Vorstenbosch van, Spek E, Colenbreder B, Wensing CJG (1984) Biol Reprod 31:565
Wachstein M, Meisel E (1956) J Histochem Cytochem 26:92
Wachtel SS (1977) Transplant Rev 33:33
Wagner TE, Yun TS (1979) Arch Androl 2:291
Wagner TE, Yun JS (1981) Arch Androl 7:251
Waibel R, Ginsberg LC, Ficsor G (1984) J Histochem Cytochem 32:63
Waites GMH (1977) In: Johnson AD, Gomes WR (eds) The testis, vol 5. Academic Press, London New York, pp 91–123
Walker MH (1971) Chromosoma 34:340
Wall RJ, Hagen DR, Foote RH (1980) In: Steinberger A, Steinberger E (eds) Testicular development, structure and function. Raven, New York, pp 455–461
Wallace E, Cooper GW, Calvin HI (1983) Gamete Res 7:389
Walt H (1981) Cell Differ 10:157
Walt H, Hedinger C (1983) Andrologia 15:34
Walt H, Saremaslani P, Heizmann CW, Hedinger C (1982) Virchows Arch Cell Pathol 38:307
Wang C, Killian G, Chapman DA (1981) Biol Reprod 25:969
Wang L, Koide SS (1982) Am J Reprod Immunol 2:277
Wang LF, Miao SY, Cao SL, Wu BY, Ke LL, Liang ZQ (1982) Chung Kuo I Hsueh Ko Hsuch Yuan Hsuch Pao 4:82
Warikoo PK, Das RP (1983) Indian J Exp Biol 21:237
Warner FD (1972) In: DuPraw EJ (ed) Advances in the cell and molecular biology, vol II. Academic Press, London New York, pp 193–235
Warner FD (1974) In: Sleigh MA (ed) Cilia and flagella. Academic Press, London New York, pp 11–37
Warner FD, Mitchell DR (1978) J Cell Biol 76:261
Warner FD, Satir P (1973) J Cell Sci 12:313
Warner FD, Satir P (1974) J Cell Biol 63:35
Warner FD, Mitchell DR, Perkins CR (1977) J Mol Biol 114:367
Warrant RW, Kim S-H (1978) Nature (London) 271:130
Wartenberg H (1981) In: Burger H, Kretser de D (eds) The testis. Raven, New York, pp 39–80
Wartenberg H, Holstein A-F (1975) Cell Tissue Res 159:435
Wartenberg H, Holstein A-F, Vassmayer J (1971) Z Anat Entwicklungsgesch 134:165
Wasco WM, Orr GA (1984) Biochem Biophys Res Commun 118:636
Watson ML (1952) Biochim Biophys Acta 8:369
Wattanaseree J, Svasti J (1983) Arch Biochem Biophys 225:892
Wauben-Penris PJ, Hoeven van der FA, Boer de P (1983) Cytogenet Cell Genet 36:547
Weaker F (1977) Cell Tissue Res 179:97
Weaker F, Cameron IL (1977) Cytobios 19:79
Weber JE, Russell LD, Wong V, Peterson RN (1983) Am J Anat 167:163
Weddington SL, McLean WS, Nayfeh SN, French FS, Hansson V, Ritzen EM (1974) Steroids 24:123
Weddington SC, Brandtzaeg P, Sletten K, Christensen T, Hansson V, French FS, Petrusz P, Nayfeh SN, Ritzen EM (1975) In: French FS, Hansson V, Ritzen EM, Nayfeh SN (eds) Hormonal regulation of spermatogenesis. Plenum, New York, pp 433–451
Weeda AJ, Cohen J (1982) Fertil Steril 37:817
Weeds NW (1975) Proc Natl Acad Sci USA 72:4110
Weignberg J, Foo V, Vogl AW (1984) Anat Rec 208:139A
Weiner C, Schlechter N, Zucker I (1984) Biol Reprod 30:507
Weitze KF (1976) Zuchthygiene 11:62
Welch JE, O'Rand MG (1985a) Biol Reprod 32 (Suppl 1):74
Welch JE, O'Rand MG (1985b) Dev Biol 109:411
Wells JN, Garbers DL (1977) Biol Reprod 15:46
Welsh MJ, Sickle van M, Means AR (1980) In: Steinberger A, Steinberger E (eds) Testicular development, structure and function. Raven, New York, pp 89–98

Welsh MJ, Ireland MC, Treisman GJ (1984) In: Catt KJ, Dufau ML (eds) Hormone action and testicular function. Ann NY Acad Sci, vol 438, pp 576–578
Welsh MJ, Ireland MC, Treisman GJ (1985) Biol Reprod 33:1050
Wendt V, Schleuning W-D, Schill W-B, Tschesche H, Leidl W, Fritz H (1976) VIII Congr Int Reprod Anim Insem Artif, Krako, vol IV, pp 947–950
Wester RC, Salisbury GW, Foote RH (1972) VII Int Congr Anim Reprod Artif Insem, München, p 429
Wheat TE, Goldberg E (1983) In: Rattazzi M, Scandalios JG, Whitt GS (eds) Isozymes: Curr Top Biol Med Res, vol VII: Molecular structure and regulation. Liss, New York, pp 113–130
Wheat TE, Goldberg E (1984) In: Catt KJ, Dufau ML (eds) Hormone action and testicular function. Ann NY Acad Sci, vol 438, pp 156–170
Wheat TE, Hintz M, Goldberg E, Margoliash E (1977) Differentiation 9:37
Wheat TE, Gonazales-Prevatt VB, Beyler SA, Shelton JA, Goldberg E (1983) J Reprod Immunol Suppl 26 (Abstr)
White IG, Wales RG (1960) Int J Fertil 5:195
White IG, Darin-Bennett A (1976) VIII Congr Int Reprod Anim Insem Artif, Krako, vol IV, pp 951–953
White SL, Schumacher GFB, Yang SL (1983) Biol Reprod 28 Suppl 1:131
Wichmann K, Kapyaho K, Sinervirta R, Janne J (1983) J Reprod Fertil 69:259
Wilkins MHF, Randall JT (1953) Biochem Biophys Acta 10:192
Williamson RA, Koehler JK, Smith WD, Stenchever MA (1984) Fertil Steril 41:103
Wilson JD (1975) In: Greep RO, Astwood EB (eds) Handbook of physiology, Sect 7: Endocrinology, vol V. Am Physiol Soc, Washington DC, p 491
Wilson RM, Griswold MD (1979) Exp Cell Res 123:127
Wimsatt WA, Krutzsch PH, Napolitano L (1966) Am J Anat 119:25
Wincek TJ, Parrrish RF, Polakoski KL (1979) Science 203:553
Wischnitzer S (1967) J Morphol 121:29
Wislocki GB (1949) Endocrinology 44:167
Withers HR, Hunter N, Barklety HT Jr, Reid BO (1974) Radiat Res 57:88
Witkin SS (1980) J Reprod Fertil 59:409
Witkin SS, Bendich A (1977) Exp Cell Res 106:47
Witkin SS, Korngold GC, Bendich A (1975) Proc Natl Acad Sci USA 72:3295
Witman GB, Carlson K, Berliner J, Rosenbaum JL (1972a) J Cell Biol 54:507
Witman GB, Carlson K, Rosenbaum JL (1972b) J Cell Biol 54:540
Witters WL, Foley CW (1976) J Anim Sci 43:159
Wolf DP (1977) J Exp Zool 199:149
Wolf DP, Boldt J, Byrd W, Bechtol KG (1985) Biol Reprod 32:1157
Wolosewick JJ, Bryan JHD (1977) Am J Anat 150:301
Wong PY, Lee WM (1983) Biol Reprod 28:206
Wong PY, Lee WM, Tsang AYF (1981) Experientia 37:69
Wong V, Russell LD (1983) Am J Anat 167:143
Wooding FBP (1973) J Ultrastruct Res 42:502
Wooding FBP (1975) J Reprod Fertil 44:185
Wooding FBP, O'Donnell JM (1971) J Ultrastruct Res 35:71
Woods MC, Simpson ME (1961) Endocrinology 69:91
Woolley DM (1970) J Reprod Fertil 23:361
Woolley DM (1971) J Cell Biol 49:936
Woolley DM (1977) J Cell Biol 75:851
Woolley DM (1979) In: Fawcett DW, Bedford JM (eds) The spermatozoon. Urban & Schwarzenberg, Baltimore Munich, pp 69–79
Woolley DM, Fawcett DW (1973) Anat Rec 177:289
Woolley DM, Richardson DW (1978) J Reprod Fertil 53:389
Working PK, Meizel S (1981) J Biol Chem 256:4708
Working PK, Meizel S (1982) Biochem Biophys Res Commun 104:1060
Working PK, Meizel S (1983) J Exp Zool 227:97
Wright WW, Frankel AI (1980) Endocrinology 107:314
Wright WW, Musto NA, Mather JP, Bardin CW (1981) Proc Natl Acad Sci USA 78:7565

Wright WW, Musto NA, Mather JP, Bardin CW (1982) In: Bardin CW, Sherins RJ (eds) The cell biology of the testis. Ann NY Acad Sci, vol 383, pp 501–510
Wright WW, Parvinen M, Musto NA, Gunsalus GL, Phillips DM, Mather JP, Bardin CW (1983) Biol Reprod 29:257
Wyker R, Howards SS (1977) Fertil Steril 28:108
Yamamoto N (1985) Acta Histochem Cytochem 18:199
Yan YC, Wang LF, Sato E, Koide SS (1983) J Reprod Immunol Suppl 17 (Abstr)
Yanagimachi R (1975) Biol Reprod 13:519
Yanagimachi R (1978a) Curr Top Dev Biol 12:83
Yanagimachi R (1978b) Biol Reprod 19:949
Yanagimachi R (1981) In: Mastroianni L Jr, Biggers JD (eds) Embryonic development in vitro. Plenum, New York, pp 82–182
Yanagimachi R, Noda YD (1970a) J Ultrastruct Res 31:465
Yanagimachi R, Noda YD (1970b) J Ultrastruct Res 31:486
Yanagimachi R, Noda YD (1970c) Am J Anat 129:429
Yanagimachi R, Suzuki F (1985) Gamete Res 11:29
Yanagimachi R, Teichman RJ (1972) Biol Reprod 6:87
Yanagimachi R, Usui N (1974) Exp Cell Res 89:161
Yanagimachi R, Noda YD, Fujimoto M, Nicolson GL (1972) Am J Anat 135:497
Yanagimachi R, Nicolson GL, Noda YD, Fujimoto M (1973a) J Ultrastruct Res 43:344
Yanagimachi R, Noda YD, Fujimoto M, Nicolson GL (1973b) Am J Anat 135:497
Yanagimachi R, Huang TTF, Fleming AD, Kosower NS, Nicolson GL (1973) Gamete Res 7:145
Yang CH (1972) Thesis, Univ Georgia, Athens
Yang CH, Srivastava PN (1973) Proc Soc Exp Biol Med 145:721
Yang CH, Srivastava PN (1974) J Reprod Fertil 37:17
Yang CH, Srivastava PN (1975) J Biol Chem 250:79
Yasuzumi G (1974) Int Rev Cytol 37:53
Yasuzumi G, Sugioka T, Nakai Y (1970a) Monit Zool Ital (NS) 4:147
Yasuzumi G, Tsubo I, Matsuzaki W (1970b) Arch Histol Jpn 31:283
Yasuzumi G, Shiraiwa S, Yamamoto H (1972) Z Zellforsch Mikrosk Anat 125:497
Yokoyama M, Chang JP (1977) Biol Reprod 17:265
Yoo BY (1982) Arch Androl 8:153
Yoo BY, Ji I, Ji TH (1982) Can J Zool 160:2970
Young LG, Goodman SA (1980) Biol Reprod 23:826
Young LG, Goodman SA (1982) Gamete Res 6:281
Young LG, Gould KG (1982) Arch Androl 8:15
Young LG, Nelson L (1968) Exp Cell Res 51:34
Young LG, Nelson L (1974) J Reprod Fertil 41:371
Young LG, Smithwick EB (1975a) J Cell Physiol 85:143
Young LG, Smithwick EB (1975b) Exp Cell Res 90:223
Young LG, Smithwick EB (1976) Exp Cell Res 102:179
Young LG, Smithwick EB (1983) J Exp Zool 226:459
Young RJ, Cooper GW (1979) In: Fawcett DW, Bedford JM (eds) The spermatozoon. Urban & Schwarzenberg, Baltimore Munich, pp 391–394
Young RJ, Cooper GW (1983) J Reprod Fertil 69:1
Zahler WL, Doak GA (1975) Biochim Biophys Acta 406:479
Zahler WL, Herman CA (1976) Fed Proc Fed Am Soc Exp 35:17B
Zahler WL, Polakoski KL (1977) Biochim Biophys Acta 480:461
Zakhariev Z, Lotova M, Kichev G, Marinov MF, Al-Hanak (1983) Vet Med Nauk 20:27
Zamboni L (1971) Fine morphology of mammalian fertilization. Harper & Row, New York
Zamboni L, Stefanini M (1968) Fertil Steril 19:570
Zamboni L, Stefanini M, Hando T (1968) VI Congr Int Reprod Anim Artif Insem, vol I, pp 229–232
Zamboni L, Zemjanis R, Stefanini M (1971) Anat Rec 169:129
Zane RS, Johnson L (1984) J Androl 5:11
Zaneveld LJD (1975) In: Hafez ESE, Thibault CG (eds) The biology of spermatozoa. Karger, Basel, pp 192–200

Zaneveld LJD (1982) In: Gerlad GI (ed) Research frontiers in fertility regulation, vol III. Northwestern Univ, Chicago, pp 1–15
Zaneveld LJD, Schumacher GFB (1972) VII Int Congr Anim Reprod Artif Insem, München, pp 432 (Abstr)
Zaneveld LJD, Williams WL (1970) Biol Reprod 2:363
Zaneveld LJD, Robertson RT, Kessler M, Williams WL (1971) J Reprod Fertil 25:387
Zaneveld LJD, Dragoje BM, Schumacher GFB (1972a) Science 177:702
Zaneveld LJD, Polakoski KL, Williams WL (1972b) Biol Reprod 6:30
Zaneveld LJD, Schumacher GFB, Travis J (1973a) Fertil Steril 24:479
Zaneveld LJD, Polakoski KL, Schumacher GFB (1973b) J Biol Chem 248:564
Zaneveld LJD, Polakoski KL, Williams W (1973c) Biol Reprod 9:219
Zaneveld LJD, Polakoski KL, Schumacher GFB (1975) In: Reich E, Rifkin DB, Shaw E (eds) Proteases and biological control, vol II. Cold Spring Harbor Conf Cell Proliferation, pp 683–706
Zao PZR, Meizel S, Talbot P (1985) J Exp Zool 234:63
Zavos PM, Goodpasture JC, Zaneveld JJ, Cohen MR (1980) Fertil Steril 34:607
Zeheb R, Orr GA (1984) J Biol Chem 259:839
Zeleny C, Faust EC (1915) J Exp Zool 18:187
Zelezna B, Cechova D (1982) Hoppe-Seylers Z Physiol Chem 363:757
Zibrin M (1971) Mikroskopie 27:10
Zimmerman KJ, Crabo G, Moore R, Weisberg S, Deihel FC, Graham EF (1979) Biol Reprod 21:173
Zimmerman RE, Nevin RS, Allen FJ, Jones CD, Goettel ME, Burck PJ (1983) J Reprod Fertil 68:257
Ziparo E, Geremia R, Russo MA, Stefanini M (1980) Am J Anat 159:385
Ziparo E, Siracusa G, Palombi F, Russo MA, Stefanini M (1982) In: Bardin CW, Sherins RJ (eds) The cell biology of the testis. Ann NY Acad Sci, Washington DC, vol 383, pp 511–512
Zirkin BR, Chang TSK (1977) Biol Reprod 17:131
Zirkin BR, Chang TSK, Heaps J (1980) J Cell Biol 85:116
Zuckerman Z, Rodriguez-Rigau LJ, Weiss DB, Chowdhury AK, Smith KD, Steinberger E (1978) Fertil Steril 30:448
Zysk JR, Bushway AA, Whistler RL, Carlton WW (1975) J Reprod Fertil 45:69

Subject Index

S. S. Guraya

Biology of Ovarian Follicles in Mammals

1985. 76 figures. XI, 320 pages
ISBN 3-540-15022-6

Contents: Introduction. - Primordial Follicle. - Follicle Growth. - Ovum Maturation. - Ovulation. - Luteinization and Steroidogenesis in the Follicle Wall During Pre-ovulatory and Ovulatory Periods. - Follicular Atresia. - References. - Subject Index.

This book presents, for the first time, a unique, up-to-date, integrated, comprehensive, and critical review of morphological (including ultrastructural), histochemical, biochemical, immunological, physiological, endocrinological and biophysical aspects of primordial follicles, follicle growth, ovum maturation, ovulation, luteinization and steroidogenesis in the follicle wall during pre-ovulatory and ovulatory periods and follicular atresia in mammals. The author summarizes the vast amount of information in the international literature on the problems of development, differentiation, growth, structure, composition, regulation of functions in various components such as oocyte, zona pellucida, granulosa and theca of follicles as well as interactions at the cellular and molecular levels between them during growth, maturation, atresia, and ovulation of follicles.

Springer-Verlag
Berlin Heidelberg New York
London Paris Tokyo

Zoophysiology

Coordinating Editor: D. S. Farner

Editors: B. Heinrich, K. Johansen, H. Langer, G. Neuweiler, D. J. Randall

Volume 15

T. Mann

Spermatophores

Development, Structure, Biochemical Attributes and Role in the Transfer of Spermatozoa

1984. 50 figures. XII, 217 pages
ISBN 3-540-13583-9

Contents: General Considerations. – Platyhelminthes, Aschelminthes and Phoronida. – Mollusca. – Annelida. – Onychophora and Myriapoda. – Insecta. – Crustacea. – Arachnida. – Chaetognatha and Pogonophora. – Sporadic Occurrence of Spermatophores and Spermatozeugmata in Vertebrata. – Concluding Remarks. – References. – Systematic and Species Index. – Subject Index.

Springer-Verlag
Berlin Heidelberg New York
London Paris Tokyo

Springer